Graphene Oxide

Graphene Oxide

Fundamentals and Applications

Edited by

AYRAT M. DIMIEV

Laboratory of Advanced Carbon Nanostructures, Kazan Federal University, Kazan, Russian Federation

SIEGFRIED EIGLER

Department of Chemistry and Chemical Engineering, Chalmers University of Technology, Sweden

WILEY

This edition first published 2017

Registered Office
John Wiley & Sons, Ltd, The Atrium, Southern Gate, Chichester, West Sussex, PO19 8SQ, United Kingdom

For details of our global editorial offices, for customer services and for information about how to apply for permission to reuse the copyright material in this book please see our website at www.wiley.com.

Library of Congress Cataloging-in-Publication Data

Names: Dimiev, Ayrat M., 1965– | Eigler, Siegfried, 1978–
Title: Graphene oxide : fundamentals and applications / Ayrat M. Dimiev, Siegfried Eigler.
Description: Chichester, West Sussex : John Wiley & Sons, Inc., 2017. |
Includes bibliographical references and index.
Identifiers: LCCN 2016018225| ISBN 9781119069409 (cloth) | ISBN 9781119069430 (epub)
Subjects: LCSH: Graphene. | Graphene–Oxidation. | Graphene–Industrial applications.
Classification: LCC TA455.G65 D56 2017 | DDC 662/.9–dc23 LC record available at https://lccn.loc.gov/2016018225

A catalogue record for this book is available from the British Library.

Set in 10/12pt Times by SPi Global, Pondicherry, India

10 9 8 7 6 5 4 3 2 1

Contents

About the Editors

Dr. Ayrat M. Dimiev

Laboratory of Advanced Carbon Nanostructures, Kazan Federal University, Kazan, Russian Federation

Ayrat M. Dimiev received his PhD degree in physical chemistry from Kazan University, Kazan, Russian Federation. After three years of holding an assistant professorship at Kazan Agricultural University, he emigrated to the USA to teach the International Baccalaureate chemistry course. In 2008 he joined the group of Professor James M. Tour at Rice University, where he started his studies in the field of carbon. His research spreads over the areas of unzipping carbon nanotubes, carbon-based dielectric composite materials, graphite intercalation compounds, and graphene oxide chemistry. His most important contributions to the field were revealing the mechanism of the stage transitions in graphite intercalation compounds, and developing the dynamic structural model of graphene oxide. In 2013, Dr. Dimiev accepted a personal invitation from AZ Electronic Materials (presently EMD Performance Materials Corp., USA, a business of Merck KGaA, Darmstadt, Germany) to help in developing their newly started carbon program. During his time at EMD Performance Materials Corp, he employed his expertise in the field to develop and commercialize new products comprising graphene oxide and other carbon nanostructures. In May 2016, Dr Dimiev returned to his alma mater in Kazan as a head of the Laboratory of Advanced Carbon Nanostructures, Kazan Federal University. Dr. Dimiev is the author of 18 recent publications in the carbon field, with an overall total number of citations exceeding 2000; he also has five recent patent applications.

Dr. Siegfried Eigler

Department of Chemistry and Chemical Engineering, Chalmers University of Technology, Sweden

Siegfried Eigler received his PhD in organic chemistry from the Friedrich-Alexander-Universität Erlangen-Nürnberg in 2006 under the guidance of apl. Prof. Dr. Norbert Jux. Subsequently he conducted industrial research at DIC-Berlin GmbH, part of DIC Corporation, Japan. The research concentrated on electrically conductive polymers and the development of novel semiconducting monomers. In 2009 he started working on the synthesis and application of graphene oxide. Two years later he became a lecturer and research associate at the Friedrich-Alexander-Universität Erlangen-Nürnberg. There, he conducted deep research on the synthesis of graphene oxide and he realized that defects in the carbon lattice could be avoided by controlling the synthesis. With this discovery he could investigate the controlled chemistry of graphene oxide and synthesized several new graphene derivatives and composites. Currently, his research focuses on advancing the controlled chemistry of graphene. Dr. Eigler has authored 27 papers in the field of carbon research and applied for one patent related to the wet-chemical synthesis of graphene that allows the control of the defect density. He accepted an offer from Chalmers University of Technology, Gothenburg, Sweden, as Associate Professor which started in 2016.

List of Contributors

Seyed Hamed Aboutalebi Condensed Matter National Laboratory, Institute for Research in Fundamental Sciences (IPM), Tehran, Iran

Young Hoon Cho Department of Energy Engineering, Hanyang University, Seoul, Republic of Korea

Ayrat M. Dimiev Laboratory of Advanced Carbon Nanostructures, Kazan Federal University, Kazan, Russian Federation

Siegfried Eigler Department of Chemistry and Chemical Engineering, Chalmers University of Technology, Gothenburg, Sweden

Mohsen Moazzami Gudarzi Department of Inorganic and Analytical Chemistry, University of Geneva, Geneva, Switzerland

Cary Michael Hayner Chief Technology Officer, SiNode Systems Inc., Chicago, IL, USA

Larisa Kovbasyuk Department of Chemistry and Pharmacy, Inorganic Chemistry II, Friedrich-Alexander-Universität, Erlangen-Nürnberg (FAU), Erlangen, Germany

Anton Lerf Emeritus Walther-Meissner-Institut der Bayerischen Akademie der Wissenschaften, Garching, Germany

Sean E. Lowe Department of Materials Science and Engineering, Monash University, Clayton, Australia

Andriy Mokhir Department of Chemistry and Pharmacy, Inorganic Chemistry II, Friedrich-Alexander-Universität, Erlangen-Nürnberg (FAU), Erlangen, Germany

Anton V. Naumov Department of Physics and Astronomy, Texas Christian University, Fort Worth, USA

Ho Bum Park Department of Energy Engineering, Hanyang University, Seoul, Republic of Korea

Ioannis V. Pavlidis Department of Biochemistry, University of Kassel, Kassel, Germany

Samuele Porro Department of Applied Science and Technology, Politecnico di Torino, Torino, Italy

Ignazio Roppolo Center for Space Human Robotics, Istituto Italiano di Tecnologia, Torino, Italy

Farhad Sharif Department of Polymer Engineering and Color Technology, Amirkabir University of Technology, Tehran, Iran

Cristina Vallés School of Materials, University of Manchester, Manchester, UK

Hee Wook Yoon Department of Energy Engineering, Hanyang University, Seoul, Republic of Korea

Yu Lin Zhong Department of Materials Science and Engineering, Monash University, Clayton, Australia

Foreword

This book presents a timely review of research on graphene oxide, which is a term representing the individual layers obtained by exfoliation of graphite oxide.

Although graphite oxide was first synthesized in the 1850s, it is a material that has attracted renewed and intense interest only in the past decade because it affords a product material, by relatively straightforward exfoliation in water, of individual layers of functionalized graphene. The functionalization of the graphene consists of hydroxyl and epoxide groups, among others, and the individual layers, for which the term "graphene oxide" has been used, are thus hydrophilic, such that they form stable dispersions in solvents like water. This has allowed stable dispersions of chemically modified graphene (of a particular type) to be readily prepared and then used in interesting ways.

It was fortunately possible that we could help with "early contributions" in this past decade, including use of graphene oxide or its modifications to make (i) electrically conducting polymer composites, (ii) thin "paper-like" material composed of stacked and overlapped platelets, and (iii) electrodes for supercapacitors. It has been gratifying to see the growth of a now relatively large body of literature about both fundamental aspects of the chemistry and properties of graphene oxide and its derivatives, and of applications or potential applications. Graphene oxide and its related product or derivative materials have been shown to be highly versatile and have been applied in a wide range of studies.

Fundamental aspects as well as applications are well covered herein, and the book should be useful for learning about graphene oxide and related materials, and thus for providing a basis for thinking about new possibilities as well.

It is perhaps of interest to speculate about the future, and I do so here only briefly. For graphene oxide, and in general chemically modified graphenes, there are many exciting possibilities. Fine control of the chemical functionalization, including with oxygen-containing groups, remains an important challenge. As greater control of the specific location and distribution of functional groups is achieved, and, when desired, of the deliberate removal of carbon atoms from the "graphene lattice", a wider range of applications, including for better sensors and in materials such as composites and filters, will emerge. There is also the fascinating possibility of control of folding or "crumpling" through thermodynamics and clever design of where functional groups are "placed" and how they interact: intra-sheet to "fix" certain morphologies, or perhaps inter-sheet and also with their surrounding environment. (This could be a type of origami but is not limited to specific types of folding.)

Rodney S. Ruoff
IBS Center for Multidimensional Carbon Materials
Ulsan National Institute of Science and Technology
March 2016

Preface

Graphene oxide (GO) has become one of the most extensively studied materials of the last decade. It has facilitated massive interdisciplinary research in the fields of chemistry, physics and materials science. Due to its unique properties, GO has been successfully tested for numerous applications. This fruitful research has resulted in an enormous number of publications. Several review articles have summarized the most recent advances. However, up to the present day, very little has been done to systematize all the published research, and to assist a non-expert audience interested in the field. This book is designed to fulfill this task. The content of each chapter and the book in general develops from basic to more complex. The material is presented in the categories typical for the classical fields of science. This makes this book unique and different from the other literature.

Today, keeping track of all the recent publications in the field is difficult even for experts. For non-experts, it is practically impossible to navigate this ocean of publications. This task is further complicated by confusion that is widespread in the modern GO field. This confusion is caused by the misuse of the main fundamental concepts, and by oversimplification and misinterpretation of GO chemistry. It is very difficult to identify trustable high-quality publications that correctly employ fundamental chemical terms, and correctly interpret experimental data. In this book, we intend to demonstrate the actual GO chemistry based on trustable publications, with correct usage of the main fundamental concepts, as they have been identified up to now.

Since the beginning of the graphene era in 2004, GO has been closely associated with graphene. At that time, GO was considered mainly as a precursor for graphene. The term "chemically converted graphene" (CCG) was introduced for reduced graphene oxide (RGO) to highlight the graphene-like nature of RGO. The misuse of the term "graphene" instead of RGO in the literature creates significant confusion among a non-expert readership. We aim to help readers to differentiate between the two by drawing a clear borderline between graphene and RGO, and by showing where they are similar, and where they are different. Additional confusion arises from the misuse of the term RGO for material obtained by annealing of GO. We highlight that those two are very different materials, and we introduce the term "thermally processed graphene oxide" (tpGO) for the latter.

Since the electrical properties of RGO are inferior of those of real graphene, GO is often considered as graphene's "younger brother", or as a low-grade graphene. This point of view was dominant up until about 2011. Later, it was demonstrated that GO is a unique and valuable material in itself, both from fundamental science perspectives and for practical applications. The main advantage of GO over the graphene counterpart is its solubility and processability in water and in several organic solvents. Another benefit of GO is due to its versatility of chemical modification to alter its properties. The ability for mass production on the scale of tonnes makes GO especially attractive for applications compared to its graphene counterpart. We intend to demonstrate all the advantages and uniqueness of GO in this book.

The book is divided into two parts. Part I focuses on the fundamentals of GO, and Part II on the applications of GO.

Part I starts with research on GO, which has a very long history. It did not start in 2006 with the work on GO reduction, as one might think by looking at the citation indexes of some publications from that period. Very serious and in-depth studies on GO chemistry were conducted throughout the entire twentieth century. Most of these studies, performed in the best old-school traditions, were in many ways advantageous when compared to some modern publications. The fundamentality of scientific thinking, the methodology of the research, and, importantly, the trustworthiness of reported data were on a level that is rather rare in the modern GO field. One could easily avoid misinterpretations of experimental results by studying those early works before even designing one's own experiments. Because of the high importance of that early research, and in an attempt to make the connection between the two eras, we begin the book with a historical retrospective of GO research done in the twentieth century (*Chapter 1*). This is written by the long-term expert in the field, one of the developers of the famous Lerf–Klinowski structural model, *Anton Lerf*.

In the modern literature, the structure of GO is greatly oversimplified. This leads to misinterpretation of the chemical reactions involving GO. *Chapter 2*, written by *Ayrat Dimiev*, aims to clarify some aspects of GO structure. In the form typical for textbooks, the mechanism of GO formation, its transformation during aqueous work-up, and the fine chemical structure of GO are methodologically described. The structure of GO is discussed with respect to its intrinsic chemical properties, such as the acidity of aqueous solutions.

The methods used for GO characterization are reviewed in *Chapter 3* in a tutorial manner. This chapter will be of particular importance for researchers entering the field. The advantages and disadvantages of different methods are highlighted. Several examples where different methods have helped to understand the structure of GO are discussed. This chapter is written jointly by the editors, *Siegfried Eigler* and *Ayrat Dimiev*.

In aqueous solutions, GO delaminates to single-layer sheets and forms colloidal solutions. From aqueous solutions, GO flakes can be transferred into the phase of low-molecular-weight alcohols; the alcoholic solutions are as stable against precipitation as aqueous ones. At certain concentrations, GO solutions form liquid crystals. The rheology of GO solutions is reviewed in *Chapter 4* by *Cristina Vallés*. Colloid chemistry, surface science, rheology and liquid chemistry of GO are discussed in this chapter.

Due to its electronic configuration, GO possesses a number of remarkable optical properties. As opposed to pristine graphene, GO exhibits photoluminescence in the ultra-violet, visible and near-infrared regions, depending on its structure. The origins of this emission and other related questions are discussed in *Chapter 5* by *Anton Naumov*.

The chemical properties of GO is the largest, most difficult and most controversial topic. In *Chapter 6*, written by *Siegfried Eigler* and *Ayrat Dimiev*, the following topics are discussed. The thermal and chemical stability of GO is reviewed first, followed by introducing wet-chemical non-covalent functionalization protocols. The covalent function-alization of GO, which is discussed next, is a very controversial topic. When well-known organic chemistry principles are applied to GO, it remains challenging to prove the successful accomplishment of reactions by analyzing the as-modified GO product. We provide an alternative interpretation for experimental results of some selected examples to demonstrate this challenge. The chemical reduction methods are summarized next, and special emphasis is given to differentiating true chemical reduction from so-called "thermal reduction". While discussing GO chemical properties, in parallel with typical GO, we discuss these properties for the oxo-functionalized graphene (oxo-G_1), a type of GO with very low density of structural defects. This sheds additional light on the role of defects in GO chemistry. Finally, additional properties of oxo-G_1 are introduced. Oxo-G_1 can act as a compound that enables the controlled chemistry for the design and synthesis of functional materials and devices.

In Part II, applications that use the reduced and non-reduced forms of GO are reviewed separately. A reduced form of GO is required where electrical conductivity is of importance. These applications exploit the graphene-like properties of RGO and tpGO.

Due to its two-dimensional character, real graphene is not available in bulk quantities by definition. It is obtained only as a substrate-supported material either by micromechanical cleavage of graphite, or by chemical vapor deposition (CVD) growth on the surface of a catalytically active metal. The electrical conductivities of RGO and tpGO are three or four orders of magnitude lower than that of real graphene due to the numerous defects or scattering centers in the former. Nevertheless, in applications where bulk forms of graphene are needed, GO derivatives are the only choice. Currently, about 90% of the studies performed with RGO and tpGO use the term "graphene" both in the title and in the abstract. We highlight that GO derivatives, and not real graphene, are used for the applications reviewed in *Chapters 7* and *8*.

Field-effect transistors and sensors are the two most promising applications that exploit the unique electronic properties of GO. RGO is also considered as one of the best candidates for fabricating transparent conductive films for many applications, due to its electrical and mechanical properties, reasonable carrier mobility, and optical transparency in the visible range. *Chapter 7*, written by *Samuele Porro* and *Ignazio Roppolo* summarizes the enormous potential for applications of GO in the above-mentioned fields.

The electrical conductivity and high surface area of tpGO have driven substantial efforts for its integration into advanced energy systems. In *Chapter 8*, the integration of GO into two major energy storage systems – lithium-ion batteries and supercapacitors – is discussed. Particular attention is given to understanding the important physicochemical properties that can be emphasized in order to achieve the greatest performance, as well as the synthetic processes used to derive these unique benefits. This chapter is written by *Cary Hayner*, CTO of SiNode Systems, a start-up company that develops a new generation of lithium-ion batteries based on the novel electrode material comprising GO.

Due to the two-dimensional character of GO flakes, and their solubility in water, GO can be cast into thin films by simple drop casting or filtration. The as-formed GO membranes exhibit unimpeded permeability to water molecules, being absolutely impermeable to other molecules and atoms. Applications of GO and RGO for selective membranes are reviewed in *Chapter 9* by *Ho Bum Park*, *Hee Wook Yoon* and *Young Hoon Cho*.

Due to the processability of GO in water and organic solvents, GO has been tested as a component in numerous composite materials. The incorporation of GO into polymers modifies electrical and thermal conductivity, lowers permeability and improves mechanical properties. This topic is covered in *Chapter 10* by *Mohsen Moazzami Gudarzi*, *Seyed Hamed Aboutalebi* and *Farhad Sharif*.

Biomedical applications and toxicity studies of GO are of utmost importance for using GO in real applications. Other materials, such as carbon nanotubes, are suspected to be toxic or carcinogenic. Therefore, the current advances in analyzing the medical properties and biomedical applications of GO are covered by *Larisa Kovbasyuk* and *Andriy Mokhir* in *Chapter 11*.

GO and its derivatives possess unique properties that grant them interest as catalysts in oxidative reactions, Friedel–Crafts and Michael additions, polymerization reactions, oxygen reduction reactions and photocatalysis. This property is reviewed in *Chapter 12* by *Ioannis Pavlidis*.

The scalable production of GO still holds the key to its commercialization. The most crucial factor for GO to be commercially viable is its cost-effectiveness. This is not a simple task to ascertain, since GO production involves lengthy purification procedures that produce significant quantities of acidic waste. The challenges facing commercial GO production are discussed in *Chapter 13* by *Sean Lowe* and *Yu Lin Zhong*.

This book is written and edited by professionals in their respective fields, and it is intended to be helpful for a very broad community, including experts broadening their field of research.

Ayrat M. Dimiev
Russian Federation
and
Siegfried Eigler
Sweden

Part I

Fundamentals

1

Graphite Oxide Story – From the Beginning Till the Graphene Hype

Anton Lerf

1.1 Introduction

The formation of graphite oxide (GO) was described for the first time by Brodie in a short note that appeared in 1855 in *Annales de Chimie* in French [1]. Another preparation method – the reaction of graphite with potassium chlorate in fuming nitric acid, now known as the "Brodie method" – and a detailed description of the composition and the chemical properties of the new compound were published in 1859 in the *Philosophical Transactions of the Royal Society of London* [2]. One year later this paper was published in both French and German translations [3, 4]. The titles of all these papers [1–4] do not give any hint of a new carbon compound. The title of the English version is as follows: "On the atomic weight of graphite". The new compound was called "oxyde de graphite" in the first publication, and "graphitic acid" in the later papers. It is also worth mentioning that Brodie himself did not cite his first work on the new compound.

The actual aim of Brodie's scientific work presented in his publications was to differentiate by means of chemical methods various forms of carbon with dissimilar properties but all called graphite. Among the reactions described in the second paper, there was also the treatment of graphite with a mixture of concentrated nitric and sulfuric acids, leading to the graphite sulfate intercalation compound. The graphite intercalation compound was described for the first time by Schafhäutl [5] (pp. 155–157) in 1840, but it was hidden in two other publications devoted almost exclusively to iron–carbon steels. This might be the reason why Brodie was not aware of this data, and did not cite it. On the other hand, the

Graphene Oxide: Fundamentals and Applications, First Edition. Edited by Ayrat M. Dimiev and Siegfried Eigler.

content of the first paper had been presented before in the *London and Edinburgh Philosophical Magazine*. In the paper of 1859, Schafhäutl [6] (pp. 300–301) complained that nobody took notice of his result.

The circumstances of the beginning of the GO story have been outlined *in extenso* because of their curiousness. In 1865 Gottschalk [7] reproduced and confirmed the results of Brodie. In his publication the term "graphitic acid" appeared for the first time in the title: "Beiträge zur Kenntnis der Graphitsäure" ("Contributions to the knowledge of graphitic acid"). GO received greater attention only due to the publication of Berthelot [8] in 1870 in which he proclaimed Brodie's procedure for the preparation of GO as a method to distinguish different forms of graphitic carbon, although Brodie had already described the different behavior of various graphite forms toward oxidation reactions earlier.

In 1898 Staudenmaier [9] discussed in detail the problems and disadvantages of the various preparation methods existing up to the end of the nineteenth century. He also described his trials to find more convenient and less dangerous preparation methods, and presented the new preparation method which is named after him to the present day.

Of utmost importance is the publication of Kohlschütter and Haenni in 1919 [10]. It marks the end of the classical research on GO, which was based on classical chemical analysis and a careful description of the reaction behavior. On the one hand, it carefully reviews and evaluates all the previous publications on the topic. Based on crystallographic considerations, reproducing the results of Brodie [2] and Weinschenk [11], the authors consider the close structural correlation between graphite and GO as evidence for a "topotactic" relation. This paper presents new data for the formation of GO, its thermal decomposition and chemical reduction, and the products of chemical reduction. Also, in this publication, the authors discarded their own previous pessimistic view that GO could be nothing other than an adsorption of CO and CO_2 at graphite surfaces.

A new period of GO research was opened by Hofmann in 1928 [12] and by Hofmann and Frenzel in 1930 [13] by applying for the first time powder X-ray diffraction (XRD) to GO. Based on these investigations and chemical considerations, Hofmann and his school of researchers gave the first structural model of GO to find general acceptance. This period of research started in 1928, continued through 1930 and 1934 with the first structural models by Thiele [14] and Hofmann *et al.* [15], and ended in 1969 with a new structural model proposed by Scholz and Boehm [16]. During this period, GO structural models were modified several times, mainly due to the debate between Hofmann and Thiele, and due to the application of new spectroscopy methods allowing the proof of assumptions about the functional groups playing a role in the chemistry of GO.

The third period of activity on GO was initiated by the first application of magic angle spinning nuclear magnetic resonance (MAS NMR) on ^{13}C by Mermoux *et al.* in 1989–1991 [17, 18]. In the extended publication [18], the authors questioned the structural model of Scholz and Boehm [16] and credited the model of Ruess [19] as the one that best fitted their data. However, later this interpretation was questioned again [20]. The interpretation of the 60 ppm signal as originating from epoxide functions revitalized the first model of Hofmann *et al.* [15], with some modifications [20]. This model has been confirmed in various studies [21], but is now again questioned by a two-component model [22].

Graphite oxide was a laboratory curiosity [23] till the discovery of graphene by Geim and Novoselov [24]. Soon after this discovery, the easy reduction of graphite oxide was considered as a cheap method to obtain graphene layers. This idea started a hype of research on GO, which has lasted up to now. Looking at many modern publications from a historical perspective, one cannot avoid the impression that many groups have fallen into traps, which previous generations of scientists have learned to circumvent. And, vice versa, reading the previous publications from the actual point of view, one discovers a lot of interesting aspects that have been described with scrutiny, later considered as unimportant, still later forgotten completely, but can now be re-evaluated on the basis of recent research.

The aim of this chapter is to sketch the storylines of various aspects of GO-related studies, which were important for understanding the peculiarities of GO. These topics include GO preparation and purification, the development of structural models, the problem of stability and decomposition, the swelling to a colloid, the acidity of GO, and the ability to intercalate very different chemical species. Some of these aspects have been on the agenda since the discovery of GO, and others came to attention triggered by progress in neighboring fields of research. This chapter is restricted mostly to the historical retrospective of GO studies. *My own thoughts and conclusions from the historical material, colored by my own experience in the field, are written in italic.*

1.2 Preparation of Graphite Oxide

1.2.1 Trials for Improving and Simplifying GO Preparation

In his paper of 1855 Brodie achieved the oxidation of graphite by adding concentrated sulfuric acid to a mixture of graphite and $KClO_3$ [1]. The subsequent treatment with water led to disintegration of the solid and a strong extension of the volume. Calcination of the dry product resulted in graphite, which was contaminated with sulfates and chlorates. At the end of the short paper, Brodie mentioned as alternative oxidants nitric acid and bichromate.

In his publication of 1859 Brodie described first the treatment of graphite with a mixture of concentrated nitric and sulfuric acids [2]. The product obtained can be assigned in modern terms as the graphite intercalation compound of sulfuric acid, as one can conclude from the described properties of the sample, especially the exfoliation phenomena (observed earlier by Schafhäutl [5, 6]). Then Brodie stresses that replacing nitric acid by potassium chlorate or potassium bichromate leads to a different product that is bright yellow or brown and decomposes easily to a graphite-like material.

The procedure given in this paper for the new compound has been used to the present day and is called "Brodie's method" [2]:

> The details of this process are as follows:– A portion of graphite is intimately mixed with three times its weight of chlorate of potash, and the mixture placed in a retort. A sufficient quantity of the strongest fumic nitric acid is added, to render the whole fluid. The retort is placed in a water-bath, and kept for three or four days at a temperature of 60 °C until yellow vapours cease to be evolved. The substance is then thrown into a large quantity of water, and washed by decantation nearly free from acid and salts. It is dried in a water-bath, and the oxidizing operation repeated with the same proportion of nitric acid and of chlorate until no further change is observed: this is usually after the fourth time of oxidation. The substance is ultimately dried, first in vacuo, and then at 100 °C. A modification of the process which may be advantageously adopted, consist in placing the substance with the oxidizing mixture in flasks exposed to sunlight. Under these circumstances the change takes place more rapidly, and without the application of heat.

The note at the end of this paragraph, that sunlight can favor the formation of GO, sounds very interesting, but seems to have been overlooked up to now.

From that time there were many trials to replace the used reagents (fuming nitric acid, concentrated sulfuric acid and potassium chlorate) by less dangerous and more convenient oxidizing reagents. Staudenmaier [9] and Kohlschütter and Haenni [10] mention explicitly the dangerous reaction of chlorate with concentrated sulfuric acid leading to ClO_2, which decomposes at temperature above 45 °C in explosions. *Is that fact the reason why Brodie did not mention this procedure in his second paper?*

Luzi [25] and later Charpy [26] applied $KMnO_4$ and $HCrO_4$ ("acide chromique") in sulfuric acid, both mentioning the tendency to decompose graphite (see section 1.1.2). Kohlschütter and Haenni [10] mentioned unsuccessful trials of oxidation with persulfuric acid, Caro's acid and ozone. Boehm *et al.* [27] studied Ce(IV) nitrate, Co(III) sulfate, NaOCl, $(NH_4)_2S_2O_8$ and OsO_4. Hofmann and Frenzel [13] and later Boehm *et al.* [27] obtained GO by the reaction of graphite suspended in a mixture of concentrated nitric and sulfuric acids with gaseous ClO_2. Boehm *et al.* [27] also obtained GO by oxidation of graphite suspended in concentrated sulfuric acid with Mn_2O_7 and by reaction of a suspension of graphite in fuming nitric acid with an O_2/O_3 mixture for 10 days.

Despite these efforts, at the present day only two further methods are of importance, the procedures described by Staudenmaier [9] and Hummers and Offeman [28], respectively.

Staudenmaier [9] used (as Luzi [25] did) exfoliated ("aufgeblähten") graphite, adding it to the mixture of concentrated nitric and sulfuric acids when the mixture was cooled to room temperature after mixing. He mentioned explicitly that the oxidation is faster the faster the $KClO_3$ has been added, but more $KClO_3$ is then necessary because of the temperature increase leading to stronger decomposition of the chlorate. Whereas he used up to 25 g of graphite for one procedure, he never observed an explosion. The green product obtained after washing and drying can then be transferred to the yellow product by a reaction with a solution of potassium permanganate in diluted sulfuric acid. *Interestingly, Staudenmaier commented on the procedure used by Luzi, but did not mention Luzi's trials with permanganate.*

Kohlschütter and Haenni [10], Hofmann and Frenzel [13] and Hamdi [29] used the "Staudenmaier method", but with some modifications, as follows. Instead of exfoliated graphite, they used powdered graphite; hence, the reaction afforded longer reaction times. At least three oxidation cycles were necessary to get a reasonable degree of oxidation. The oxidation with permanganate solution was dropped without giving an explicit reason. Hofmann and Frenzel [13] and Hamdi [29] also found the process hazardous.

Hummers and Offeman [28] for the first time successfully applied permanganate as an oxidant for the formation of GO: powdered graphite flakes and solid sodium nitrate were suspended in concentrated sulfuric acid and then the permanganate added in portions, so that the temperature can be kept below 20 °C. Then the temperature of the suspension was brought to 35 °C, and kept at that temperature for 30 minutes. The now pasty suspension was then diluted by adding water, causing a temperature increase up to 98 °C. After 15 minutes the mixture was diluted with more water and the residual permanganate reduced with hydrogen peroxide.

Boehm and Scholz [30] discussed for the first time the drawbacks and advantages of the three preparation methods for GO. *(Almost no attention has been paid (only 11 citations in 50 years) to this important paper.)* The following most important conclusions should be mentioned:

The GO samples obtained via "Brodie's method" are the purest and the most stable ones (see section 1.4.1).

The purification of GO samples prepared according to the "Staudenmaier method" and the "Hummers–Offeman method" is much more tedious; especially, the Hummers–Offeman samples are contaminated with a considerable amount of sulfur, probably bound to carbon as sulfonic acid or as esters of sulfuric acid.

Replacing $KClO_3$ by $NaClO_3$ prevents the formation of insoluble $KClO_4$, which is hard to remove by washing with water.

Warning is mentioned to allow temperature increase during pouring the sample into water at the end of the oxidation reaction because of decomposition.

The chemical composition of the various samples shows a great variation, but there is a trend in the degree of oxidations: the C/O ratio decreases in the order Brodie > Staudenmaier > Hummers–Offeman.

The principal routes of work-up have been established since the early days of GO research. It is a tedious process, which can lead to changes in samples exposed to light or to water for too long. The work-up always starts with a strong dilution of the acids. In the first step, it is recommended to pour the reaction mixture into a huge amount of excess water to keep the temperature as low as possible. Since the particles of GO are very small, the sedimentation takes some time. During the washing process, the sediment volume increases strongly and the time for sedimentation becomes longer. In order to shorten the process, the GO is precipitated after some dilution/sedimentation cycles by adding dilute HCl. The obtained precipitate can be separated from the solvent by filtration or centrifugation. For the cleaning steps, dialysis or electrodialysis [29] has also been applied.

A very strange purification was recommended by Thiele [14]: he removed the oxidation mixture ($KClO_3/H_2SO_4/HNO_3$) by repeated boiling with concentrated nitric acid; the nitric acid is then removed by washing with acetic acid/acetic acid anhydride, and at the end with alcohol or ether.

Obtaining dry GO is also a tricky business. It is almost impossible to remove the solvent water completely. The minimum water content (5–10%) can be achieved without decomposition at room temperature in vacuum and in the presence of P_2O_5. An alternative method is freeze-drying [30].

Apart from these chemical preparation methods, it is possible to obtain graphite oxide also by electrochemical oxidation. This was shown for the first time by Thiele [31] in 1934. He applied a very high current density and obtained in concentrated sulfuric acid first the blue phase, which is the sulfuric acid intercalation compound, and then a so-called primary oxide, graphite oxide and perhaps humic acid. Later Boehm *et al.* [27] and Besenhard and Fritz [32] obtained GO under controlled conditions in 70% perchloric acid. However, the degree of oxidation is considerably lower than for GO samples prepared via chemical oxidation. In the case of the concentrated sulfuric acid, beyond the first-stage phase $C_{24}(HSO_4)(H_2SO_4)_2$, there is a two-phase region in which there is a superposition of GO formation and O_2 evolution. The processes in this range depend strongly on the current density, indicating that the processes involved are much slower than the formation of the graphite sulfate intercalation [33].

1.2.2 Over-Oxidation of Graphite

Whereas graphite treated with boiling sulfuric acid loses only a small quantity of carbon as CO_2 [7, 34], most of it is lost when oxidizing agents such as nitric acid or $K_2Cr_2O_7$ are added to the boiling mixture [7]. Even Schafhäutl observed that repeated treatment of a piece of graphite with boiling sulfuric acid and the addition of nitric acid leads to a continuous loss of the material and finally to a complete disappearance [6].

Gottschalk [7] showed that the graphite-like carbon obtained by thermal decomposition of GO (prepared via the "Brodie method") reacted faster during the Brodie reaction the more oxidation/decomposition cycles had been carried out: after the third oxidation, only traces of GO were found. In every oxidation cycle the quantity of graphite is reduced, and after the fourth oxidation step the graphite is gone completely.

Charpy [26] showed later that pristine graphite can be transferred to GO by bichromate or permanganate in sulfuric acid, but these reactions lead easily to over-oxidation: at room temperature one can get GO with nearly no loss of graphite. By oxidation with permanganate at 45 °C, a small part of the graphite is lost, whereas at 100 °C already 50% is oxidized to CO_2. In the case of chromate, 50% is lost at 45 °C and all graphite is transferred to CO_2 at 100 °C.

In their GO preparation experiments in which they used a modified "Staudenmaier method", Kohlschütter and Haenni [10] described that the amount of GO obtained decreases with every step of oxidation (the yield of GO being 95% after the first oxidation and only 54% after the fifth repetition of oxidation) till nothing is left. The amount of GO obtained decreases also with the time that graphite or graphite oxide is in contact with the oxidizing medium.

Hofmann and Frenzel [13] found that the C/O ratio of the obtained GO decreases (thus the oxygen content increases) with decreasing size of the graphite crystallites. Very interesting are the results of Luzi [25]. He demonstrated that, besides the insoluble GO, some soluble by-products, especially mellitic acid, are formed, even in chlorate/nitric acid mixtures. The quantity of this by-product increases with subsequent oxidation steps. According to Thiele [14], up to 50% of mellitic acid is formed when GO is prepared via the procedure of Staudenmaier. To avoid it, the portions of chlorate should be small, the time intervals of additions longer, and the reaction temperature should be kept below 20 °C. Thiele [14] stated also that GO can be transformed to humic acid. In a later paper Thiele [35] mentioned that under special circumstances (*not formulated*) humic acids can be obtained as an intermediate state of the total oxidation of graphite to CO_2 by application of alkaline oxidation agents. In the same paper [35] he stated that humic acid is formed when an aqueous suspension of GO is left in air.

Ruess [19] remarked upon an increasing tendency to form mellitic acid or to destroy the graphite structure with decreasing particle size of the starting graphite. *This tendency is confirmed by my own experience. We have obtained mostly mellitic acid with the Hummers–Offeman method, when we applied the optimal reaction conditions for getting GO from crystalline graphite (particle size 500 μm) to a graphite powder with particle diameters of < 100 μm (unpublished results).* And even Hummers and Offeman [28] recommended in their work-up process to filter the suspension while it is still warm to avoid precipitation of mellitic acid. In their GO preparation Hummers and Offeman used powdered graphite flakes with particle size 325 mesh (44 μm).

On the basis of these very early results in GO research, I have the feeling that Rourke et al. [22] *do not really present a new structural model, but investigate intermediate states between the formation of GO and mellitic acid.*

1.2.3 Formation Mechanism – First Approximation

Already in 1919 Kohlschütter and Haenni [10] (p. 125) had recognized the essential role of nitric acid to open the graphite interior for the access of oxidants without acting as an oxidant for the formation of GO.

In 1930 Hofmann and Frenzel [13] showed for the first time by the application of XRD that the layer distance increases from 3.4 Å to 8 Å when graphite reacts with a mixture of sulfuric acid and nitric acid. *Thus, they confirmed the suggestion of Kohlschütter and Haenni* [10] *that nitric acid supports the opening of the interlayer space in graphite.* The product obtained is the "blue phase" described first by Schafhäutl [5, 6] and Brodie [2–4] and has been identified as a graphite sulfate [36]. Since this compound is highly sensitive to water, the pure first-stage compound (intercalation in every interlayer space) $C_{24}^{+}(HSO_4)^{-}(H_2SO_4)_2$ with the maximum content of sulfuric acid can be obtained by chemical or electrochemical oxidation of graphite in concentrated sulfuric acid [36, 37]. In a chemical oxidation with nitric acid in sulfuric acid (water content of 20%), the redox potential of the nitric acid allows only the preparation of a second-stage intercalation compound [37]. Although nitric acid is a strong oxidant, it is hard to get a pure first-stage nitric acid intercalation compound. In concentrated nitric acid (66 wt.%), only a mixture of higher stages can be obtained, and even in fuming nitric acid only a second-stage phase [36]. Only the addition of a small amount of permanganate allows the formation of the first-stage compound [38].

Thiele [39] as well as Rüdorff and Hofmann [36] mention that other oxidants such as $KClO_3$ or $KMnO_4$ or CrO_3 could take over the role of nitrate in the formation of graphite salts. However, they do not give experimental details. In 1992 Avdeev *et al.* [40] gave a physicochemical basis for the rational choice of oxidants allowing the formation of graphite sulfate salts. A detailed study of the system $KMnO_4/H_2SO_4/C$ was presented only in 2005 by Sorokina *et al.* [41]. They showed that the graphite sulfate salt is only formed when the MnO_4^- : C molar ratio is ≤ 1 : 1 [41]. The formation of a CrO_3/H_2SO_4 graphite salt is even more complex [42].

Thus, the unfortunate experiences of Brodie [1–4], Gottschalk [7], Charpy [26], Luzi [25] and Kohlschütter and Haenni [10] are not surprising (see section 1.2.1). From these experiences one can conclude that it seems to be unfortunate when the oxidant for the formation of GO has to act also as the oxidant for the formation of the graphite salt. In that case the opening of the interlayer galleries and the oxidation to GO start simultaneously and can, thus, lead to over-oxidation. The original formulation of Kohlschütter and Haenni [10] (p. 126) sounds, therefore, very modern:

> setzung der Graphitsäurebildung erfüllen. **Nur wenn die Oxydation gewissermaßen aus dem Innern heraus bzw. in der ganzen Masse zugleich ansetzt, kommt es zur Bildung von Graphitsäure.** Greift das Oxydationsmittel von außen her an, so geht die Oxydation bis zu den einfachsten Produkten, da die Einwirkungsdauer für eine kleine Menge Graphitsubstanz verhältnismäßig lang ist, wenn die Reaktion nur Schicht um Schicht fortschreiten kann, und der schließliche Effekt immer — auch bei der Chloratmischung! — von der Zeit abhängt, während deren das Reagens wirkt. Infolgedessen bleibt bei Oxydationsmitteln, die weniger leicht als Salpetersäure in die Substanz eindringen, die Oxydation vielfach nicht bei Graphitsäure stehen, und wird ander-

(English translation: "Only when the oxidation starts simultaneously in the total mass from the interior does the formation of graphitic acid occur. If the oxidizing agent attacks [the graphite] from outside, the oxidation continues to the simplest products, since the time of action for the small amount of graphite is too long, when the reaction can proceed only layer by layer …")

To get optimal access of the oxidant to the interlayer space and to get GO monolayers in the colloidal state, it must be guaranteed that a pure first-stage intercalation compound has been obtained. This corresponds quite well with the statement of Boehm et al. [27] *that the water content in sulfuric acid should not exceed 15% for the preparation of GO of high quality. However, it is hard to recognize the ideal endpoint of the intercalation reaction because of inter-stratification phenomena. Looking at the modern preparation methods of GO, I get the feeling that the importance of the intercalation process for optimal GO formation is mostly overlooked and the time before the addition of the oxidants is too short for optimal intercalation.*

If the interlayer space is free for the access of the oxidizing agent, two questions arise: What are the oxidizing species? And how do they enter into the interlayer space? It is generally assumed that ClO_2 *and* Mn_2O_7 *formed in concentrated sulfuric/nitric acid mixtures from* $KClO_3$ *and* $KMnO_4$ *are the active species. In the case of* $KMnO_4$ *in concentrated sulfuric acid, the reactive species could also be the* MnO_3^+ *cation* [43]. *However, it is hard to imagine that this cation enters the interlayer space constrained by two positively charged carbon layers. To enter into the interlayer space,* ClO_2 *and* Mn_2O_7 *have to replace the sulfuric acid molecules that form together with the* HSO_4^- *ions a densely packed arrangement in the interlayer space in the precursor intercalation compound. This exchange process has to occur against a huge excess of sulfuric acid in the surroundings of the graphite salt. Therefore, it seems to me more probable that the ion exchange of the monovalent chlorate and the permanganate ion against the* HSO_4^- *ion is the essential next step in the GO formation. (The possibility of ion exchange in the graphite salts was mentioned by Boehm* et al. [27] *and demonstrated by Rüdorff and Hofmann* [36].*) Both ions could be present to some extent since the* pK_a *values of sulfuric acid, chloric acid and permanganate acid are almost of the same order of magnitude (*H_2SO_4/HSO_4^-*, −3;* $HMnO_4/MnO_4^-$*, −2.25;* $HClO_3/ClO_3^-$*, −2.7; provided these values have any meaning under conditions in which the pH scale is not really valid). However, one essential problem is left: permanganate and chlorate act differently on organic compounds; but in the case of GO formation both must act very similarly.*

1.3 Discovery of Essential Functional O-Containing Groups and its Relation to the Development of Structural Models

1.3.1 Analytical Composition of Graphite Oxide

Throughout the history of GO research, the composition of GO was determined by combustion analysis to determine the C and H contents. The oxygen content corresponds to the difference from 100%. Only Thiele [14] carried out reduction reactions of GO to determine the amount of reducing agent necessary for full reduction, to get the C/O ratio. However, there was general agreement at that time that the reduction is never complete even under strong conditions (e.g. up to 800 °C). To get the correct C/O ratio one has to make a correction for the ash left after burning. A serious problem for the determination of the correct

C/O ratio is the amount of hydration water left in the sample after careful drying without the risk of deterioration. Boehm and Scholz [30] recommend using samples for analysis with 10–17% residual water because this corresponds to the water content of GO in the normal water vapor pressure in the laboratory atmosphere; thus, the samples do not change the weight so much during weighing.

For samples prepared via "Brodie's method", the carbon and hydrogen contents are fairly constant over the time of GO research. Brodie reported a carbon content of 60–61% and a hydrogen content of 1.75–1.91%; Gottschalk [7] reported similar values, but questioned whether these values mark an endpoint of drying. However, Boehm and Scholz [30] confirmed, despite some scatter, the values of Brodie [2].

For their GO prepared via the modified "Staudenmaier method", Kohlschütter and Haenni [10] found after the first oxidation a carbon content of 59% and a hydrogen content of 1.91%. The carbon content decreases with the number of oxidation/purification cycles to 55%. Such strong alterations have not been reported since. Boehm and Scholz [30] reported values between 56% and 58% with an average of 57%.

Hummers and Offeman [28] reported a carbon content of 47% for their product and a carbon content of 52% for the sample prepared according to the "Staudenmaier method". These values are the lowest to be found in publications. Boehm and Scholz [30] found the carbon content in the range from 56% to 63.7% for the Hummers–Offeman GO product.

Boehm and Scholz [30] stated that there is always a large scatter in the analytical results of various samples, but they are convinced that there is a trend in the degree of oxidation: the value of the C/O ratio is highest for the Brodie samples (C/O = 3.5), slightly lower for the Hummers–Offeman samples (3.3) and lowest for the Staudenmaier materials (2.9). In contrast, Thiele [14] stated *(without giving experimental proof)* that the number of oxidation cycles and the method of preparation have no serious influence on the analytical results. On the other hand, he gave the ratio of C : O : H of 2 : 1 : 1 for his own samples prepared via the "Staudenmaier method". This would correspond to the full oxidation of all aromatic double bonds.

Summing up all the analytical data in the Boehm and Scholz publication [30] *and the analytical work done by all scientists working on GO since Brodie, I conclude that carbon contents lower than 55% (corresponding to a C/O ratio below 2.5) indicate over-oxidation of GO, and that carbon contents higher than 63% (C/O ratio above 3.3) are caused by incipient decomposition due to drying the samples at about 100 °C.*

1.3.2 Creation of the Structural Model from 1930 till 2006

Based on his analytical result and especially on the unusually high hydrogen content, in 1930 Thiele proposed that oxygen is attached to the carbon grid in the form of OH groups [14]. Since the carbon atoms form a honeycomb arrangement, he gave a sum formula $C_6O_3H_3$. To be in agreement with this sum formula, Thiele had to assume that the OH functions are covalently bound to every second C atom on both sites of a carbon bilayer, which are interconnected by covalent C–C bonds. A sketch of this structural model is shown in Figure 1.1. Thiele considered the possibility to transform GO into humic acid as an additional hint for the presence of OH groups in GO.

Almost 60 years after Thiele, Nakajima *et al.* [44, 45] presented a very similar structural model of GO (see Figure 1.1, right), without giving any reference to Thiele's paper. Nakajima *et al.* proposed a slightly different sum formula ($C_8(OH)_4$) instead of $C_6(OH)_3$) and based their model on the similarity of XRD data to those of a carbon fluoride compound with the same composition. Neither of these models found attention in the community.

An alternative model for the structure of GO was presented by Hofmann *et al.* [15] in 1934. The starting point of the model construction was based on the following considerations. As a maximum, all three double bonds per carbon hexagon could be replaced by three oxygen atoms bridging two neighboring carbon atoms (forming epoxide units). Since the most frequently found C/O ratio is 3, some double bonds have to be left unoxidized. In the oxygen removal from some ethylene oxide compounds under re-formation of a double bond via HI treatment, Hofmann *et al.* [15] saw an analogy to the ease of oxygen removal in GO. This analogy was considered as a hint for the presence of epoxide functions in GO. From the fact that the heat of combustion of GO is quite similar to that of graphite, Hofmann *et al.* concluded that the oxygen is only weakly bound to the carbon layers in GO [15]. An additional proof is presented by the authors via the first single-crystal XRD analysis of GO.

For the calculation of the intensities of the reflections, four different arrangements of oxygen atoms has been considered (Figure 1.2, top, a–d). The observed intensities agree best with the calculated intensities for model a) under the additional assumptions that every third C layer is at the same position, and that the oxygen atoms are statistically distributed between the possible sites. This arrangement is only possible if the carbon grid remains flat

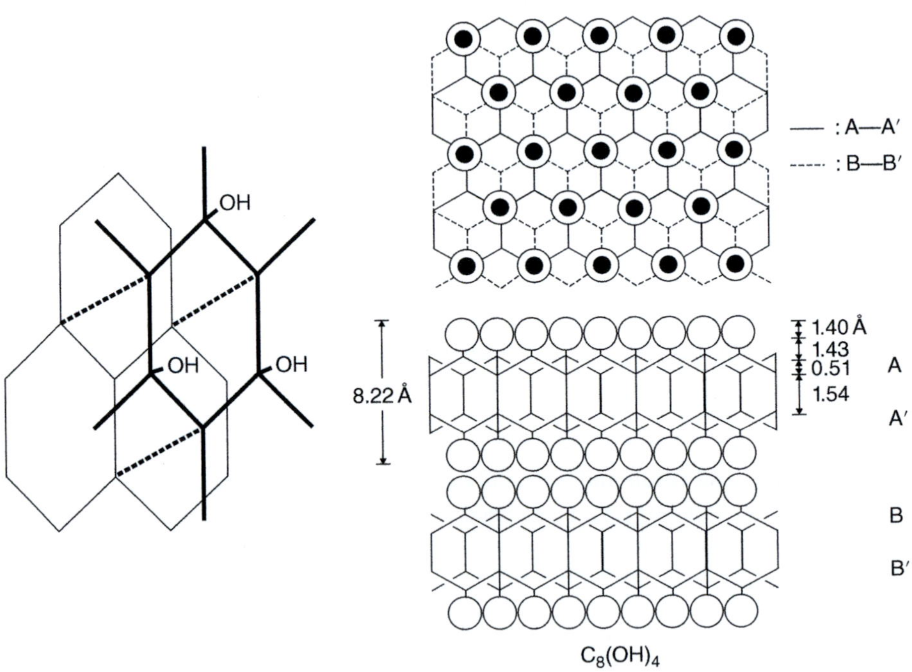

Figure 1.1 *The structural models of Thiele [14] (left) and Nakajima* et al. *[44, 45] (right). (left) Reproduced from [14] with permission of Wiley-VCH. (right) Reproduced from [44] with permission of Elsevier*

as in graphite. The authors did not see any evidence for the presence of OH groups covalently attached to the carbon grid. They ascribe the hydrogen exclusively to residual "bound" water, without specifying the nature of this bonding state [15]. *The unusually high hydrogen content in Thiele's samples can be explained by the retention of the organic solvents (acetic acid, acetic anhydride, ether) used by Thiele in his work-up process (described in section* 1.2.1*).* In a subsequent paper, Hofmann and König [46] devoted a small section to sum the formula of Thiele, which they discarded again. This conclusion has been confirmed also by a very thorough study of Thiele's purification method carried out by Ruess [47]. In his experiments Ruess observed some evidence for a covalent attachment of the acetyl groups from the acetic acid anhydride to the OH groups (see section 1.4.6).

In the 1937 paper of Hofmann and König [46] there is an interesting turn. They confirm the cation exchange capacity in slightly basic medium determined by Thiele [35] and conclude that the amount of carboxyl groups in GO will not be enough to explain the acidity of GO. Thus, OH groups like those in humic acid may fill the gap. In addition, they suggest that the OH groups may be formed by hydrolysis of the epoxide groups.

In 1947 Ruess pointed out in his paper [19] that GO is not as reactive as organic epoxides and does not show the typical ring opening reactions with ammonia, HBr or HCl forming vicinal hydroxyamines or halogenohydroxides. He questioned also the idea that hydroxyl groups can be formed by hydrolysis of the epoxide functions. Based on these chemical arguments, the covalent bonding of OH to the carbon grid and a re-evaluation of Hofmann *et al.*'s [15] and his own XRD data, Ruess [19] presented a new structural model. He claimed the best fit between the observed and calculated structure factors he had obtained, shown in Figure 1.3.

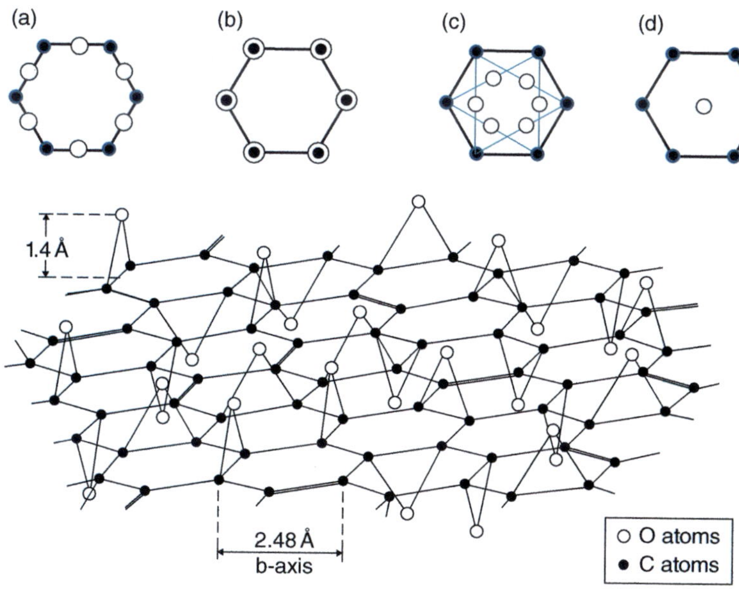

Figure 1.2 *The structural model of Hofmann* et al. *[15]. Reproduced from [15] with permission of Wiley-VCH*

The carbon grid is now corrugated in a form typical for an arrangement of cycloalkanes with sp^3-hybridized carbon atoms. This structure no longer allows epoxide functions, and is therefore replaced by 1,3-ether functions. According to Ruess's determination of non-removable hydrogen atoms, there is about one OH group per six carbon atoms in the structure. The remaining oxygen atoms form the ether bridges. Their number and the number of double bonds vary with the degree of oxidation.

However, a single type of OH groups – particularly tertiary OH functions – is hardly compatible with the variety of reactions of the GO hydroxyl functions (see section 1.4.5) [16, 46, 48, 49]. Therefore, Clauss *et al.* [48] suggested that an OH function in the neighborhood of a C–C double bond (enols) could be much more acidic. However, in that case C–C bonds must have been broken, and the arrangements shown in Figure 1.4 (left) could be present in GO.

The transition from an enol to a keto function could explain the transition from the bright to the dark state of GO. However, this tautomerism could not be clearly established by the authors in an infrared (IR) spectrum. IR spectra by Clauss *et al.* [48] and earlier by Hadži and Novak [50] clearly confirmed the presence of carboxyl and hydroxyl groups but gave no indications for the 1,3-ether functions. Trials to split these ethers were unsuccessful also [16, 48]. Hadži and Novak assigned a weak feature at a wavenumber of $980\,cm^{-1}$ as a hint for the presence of epoxide functions. In a later paper Scholz and Boehm [16] assigned the signal at $1720\,cm^{-1}$ to a carbonyl function, which in alkaline solution undergoes a keto–enol tautomerism or forms a geminal diol. Another vibration mode at $1620\,cm^{-1}$ was interpreted as originating from a carbonyl function of an extended quinonoid system. This assignment was the basis of the new structural model of Scholz and Boehm [16] shown in Figure 1.4 (right). Note that this assignment was later shown to be incorrect [51]. *I think that the model published by Szabó* et al. [51] *combines structural elements of both Scholz and Boehm* [16] *(broken C–C bonds, quinone groups) and Ruess* [19] *(1,3-ethers).*

In 1962 Aragon de la Cruz and Cowley [52] found the intensity distribution of the spots in their electron diffraction pattern to be compatible with epoxide functions over C–C double bonds and OH groups on top of carbon atoms in a random distribution, in agreement with a modified Hofmann model. However, the authors conceded that this interpretation

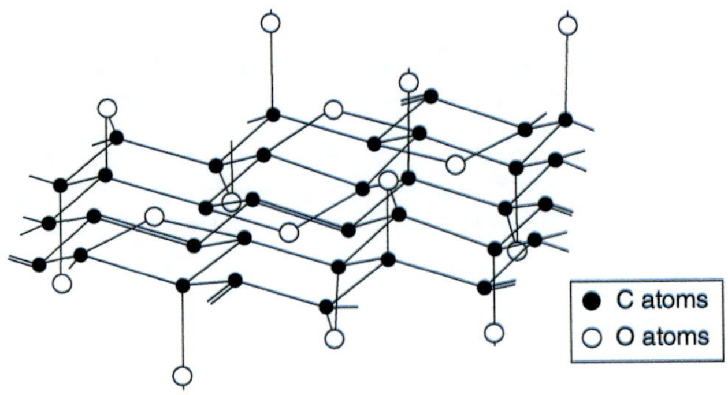

Figure 1.3 The structural GO model according to Ruess [19]. However, this figure is taken from Ref. [48] because it shows clearly the double bonds left unoxidized. Reproduced from [48] with permission of Wiley-VCH

suffers from some uncertainty. And, indeed, Scholz and Boehm [16] stated that a similar electron density distribution would be found for carbonyl functions in neighboring carbon layers lying on top of each other. The electron density distribution could also originate when the layers were built up from small areas oriented in all three *a*-directions [16].

Summing up the results of this era of investigation, it is disappointing that the application of the most modern methods at that time did not allow the structure of GO to be unambiguously clarified. With respect to the XRD, this is mainly due to the fact that, for GO, one can observe only two 001 reflections and a few *hk*0 reflections, but no *hkl* reflections.

A new era in GO studies started with the possibilities offered by MAS NMR. NMR monitors mainly the electronic structure of the first coordination sphere and is, therefore, less sensitive to the long-range order in a crystal. Thus, it is the ideal method for investigating almost amorphous materials. The first NMR experiments on GO were published by Mermoux *et al.* in 1989 and 1991 [17, 18]. The ^{13}C NMR spectrum shows just three dominant signals, as shown later in Figure 1.6(a), with chemical shifts of 60, 70 and ~130 ppm. The three signals of an as-prepared GO have been assigned to ether functions, to alcohol functions and to aromatic or conjugated double bonds, respectively. Therefore, Mermoux *et al.* [18] ruled out the structural models of Thiele [14], Nakajima *et al.* [44, 45] and Scholz and Boehm [16]. Whereas in their first paper [17] Mermoux and Chabre assigned the 60 ppm signal to epoxide functions, in the following paper [18] they did not differentiate between epoxide (1,2-ether) and 1,3-ether functions and considered their spectra as confirmation of Hofmann *et al.*'s [15] and Ruess's models [19]. At the end the authors preferred

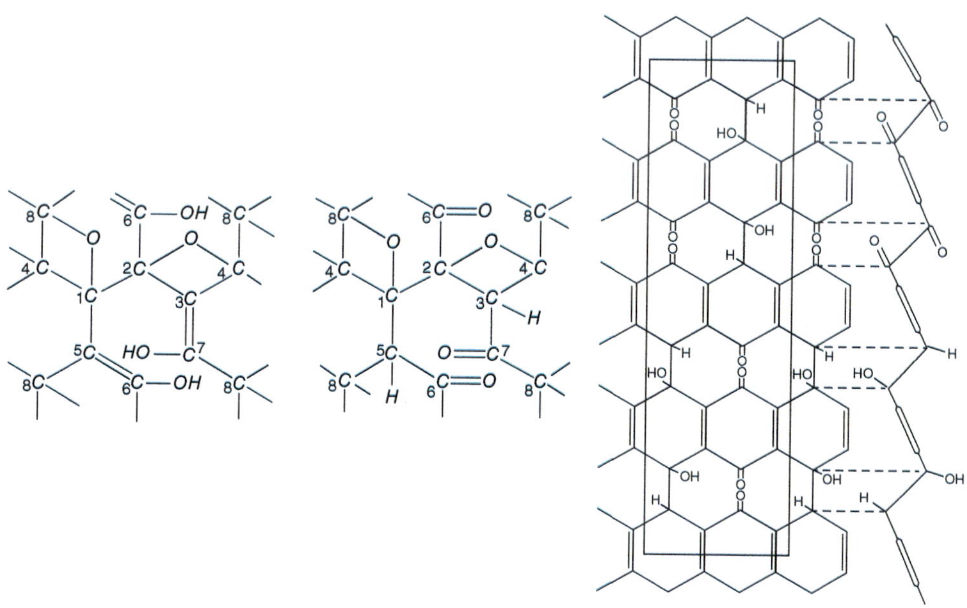

Figure 1.4 *Possible keto–enol tautomerism in graphite oxide (left) according to Clauss et al. [48], and the new structural model (right) of Scholz and Boehm [16]. (left) Reproduced from [48] with permission of Wiley-VCH. (right) Reproduced from [16] with permission of Wiley-VCH*

a Ruess-type model, but name the 1,3-ethers as epoxides as well. In subsequent papers [53, 54] the position of the authors is not clear. On the one hand they assign the 60 ppm peak to epoxides, but on the other hand they argue only based on the model of Ruess, despite Ruess [19] clearly stating that epoxides are not compatible with corrugated layers of sp^3 carbons.

The assignment of the 60 ppm signal to epoxide groups in our studies signifies the return to the structural model of Hofmann with some modifications [20, 55]. Part of the epoxide groups must be replaced by tertiary OH groups; and the two functional groups are more or less randomly distributed over both sites of the carbon grid. Since the signal at 130 ppm can be ascribed to aromatic carbon atoms, patches of aromatic rings of varying size are distributed between the oxygen-carrying areas. These aromatic areas and the epoxide groups keep the carbon grid almost flat; some deviations may result from the sp^3-hybridized carbon atoms carrying the OH groups. A sketch of the resulting structure is shown in Figure 1.5.

To be honest, one should be aware that the interpretation of the NMR spectra might not be the end of the story. We cannot differentiate if isolated or conjugated double bonds contribute to the signal at 130 ppm. In addition, NMR is a local probe and, thus, we do not know whether this signal is caused by isolated aromatic rings or by polycyclic arrangements as we have shown it in the sketch of our model. Thus, the NMR data could also be compatible with a distribution of aromatic rings following Clar's concept of aromaticity (as shown in figure 1 of Ref. [56]*). The signals at 60 and 70 ppm overlap partially. A deconvolution analysis showed that perhaps four signals are hidden in the experimental peak. Going through NMR data collections (e.g. Ref.* [57]*), one can find structural features that could be compatible with the structural models discussed with almost similar chemical shifts, i.e. allylic alcohol. However, it is clear that in pristine GO there is little evidence for the presence of enol or phenol functions.*

1.3.3 Considerations for the Formation Mechanism – Second Approximation

As discussed in section 1.2.3, the intercalation of sulfuric acid and/or nitric acid into graphite is the first step of GO formation [10, 13, 16]. Recently, Dimiev and Tour [58] confirmed elegantly by means of modern scientific methods (XRD, Raman spectroscopy) all the observations on GO formation described from the early days of GO research by Brodie [2], Gottschalk [7], Staudenmaier [9], Luzi [25], Kohlschütter and Haenni [10] and Hofmann and Frenzel [13], just to mention the most important workers.

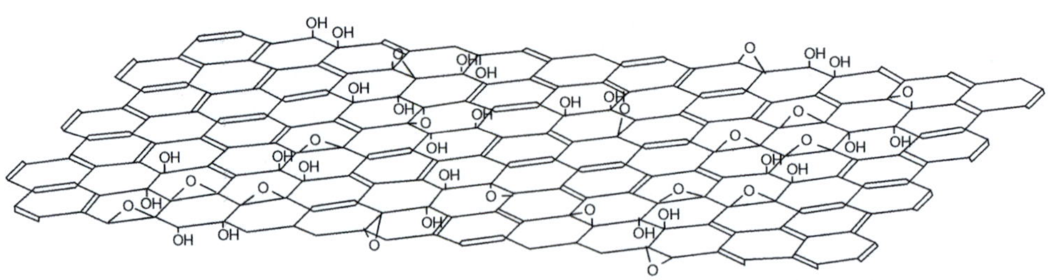

Figure 1.5 Lerf–Klinowski model of the as-prepared GO structure [55]. Reproduced from [55] with permission of Elsevier

Since it is hard to guarantee a pure first-stage phase throughout the intercalated graphite particles, it cannot be ruled out that there are multi-layer assemblies of a second-stage phase. In that case the oxidation could occur only in any second interlayer gallery, resulting in an assembly according to Thiele's model [14]. The real arrangement must not follow the ideal and very regular arrangement in Thiele's model, but there could be some covalent linkages of C atoms (carrying OH groups) between the neighboring C layers. Besides the OH groups, there could also be epoxides or other functional groups.

Hofmann *et al.* [15], Boehm *et al.* [16, 48] as well as Dimiev and Tour [58] refrain from discussing the molecular mechanism of graphite oxidation. This is understandable because there are no low-molecular-weight model systems with large polycyclic aromatic hydrocarbons (PAHs) where the reaction behavior of hydrogen-free carbon atoms has been investigated. Therefore, considerations on the mechanism of GO formation must start from knowledge of the reaction behavior of low-molecular-weight PAHs.

Assuming epoxides as an important structural element in GO, one is faced with the fact that neither ClO_3^- nor MnO_4^- are known to oxidize PAHs to epoxides [59]. MnO_4^- reacts with isolated double bonds to form *cis*-diols [59, 60]. This sort of reaction could happen only at the peripheral carbon atoms of the graphite layers, and the subsequent reaction from diols to ketones could occur with breaking of C–C bonds. Whether isolated double bonds can undergo the same reactions is not clear. The proposed mechanism of diol formation has been proved experimentally only for permanganate oxidation in neutral or basic solutions [60]. However, it cannot be ruled out completely that this diol–ketone reaction can happen also in concentrated sulfuric acid, as has been discussed for the unzipping of carbon nanotubes [61].

In the destructive oxidation of PAHs, the permanganate ion is bound via an oxide ion to the carbon under abstraction of a proton, the formation of a carbonium ion and the reduction of manganese to the Mn(v) state [62]. At the end of the reaction, the manganese is liberated and a quinone is formed. This process could work again only for peripheral carbon atoms or at defect sites in the graphite layers. Such defect sites could be the nuclei for the subsequent oxidation of the whole layers. The product of this reaction would be compatible with the structural model of Scholz and Boehm [16] and Szabó *et al.* [51]. However, according to the NMR spectra, quinones are at best minority species in pristine GO *(I guess less than 10%)*.

If one assumes that *cis*-diols are formed, then there is no mechanism explaining their transformation to 1,3-ethers postulated by Ruess [19]; the water abstraction to epoxide groups by concentrated sulfuric acid can also be questioned. Whether these reactions can occur in the interlayer space of GO is also unclear. Even less clear is how these functional groups could be formed with the other important oxidant ClO_2 (or ClO_3^-, $HClO_3$?). Their oxidation power is so high that low-molecular-weight compounds are completely destroyed. In addition, when oxidants diffuse from outside into the interior of graphite salt crystals, one would expect that oxidation occurs at the site of entry. The removal of the reaction by-products and the different chemical nature of the carbon surfaces could obstruct the progress of oxidation toward the center of the crystals.

Therefore, it seems necessary to think about a common mechanism for both oxidants. If in the case of the chlorate and permanganate ions the ion exchange process were faster than the oxygen liberation, the process of oxidation could start in the interior of the graphite salt crystals. If this liberated oxygen is in the $^1O_2$ state and cyclohexadiene units exist due to the disturbances in the graphite π-electron system (positive electron holes or the carbocations), it could be attached to the graphite sheets as peroxide functions, which can *thermally rearrange to diepoxides in analogy to reactions discussed in Ref.* [63]. *Starting from these oxidized nuclei, new sites would be created to continue the reaction. The epoxides could be hydrolyzed to hydroxyl groups. It should be recalled that Boehm postulated that some residual water in the concentrated acids is necessary for GO formation* [27].

However, one should keep in mind that these considerations are deduced from the solution chemistry of PAHs, and it is not clear whether these reactions can occur in the same way in the constrained space of the interlayer galleries and in an electrostatic field created by the interaction of the intercalated anions with the positively charged graphite planes. In addition, there is a highly protic solvent in the surrounding of the crystals which is also present as a solvent in the interlayer space.

1.4 Properties of Graphite Oxide

1.4.1 Thermal Degradation and its Products

In the 1850s Brodie [2] described the vigorous decomposition of GO during heating. To control the process better, he carried out the decomposition reactions in purified naphtha. In the temperature range of 100–200 °C considerable amounts of water and CO_2 are liberated. The black residue obtained after heating GO in this fluid for a few hours at 240 °C still contains some oxygen, and Brodie determined its composition as $C_{22}H_2O_4$. By heating this sample in a stream of nitrogen, more CO_2 and CO were liberated, but even after heating the sample to red heat some oxygen and hydrogen was left in the sample.

Kohlschütter and Haenni [10] as well as other workers before (e.g. Gottschalk [7], Berthelot [8] and Staudenmaier [9]) confirmed the observations of Brodie and stated that the form of the black decomposition products is in close correlation with the form of the graphitic acid and the starting graphite materials. The decomposition residues become more voluminous the more often the oxidation and cleaning cycle is repeated [10]. Kohlschütter and Haenni [10] found also that the temperature of deflagration is the lower the slower the materials are heated up. Decomposition under pressure resulted in a more graphite-like residual compound with higher density. When the samples are heated in concentrated sulfuric acid, the samples decompose again to graphite-like material, but do not deflagrate. Hofmann found by XRD investigation *(this is the first application of XRD in the GO research carried out in 1928)* that the decomposition products obtained under pressure or in sulfuric acid are more crystalline than the soot-like decomposition products obtained in decomposition products at normal pressure [12].

Hofmann *et al.* [15] stressed later that no oxygen can be detected in the gaseous decomposition products and concluded that the liberated oxygen had reacted with carbon at the crystal edges, forming CO_2 and CO. Later De Boer *et al.* [64] reported that they discovered a considerable amount of oxygen due to the decomposition of the proposed peroxides present in GO. This conflicting result was not confirmed by Boehm and coworkers [16, 65]. Boehm and Scholz [66] showed in addition that the deflagration point of GO decreases in the order $GO_{Brodie} > GO_{Staudenmaier} > GO_{Hummers–Offeman}$. The low decomposition temperature of the Hummers and Offeman samples could be caused by catalysis from residual manganese ions. This conclusion is based on the observation that a GO sample soaked in an aqueous $FeCl_3$ solution explodes under fire. In general, any impurity leads to lowering of the deflagration point [66].

Hofmann *et al.* [15] found that the interlayer distance decreases almost continuously during decomposition. At a temperature slightly above the temperature of deflagration (320 °C), the layer distance amounts to 4.05 Å, and after heating up to 850 °C, it is 3.38 Å, almost identical to the layer distance of graphite. This can be explained only when the gaseous decomposition products leave the interlayer spaces almost randomly and when there is an almost random interstratification of completely empty and partially oxygen-loaded interlayer spaces.

Decomposition of GO can occur at 70 °C [66]. Very slow decomposition of GO with the liberation of CO_2 and CO can be observed already at 50 °C [67]. Hofmann and Holst [49] showed that the ion exchange capacity decreases irreversibly when the samples have been dried at 130 °C.

To the best of my knowledge, Matuyama carried out the first thermogravimetric and differential thermal analysis (TG-DTA) measurement in 1954 [68]. He chose a heating rate of 42 °C h^{-1} to avoid deflagration. The weight loss starts slightly below 50 °C and reaches a plateau at ~100 °C. This weight loss is mainly caused by dehydration. The greatest weight loss occurs in the range 150–200 °C and amounts to about 40%. Up to 300 °C the weight loss continues smoothly. Martín-Rodríguez and Valerga-Jiménez [69] extended the TG-DTA analysis up to 800 °C and determined the evolved gases by gas chromatography. They confirmed the results of the previous works. In a most recent TG-DTA study of the thermal decomposition of GO coupled with a mass spectroscopic analysis of the evolved gases, new (up to now unknown) gaseous species with molar masses of 43, 46, 59 and 60 have been detected [70]. Hofmann *et al.* had already mentioned the liberation of "carbon like oxygen containing compounds" beside CO_2 and CO as decomposition products [15].

1.4.2 Chemical Reduction Reactions

Again, Brodie [2] was the first to describe that the treatment of GO with reducing agents such as potassium or ammonium sulfide, Cu(I) chloride or Sn(II) chloride leads to a black residue with the appearance of graphite. However, he could not purify the obtained solids and, thus, a further study was given up. Gottschalk [7], Staudenmaier [9] and Kohlschütter and Haenni [10] using Cu(I) chloride and Fe(II) sulfate confirmed the graphite-like appearance of the reduction products. These authors stated also that the reduction product can be transferred again to GO.

Thiele's conclusion [14] that GO can be quantitatively reduced to graphite was not confirmed by Hofmann *et al.* [15]. They found that the amount of oxygen that can be removed depends on the reducing agent used: with Fe(II) chloride, ~68% oxygen can be removed; with hydrazine hydrate in aqueous medium, 82%; and with H_2S in water, 91%. In the latter case Hofmann and Frenzel assumed that sulfur replaces oxygen by forming the sulfur analog of the epoxides, which decompose already at room temperature [71]. Lerf *et al.* treated GO with thiourea, an agent sensitive to epoxides; however, they could not get an NMR spectrum because the sample exploded in the rotator after starting the measurement [20]. Only more recently, in 2003, have Bourlinos *et al.* [72] introduced new reducing agents: $NaBH_4$ and hydroquinone. Especially the latter is of interest because the graphite obtained shows a higher crystallinity.

When GO is reduced with hydrazine, first mentioned by Thiele [73], the residual solid retains a lot of water (>30%) although the layer distance of 3.5 Å is only slightly larger than that of the graphite [46]. Most of the water can be removed by application of mechanical pressure on the sample. More water can be removed then by heating the sample to 110 °C, and it requires temperatures up to 350 °C to remove all the water. Hofmann and König [46] explained this surprising fact in the following way: since an aqueous solution of hydrazine is alkaline, GO disperses to piles with a few GO monolayers and these piles are further propped apart by the reduction processes; after reduction the lamellae are deposited randomly to a dense felt where the water is occluded in hollow spaces. The loss of water during heating is again a continuous process and is accompanied by a decrease of the linewidth of the 00 *l* Bragg peaks indicating an increased ordering of the graphite layers in the direction of the crystallographic *c*-axis. Thiele reported that the specific conductivity is in the order of that of graphite [35].

Whereas Hofmann *et al.* [15] considered the deoxygenation of molecular epoxides with hydrogen iodide under the formation of double bonds as a model case for the deoxygenation of GO, Ruess [19] and later Scholz and Boehm [16] mentioned that hydrogen iodide does not react with the epoxide or the 1,3-ethers in GO. However, De Boer and van Doorn [74] observed a reduction of GO with sodium iodide and ascribed the amount of liberated iodine to the reduction of a hydroperoxide function and to a ketone function which is involved in a keto–enol equilibrium.

By means of ^{13}C NMR spectroscopy it has been shown that iodide treatment affects GO considerably (see Figure 1.6). Following the suggestion of Hofmann *et al.* [15], Lerf *et al.* [20] assumed that iodide could lead to a selective abstraction of the epoxide oxygen. However, the intensity of the peak assigned to epoxide groups (60 ppm) decreased simultaneously with the intensity of the 70 ppm signal (Figure 1.6a). Simultaneously, new features appear

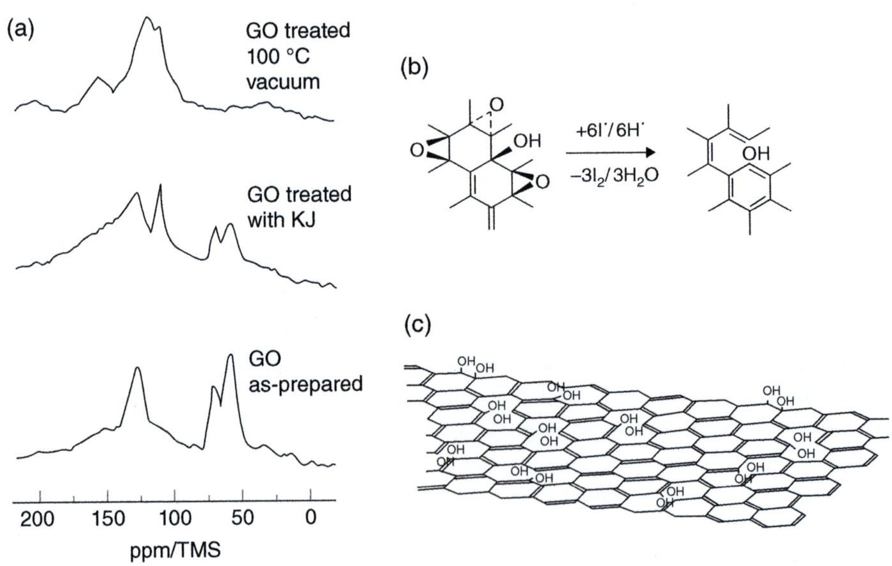

Figure 1.6 (a) ^{13}C MAS NMR spectra of GO modified by thermal treatment and reduction with KI [20, 55]. (b) Proposed reorganization of the double bond system after liberation of oxygen [20, 55]. (c) Modified structure of GO containing only phenols and double bonds after complete removal of epoxide groups [55]. Reproduced from [55] with permission of Elsevier

in the NMR spectrum: a peak at ~110 ppm and another at ~160 ppm. The simultaneous appearance of these signals has been interpreted as the formation of phenols in GO.

A complete disappearance of the epoxide and OH functions and a still higher portion of signals typical of phenols were observed when GO was treated at 100 °C in vacuum. The transformation of OH functions can only occur when the electron system has been reorganized completely after removal of oxygen and breaking of some C–C single bonds (Figure 1.6c). In an extreme case, only phenol groups and the rearranged π electron system could be left (Figure 1.6c).

These results demonstrate a relation between the structural models of Hofmann et al. [15] *and Lerf and Klinowski* [20, 55] *and the slightly modified model of Scholz and Boehm* [16] *and Szabó* et al. [51] *due to the proposed chemical transformations. The Scholz–Boehm model can now be considered as a verification of Thiele's suggestion that oxygen could enter the carbon grid as a first step of GO transformation to humic acid* [14, 35]. *Pallmann* [75] *and Hamdi* [29] *compared the colloidal dispersion, the ion exchange capacity and the acidity of GO, humic acid and different forms of lignin (see section* 1.4.5) *and found that lignin takes an intermediate state between humic acid and GO.*

1.4.3 Reactions with Acids and Bases

GO was shown to be very stable against acids, at least against oxidizing acids. In his cleaning process Thiele treated GO with boiling concentrated nitric acid without evidence for destruction [14]. Several authors added ClO_2 or nitric acid to aqueous suspensions of GO to keep the samples bright-colored [7, 9, 30, 48].

Aqueous alkali/alkaline earth–alkali hydroxide solutions can interact with GO in different ways:

- Dilute NaOH solutions are essential for the redispersion of acid-precipitated GO or to cause osmotic swelling of GO (see section. 1.4.4).
- Alkali and alkaline earth acetate solutions deprotonate GO, leading to GO–alkali salts (see section 1.4.5) and these neutralization reactions seem to be reversible.
- OH^- ions can initiate more or less reversible alterations of GO.

Only the latter phenomenon will be discussed here. GO suspensions in dilute alkaline solutions (<0.1 M) turn black within minutes, but NMR spectra do not show significant changes even when the samples are exposed to the solution for a day at room temperature (unpublished results). In contrast, Dimiev *et al.* [76] found a significant decrease of the peak intensity of the alcohol and epoxide peaks, but no other significant changes in the spectrum. When a GO–ethylate sample is treated with a 1 M NaOH solution at 70 °C for 3 h, the NMR spectrum looks similar to the NMR spectrum of the thermally treated sample shown in Figure 1.6(a) [20]. The only difference is that the sharp signals are superimposed on a very broad background with a maximum at about 130 ppm, perhaps indicating that the changes of the samples are more serious than in the case of the thermally treated sample. Fan *et al.* [77] observed, by treating GO with 18 M NaOH solution at 80 °C, a spectrum similar to ours, which we have obtained for a sample treated at 100 °C in vacuum.

Therefore, I think that Fan et al. *observed a "base-assisted thermal decomposition". Recently, Dimiev* et al. [76] *proposed chemical rearrangements in the GO structure,* *triggered by the nucleophilic attack of OH^- ions. The proposed mechanism explains for the first time the liberation of CO_2 during the slow decomposition of GO in basic solutions. It is also a model process to explain the lack of oxygen liberation during thermal deflagration of GO* [15, 65] *as well as the evolution of CO_2 in the thermal decomposition of GO at 50 °C* [66], *if one assumes that the strongly bound water can act as a nucleophile, perhaps after deprotonation. In that case it could be that the changes of GO caused by soaking GO in water* [35] *or by exposure to sunlight* [67] *follow the same mechanism. The presence of Ba salts, which can act as a trap of the liberated CO_2 due to the formation of insoluble $BaCO_3$, may enhance the rate of decomposition* [35, 49].

The iodide reaction [20] *leads to very similar products as the thermal decomposition, but it acts as a reducing agent. Therefore, the question arises how it can be combined with the reaction scheme of Dimiev* et al. [76].

1.4.4 "Osmotic Swelling": Hydration Behavior and Colloid Formation

The hygroscopic properties of GO have been described already by Brodie [2]. Gottschalk [7] mentioned the presence of "Hydratwasser" casually. Kohlschütter and Haenni [10] mentioned the weight gain of dried GO exposed to air and that the water taken up can be removed easily during mild heating. They mentioned also that the water is liberated in vacuum or by keeping the samples over concentrated sulfuric acid, but that it is hard to get constancy in weight even for long time runs.

Hofmann and Frenzel [13] found by means of powder XRD that the water taken up in GO props open the graphite oxide sheets. A sample containing about 95% water showed a layer distance of 11.3 Å; an air-dried sample with ~15% water, a distance of 7.84 Å; and a thoroughly dried sample with 7–8% water, a distance of 6.4 Å. The uptake and the release of water is a reversible process. From the layer expansion of ~6 Å the authors assumed that two layers of water are inserted between the GO monolayers when GO is soaked in water. Since the *hk*0 reflections are not changed by the water uptake, the authors call the phenomenon a "one-dimensional swelling".

Hofmann *et al.* [15] demonstrated for the first time the correlation between the water vapor pressure in the air with the percentage of water taken up and with the layer distances of the hydrated product. This correlation was later confirmed by Derksen and Katz [78] and Ruess [19]. Derksen and Katz [78] as well as Hofmann *et al.* [15] found that the isotherm of water adsorption shows a sigmoidal curve as for other known water adsorption processes. The maximum layer distance in fully hydrated GO decreases with temperature from 11.3 Å to 9.95 Å at 80 °C [78]. They found also, like Hofmann *et al.*, that the layer distance increases almost linearly with the amount of water taken up during hydration. This effect has been confirmed later several times [79, 80]. It can be explained only in terms of random interstratification [15, 81, 82]. Hofmann *et al.* [15] mentioned that partially decomposed GO samples with C/O ratios up to 5 do not lose their swelling behavior in water.

According to the latter authors [15], shaking the fully hydrated GO in alkaline and ammonia solutions leads to a complete delamination of GO. Derksen and Katz [78] gave the layer distance of GO as a function of the pH. They showed the increase of the layer distance from 10.7 Å to 11.4 Å with decreasing proton concentration and a layer distance of 12.5 Å at pH 13. In the range $6 < pH < 13$ no layer distances are given because of the colloidal dispersion of GO. Clauss *et al.* [48] showed that the layer distance in 0.05 M and 0.03 M NaOH solution increases from 11.3 Å to 12.5 Å. In 0.02 M NaOH solutions the colloidal dispersion of the GO occurs discontinuously. The authors mentioned also that disintegration occurs at the same layer charge density (1–2 sodium ions per 100 Å^2) as for montmorillonites [48, 83].

Whereas the swelling behavior during the work-up of GO has been well known since the discovery of GO by Brodie, Kohlschütter and Haenni [10] described this process as colloidal dispersion ("Kolloidisierung") of GO for the first time and mentioned the precipitation with strong acids. Thiele was then the first who described the fundamental colloidal properties of GO [84], e.g. the movement of the colloidal particles towards the anode in an electrophoresis experiment, and the increase of the flocculation tendency according to the Hardy–Schulze rule in the series of flocculants $H < K < Ca < Al < Ce$.

Hofmann *et al.* used colloidal dispersions of GO to prepare membranes and measured their properties [85, 86]. These membranes were permeable for water and cations but not for oxygen and nitrogen. An application is difficult because of their instability due to the strong swelling. The fractal dimension of such "membranes" (*single-* or *multi-layer aggregates*) determined by light scattering has been determined to be 2.5. This has been interpreted by crumpling of the GO layers [87, 88]. Other workers disproved this finding and corrected the value of the fractal dimension to 2.15, indicating flat GO monolayers [89].

Boehm *et al.* [90, 91] succeeded in preparing extremely thin graphite lamellae by reducing dispersed GO via reduction with hydrazine or by thermal treatment (Thiele mentioned this possibility [84]). They found monolayers and aggregates up to four layers. Reduced samples in suspension have a specific surface area of ~800 m^2 g^{-1} determined by means of the methylene blue method, and only ~100 m^2 g^{-1} when measured with the Brunauer–Emmett–Teller (BET) method. Thus, Boehm *et al.* were the first to prepare graphene layers. In the new era, Stankovich *et al.* [92] simply used this method in the context of graphene formation. They showed also that hydrazine attacks the epoxide functions and transfers them to the nitrogen analog (aziridine), which then splits off the nitrogen.

Colloidal dispersions of GO have been used by Matsuo *et al.* to prepare GO–polyethylene oxide nanocomposites by flocculation of GO with the polymer [93, 94]. Later, synthesis of other nanocomposites, e.g. poly(vinyl acetate), polyacrylamide, poly(vinyl alcohol) [95–97] (and references therein) were reported. Almost at the same time Kotov *et al.* [98] described highly organized GO–polymer nanocomposites by means of layer-by-layer deposition using GO colloidal dispersions. Only a few research groups have used this elegant method [99–101].

1.4.5 GO Acidity

Again, Brodie [2] in 1859 as well as Gottschalk [7] in 1865 described the acidic properties of GO by the reaction with litmus. Thiele was the first to start a thorough investigation of the acidic properties of GO in 1937 [35]. He found that an aqueous NaCl solution becomes strongly acidic when GO is soaked in it. Thiele explained this observation by a release of protons from GO and a simultaneous uptake of sodium ions with the formation of a graphitic acid sodium salt. In sodium hydroxide solutions, Thiele determined a reversible uptake of sodium up to 800 mval/100 g of GO (mval is the same as mmol).

Almost simultaneously Hofmann and König [46] found that the cation exchange capacity from sodium acetate solutions of increasing concentrations approaches a limit of ~150 mval/100 g GO. The most obvious explanation for the acidity would be that there are COOH groups at the external edges of the GO particles. A rough estimate of the possible number of such acidic groups would result in an uptake of sodium in the order of 1 mval/100 g GO [46]. The assumption of a partial destruction of the graphite layers during exhaustive oxidation could lead to a strong increase in the number of COOH groups when the edges of the interior holes are also fully covered with them, as shown in Figure 1.7. Even in such a case, the high cation exchange capacity cannot be explained. Therefore, Hofmann and König concluded that, apart from the carboxylic groups, some acidic OH groups must be present in GO.

Hofmann and Holst [49] observed later that the cation exchange capacity determined with sodium and calcium acetate varies strongly in the range of 27–184 mval/100 g GO depending on the degree of oxidation. GOs with higher C/O ratios show lower values of the exchange capacity, but there is no systematic correlation. The exchange capacity of GO is much higher when the experiments are carried out with sodium hydroxide. The limiting value is scattered in the range 500–600 mval/100 g GO depending on the nature of the GO samples used, in fair agreement with the data of Thiele [35].

To get more insight into the acidic properties of GO, Hofmann and Holst [49] carried out methylation reactions. The methylation with methanol and HCl is known to be sensitive for carboxyl groups, whereas diazomethane reacts with different OH groups including carboxyl OH and phenolic OH groups. Only ~30 mval of CH_3O/100 g GO have been obtained for the first reaction independent of the pretreatment of GO (wet, dried at 100, 130 or 180 °C). Thus, only the carboxyl groups at the edges have been methylated with this method. The methylation of GO dried at 100 °C with diazomethane in ether leads to ~500 mval CH_3O/100 g GO, in fair agreement with the values of the sodium ion exchange

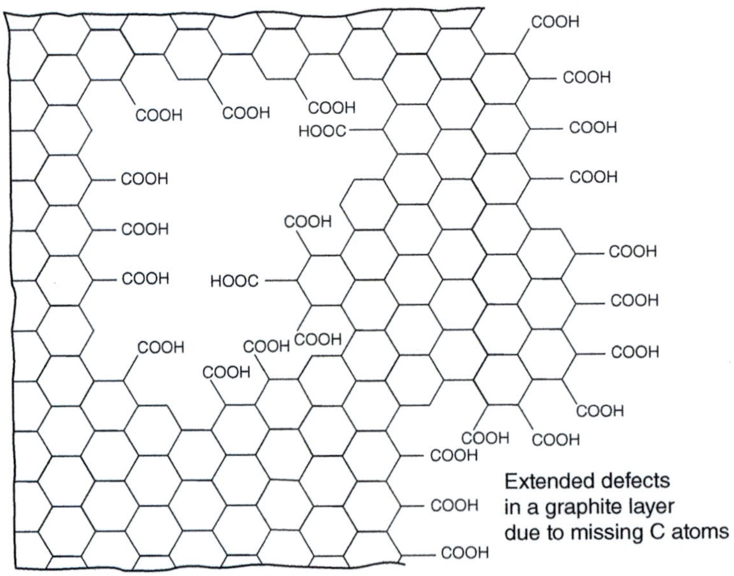

Figure 1.7 Defective graphite layer [46]. Reproduced from [46] with permission of Wiley-VCH

capacity. This value is strongly reduced for a GO dried at 180 °C. Ruess [19] carried out acetylation reactions to transfer carboxyl and acidic OH groups into acetic esters by treating GO with acetic acid anhydride and a trace of sulfuric acid in closed ampoules at 70 °C. The amount of acetic acid bound to OH groups in GO is again in the range 500–600 mval/100 g GO. When he carried out the acetylation reaction in a dioxane solution, the amount of inserted acetate was in the order of 800 mval/100 g GO, the highest values of the cation exchange capacity (CEC) found by Thiele [35]. Ruess explained the difference between the two methods by the enhanced diffusion of the reactant to the acidic sites, because dioxane is co-intercalated and props up the interlayer expansion to 9–10 Å. He suggests that the acetylation reaction without an additional solvent and the methylation reaction with diazomethane in ether were not finished because of retarded diffusion in the constrained interlayer space; in these cases the layer expansion amounts only 3–4 Å.

Pallmann [75] and Hamdi [29] confirmed the values of the exchange capacity determined by Hofmann and Holst [49]. Pallmann as well as Hamdi compared the cation exchange capacity of GO with those of humic acid, lignin and oxidized lignin, and they found a correlation between the C/O ratio and the cation exchange capacity: it decreases the higher the carbon content. The partially destroyed samples of Hofmann and Holst [49] fit quite well into this correlation.

Hamdi [29] carried out also the first titration experiments monitoring the pH change in the supernatant solution versus the applied quantity of mval NaOH/g of GO (Figure 1.8, left). The titration curve shows two well-pronounced inflection points. If they are not just experimental artifacts, the one at a pH of 10.5 corresponds to the number of OH groups that can be deprotonated with NaOH solutions, with methylation or esterification. The pH of the other inflection point is in the order of the pK_a values of phenols. The maximum deprotonation that can be achieved with acetate solutions (~100 mval/100 g GO) leaves no trace in the pH curve. Of the recently published titration experiments [76, 102, 103], one [103] seems to be compatible with the previous data of Hamdi [29], whereas the other two show a continuous featureless increase of pH with increasing amounts of NaOH. All four titration curves start at a pH of ~ 4.

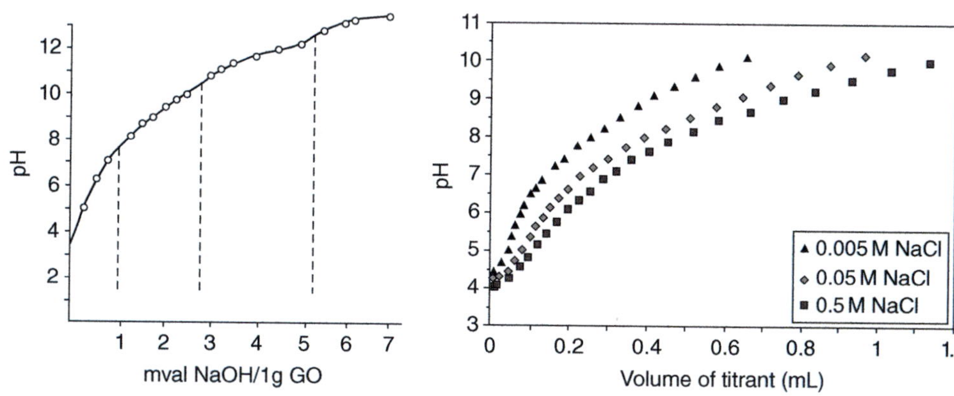

Figure 1.8 Titration curves of Hamdi [29] (left) and Szabó et al. [102] (right). (left) Reproduced from [29] with permission of Springer (right). Reproduced from [102] with permission of Elsevier

The amount of exchangeable protons is roughly half of the expected value when one assumes that all hydrogen determined by chemical analysis (including the hydrogen of not extractable water, which amounts to 7–8%) is bound as OH to the carbon grid. When the carefully dried GO samples are treated with sodium ethylate, it is possible to deprotonate also those OH groups which cannot be exchanged by highly concentrated NaOH solutions [16, 48].

Scholz and Boehm [16] *collected all information available about deprotonation (titration) experiments for the three different types of GOs (Brodie, Staudenmaier and Hummers–Offeman), showing no clear distinction between them. The scatter of available data could be due to the properties of the GOs (particle size, degree of oxidation and degree of decomposition), the equilibration time, the deprotonation methods applied and the skill of the workers. The results of the exchange experiments of Thiele, Hofmann and Boehm indicate that there are four different types of OH groups with differing acidities*:

- *OH deprotonated with Na ethylate*
  pK_a values 14–17 — *tertiary alcohols (sp^3 C)*
- *OH deprotonated with NaOH*
  pK_a values 10–12 — *enols, vicinal diols*
  pK_a values 7–9 — *phenols*
- *OH deprotonated with Na acetate*
  pK_a values < 8 — *carboxyl groups, phenols, water (?)*

For pH < 12, this division seems to be in fair agreement with the inflection points of the titration experiments of Hamdi and the most recent ones of Konkena and Vasudevan [103]. *The titration curves found by Szabó* et al. [102] *were greatly affected by the solution conditions: increasing the pH and the ionic strength promotes the dissociation of acidic surface sites as weaker functional groups progressively participate in the ion exchange process, and the electrolyte provides an effective shielding for the surface charge. The acidity and surface charge characteristics were also found to depend on the degree of oxidation. Thus, their study points out that additives such as acids, bases and salts affect the surface charge behavior, which, in turn, may reflect the colloidal stability of their aqueous dispersions.*

In addition, the question arises whether it is really possible to define discrete pK_a values for OH groups in the constrained space of the interlayer galleries in which an increasing electric field is created by the deprotonation of the OH groups and the concomitant charging of the GO layers partially shielded by the uptake of cations. In addition, the NMR experiments (dipolar dephasing, short contact time measurements) of Mermoux et al. [18] *and He* et al. [104] *show that the protons of the OH groups and of the bound water molecules are in strong van der Waals interactions to the neighboring O atoms and to the carbons on which the atoms are attached. This may enhance their acidity so that they take part also in the deprotonation reactions.*

1.4.6 Intercalation and Functionalization Reactions

In 1962, Slabaugh and Seiler intercalated ammonia into GO for the first time [105]. They carried out the reaction at temperatures between −25 and −45 °C and they found an exceptionally high ammonia uptake of ~170 mg/g GO, which corresponds to the total amount of OH groups determined by Clauss *et al.* [48]. Later Seredych and Bandosz found an ammonia uptake of ~18 mg/g GO prepared by the "Brodie method". These authors explained the uptake by the neutralization of the carboxylic groups and the deprotonation of interlamellar weakly acidic OH groups [106]. The $GO_{Hummers-Offeman}$ takes up even more NH_3 (~60 mg/g GO) because it forms $(NH_4)_2SO_4$ with the sulfate groups covalently bound to GO (first proposed by Boehm and Scholz [30] and later by Titelman *et al.* [107]) and detached from the GO by the action of the superoxide radical anion present in GO [108].

Whereas Kohlschütter and Haenni [10] and earlier authors discussed the solubility and insolubility of GO, Hofmann *et al.* [15] described for the first time reactions of solvents with GO in terms of swelling. Kohlschütter and Haenni mentioned the insolubility of GO in alcohols, ether, benzene, toluene, xylol, acetone, carbon disulfide and chloroform. Hofmann *et al.* [15] stated that GO swells in many polar liquids, e.g. ethanol, acetic acid, acetone and red fuming nitric acid, but they do not give layer distances. Derksen and Katz [78] gave for the first time the layer distance of GO swollen in ethanol, which is 9.3 Å. These authors show also that the admixture of ethanol or tannin to water reduces the layer distance with respect to the fully hydrated GO.

Ruess [19] discovered that dioxane is the only swelling agent that leads to some ordering in the GO compound, as indicated by the appearance of up to six $00l$ reflections (normally only one or two $00l$ reflections are observed). Therefore, he used this intercalation compound to improve his structural analysis.

"Interlamellar sorption complexes" of a great number of organic compounds were studied for the first time by Cano-Ruiz and MacEwan [109, 110] following the research program of MacEwan first applied to montmorillonite and halloysite [111]. They showed that non-polar molecules like alkanes, PAHs, CS_2 or long-chain alcohols are not taken up in the interlayer space. For various other molecules, the insertion into the interlayer galleries can be questioned. If the so calculated expansions are ≤3 Å, an uptake of organic molecules in the interlayer space is highly improbable because it affords an expansion of at least 3.5 Å, the smallest possible dimension for an organic molecule. Layer expansions of 4–5 Å have been found for various aliphatic alcohols, glycol, glycerol, other diols, phenol, aliphatic and aromatic amines, and aliphatic and aromatic nitro-compounds. Significantly higher layer expansions have been described for aniline or pyridine and long-chain aliphatic alcohols and amines. Intercalation of amines is preferred with respect to the uptake of other molecules because they undergo acid–base reactions like ammonia with particular OH groups in the interlayer space forming ammonium cations. This effect leads to a higher uptake of amines over that of the alcohols. Recently, Bourlinos *et al.* have shown that amines are intercalated easily in GO, but not ammonium ions [72]. They show also that tertiary amines are hardly inserted in GO [72].

Subsequent papers were exclusively devoted to the uptake of long-chain alkyl alcohols and amines (up to 16 C atoms) [112, 113] (and references therein). These compounds found increased attention after their discovery by MacEwan [111] and Jordan [114, 115] in 1948–1959, because of the membrane-like arrangements of the alkyl compounds in the interlayer space. Krüger-Grasser continued this work, but used a sodium-loaded GO for her studies; thus she transferred GO to a montmorillonite-like system [116]. Correspondingly, she found much higher layer distances (up to 50 Å) and could also insert long-chain alkyl alcohols. In addition, mixtures of alcohols and amines have been intercalated. These systems allow the selective adsorption of hydrophobic solvents from their mixtures [117]. Matsuo *et al.* [118] used the sodium salt of GO to insert quaternary ammonium surfactants $[R_3NC_nH_{2n+1}]^+$ ($R = CH_3$, C_2H_5; $n = 12, 14, 16$)

and $[C_{16}H_{33}NC_5H_5]^+$. In subsequent papers they used them for selective adsorption of PAHs (e.g. [119, 120]).

Due to the reactive oxygen atoms in the interlayer space of GO, the uptake of organic molecules (intercalation) can be accompanied by the reaction of some of the reactive species taken up. The first examples of this functionalization were realized with two very reactive organic species – diazomethane and acetic acid anhydride – both reacting with the OH groups with the formation of a methyl ether and an acetic acid ester, respectively [19, 47, 49]. Both reactions have been used to determine acidic sites on GO (see section 1.4.5). Aragón de la Cruz and MacEwan [113] and later Krüger-Grasser [116] intercalated long-chain alkyl compounds into these functionalized GO. However, the functionalization leads to a reduced uptake of these compounds.

Later the functionalization was used to identify the functional groups on GO by specific reactions [20, 55, 104, 121]. While in the cases of acetylation, ethylation and the reaction with an isocyanate, the specific functionalization could be confirmed by NMR, most of the other reagents studied did not lead to detectable changes in the NMR spectrum. *In all the studies a new peak at about 110 ppm has been assigned to the formation of phenols. Protonation/deprotonation reactions and additions of water or OH groups like those proposed by Dimiev* et al. [76] *could also induce phenol formation near defect sites apart from epoxide ring opening reactions.*

Bourlinos *et al.* [72] suggested that nucleophilic attack on epoxides and covalent bonding is the main mechanism for the uptake of primary aliphatic amines by GO. This conclusion has been drawn mainly from the observation that the treatment of amine–GO complexes with hot ethanol or aqueous NaOH solution does not reduce the layer distance significantly.

Bourlinos *et al.* [72] and Matsuo *et al.* [121] were able to silylate the hydroxyl groups of GO, forming C–O–Si bonds. The product is stable up to 300 °C. Since that time the functionalization of GO has become a new field of research.

1.4.7 Functional Groups, their Reactions and their Relation to GO Formation and Destruction

The thermal decomposition behavior of GO is indicative of a reactive oxygen function. The 1,3-ethers of Ruess are not as reactive as the 1,2-ethers. Peroxide groups would be even more reactive, but the evidence for peroxides in GO is weak. The H_2S, the thiourea and probably the iodide reactions seem to be indicative for epoxide functions.

The argument of Ruess that typical nucleophilic addition reactions do not occur in GO is not a convincing argument against epoxides. Nucleophilic ring opening reactions on epoxides start with the addition of the nucleophile from the opposite site of the epoxide oxygen. But both sites of the carbon grid are covered with a more or less densely packed oxygen layer, which is negatively polarized. So the access of the nucleophiles to the carbon atoms is restricted to defects in the oxygen coverage. This may retard the reaction considerably. And indeed steric hindrance by bulky substituents on the opposite site of the epoxide oxygen prevents any nucleophilic reaction [122].

Hydronium ions in the interlayer space could induce ring opening of epoxides via a proton transfer to the epoxide oxygen followed by the covalent bonding of a hydroxyl group to the carbocation [59, 60, 123]. *Electrophilic reactions induced by protons can also induce the transformation of epoxides to phenols, at least for low-molecular-weight PAHs. The transformation to phenols in GO would afford C–C bond breaking, which seems improbable, but it could happen at defects in the carbon grid. Protons could also induce an oxygen walk on the carbon surface, another interesting aspect for GO chemistry* [59]. *Perhaps electrophilic reactions could be of much more importance in GO than nucleophilic reactions.*

1.5 Epilogue

Going through the publications of 150 years or so, it is interesting to see how the mode of scientific work and the style of writing publications have changed, what has been cited in later generations, what has been omitted consciously or subconsciously, and what has been neglected because it could not be understood in later times. These topics would be an interesting case for a study in the sociology of science. I would like to give just a few comments for such a study.

Knowledge of the chemical properties of GO was highly developed by 1919, and it was excellently reviewed by Kohlschütter and Haenni [10]. This remarkable paper and all the previous publications had practically no figures. The experimental conditions were not described separately. The manipulations done were integrated in the very precise and detailed description of the observations made. The experiments were focused on the reaction behavior with respect to other chemical substances, temperature or in vacuum. The observations were mostly qualitative in nature. Analytical results were restricted to the determination of the C/H contents via combustion analysis, as in organic chemistry. The procedure applied was a standard method and not further described.

Worth noting is the debate between Thiele [14, 31, 35, 39] and Hofmann [12, 13, 15, 46, 49]. Thiele was the first to make a statement concerning the functional groups in GO and presented the first structural model. He carried out for the first time the ion exchange reactions, the electrochemical intercalation of sulfuric acid and the electrochemical preparation of GO, and discussed the relation to humic acids. Hofmann in 1934 presented an alternative structural model and criticized Thiele's model with a few pejorative remarks. In 1937, shortly after the paper by Thiele, he presented a confirmation of Thiele's data concerning the ion exchange and admitted the existence of OH groups in GO besides the epoxide functions. He never corrected his structural model; modifications of it were given by his young coworkers, first Ruess [19] and later Boehm [16, 27, 30, 48], maybe because his scientific interests shifted towards clay minerals. It is interesting to look at the style of the publications. Thiele published in the style of previous workers; his publications are narrative and many important experimental details are not reported; exceptions from this attitude were the description of his reduction and his ion exchange experiments. Hofmann was more modern; although he was also narrative in large parts of his publications, he gave the experimental details in a separate part of the publications; one can also recognize in his arguing the influence of physical chemistry.

The research on GO received some stimuli from other fields of research: the invention of XRD, colloid chemistry, ion exchange processes and soil science (i.e. the relation between GO and humic acids). An important aspect was the phenomenon of swelling. In the 1920s macromolecular chemistry was established and it was found that especially biological macromolecules show a remarkable swelling behavior, which was investigated with XRD [78]. This triggered the corresponding work of Hofmann and his coworkers. In at least one of his publications he called the isolated graphite oxide layers "macromolecules". Because these macromolecules were layered materials, the swelling could occur in just one dimension. This one-dimensional swelling was also discovered by Hofmann almost simultaneously for the low-charged clay minerals.

The research on GO was for a long time a specialty of solid-state chemists. The most important work was done by researchers in France and Germany. Therefore, many important publications on GO were written in German. This is a problem for estimating the value of previous work in the field today.

In 2015 one can say the following about GO from the point of view of scientific activities over the 150 years from 1855 to 2005: GO is a metastable non-stoichiometric solid carbon compound whose structure has not been fully established. It is a tricky intermediate state between over-oxidation and self-destruction. While all the information was available, nobody had drawn this conclusion in 1970. Only under careful adjustment of the preparation conditions, starting with the choice of the graphite (crystalline, powders, particle size), the control of the water content of the acids applied, the temperature and time regime, exclusion of sunlight during work-up, and the drying conditions, can one obtain GO with reproducible properties. A useful guide for the choice of the reaction conditions is the almost neglected publication of Boehm and Scholz [30]. The sample with a residual water content of about ~8% should have a C/O ratio of 3 ± 0.3 and in the NMR spectrum there should be only three peaks at ~60 ppm, ~70 ppm and ~130 ppm.

Comparing recent work on GO and the historical material, a statement of an older colleague of mine came to my mind: "You can redo the same thing in 25 years, and almost nobody will recognize it." I also feel I should mention a statement by the English mathematician and philosopher, A.N. Whitehead, who wrote in a footnote of his most famous book *Process and Reality*: "All philosophical concepts are just footnotes to Plato." So, in a manner of speaking, most workers active in the field of GO research since 2005 are just preparing footnotes to the seminal works of Kohlschütter, Hofmann and Boehm, and I must confess that my own work is also nothing else. I realized that I have overlooked also many interesting aspects in the previous works when we wrote our own papers. I fear this is quite normal, because you can recognize and understand the importance of details in publications only when you have your own experimental experience in a new field into which you have jumped.

References

[1] Brodie, M.B.C., Note sur un nouveau procédé pour la purification et la désagrégation du graphite. *Ann. Chim. Phys.* **1855**, *45*, 351–352.

[2] Brodie, B.C., On the atomic weight of graphite. *Phil. Trans. R. Soc. Lond.* **1859**, *149*, 249–259.

[3] Brodie, B.C., Sur le poids atomique du graphite. *Ann. Chim. Phys.* **1860**, *59*, 466–472.

[4] Brodie, B.C., Über das Atomgewicht des Graphits. *Ann. Chem. Pharm.* **1860**, *114*, 6–24.

[5] Schafhäutl, C., Über die Verbindungen des Kohlenstoffs mit Silizium, Eisen und anderen Metallen, welche die verschiedenen Gattungen von Roheisen, Stahl und Schmiedeeisen bilden. *J. Prakt. Chem.* **1840**, *21* (1), 129–157.

[6] Schafhäutl, C., Über weißes und graues Roheisen, Graphitbildung u.s.w. *J. Prakt. Chem.* **1859**, *76* (1), 257–310.
[7] Gottschalk, F., Beiträge zur Kenntnis der Graphitsäure. *J. Prakt. Chem.* **1865**, *95* (1), 321–350.
[8] Berthelot, M., Recherches sur les états du carbone. *Ann. Chim. Phys.* **1870**, *19*, 392–417.
[9] Staudenmaier, L., Verfahren zur Darstellung von Graphitsäure. *Ber. Dtsch. Chem. Ges.* **1898**, *31* (2), 1481–1487.
[10] Kohlschütter, V.; Haenni, P., Zur Kenntnis des graphitischen Kohlenstoffs und der Graphitsäure. *Z. Anorg. Chem.* **1919**, *105* (1), 121–144.
[11] Weinschenk, E., Über den Graphitkohlenstoff und die gegenseitigen Beziehungen zwischen Graphit, Graphitit und Graphitoxid. *Z. Kristallogr.* **1897**, *28* (3), 291–304.
[12] Hofmann, U., Über Graphitsäure und die bei ihrer Zersetzung entstehenden Kohlenstoffarten. *Ber. Dtsch. Chem. Ges.* **1928**, *61* (2), 435–441.
[13] Hofmann, U.; Frenzel, A., Quellung von Graphit und die Bildung von Graphitsäure. *Ber. Dtsch. Chem. Ges.* **1930**, *63* (5), 1248–1262.
[14] Thiele, H., Graphit und Graphitsäure. *Z. Anorg. Allg. Chem.* **1930**, *190* (1), 145–160.
[15] Hofmann, U.; Frenzel, A.; Csalán, E., Die Konstitution der Graphitsäure und ihre Reaktionen. *Liebigs Ann. Chem.* **1934**, *510* (1), 1–41.
[16] Scholz, W.; Boehm, H.P., Betrachtungen zur Struktur des Graphitoxids. *Z. Anorg. Allg. Chem.* **1969**, *369* (3–6), 327–340.
[17] Mermoux, M.; Chabre, Y., Formation of graphite oxide. *Synth. Met.* **1989**, *34*, 157–162.
[18] Mermoux, M.; Chabre, Y.; Rousseau, A., FTIR and ^{13}C NMR study of graphite oxide. *Carbon* **1991**, *29* (3), 469–474.
[19] Ruess, G., Über das Graphitoxyhydroxyd (Graphitoxyd). *Monatsh. Chem.* **1947**, *76* (3), 381–417.
[20] Lerf, A.; He, H.; Forster, M.; Klinowski, J., Structure of graphite oxide revisited. *J. Phys. Chem. B* **1998**, *102* (23), 4477–4482.
[21] Dreyer, D.R.; Park, S.; Bielawski, C.W.; Ruoff, R.S., The chemistry of graphene oxide. *Chem. Soc. Rev.* **2010**, *39* (1), 228–240.
[22] Rourke, J.P.; Pandey, P.A.; Moore, J.J.; *et al.*, The real graphene oxide revealed: stripping the oxidative debris from graphene-like sheets. *Angew. Chem. Int. Ed.* **2011**, *50* (14), 3173–3177.
[23] Boehm, H.P., Graphene – how a laboratory curiosity suddenly became extremely interesting. *Angew. Chem. Int. Ed.* **2010**, *49* (49), 9332–9335.
[24] Geim, A.K.; Novoselov, K.S., The rise of graphene. *Nature Mater.* **2007**, *6* (3), 183–191.
[25] Luzi, W., Beiträge zur Kenntnis des Graphitkohlenstoffs. *Z. Naturwiss.* **1891**, *64*, 224–269.
[26] Charpy, G., Sur la formation de l'oxyde graphitique et la définition du graphite. *C. R. Hebd. Séances Acad. Sci.* **1909**, *148*, 920–923.
[27] Boehm, H.P.; Eckel, M.; Scholz, W., Über den Bildungsmechanismus des Graphitoxids. *Z. Anorg. Allg. Chem.* **1967**, *353* (5–6), 236–242.
[28] Hummers, W.S.; Offeman, R.E., Preparation of graphite oxide. *J. Amer. Chem. Soc.* **1958**, *80* (6), 1339–1339.
[29] Hamdi, H., Zur Kenntnis der kolloidchemischen Eigenschaften des Humus. Dispersoidchemische Beobachtungen an Graphitoxid. *Kolloid Beihefte* **1942**, *54*, 554–643.
[30] Boehm, H.P.; Scholz, W., Vergleich der Darstellungsverfahren für Graphitoxyd. *Liebigs Ann. Chem.* **1966**, *691* (1), 1–8.
[31] Thiele, H., Die Quellung des Graphit an der Anode und die mechanische Zerstörung von Kohleanoden. *Z. Elektrochem.* **1934**, *40* (1), 26–33.
[32] Besenhard, J.O.; Fritz, H.P., Uber die Reversibilität der elektrochemischen Graphitoxydation in Säuren. *Z. Anorg. Allg. Chem.* **1975**, *416* (2), 106–116.
[33] Besenhard, J.O.; Wudy, E.; Möhwald, H.; *et al.*, Anodic oxidation of graphite in H_2SO_4: dilatometry–in-situ X-ray diffraction–impedance spectroscopy. *Synth. Met.* **1983**, *7* (3–4), 185–192.
[34] Marchand, R.F., Über die Einwirkung der Schwefelsäure auf die Holzkohle. *J. Prakt. Chem.* **1845**, *35* (1), 228–231.
[35] Thiele, H., Über Salzbildung und Basenaustausch der Graphitsäure. *Kolloid-Z.* **1937**, *80* (1), 1–20.
[36] Rüdorff, W.; Hofmann, U., Über Graphitsalze. *Z. Anorg. Allg. Chem.* **1938**, *238* (1), 1–50.

[37] Inagaki, M.; Iwashita, N.; Kouno, E., Potential change with intercalation of sulfuric acid into graphite by chemical oxidation. *Carbon* **1990**, *28* (1), 49–55.
[38] Sorokina, N.E.; Shornikova, O.N.; Avdeev, V.V., Stability limits of graphite intercalation compounds in the systems graphite–$HNO_3(H_2SO_4)$–H_2O–$KMnO_4$. *Inorg. Mater.* **2007**, *43* (8), 822–826.
[39] Thiele, H., Über die Quellung von Graphit. *Z. Anorg. Allg. Chem.* **1932**, *206* (4), 407–415.
[40] Avdeev, V.V.; Monyakina, L.A.; Nikol'skaya, I.V.; *et al.*, The choice of oxidizers for graphite hydrogensulfate chemical synthesis. *Carbon* **1992**, *30* (6), 819–823.
[41] Sorokina, N.E.; Khahkov, M.A.; Avdeev, V.V.; Nikol'skaya, I.V., Reaction of graphite with sulfuric acid in the presence of $KMnO_4$. *Russ. J. Gen. Chem.* **2005**, *75* (2), 162–168.
[42] Skorwroński, J.M., Distribution of intercalates in CrO_3–H_2SO_4–graphite and CrO_3–$HClO_4$–graphite bi-intercalation compounds. *Synth. Met.* **1998**, *95* (2), 135–142.
[43] Royer, D.J., Evidence for the existence of the permanganyl ion in sulfuric acid solutions of potassium permanganate. *J. Inorg. Nucl. Chem.* **1961**, *17* (1–2), 159–167.
[44] Nakajima, T.; Mabuchi, A.; Hagiwara, R., A new structural model of graphite oxide. *Carbon* **1988**, *26* (3), 357–361.
[45] Nakajima, T.; Matsuo, Y., Formation process and structure of graphite oxide. *Carbon* **1994**, *32* (3), 469–475.
[46] Hofmann, U.; König, E., Untersuchungen über Graphitoxyd. *Z. Anorg. Allg. Chem.* **1937**, *234* (4), 311–336.
[47] Ruess, G., Zur Formel des Graphitoxyds. *Kolloid-Z.* **1945**, *110* (1), 17–26.
[48] Clauss, A.; Plass, R.; Boehm, H.P.; Hofmann, U:, Untersuchungen zur Struktur des Graphitoxyds. *Z. Anorg. Allg. Chem.* **1957**, *291* (5–6), 205–220.
[49] Hofmann, U.; Holst, R., Über die Säurenatur und die Methylierung von Graphitoxid. *Ber. Dtsch. Chem. Ges.* **1939**, *72* (4), 754-771.
[50] Hadži, D.; Novak, A., Infrared spectra of graphitic oxide. *Trans. Faraday Soc.* **1955**, *51*, 1614–1620.
[51] Szabó, T.; Berkesi,O.; Forgó, P.; *et al.*, Evolution of surface functional groups in a series of progressively oxidized graphite oxides. *Chem. Mater.* **2006**, *18* (11), 2740–2749.
[52] Aragón de la Cruz, F.; Cowley, J.M., Structure of graphitic oxide. *Nature* **1962**, *196* (4853), 468–469.
[53] Blumenfeld, A.L.; Muradyan, V.E.; Shumilova, B.; *et al.*, Investigation of graphite oxide by means of ^{13}C NMR and ^{1}H spin lattice relaxation. *Mater. Sci. Forum* **1992**, *91–93*, 613–617.
[54] Hontoria-Lucas, C.; Lopez-Peinado, A.J.; Lopez-Gonzalez, J.deD.; *et al.*, Study of oxygen containing groups in a series of graphite oxides: physical and chemical characterization. *Carbon* **1995**, *33* (11), 1585–1592.
[55] He, H.; Klinowski, J.; Forster, M.; Lerf, A., A new structural model for graphite oxide. *Chem. Phys. Lett.* **1998**, *287* (1–2), 53–56.
[56] Fujii, S.; Enoki, T., Clar's aromatic sextet and π-electron distribution in nanographene. *Angew. Chem.* **2012**, *124* (29), 7348–7353.
[57] Kalinowski, H.O.; Berger, S.; Braun, S., *^{13}C NMR-Spektroskopie*, Georg Thieme, Stuttgart, **1983**.
[58] Dimiev, A.M.; Tour, J.M., Mechanism of graphene oxide formation. *ACS Nano* **2014**, *8* (3), 3060–3068.
[59] Boyd, D.R.; Jerina, D.M., Arene oxides-oxepins. In *The Chemistry of Heterocyclic Compounds*, ed. A. Hassner, vol. *42*, pp. 197–282. John Wiley & Sons, Inc., New York, **1985**.
[60] Fatiadi, A.J., The classical permanganate ion: still a novel oxidant in organic chemistry. *Synthesis* **1987**, *1987* (2), 85–127.
[61] Kosynkin, D.V.; Higginbotham, A.L.; Sinitskii, A.; *et al.*, Longitudinal unzipping of carbon nanotubes to form graphene nanoribbons. *Nature* **2009**, *458* (7240), 872–877.
[62] Forsey, S.P.; Thomson, N.R.; Barker, J.M., Oxidation kinetics of polycyclic aromatic hydrocarbons by permanganate. *Chemosphere* **2010**, *79* (6), 628–636.
[63] Foster, C.H.; Berchtold, G.A., Addition of singlet oxygen to arene oxides. *J. Org. Chem.* **1975**, *40* (25), 3743–3746.

[64] De Boer, J.H.; van Doorn, A.B.C., Graphitic oxide III. The thermal decomposition of graphitic oxide. *Proc. K. Ned. Akad. Wetensch. B* **1958**, *61*, 17–21.
[65] Scholz, W.; Boehm, H.P., Die thermische Zersetzung von Graphitoxyd. *Naturwissenschaften* **1964**, *51* (7), 160.
[66] Boehm, H.P.; Scholz, W., Der "Verpuffungspunkt" des Graphitoxyds. *Z. Anorg. Allg. Chem.* **1965**, *335* (1–2), 74–79.
[67] Scholz, W.; Boehm, H.P., Die Ursache der Dunkelfärbung des hellen Graphitoxids. *Z. Anorg. Allg. Chem.* **1964**, *331* (3–4), 129–132.
[68] Matuyama, E., Pyrolysis of graphitic acid. *J. Phys. Chem.* **1954**, *58* (3), 215–219.
[69] Martín-Rodríguez, A.; Valerga-Jiménez, P., Thermal decomposition of the graphite oxidation products. *Thermochim. Acta* **1984**, *78* (1–3), 113–122.
[70] Barroso-Bujans, F.; Alegría, A.; Colmenero, J., Kinetic study of the graphite oxide reduction: combined structural and gravimetric experiments under isothermal and nonisothermal conditions. *J. Phys. Chem. C* **2010**, *114* (49), 21645–21651.
[71] Hofmann, U.; Frenzel, A., Die Reduktion von Graphitoxyd mit Schwefelwasserstoff. *Kolloid-Z.* **1934**, *68* (2), 149–151.
[72] Bourlinos, A.B.; Gournis, D.; Petridis, D.; *et al.*, Graphite oxide: chemical reduction to graphite and surface modification with primary aliphatic amines and amino acids. *Langmuir* **2003**, *19* (15), 6050–6055.
[73] Thiele, H., Über die Quellung von Graphit. *Z. Anorg. Allg. Chem.* **1932**, *206* (4), 407–415.
[74] De Boer, J.H.; van Doorn, A.B.C., Graphitic oxide IV. Some chemical properties. *Proc. K. Ned. Akad. Wetensch. B* **1958**, *61* (3), 160–169.
[75] Pallmann, H., Dispersoidchemische Probleme in der Humusforschung. *Kolloid-Z.* **1942**, *101* (1), 72–81.
[76] Dimiev, A.M; Alemany, L.B.; Tour, J.M., Graphene oxide. Origin of acidity, its instability in water, and a new dynamic structural model. *ACS Nano* **2013**, *7* (1), 576–588.
[77] Fan, X.; Peng, W.; Li, Y.; *et al.*, Deoxygenation of exfoliated graphite oxide under alkaline conditions: a green route to graphene preparation. *Adv. Mater.* **2008**, *20* (23), 4490–4493.
[78] Derksen, J.C.; Katz, J.R., Untersuchung über die intramicellare Quellung der Graphitsäure. *Rec. Trav. Chim.* **1934**, *53* (7), 652–669.
[79] Lerf, A.; Buchsteiner, A.; Pieper, J.; *et al.*, Hydration behavior and dynamics of water molecules in graphite oxide. *J. Phys. Chem. Solids* **2006**, *67* (5–6), 1106–1110.
[80] Barroso-Bujans, F.; Cerveny, S.; Alegría, A.; Colmenero, J., Sorption and desorption behavior of water and organic solvents from graphite oxide. *Carbon* **2010**, *48* (11), 3277–3286.
[81] Hendricks, S.B.; Jefferson M.E., Structures of kaolin and talc-pyrophylite hydrates and their bearing on water sorption of the clays. *Amer. Miner.* **1938**, *23* (12), 863–875.
[82] Hofmann, U.; Hausdorf, A., Kristallstruktur und innerkristalline Quellung von Montmorillonit. *Z. Kristallogr. Mineral. Petrogr. A* **1942**, *104* (1–6), 265–293.
[83] Weiss, A., Die innerkristalline Quellung als allgemeines Modell für Quellungsvorgänge. *Ber. Dtsch. Chem. Ges.* **1958**, *91* (3), 487–502.
[84] Thiele, H., Die Mizellarstruktur der Graphitsäure. *Kolloid-Z.* **1931**, *56* (2), 129–138.
[85] Clauss, A.; Hofmann, U.; Weiss, A., Membranpoteniale an Graphitoxydfolien. *Z. Elektrochem.* **1957**, *61* (10), 1284–1290.
[86] Boehm, H.P.; Clauss, A.; Hofmann, U., Graphite oxide and its membrane properties. *J. Chim. Phys.* **1961**, *58* (12) 141–147.
[87] Hwa, T.; Kokufuta, E; Tanaka, T., Conformation of graphite oxide membranes in solution. *Phys. Rev. A* **1991**, *44* (4), R2235–R2238.
[88] Wen, X.; Garland, C.; Hwa, T.; *et al.*, Crumpled and collapsed conformations in graphite oxide membranes. *Nature* **1992**, *355* (6359), 426–428.
[89] Spector, M.S.; Naranjo, E.; Chirovolu, S.; Zasadzinski, J.A., Conformation of tethered membrane: crumpling in graphite oxide? *Phys. Rev. Lett.* **1994**, *73* (21), 2867–2870.
[90] Boehm, H.P.; Clauss, A.; Fischer, G.O.; Hofmann, U., Dünnste Kohlenstoff-Folien. *Z. Naturforsch. B* **1962**, *17* (b), 150–153.

[91] Boehm, H.P.; Clauss, A.; Fischer, G.O.; Hofmann, U., Surface properties of extremely thin graphite lamellae. In *Carbon: Proceedings of the Fifth Conference*, vol. *1*, pp. 73–80. Pergamon Press, Oxford, **1962**.

[92] Stankovich, S.; Dikin, D.A.; Piner, R.D.; *et al.*, Synthesis of graphene-based nanosheets via chemical reduction of exfoliated graphite oxide. *Carbon* **2007**, *45* (7), 1558–1565.

[93] Matsuo, Y.; Tahara, K.; Sugie, Y., Synthesis of poly(ethylene oxide)-intercalated graphite oxide. *Carbon* **1996**, *34* (5), 672–674.

[94] Matsuo, Y.; Tahara, K.; Sugie, Y., Structure and thermal properties of poly(ethylene oxide)-intercalated graphite oxide. *Carbon* **1997**, *35* (1), 113–120.

[95] Liu, P.; Gong, K.; Xiao, P.; Xiao, M., Preparation and characterization of poly(vinyl acetate)-intercalated graphite oxide nanocomposite. *J. Mater. Chem.* **2000**, *10* (4), 933–935.

[96] Xu, J.; Hu, Y.; Song, L.; *et al.*, Preparation and characterization of polyacrylamide intercalated graphite oxide. *Mater. Res. Bull.* **2001**, *36* (10), 1833–1836.

[97] Xu, J.; Hu, Y.; Song, L.; *et al.*, Preparation and characterization of poly(vinyl alcohol)/graphite oxide nanocomposite. *Carbon* **2002**, *40* (3), 445–467.

[98] Kotov, N.A.; Dékány, I.; Fendler, J.H., Ultrathin graphite oxide–polyelectrolyte composites prepared by self-assembly: transitions between conductive and non-conductive states. *Adv. Mater.* **1996**, *8* (8), 637–641.

[99] Cassagneau, T.; Fendler, J.H., High density rechargeable lithium-ion batteries self-assembled from graphite oxide nanoplatelets and polyelectrolytes. *Adv. Mater.* **1998**, *10* (11), 877–881.

[100] Kovtyukhova, N.I.; Ollivier, P.J.; Martin, B.R.; *et al.*, Layer-by-layer assembly of ultrathin composite films from micron-sized graphite oxide sheets and polycations. *Chem. Mater.* **1999**, *11* (3), 771–778.

[101] Cassagneau, T.; Fendler, J.H., Preparation and layer-by-layer self-assembly of silver nanoparticles capped by graphite oxide nanosheets. *J. Phys. Chem. B* **1999**, *103* (11), 1789–1793.

[102] Szabó, T.; Tombácz, E.; Illés, E; Dékány, I., Enhanced acidity and pH-dependent surface charge characterization of successively oxidized graphite oxides. *Carbon* **2006**, *44* (3), 537–545.

[103] Konkena, B.; Vasudevan, S., Understanding aqueous dispersibilty of graphene oxide and reduced graphene oxide through pK_a measurements. *J. Phys. Chem. Lett.* **2012**, *3* (7), 867–872.

[104] He, H.; Riedl, T.; Lerf, A.; Klinowski, J., Solid-state NMR studies of the structure of graphite oxide. *J. Phys. Chem.* **1996**, *100* (51), 19954–19958.

[105] Slabaugh, W.H.; Seiler, B.C., Interactions of ammonia with graphite oxide. *J. Phys. Chem.* **1962**, *66* (3), 396–401.

[106] Seredych, M.; Bandosz, T.J., Removal of ammonia by graphite oxide via its intercalation and reactive adsorption. *Carbon* **2007**, *45* (10), 2130–2132.

[107] Titelman, G.I.; Gelman, V.; Bron, S; *et al.*, Characteristics and microstructure of aqueous colloidal dispersions of graphite oxide. *Carbon* **2005**, *43* (3), 641–649.

[108] Petit, C.; Seredych, M.; Bandosz, T.J., Revisiting the chemistry of graphite oxides and its effect on ammonia adsorption. *J. Mater. Chem.* **2009**, *19* (48), 9176–9185.

[109] Cano-Ruiz, J.; MacEwan, D.M.C., Interlamellar sorption complexes of graphitic acid with organic substances. *Nature* **1955**, *176* (4495), 1222–1223.

[110] Cano-Ruiz, J.; MacEwan, D.M.C., Interlamellar organic complexes of graphitic acid. A preliminary study. In *Proceedings of the Third International Symposium on the Reactivity of Solids*, Madrid, **1956**, pp. 227–242.

[111] MacEwan, D.M.C., Complexes of clays with organic compounds. I. Complex formation between montmorillonite and halloysite and certain organic liquids. *Trans. Faraday Soc.* **1948**, *44*, 349–367.

[112] Aragón, F.; Cano-Ruiz, J.; MacEwan, D.M.C., β-type interlamellar sorption complexes. *Nature* **1959**, *183* (4663), 740–741.

[113] Aragón, F.; MacEwan, D.M.C., Sorption of organic molecules by graphitic acid and methylated graphitic acid: a preliminary study. *Kolloid-Z.* **1965**, *203* (1), 36–42.

[114] Jordan, J.W., Organophilic bentonites. I. Swelling in organic liquids. *J. Phys. Chem.* **1949**, *53* (2), 294–306.

[115] Jordan, J.W.; Hook, B.J.; Finlayson, C.M., Organophilic bentonites. II. Organic liquid gels. *J. Phys. Chem.* **1950**, *54* (8), 1196–1208.

[116] Krüger-Grasser, R., Interkalatkomplexe von Graphitoxiden. Ph.D. thesis, Munich, 1980.
[117] Dékány, I.; Krüger-Grasser, R.; Weiss, A., Selective liquid sorption properties of hydrophobized graphite oxide nanostructures. *Colloid Polym. Sci.* **1998**, *276* (7), 570–576.
[118] Matsuo, Y.; Niwa, T.; Sugie, Y., Preparation and characterization of cationic surfactant-intercalated graphite oxide. *Carbon* **1999**, *37* (6), 897–901.
[119] Matsuo, Y.; Hatase, K.; Sugie, Y., Selective intercalation or aromatic molecules into alkyltrimethylammonium ion-intercalated graphite oxide. *Chem. Lett.* **1999**, *28* (10), 1109–1110.
[120] Matsuo, Y.; Watanabe, K.; Fukutsua, T.; Sugie, Y., Characterization of n-hexadecylalkylamine-intercalated graphite oxides as sorbents. *Carbon* **2003**, *41* (8), 1545–1550.
[121] Matsuo, Y.; Fukunaga, T.; Fukutsua, T.; Sugie, Y., Silylation of graphite oxide. *Carbon* **2004**, *42* (10), 2117–2119.
[122] Thiergardt, R.; Rihst, G.; Hugt, P.; Peter, H.H., Cladospirone bisepoxide: definite structure assignment including absolute configuration and selective chemical transformation. *Tetrahedron* **1995**, *51* (3), 733–742.
[123] Bruice, T.C.; Bruice, P.Y., Solution chemistry of arene oxides. *Acc. Chem. Res.* **1976**, *9*, 378–384.

2

Mechanism of Formation and Chemical Structure of Graphene Oxide

Ayrat M. Dimiev

2.1 Introduction

Graphene oxide (GO) has been attracting the recurring interest of the chemical community since it was first synthesized almost 160 years ago. The fine chemical structure of this wonder material remains ambiguous up to the present day despite the efforts of generations of chemists to propose a suitable structural model. The mechanism of GO formation is even less studied and understood. As the reader might have learned from the previous chapter, very serious studies on GO chemistry have been conducted throughout the entire twentieth century. Unfortunately, today a large part of those early studies has been forgotten, because they were written in German and French. In the author's opinion, those early studies, performed in the best old-school traditions, were in many ways advantageous when compared to some modern publications. The fundamentality of scientific thinking, the methodology of the research and, importantly, the trustworthiness of the reported data were on a level that is rather rare in the modern GO field. The lack of advanced instrumentation was the main factor that prevented the old-time researchers from arriving at the conclusions we can make today. Another factor is that experimental data in the field has been accumulating gradually throughout the years. Today, one can build a theory on the structure of GO simply by reading and re-evaluating previously published research, in the way Anton Lerf did in Chapter 1. At the same time we should admit that for several reasons the early researchers were wrong in their views on GO chemistry. Even the most recent and broadly accepted model from Lerf himself contains some drawbacks, and cannot account for all the experimental observations.

Graphene Oxide: Fundamentals and Applications, First Edition. Edited by Ayrat M. Dimiev and Siegfried Eigler.

The "graphene era", started in 2004 by works of Novoselov and Geim, sparked renewed interest in GO, first as a potential graphene precursor, and later as a unique and interesting material in itself. However, in contrast to the twentieth century, the major part of modern GO-related studies is applied research: GO has been tested for numerous potential applications. The fundamental aspects of this material have gained significantly less attention. In modern application-driven research, fundamental science is considered a luxury, and is the realm of a few enthusiasts who can afford such a luxury. As a result, today one can observe a substantial decrease in the quality of GO-related publications caused by underestimating the importance of the fundamentals. One should never forget that fundamentals are paramount for successful applications. Understanding the mechanism of GO formation is critical for developing cost-effective production methods for products with target properties, while understanding the actual chemical structure is critical for developing meaningful functionalization protocols.

In this chapter, the most recent view on GO formation and its chemical structure will be described. However, this chapter is not a review article; we do not pursue the goal of referencing all the related publications. For those interested in a broader coverage of the published literature, we recommend the excellent review articles by Loh *et al.* [1], Chua and Pumera [2], Dreyer *et al.* [3], and Eigler and Hirsch [4]. In this chapter, we discuss only selected publications that directly contribute to the topic. In the form typical of textbooks, we methodologically describe the mechanism of GO formation, its transformation during aqueous work-up, and finally arrive at the fine chemical structure of GO. In large part, this chapter is based on the author's personal experimental studies, and reflects the author's personal views on the topic. Open questions are clearly indicated. The important thoughts and conclusions are underlined.

2.2 Basic Concepts of Structure

Graphene oxide (GO) is a two-dimensional (2D) material that is derived from the parent graphene backbone by introducing several oxygen functionalities. The oxygen atoms are covalently bonded to carbon atoms, converting them from the sp^2-hybridized state in the parent graphene into the sp^3-hybridized state. In a typical GO, the number of carbon atoms bonded to oxygen exceeds the number of intact sp^2-hybridized carbon atoms. This makes GO very different from the parent graphene. On the one hand, these oxygen functionalities can be considered as defects introduced into the otherwise ideal graphene plane. These defects convert electrically conductive graphene into an insulator. On the other hand, the oxygen functionalities provide GO with many unique properties that the parent graphene does not possess. One of these properties is hydrophilicity, i.e. the ability to be dissolved and to form stable colloid solutions in water and in some low-molecular-weight alcohols. Another advantage is opening a tunable bandgap that is responsible for unique optical and electronic properties.

The oxygen functionalities decorate the carbon grid from both sides. From this perspective, every single GO layer can be considered as a triatomic-layer 2D material where the carbon layer is sandwiched between two layers of oxygen atoms. This simplified model is not exactly correct, because oxygen atoms do not fully cover the carbon grid. The C/O atomic ratio varies to some extent, and for sufficiently oxidized GO samples it is approximately 2 : 1. In the oxygen-free areas, called graphitic or graphenic domains, the GO flakes are only one atomic layer thick. Thus, it will be more correct to state that every GO layer is one atomic layer thick with oxygen functionalities extending out of the plane in certain locations. Atomic force microscopy (AFM) is not able to resolve oxidized domains from graphitic domains and provides only the average thickness for a GO flake. The reported flake thicknesses vary from 0.8Å through 1.2Å. For comparison, the measured thickness of a single-layer intact graphene on a substrate is 0.5–0.6Å.

While GO is only one carbon layer thick, the lateral size of the flakes can vary from hundreds of nanometers through the tens and even hundreds of micrometers. In this truly 2D form, GO can exist only in solution (Figure 2.1a), where it is completely exfoliated to single-layer sheets, or being placed on a substrate surface (Figure 2.1b). In these two cases, GO flakes are in totally different environments. In solutions, GO flakes are fully surrounded by solvent molecules (water or alcohols). When on a substrate, GO flakes are in contact with a substrate surface on one side and with air on the other side. Most likely, the water molecules are still strongly physisorbed onto the GO surface on both sides, even when it is on a substrate surface. Note that, in aqueous solutions, GO flakes are negatively charged with the formation of an electrical double layer at the GO/water interface.

In the bulk form, the 2D character of GO is not really pronounced. The bulk form of GO is a lamellar paper-like solid material that was historically referred to as "graphite oxide". Solid graphite oxide is the product formed by restacking of the previously exfoliated single-layer GO sheets. Graphite oxide can also be obtained after oxidation of graphite

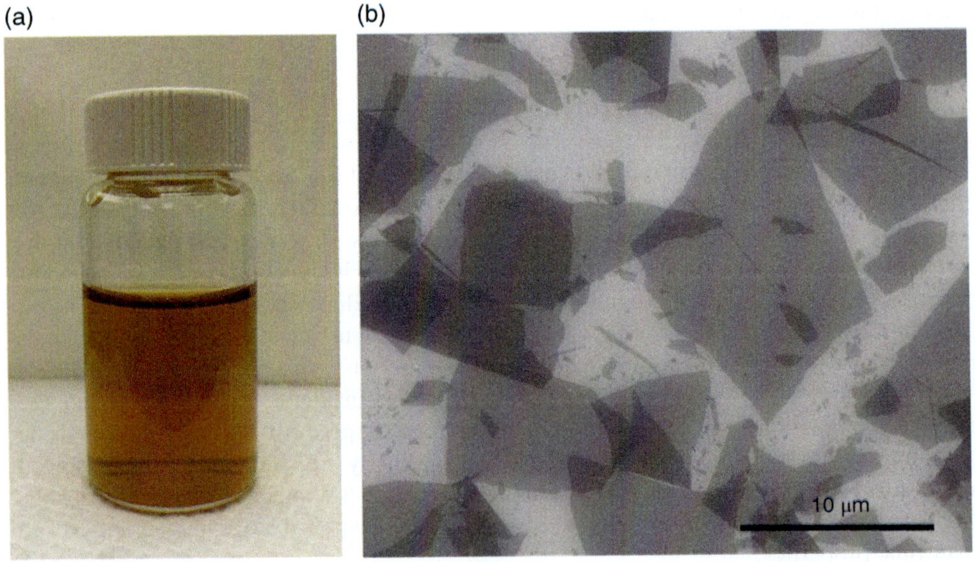

Figure 2.1 *Graphene oxide in its true 2D single-layer form. (a) Photograph of GO aqueous solution; the solution color may vary from yellow to brown. (b) Scanning electron microscope image of GO flakes on a Si/SiO_2 wafer. The number of layers can be distinguished by their opacity. All the flakes on this image are single-layered. The image is darker where the flakes are folded or overlapped, making double-layered structures*

without applying a delamination step. This substance has a loosely ordered structure in the *c*-axis direction with a repeat distance of 6–9 Å between the constituent GO layers, depending on the amount of absorbed water. Despite a weak interaction between the flakes, upon dissolution in water, the GO flakes easily separate from each other to form a colloid solution of single-layer GO sheets (Figure 2.1a). From these perspectives, solid graphite oxide can be considered simply as an accumulation of single GO sheets. From another point of view, we have to stress that the fine chemical structure of GO depends on the form of the material and on its environment. To some extent, the nature of the functional groups of a GO flake suspended in an aqueous solution (Figure 2.1a) is different from that of the same flake placed on a solid substrate (Figure 2.1b), or that sandwiched between neighbors in the form of bulk graphite oxide. This is one of the reasons why the fine chemical structure of GO is still not fully understood.

2.3 Preparation Methods

GO is produced from graphite by oxidation with strong oxidants in concentrated acid media. Depending on the choice of the oxidant and the acidic medium, today one can differentiate three major methods of GO preparation: Brodie's method [5], Staudenmaier's method [6] and the Hummers method [7]. Brodie's method uses potassium chlorate ($KClO_3$) as an oxidant in fuming nitric acid medium. This method is not effective, since it is not complete in one step and in one vessel. The partially oxidized product from the first reaction needs to be isolated, purified and subjected to a new oxidation cycle several times until sufficiently oxidized product is obtained. Staudenmaier's method uses $KClO_3$ in a mixture of concentrated nitric and sulfuric acids. It is similar to Brodie's method in the sense that it requires multiple portions of $KClO_3$ to be added to the reaction mixture during the course of oxidation. This is needed because it is difficult to maintain a high concentration of the oxidizing agent in an acidic medium for a long time. The added $KClO_3$ portion quickly decomposes in a reaction, which is parallel to the target reaction, i.e. oxidizing graphite. The Hummers method uses potassium permanganate ($KMnO_4$) as on oxidant in concentrated sulfuric acid medium. Hummers and Offeman claimed that their reaction completes in two hours, and thus it is more effective compared to Brodie's and Staudenmaier's methods. Indeed, this method is more effective compared to the two predecessors; however, it is that fast only with the small-particle-size powdered graphite source that was used by Hummers and Offeman. When large-particle-size graphite is subjected to oxidation by exactly the same Hummers method, an incompletely oxidized graphite–GO hybrid is obtained.

As one might have learned from the previous chapter, credit for the so-called Hummers method should in fact be given to Charpy [8], who had successfully used the combination of potassium permanganate and sulfuric acid fifty years before Hummers and Offeman published their report. In this chapter, we will use the term "Charpy–Hummers approach" whenever the combination of potassium permanganate and sulfuric acid is involved. We will use the term "Hummers method" strictly for the procedure described by Hummers and Offeman [7].

The GOs prepared by the three methods vary slightly in chemical composition [9–11]. The samples prepared by the Charpy–Hummers approach are normally more oxidized, while the samples prepared by Brodie's method are less oxidized. This is often attributed to the fact that potassium permanganate, used in the Hummers method, is a stronger oxidizing agent than potassium chlorate. However, the actual situation is not as simple. The oxidation level of GO depends not only on the nature of the oxidants, but most importantly on the amount of the oxidant (graphite to oxidant ratio) used for oxidation. GOs with different oxidation levels can be prepared by all three methods by varying the amount of oxidant used, and the time allowed for reaction. Here, we must differentiate between an oxidant, i.e. the chemical compound used for reaction, and an oxidizing agent, i.e. the actual species that interacts with graphene to convert it to GO. The nature of the actual oxidizing agents is still unknown for all three oxidation protocols, mainly due to the lack of research targeting this topic. Today, one can speak only about effectiveness of the compounds, or combinations of compounds, used for oxidation.

Of the three methods, the Charpy–Hummers approach is the one most widely used today because of the shorter reaction time and the relative ease of GO preparation. Also, this method avoids having to deal with the toxic fumes of nitric acid and chlorine that are formed in large quantities with the Brodie and Staudenmaier methods. The Charpy–Hummers approach has two major modifications [12, 13] that are used almost as often as the original Hummers method [7]. The Kovtyukhova modification [12] includes a so-called pre-oxidation step where graphite is exposed to a mixture of potassium persulfate, diphosphorus pentaoxide (phosphorus(v) oxide) and sulfuric acid. While this step is called "pre-oxidation", its actual role is the intercalation-assisted expansion of graphite. This expansion affords better access of the oxidizing agent into the graphite galleries during the following oxidation step with potassium permanganate. As a consequence, higher oxidation rates are afforded. The modification proposed by Marcano *et al.* [13] includes the use of more oxidant (twice as much as in the original Hummers method), and the addition of phosphoric acid to the sulfuric acid medium. The use of more oxidant affords higher oxidation levels of GO product, while the addition of phosphoric acid, as was claimed by the authors, affords less structurally damaged GO product.

An absolutely new preparation method for GO was proposed very recently by Peng *et al.* [14]. This method uses potassium ferrate as an oxidant and concentrated perchloric acid as an acidic medium. After decades of dealing with the same oxidants and the same acids, this new combination of reactants is very refreshing, and the entire preparation method is very elegant. The authors claim significantly higher reaction rates compared to all the previously known methods, and an eco-friendly technology, since no manganese is present in the acidic waste. In addition to the fact that this method might be very attractive for industrial-scale GO production, this finding also provides additional experimental data for understanding the mechanism of GO formation in general. In the very near future one can expect feedback on the method's effectiveness from a broader community.

Regarding the production cost, the oxidation process itself is not the most critical step in GO production. The bottleneck is the purification steps that follow the oxidation. Purification of as-prepared GO is performed by lengthy washing procedures in water. After each washing cycle, the GO product and washing waters are separated by centrifugation. Upon washing and removing acids, GO product exfoliates to single-layer sheets, and forms a very stable and voluminous colloid solution, or a gel at higher concentrations. This introduces additional difficulties into purification protocols. Alternatively, dialysis can be used for purification. However, this method takes even longer than multiple washing–separation cycles, and it is usable for small portions of GO only.

2.4 Mechanism of Formation

2.4.1 Theoretical Studies and System Complexity

Despite recent progress in understanding GO chemistry and structure, the mechanism of its formation has received significantly less attention. Until recently, most reported studies in this field were theoretical [15–18] and focused on the act of introducing oxygen atoms into the graphene lattice with the formation of C–O covalent bonds. Back in 2006, by utilizing a density functional theory, Li *et al.* [15] proposed a mechanism to explain the observed fault lines and cracks on GO. The authors showed that strain, generated by cooperative alignment of epoxy groups on opposite sites of the hexagon ring, can initiate the formation of cracks on GO. Sun *et al.* [16] arrived at similar conclusions, suggesting that oxidation of graphene is accompanied by cutting of graphene sheets. This process is limited by oxygen diffusion and driven by the local strain induced by the oxygen functional groups. The authors show that as-formed epoxy groups and their diffusion play the central role in the nucleation and growth of extended linear defects and in the subsequent cutting of graphene. Shao *et al.* [17] suggested that oxidation begins from the existing defects on the graphene plane or from the flake edges. The authors suggested that oxidation of graphene begins with the formation of phenolic groups. The phenolic groups form first on the edges and defects, and then propagate to the basal planes. In the next step, phenolic groups transform to either epoxides or ketones. Importantly, Shao *et al.* further suggested the rupture of the C–C bonds, or even the loss of carbon atoms with the formation of point defects. Note that the mechanism suggested by Shao *et al.*, involving primary formation of phenols, is very different from the mechanisms proposed by Li *et al.* and Sun *et al.*, where epoxy groups are the primary functionalities.

Theoretical studies, published to the present day, consider both graphene and an oxidizing agent as freestanding species with no interaction with their surroundings. However, in fact, GO is produced from bulk graphite, where individual graphene layers are closely aligned and stacked. In order to attack graphene layers, the oxidizing agent needs to first penetrate between those layers. As will be shown below, at the time when the oxidizing agent attacks graphene, the latter is in intimate contact with sulfuric acid. These factors have not been yet taken into account in the theoretical studies. This is true even for the most recent studies, such as the one by Boukhalov [18]. As an advantage over the earlier theoretical works, Boukhalov demonstrated a significant difference in the oxidation mechanisms between single-layer graphene and graphite. This is a significant step forward that makes the studied system closer to reality, compared to previous publications. Unfortunately, the chosen standing points in this work were very different from the actual experimental conditions. Thus the study was based on the assumption that, in Brodie's method, water molecules penetrate graphite galleries first, and assist in the oxidation that occurs next. This conclusion is in fact misleading. In reality, in all the known methods, graphite is first oxidized in water-free conditions, and as the next step the oxidized graphite is exposed to water.

While theoretical studies proved to be beneficial in understanding the structure of GO and its thermal decomposition, concerning the mechanism of GO formation, they are still far away from the real system. Due to the complexity of the real system, there is a long road ahead for theoretical studies, before the actual complex reaction conditions can be modeled in a reliable way. At the present day, mostly experimental work has contributed to the understanding of this complex phenomenon. Little by little experimental data are accumulated and the whole picture of GO formation is shaping up.

2.4.2 Step 1: Formation of Stage-1 H_2SO_4-GIC Graphite Intercalation Compound

Work on the reaction mechanism should ultimately reveal the full content of the black box shown in Figure 2.2.

Since the Charpy–Hummers approach and its modifications are the most widely used methods for GO production today, we will first focus our discussion on this method, and then we will generalize our discussion on other preparation protocols. In our recent experimental study, we tried to reflect the complexity of the actual reaction conditions between graphite and the oxidizing medium [19]. In the course of conversion of bulk graphite to GO, we single out three distinct independent steps. The reaction can be stopped at any step, and the corresponding intermediate products can be isolated, characterized and stored under appropriate conditions. The first step is the conversion of graphite to the sulfuric acid–graphite intercalation compound (H_2SO_4-GIC), which can be considered as the first intermediate. The second step is conversion of the GIC to the oxidized form of graphite that was defined as "pristine graphite oxide" (PGO), constituting the second intermediate. The third step is conversion of PGO to GO by the reaction of PGO with water.

The first step, the formation of H_2SO_4-GIC, begins immediately upon exposing graphite to the acidic oxidizing medium, and it comes to completion within a few minutes. The GIC formation is manifested by the characteristic deep-blue color acquired by graphite flakes. The nature of the intercalation compound formed is the stage-1 GIC, where every graphene layer alternates with a layer of intercalant. The characteristics of the GIC intermediate formed in the course of GO production are identical to those of H_2SO_4-GICs produced by different methods electrochemically or chemically [20–24]. The stoichiometry of the stage-1 H_2SO_4-GIC can be represented by the formula $C_{(21-28)}{}^{+} \cdot HSO_4{}^{-} \cdot 2.5H_2SO_4$ [20, 23]. The interlayer galleries in the stage-1 H_2SO_4-GIC are closely packed with H_2SO_4 molecules

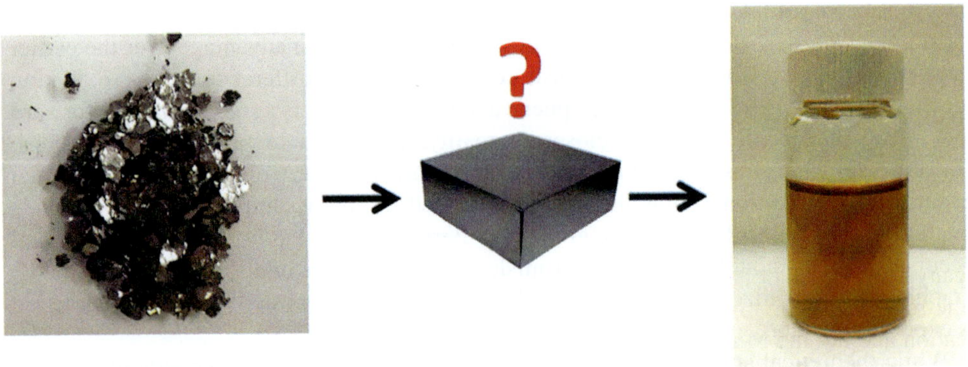

Figure 2.2 *The conversion of bulk graphite to GO is a black box. Researchers studying the mechanism of this conversion must reveal and fully describe all the steps and underlying mechanisms that lead to this transformation*

and HSO_4^- ions, which do not form any orderly structure [20, 23]. The second step, conversion of the GIC into PGO, is significantly slower; it takes several hours or even days depending on the graphite source. Thus, the stage-1 GIC can be considered as a real intermediate that can be isolated and characterized. The ability of acidic oxidizing media to intercalate graphite with the formation of stage-1 GIC is the first necessary condition for successful oxidation of graphite. This occurs with all the known GO preparation methods. All the used combinations of reactants, i.e. $KClO_3/HNO_3$, $KClO_3/H_2SO_4$, $KMnO_4/H_2SO_4$ and the new $K_2FeO_4/HClO_4$, afford the formation of respective stage-1 GICs. Intercalation increases the distance between the graphene layers in graphite, making the interlayer galleries accessible for the oxidizing agent. While the intercalation phenomenon was well known to scientists in the past, in the new graphene era, the significance of this step in the course of GO formation is often neglected and many researchers do not even take this significant step into consideration, while discussing the reaction conditions.

2.4.3 Step 2: Transformation of Stage-1 H_2SO_4-GIC to Pristine Graphite Oxide

The second step, the conversion of the stage-1 GIC to PGO, is the most intriguing step. It involves the insertion of the oxidizing agent into the pre-occupied graphite galleries. Careful study of this conversion by Raman microscopy has lifted the veil on this transformation. Figure 2.3(a) shows a partially oxidized graphite flake in the course of its transformation from stage-1 GIC to PGO.

It is apparent that the transformation of stage-1 GIC to PGO has clear edge-to-center propagation. The spectra taken from the blue-colored areas in the middle of the flake (Figure 2.3b) show the presence of the stage-1 GIC only. The Raman spectrum from the light-yellow-pearl-colored area along the flake perimeter (Figure 2.3c) confirms the presence of GO phase only. The spectrum acquired from the dark-looking bordering area between the blue and yellow areas (Figure 2.3d) is simply the sum of the two spectra discussed above. Thus, on the border, there is a mixture of the two phases: stage-1 GIC and GO. No traces of stage-2 GIC, higher stages or pure graphite were ever detected. These observations suggest that PGO is formed directly from the stage-1 GIC without any additional rearrangements in the graphite structure.

The actual nature of the oxidizing agent in the Hummers method is as elusive as in the methods that use $KClO_3$. Not only is the identity of the specific oxidizing agent species that attacks the graphene layers unknown, but also even the nature of the $KMnO_4/H_2SO_4$ solution itself has not been systematically studied. Some researchers suggest that the oxidizing agent is manganese heptaoxide (Mn_2O_7) [25]. Indeed, the green-colored Mn_2O_7 can be isolated from the $KMnO_4/H_2SO_4$ solution of similar color [26]. However, it was also shown that, in concentrated sulfuric acid medium, Mn(VII) exists in the form of the planar permanganyl (MnO_3^+) cation, which might be closely associated with hydrogen sulfate (HSO_4^-) in the form of MnO_3HSO_4 [27, 28]. In the author's opinion, it is more likely that the MnO_3^+ cation and not the Mn_2O_7 molecule is the real oxidizing agent.

In order to diffuse between the graphene layers, the oxidizing agent needs to either replace existing intercalant molecules or insert between them. The clear edge-to-center front-like propagation of the reaction (Figure 2.3a) is an indication that the rate of diffusion of the oxidizing agent into the graphite interlayer galleries must be lower than the rate of

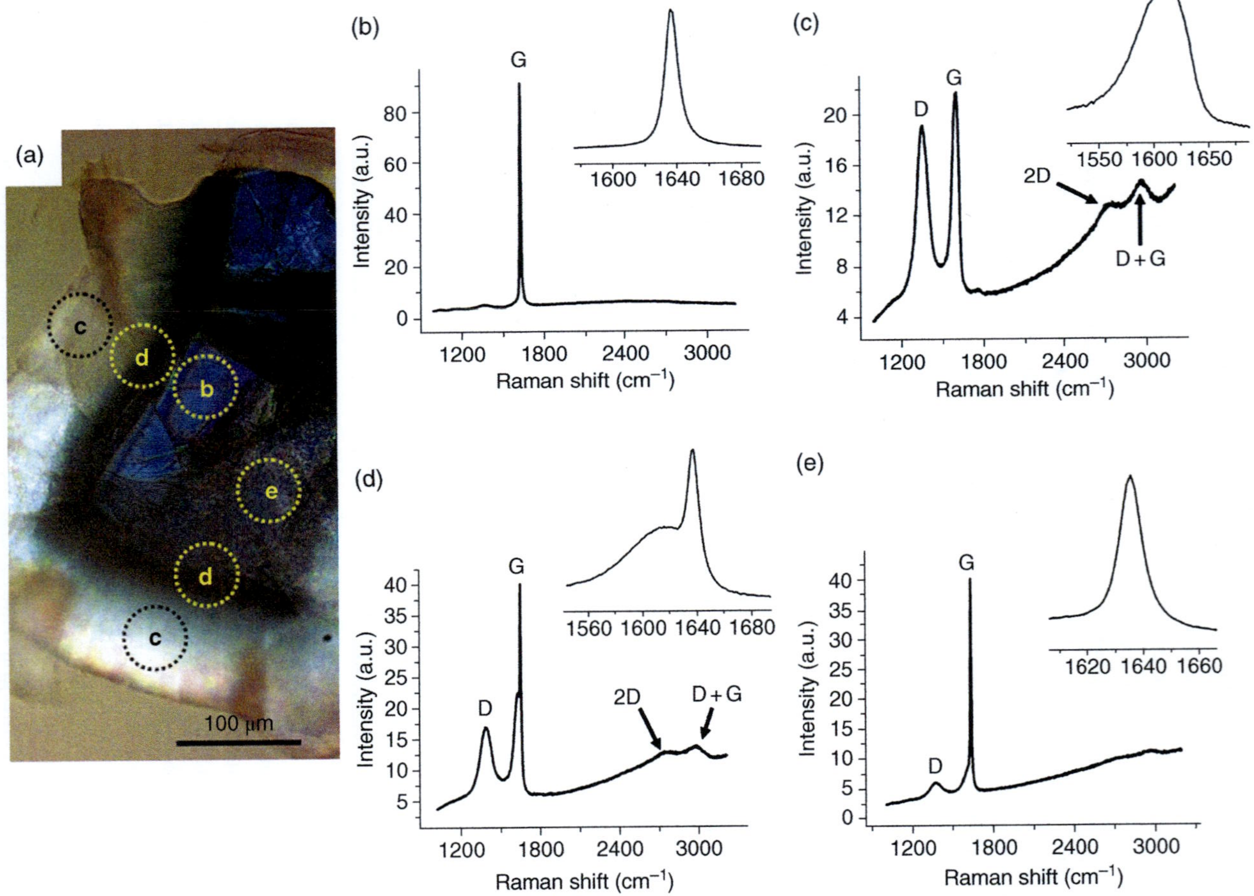

Figure 2.3 *A graphite flake in the course of its transition from stage-1 GIC to PGO. (a) Map of the flake showing spots of acquisition; the circles labeled "b", "c", "d" and "e" indicate the four typical areas on the flake surface where the spectra were acquired. (b)–(e) Typical Raman spectra acquired from the corresponding spots labeled in (a). The insets represent the x-axis expansion in the G-band area. The spectra taken from the blue-colored areas in the middle of the flake (b and e) show the presence of the stage-1 GIC only. The spectra taken from the light-yellow-colored areas (c) on the flake's edge represent GO. The spectrum acquired from the dark-looking bordering area between the blue and yellow areas (d) shows the presence of two phases: stage-1 GIC and PGO. A 514 nm laser was used for excitation. Reproduced from [19] with permission of ACS*

the chemical reaction itself. As soon as the oxidizing agent diffuses between the graphene layers, it must quickly react with nearby carbon atoms. Otherwise, the GO phase would gradually form over the entire flake body. Thus, the second step of PGO formation is diffusion-controlled, where the oxidizing agent replaces the acid intercalant. This is the rate-determining step of the entire GO formation process. This conclusion explains why small-particle-size graphite reacts faster than large-particle-size graphite: the diffusion pathway is much shorter for the smaller graphite flakes and particles. It also explains why only partially oxidized GO product (GO shell and graphite core) is obtained when a large-particle-size graphite is subjected to oxidation according to the exact Hummers protocol with the two hour reaction time limit: there is simply not enough time for an oxidizing agent to penetrate between the graphene layers.

Note that this interpretation of the GO formation reaction is not consistent with the existence of two different types of domains in GO (see section 2.6). If the process is indeed limited by diffusion, it is unclear how the graphitic domains survive during the slow continuous front-like diffusion of the oxidizing agent into the stage-1 GIC galleries. It is possible, therefore, that there are additional still unknown factors controlling this process.

2.4.4 Pristine Graphite Oxide Structure

The Raman data, presented and discussed above, are limited by the laser spot size and the probing skin depth. Unlike Raman microscopy, X-ray diffraction (XRD) provides information about the bulk quantities, and reveals all the phases present in the specimen. In this experiment, 4 wt equiv (weight equivalents) of $KMnO_4$ were added sequentially to the graphite/H_2SO_4 slurry, and the corresponding samples were subjected to XRD analysis after complete consumption of each weight equivalent of $KMnO_4$. The four samples were referred to as "transition forms" 1 to 4 (TF-1, -2, -3 and -4), respectively. The samples were protected from moisture to prevent decomposition of their unique structure.

The graphite sample taken from the reaction mixture 20 min after addition of the first weight equivalent of $KMnO_4$ exhibits a diffraction pattern typical for the stage-1 H_2SO_4-GIC (Figure 2.4). The 002 diffraction line at 22.3° 2θ angle along with the 003 and 004 signals at 33.7° and 45.2° can be unambiguously assigned to the stage-1 H_2SO_4-GIC with interlayer distance (d_i) of 7.98 Å. The stage-1 GIC is the only phase present in the sample. After the first added $KMnO_4$ portion is consumed (TF-1), the stage-1 GIC signals are still observable, though they are less intense. The broad curve in the 15° to 28° 2θ region is indicative of the formation of an amorphous phase. New weak signals appear at 21.6°, 11.4° and 11.7° 2θ diffraction angles (Figure 2.4a,b).

After addition and consumption of the second weight equivalent of $KMnO_4$ (TF-2), the signals associated with the stage-1 GIC are no longer present. This is consistent with the fact that the uniformly blue-colored areas like on Figure 2.3(a) were no longer observable in graphite flakes with optical microscopy. The new strong signal at 9.7° 2θ (9.12 Å) is observed in the sample taken after the addition and consumption of the third weight equivalent of $KMnO_4$ (TF-3). This signal becomes even stronger in TF-4 (Figure 2.4c). This signal is associated with the formation of PGO.

The as-prepared PGO is highly ordered along the *c*-axis, with a repeat distance of 9.12 Å, corresponding to the 9.7° 2θ. The 1.14 Å increase in spacing compared to the stage-1 GIC (d_i = 7.98 Å) is likely due to insertion of oxygen atoms within the oxidized domains in

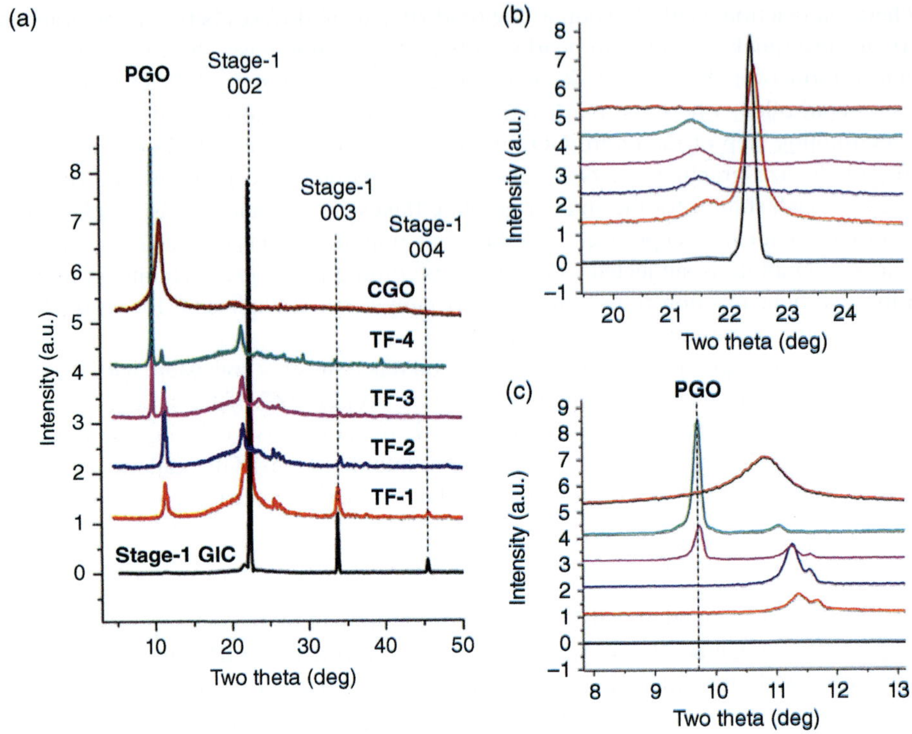

Figure 2.4 X-ray diffraction data. (a) X-ray diffraction patterns for stage-1 GIC (black), CGO (brown line) and four transition forms (red, blue, pink and dark cyan lines). The labels "TF-1" to "TF-4" represent the four consecutive transition forms obtained by consumption of 1, 2, 3 and 4 wt equiv of $KMnO_4$, respectively. (b,c) The x-axis expansions of (a) in the 20° to 24° and 8° to 13° 2θ diffraction angle regions. Reproduced from [19] with permission of ACS

addition to the existing sulfuric acid molecules. This structure is relatively stable and can exist for several months with no visible change if not exposed to large amounts of water. Thus, PGO can be considered as the second intermediate in the course of GO production.

Interestingly, the step of PGO formation and the unique PGO structure is not something specific to only the Charpy–Hummers method. Back in 1989, a green-colored intermediate with a structure very similar to that of PGO was reported by Mermoux and Chabre [29] in the course of graphite oxidation by Brodie's method. A repeat distance of about 8 Å was reported for that product. Unfortunately, Mermoux and Chabre missed the first intermediate, the stage-1 HNO_3-GIC. They simply did not perform the XRD study of the graphite sample immediately after introducing $KClO_3$ into the graphite–HNO_3 mixture. However, there are numerous studies showing that the stage-1 HNO_3-GIC with a repeat distance of 7.84 Å is formed when graphite is exposed to a mixture of concentrated HNO_3 with strong oxidants such as $KMnO_4$ and $KClO_3$ [30]. Interestingly, there was only 0.16 Å increase in the repeat distance of Mermoux and Chabre's PGO compared to the 7.84 Å for the corresponding stage-1 HNO_3-GIC. It is unclear how to account for such an insignificant increase in the repeat distance with oxidation. This might be related to (i) lower oxidation level of Brodie's GO, and (ii) planar geometry of HNO_3 molecules, allowing better accommodation of the oxygen functional groups, and thus denser packing of the interlayer galleries in Mermoux and Chabre's PGO compared to our PGO.

There are no publications reporting the formation of PGO during the oxidation of graphite with the Staudenmaier method. This method is similar to the Charpy–Hummers method in that the reaction occurs in sulfuric acid medium. This is why with a high degree of confidence we can assume that the reaction proceeds through the same three steps as in the case of the other two methods. Thus, the above-described mechanism of GO formation, involving the two intermediate structures, stage-1 GIC and PGO, is a general pathway for all three major methods of GO production in acidic media.

2.4.5 Step 3: Delamination of Pristine Graphite Oxide

The third step of GO formation, i.e. the conversion of PGO to GO, is afforded by the exposure of PGO to water during quenching and washing procedures. This involves the loss of the *c*-axis interlayer registry and exfoliation of PGO into single-atomic-layer sheets. Many synthetic protocols specify that sonication is needed to disperse as-prepared PGO in aqueous solutions. In fact, sonication is only needed for not completely oxidized GO samples prepared from large-particle-size graphite under the original Brodie and Hummers protocols. Many researchers simply do not provide sufficient time for the oxidizing agent to diffuse between the graphene layers of stage-1 GIC, and subsequently obtain only partially oxidized GO. When as-prepared GO is sufficiently oxidized, it spontaneously delaminates into single-layer sheets with simple stirring in water.

Figure 2.5 summarizes and schematically represents the three steps constituting the process of conversion of bulk graphite into GO.

2.5 Transformation of Pristine Graphite Oxide Chemical Structure Upon Exposure to Water

The loss of the *c*-axis interlayer registry and exfoliation into single-atomic-layer sheets is not the only process that occurs upon exposure of PGO to water. The chemical composition of PGO also undergoes very serious transformation. As-prepared PGO has bright-yellow color, even when the reaction is already quenched with water and PGO flakes are partially exfoliated. Upon prolonged exposure to water during washing procedures, while the solution becomes more and more neutral, the bright yellow color gradually darkens and transforms to brown. It was straightforward for us to suggest that chemical reactions between PGO and water must be behind the visual observations [31]. To investigate the chemical structure of as-prepared PGO, one needs to free it from sulfuric acid, but also to avoid exposing the material to water. For this purpose we conducted a series of experiments [31], where as-prepared PGO was quenched and washed by non-aqueous organic solvents capable of dissolving H_2SO_4, but not possessing nucleophilic character to prevent chemical transformation of PGO to GO. The as-purified organic-solvent-washed GO samples (OS-GO) were compared with those obtained by washing with 33% HCl, and the one washed with water (conventional GO). Quenching and washing with the organic solvents yields products with very light colors that vary from light yellow through bright yellow. The

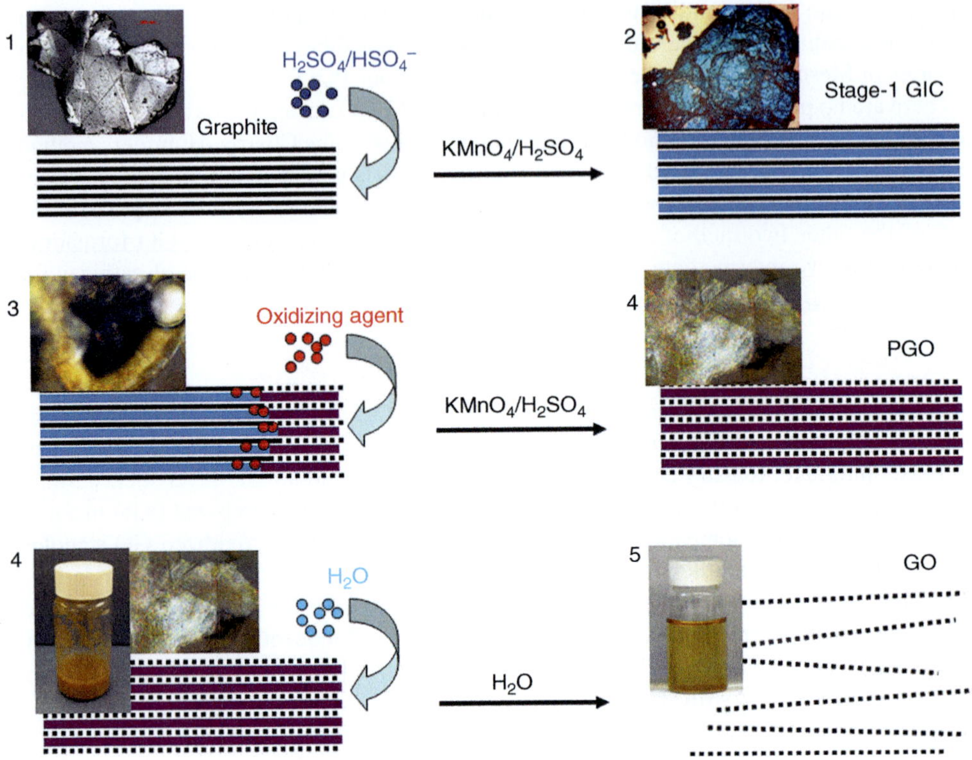

Figure 2.5 Schematics of conversion of bulk graphite into GO with corresponding micrographic images of sample appearances at each phase. The three steps signify formation of the two intermediate products (stage-1 GIC and PGO) and the final GO product. The solid black lines represent graphene layers; dotted black lines represent single layers of GO; wide blue lines represent H_2SO_4/HSO_4^- intercalant; wide purple lines represent a layer of the mixture of H_2SO_4/HSO_4^- intercalant with reduced form of oxidizing agent. Reproduced from [19] with permission of ACS

HCl-washed GO was light brown. This preservation of the original light color for OS-GO versus color darkening for conventional GO confirms that, during the water-based purification, a chemical reaction occurs between PGO and water.

In contrast to water, all the tested OS-GO samples contained 1.2–6.0 at.% sulfur according to X-ray photoelectron spectroscopy (XPS) data. The presence of sulfur from 0.5% to 2% even in water-washed GO samples had been reported by several researchers earlier. Sulfur is very difficult to remove even with extensive water washings, suggesting that the sulfur-containing impurities are covalently bound to or strongly physisorbed onto GO. To account for this observation, Petit *et al.* suggested that sulfur is covalently bonded to carbon atoms by C–S bonds, and exists in the form of sulfones [32]. This suggestion was in line with the ideas expressed earlier by Boehm and Scholz suggesting the presence of either sulfones or sulfates [9].

The Fourier-transform infrared (FTIR) spectra of all the light-colored OS-GO samples (Figure 2.6) contain the two enhanced bands at 1417 and 1221 cm^{-1}, which are barely

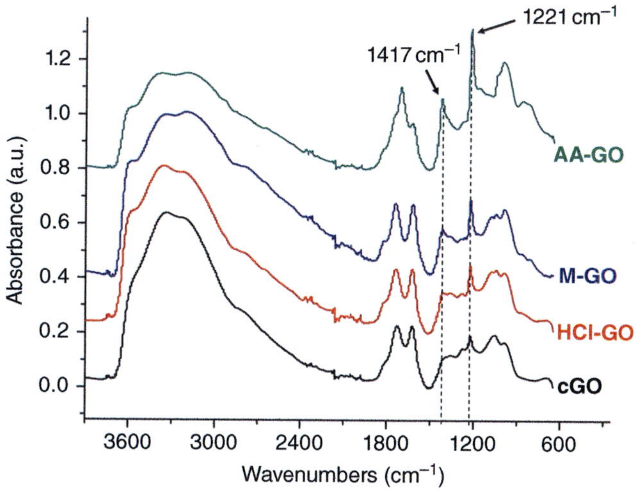

Figure 2.6 FTIR spectra of the four different GO samples. The samples are obtained by purification of as-synthesized PGO samples with acetic acid (AA-GO), methanol (M-GO), 33% hydrochloric acid (HCl-GO) and water (cGO). The two absorption bands at 1417 and 1221 cm^{-1} are assigned to the symmetric and asymmetric stretching of the S=O bonds in covalent sulfates. The intensity of these two bands increases with the sulfur content in GO, and decreases with increasing aqueous/nucleophilic character of the solvent used for purification. Reproduced from [31] with permission of ACS

present, if at all, in the spectrum of the water-washed GO sample. The intensities of the two bands increase with the sulfur content in GO samples. This trend is opposite to the nucleophilic character of the solvents used for GO purification. From all the possible sulfur-containing compounds, only disubstituted organic sulfates exhibit absorption at both 1417 and 1221 cm^{-1} [33, 34]. Sulfones, sulfuric acid and inorganic sulfates do not absorb at 1417 cm^{-1}. Thus, sulfones were excluded, and the two bands at 1417 and 1221 cm^{-1} were assigned to the symmetric and asymmetric stretching of the S=O bond in disubstituted cyclic organic sulfates.

Most likely, the organic sulfates are formed during the second step of GO formation, i.e. when stage-1 GIC is converted to PGO. If one assumes that epoxides are indeed the first functional groups formed during the oxidation, then the reaction mechanism in Schemes 2.1 and 2.2 can be envisioned. Sulfuric acid or hydrogen sulfate present in the interlayer galleries of PGO attack newly formed epoxides (**1**) according to Scheme 2.1. The reaction mechanism is typical of epoxide ring opening. The intermediate product is monosulfate (**2**). The monosulfate can attack the adjacent epoxide and convert to the cyclic sulfate (**3**).

During the third step of GO production (exposing PGO to water), the reverse reactions occur: cyclic sulfates hydrolyze with the formation of tertiary alcohols (Scheme 2.2). The first step of cyclic sulfate hydrolysis occurs with C–O bond cleavage and results in the formation of the monosulfate (**4**) [35]. The second step (hydrolysis of the monosulfate) occurs mainly with S–O bond cleavage and results in the formation of a 1,2-diol (**5**).

The low rate of sulfate hydrolysis under acidic conditions explains why it is difficult to remove sulfur-containing by-products from a GO sample even when washing with water.

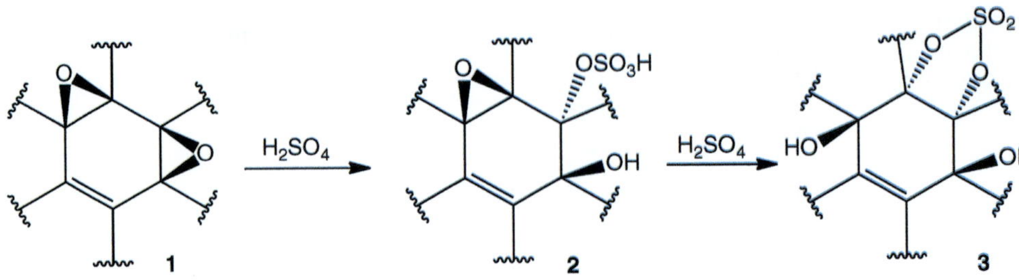

Scheme 2.1 The sulfate ester (**2**) formed in the first step is the intermediate product. The sulfate ester can attack the neighboring epoxide group, resulting in a 1,2-cyclic sulfate (**3**). 1,3-Cyclic sulfates (not shown) can also be formed by the reaction. Note that one hydroxyl group is formed for every sulfate ester, and two hydroxyl groups form for every cyclic sulfate

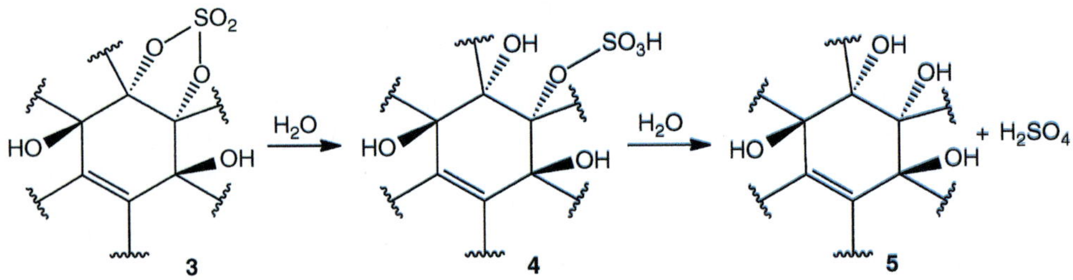

Scheme 2.2 The hydrolysis of cyclic sulfate (**3**) occurs in two steps. The first step occurs with the C–O bond cleavage and results in the formation of the monosulfate (**4**). The second step (hydrolysis of the monosulfate) results in the formation of a 1,2-diol (**5**)

Sulfuric acid itself is easily washed away with a few washings, but the organic sulfates are removed only via hydrolysis, the rate of which is slow under acidic conditions, so that longer reaction times are needed to remove more of those moieties.

Shortly after our work, Eigler *et al.* demonstrated that even in the water-washed GO samples covalent sulfates might be present in quantities up to 5%, i.e. one sulfate group per every 20 carbon atoms [36]. However, to the present day, this is the only work claiming such a high content of sulfur in water-washed GO samples. In our personal experience, the sulfur content in moderately washed GO samples was between 1.5 and 2.5 at.%. In thoroughly water-washed samples, the sulfur content is normally less than 1 at.%. Thus it is reasonable to state that the content of covalent sulfates is largely dependent on the purification procedures.

Concerning the chemical composition of PGO, it was shown to be notably different from the composition of typical water-washed GO [31]. The major functional groups in the PGO structure are epoxides. Their content substantially exceeds the content of the tertiary alcohols. The content of tertiary alcohols, in turn, only insignificantly exceeds the content of organic sulfates. This observation suggests that the first functional group formed on graphene during the oxidation of graphite is indeed epoxide, which is converted to organic sulfate by reacting with sulfuric acid, and to tertiary alcohol by reacting with water.

Unlike the first two steps of GO formation described in section 2.4, the third step does not have clear indicators to signal when this step is actually complete. The color of the GO dispersion changes gradually; the same is true of the pH of the washing waters. After some time of exposing PGO to water, this process can be considered not only as a step of GO formation (exfoliation and purification), but also as a chemical reaction of as-prepared GO with water. There is no clear borderline between the two processes: the washing–exfoliation on the one hand and the chemical modification of the functional groups on the other. The chemical reactions of GO with water will be reviewed in more detail in section 2.6 and in Chapter 6, describing the chemical properties of GO.

2.6 Chemical Structure and Origin of Acidity

2.6.1 Structural Models and the Actual Structure

Now let us discuss the fine chemical structure of a single-layer GO regardless of the organic sulfates. In Chapter 1, Anton Lerf discussed all the structural models that were historically proposed throughout the decades from 1939 through 2006. For our further discussion, two models will be of particular importance: the now-famous Lerf–Klinowski (LK) model [37] proposed in 1998, and the Szabó–Dékány (SD) model [38], proposed in 2006. While Lerf has reviewed the earlier models in detail, he was modestly laconic about his own development. Regarding the SD model, in Chapter 1 Lerf wrote one single sentence:

> I think that the model published by Szabó *et al.* combines structural elements of both Scholz and Boehm (broken C–C bonds, quinone groups) and Ruess (1,3-ethers).

This is not sufficient to understand the essence of this model. Both the LK and SD models are significantly more advanced compared to their predecessors, and more accurately reflect the actual GO structure. Because of the importance of these two models in understanding GO structure, we begin our discussion by reviewing the LK and SD models. Indeed, the SD model is the further development of the model proposed earlier by Clauss [39], and further modified by Scholz and Boehm [40]. As such, it inherits and further develops the studies of generations of chemists, who attempted to propose a stoichiometric and orderly arranged GO structure. In contrast to the SD model, the LK model rejects any orderly structure, and proposes an irregular chaotic arrangement of the functional groups. From these perspectives, the LK model was a revolutionary breakthrough in GO studies; amazingly, it was very close to reality.

The LK model (Figure 2.7) concludes that GO planes consist of two different types of randomly distributed domains: (i) areas of pure graphene with sp^2-hybridized carbon atoms, which have survived the oxidation, and (ii) areas of oxidized and thus sp^3-hybridized carbon atoms. The oxidized GO domains contain epoxy and hydroxyl (tertiary alcohol) functional groups. The edges of the flakes are terminated by carboxyl and hydroxyl groups.

The SD model (Figure 2.8) represents GO as a periodic ribbon-like structure of aromatic and non-aromatic (cyclohexane) stripes. The oxygen functionalities that are thought to be present on basal planes (cyclohexane stripes) are hydroxyls and four-membered-ring 1,3-ethers. The SD model suggests that ketones and quinones are formed where C–C bonds

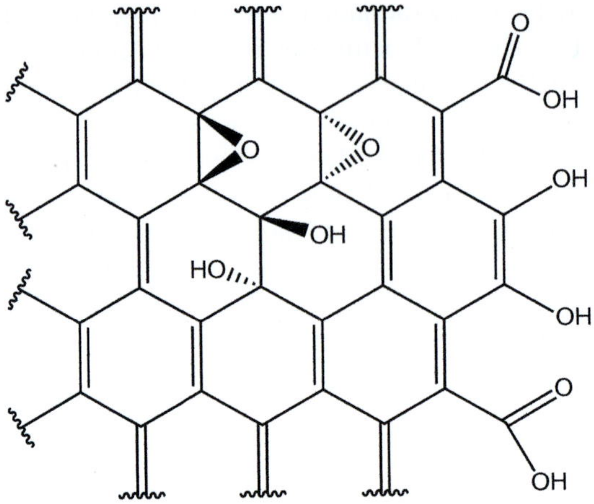

Figure 2.7 Simplified version of the Lerf–Klinowski GO structural model. GO flakes contain epoxides and tertiary alcohols on basal planes; carboxyl and hydroxyl groups terminate flake edges

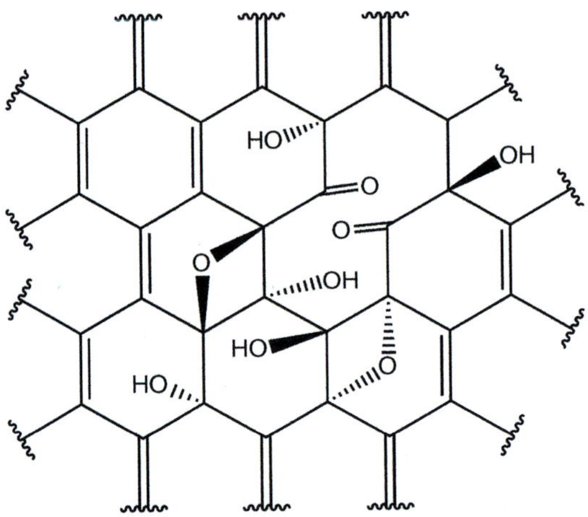

Figure 2.8 Simplified version of the Szabó–Dékány GO structural model. GO flakes contain tertiary alcohols and 1,3-ethers on basal planes. Ketones are formed where C–C bonds are cleaved

have been cleaved. This idea of C–C bond cleavage was first proposed by Clauss [39] and was later supported by Scholz and Boehm [40].

One can see that the LK and SD models are substantially different. This fact demonstrates that, at the time the SD model was proposed (2006), the situation with understanding GO structure was far from coming to mutual agreement. For some reason, the SD model did not receive as much attention as the LK model. As will be shown below, the ideas proposed in the SD model and in its earlier predecessors deserve more attention.

The major breakthrough in understanding the real GO structure came in 2010 with the two studies reporting high-resolution transmission electron microscopy (HRTEM) images of GO and RGO with atomic-level resolution [41, 42]. While the work by Gomez-Navarro *et al.* [41] demonstrated the images of RGO only, the work by Erickson *et al.* [42] demonstrated the structure of both RGO and GO. The images confirmed the structure of GO with two different types of domains, as Lerf *et al.* had hypothesized. Two different types of domains were clearly observable on the HRTEM images of GO (Figure 2.9), with the only

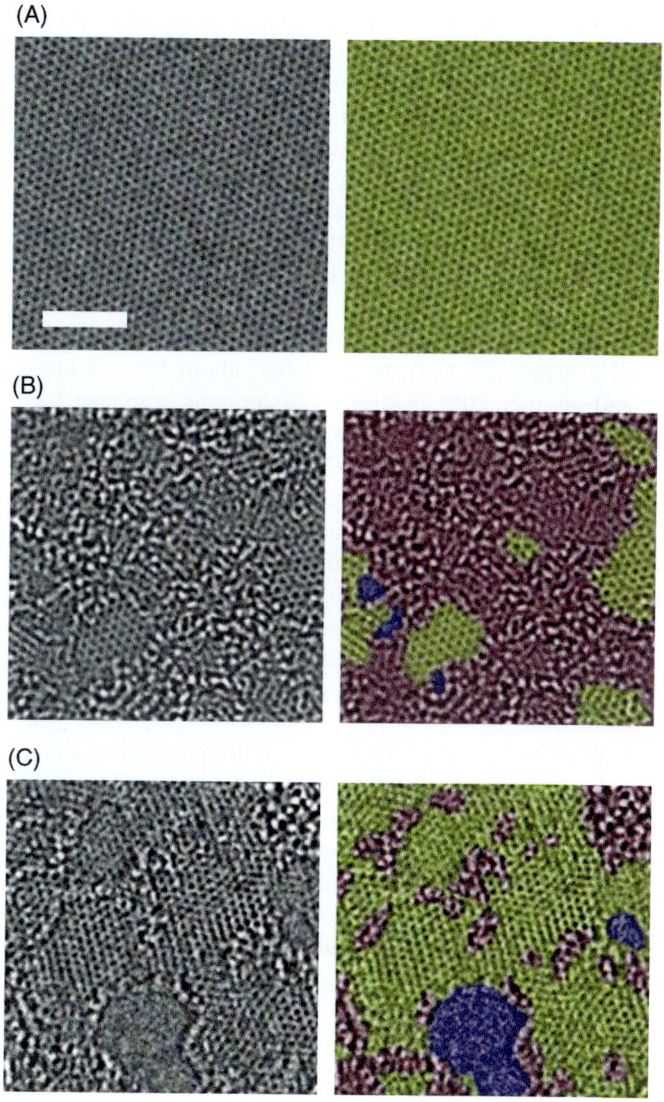

Figure 2.9 *Aberration-corrected transmission electron microscope images of (A) graphene, (B) GO and (C) RGO. The scale bar, denoting 2 nm, is valid for all images. On the right are the same images as on the left, where different areas are color-coded: graphitic domains in yellow-green, holes in blue and oxidized domains in red-purple. Reproduced from [42] with permission of John Wiley & Sons*

difference being that the size of the domains was larger than predicted by Lerf *et al.* As is evident from the HRTEM images (Figure 2.9B), oxidized domains form a continuous network, while graphitic domains form isolated islands. In addition to intact graphene domains and oxidized domains, as predicted by the LK model, nanometer-sized holes were observed in GO flakes, which was a new finding. However, it is not very clear if these holes pre-existed on the GO planes, or were generated by the electron beam.

Unfortunately, one cannot claim that the two published HRTEM studies have completely resolved the mystery of GO structure. Despite the fact that the GO lattice was observed at the atomic level, a more precise chemical structure has not arisen. The chemical identity of the oxygen functional groups cannot be determined from the provided HRTEM images. Keeping in mind the confirmed existence of the two types of domains, today, the term "chemical structure of GO" signifies the fine chemical structure of oxidized domains that constitute more than two-thirds of the entire GO flake body. Even more important for understanding GO chemistry is the chemical structure of interfaces between the two types of domains, and the structure of the edges of the numerous holes.

Another question is whether the presented HRTEM images represent the true oxidation level and the structure of GO in general. One can easily see that RGO presented by Erickson *et al.* [42] (Figure 2.9) is significantly more disordered compared to that presented by Gomez-Navarro *et al.* [41]. In the latter, about 60% of the carbon plane is intact graphene, and another 20% is slightly disordered graphene (see Figure 3.25), while in the former the graphenic areas constitute less than 50%. Apparently, the two studies were dealing with two very different GO samples. The presented HRTEM images are simply screenshots of a few particular locations of a particular GO flake(s). These images are not statistically sufficient to claim a similar structure even for the entire batch of a given GO product. They are definitely not sufficient to claim the structure of GO as a substance in general. Today, we know that there is significant flake-to-flake variation in oxidation levels even within the same batch of prepared GO samples [19, 43]. Thus the observed structures cannot be extended with confidence for GO in general. Additional HRTEM studies of different GO samples are highly desired to better understand the GO structure.

Confirmation of the existence of the two types of domains raised the popularity of the LK model even further. The correctness of the two-type-domain structure was automatically extended to the entire model, including the chemical identity of the functional groups. Simultaneously, the importance of other models was downgraded. As often happens in science, focusing on one single point of view and ignoring alternatives might play a negative role in scientific development. This is exactly what happens in GO-related studies today. One should remember that the publication of Lerf *et al.* [37] was one out of many studies in the chain of works on the structure of GO started in 1939. As Lerf himself modestly noted in the previous chapter:

> So ... most workers active in the field of GO research since 2005 are just preparing footnotes to the seminal works of Kohlschütter, Hofmann and Boehm, and I must confess that my own work is also nothing else.

However, in the modern GO-related literature, the LK model is considered almost as an axiom. At the same time, several valuable ideas proposed in the earlier models remain largely forgotten.

Misinterpretation of GO structure leads to confusion in understanding the nature of reactions involving GO. This confusion, unfortunately, has only increased in recent years. Despite the existence of the high-quality early publications (see Chapter 1), and several serious studies on GO structure published in the past few years, they have been appreciated only by a small circle of experts. The LK model, in turn, has been largely oversimplified and misinterpreted. The statement that "epoxides and hydroxyls exist on basal planes, and carboxyls decorate flake edges" is a typical cliché in almost every GO-related paper today. Many researchers refer to the LK model without even having read the actual paper, but trust generalized expressions. Being focused on the final applications of GO, such as devices, films, hydrogels, etc., the authors do not provide a proper description of the precursor material they work with. Newer publications refer to previously misinterpreted chemical structures, and this is growing as a chain reaction. As a consequence, the current literature on GO is already dominated by oversimplified misleading interpretations. Those publications create a parallel reality that greatly confuses a non-expert audience regarding actual GO structure.

In this chapter, we will highlight several experimental facts about GO chemistry that do not fit into the broadly accepted simplified views on GO structure.

Let us start our discussion of GO structure by reviewing a ^{13}C solid-state nuclear magnetic resonance (SSNMR) spectrum of GO that provides the most reliable data on existing functional groups (Figure 2.10). With 100% confidence, one can assign the signal at 134 ppm as originating from carbon atoms of graphitic domains. Concerning the oxidized domains, at the present day, there is no solid indisputable evidence for the existence of any

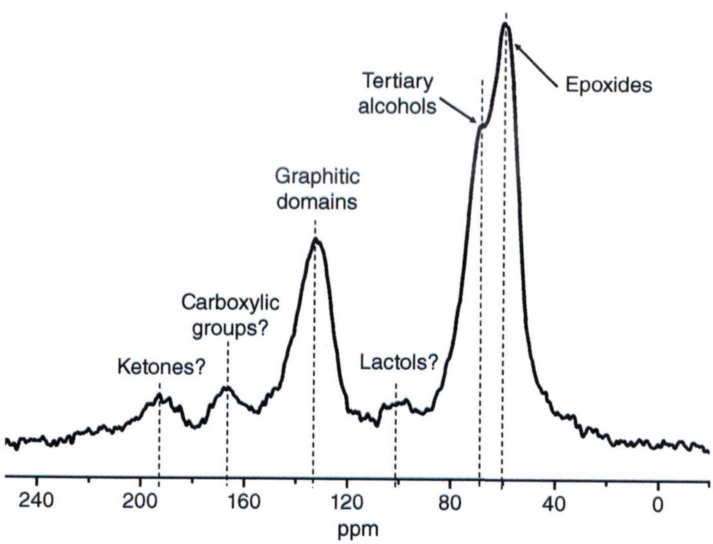

Figure 2.10 *Typical direct-pulse ^{13}C SSNMR spectrum for GO. The spectrum is taken from the author's personal research files. The signals at 60, 70 and 134 ppm are assigned to epoxides, tertiary alcohols and aromatic carbons of graphitic domains, respectively. The signals at 101, 167 and 193 ppm might be assigned, respectively, to lactols/gem-diols, carboxylic groups and ketones*

functional groups that can be accepted with 100% confidence. With high certainty, one can accept the existence of the major functional groups situated on the basal planes: tertiary alcohols and epoxides. The signals at 60 ppm and 70 ppm most likely originate from the epoxides and tertiary alcohols, respectively. However, the 60 ppm signal might originate from 1,3-ethers as well, as Szabó *et al.* [38] suggested. While the actual species corresponding to the 60 ppm signal are indeed most likely epoxides, there is no solid experimental evidence ruling out the 1,3-ethers. Reactions specific to epoxides (ring opening) have never been unambiguously demonstrated for GO.

The assignment of the weak signals of the NMR spectrum, however, is even less straightforward. The 101 ppm peak could originate from the sp^3-hybridized carbon atoms connected to two oxygen atoms by single bonds. Gao *et al.* assigned this peak to the C atoms in the five- and six-membered rings of lactols [44]. Alternatively, Dimiev *et al.* assigned this peak to the carbon atoms of *gem*-diols [31]. Neither assignment has been confirmed chemically or by the use of additional characterization methods. While Gao *et al.* place lactols on the flake's edges, Dimiev *et al.* place *gem*-diols at the points of C–C bond cleavage.

The minor peaks at 167 and 192 ppm, attributed to carbonyl-containing groups, originate from a relatively small number of carbon atoms. At the same time the role of these functional groups is extremely important for understanding GO chemistry and structure. Let us take a closer look at these minor oxygen functionalities. We will begin with one of the most controversial functionalities: carboxyl groups that according to the LK model are situated on the edges of GO flakes.

Note that Lerf *et al.* in their famous work [37] did not provide any experimental evidence for the presence of carboxyl groups; the conclusion was made based on the previously published infrared (IR) data. It was only speculation about a functional group that might possibly terminate the flake edges. For some reason, this suggestion by Lerf *et al.* was later interpreted as an important integral part of the LK model. Today, despite the lack of solid experimental evidence, the presence of carboxyl groups in GO is one of the common beliefs, and often experimental data are interpreted from the perspective of a chemical reaction involving carboxyl groups. This is especially true for the reactions of GO with amines. What is often interpreted as the formation of amide bonds most likely is an electrostatic interaction between negatively charged GO and positively charged ammonium. This type of reaction is discussed in Chapter 6.

Here we provide some arguments showing the lack of solid evidence for the existence of significant quantities of carboxyl groups in the GO structure. To begin with, the spectroscopic data either are inconclusive or suggest against carboxyls. The absorbance band at ~1730 cm^{-1} in the GO FTIR spectra that was interpreted by Lerf *et al.* as the C=O bond stretch of carboxyls can be assigned to any carbonyl group, including ketones and lactones. Most importantly, the ^{13}C SSNMR spectra of GO do not contain the signal at ~180 ppm, where the typical carboxyl carbon should appear. The two signals at 167 and 193 ppm are both 13 ppm from the theoretical 180 ppm value. The 193 ppm signal is most likely originated by ketones: alpha- and beta-unsaturated ketones are predicted to give a signal at 190 ppm. Subsequently, carboxyls might be associated with the 167 ppm signal. However, what causes the 13 ppm shift? Up to the present day these questions remain open. The only spectroscopic method suggesting carboxyls is X-ray photoelectron spectroscopy (XPS) (see section 3.1). The shoulder at ~289 eV in the C 1s XPS spectrum (see Figures 3.11 and 3.12) is most likely to have originated from carboxyl groups. The remaining question is what is the actual content and the location of carboxyls on the GO platform?

2.6.2 Origin of Acidity and the Dynamic Structural Model

To answer questions related to GO structure, one needs not only to interpret the spectroscopy data (these data can be interpreted in different ways) but also to explain the GO chemical properties from a structural perspective. One of the most important and unique properties of GO is the acidity of its aqueous solutions. Understanding the origin of GO acidity can greatly help in understanding its chemical structure. This was the approach we took in our studies and reported in two consecutive publications [31, 45]. Depending on the concentration, the pH of GO solutions varies from 2.0 through 4.0. The formal pK_a values calculated from this pH are in the range of 3.93–3.96, suggesting that GO is a stronger acid than commonly known carboxylic acids with one functional group in a molecule [45]. When GO flakes are separated from water in which they were dispersed, the water does not possess any acidic character. This signifies that hydronium cations are tightly associated with GO flakes, and do not exit the boundaries of the diffusion layer of counter-ions at the GO/water interface. There are no negatively charged ions in the bulk of solution to equilibrate the positive charge of hydronium cations; they can be associated only with negatively charged GO planes.

GO demonstrates extremely high cation exchange capacity (CEC), a property that has been known for decades. Based on the early studies, 100 g of GO contains 500–800 mmol of active acidic sites that can take part in cation exchange reactions with bases [45, 46]. This translates into approximately one acidic site for every six to eight carbon atoms. Both the LK and SD models fail to account for such a high acidity of GO; the two structural models do not suggest any functional groups that can account for such a high CEC. The demonstrated GO acidity and Lerf's suggestion regarding the presence of carboxyl groups on the flakes' edges have led many modern researchers to link the two statements, and to conclude that carboxyl groups are the entities responsible for GO's CEC. Indeed, the classical representation for the cation exchange reaction involves conversion of carboxyls to carboxylates. However, it is unlikely that the small number of carboxylic acid moieties situated on the edges of GO flakes, as proposed by the LK model, can account for such a high CEC. Even the edges of the holes cannot host so many carboxyls. Despite the fact that this was first noted back in 1909 by Hofmann and König [47], it is largely ignored in the modern literature. Carboxylic acids are still considered as the main moieties responsible for the acidic properties.

Simple geometrical considerations suggest that more or less stable two-dimensional structures containing one carboxyl group per six carbon atoms can hardly be envisioned. Thus, the number of carboxyl groups on the GO platform (if any) should be very small, and ordinary carboxyls cannot be the groups accountable for GO's acidic properties. In our study from 2012, we singled out two sources of GO acidity [45]. The low pK_a values and roughly one-third of the CEC were attributed to organic sulfates, and the other two-thirds of the CEC was attributed to oxygen-containing groups, or more exactly to reactions involving oxygen-containing groups running in basic conditions. Note that organic sulfates are present only in GO prepared in sulfuric acid medium. At the same time, acidity is a property specific to all GO samples, regardless of the preparation method. Thus, the main contribution to acidic properties is made by oxygen functionalities. Considering the magnitude of the CEC, these are the major functionalities, i.e. the groups located on the basal planes. The only structural model that attempted to account for GO acidity was the one proposed by Clauss back in 1957 [39]. Clauss suggested the presence of acidic enolic groups at the points of the C-C bond cleavage. Note, Clauss derived enolic groups from ketones according to the keto–enol tautomerism; however, no evidence was ever provided in support of this statement. Interestingly, in the modern era, no attempts have been made to explain the acidity from the structural perspective. A few acidity-related studies and some structure-related studies have been carried out separately. We were probably the first since Clauss to try to combine the two aspects together [31, 45].

It was discovered that in aqueous solutions, especially in moderately basic conditions, GO constantly generates hydrogen cations [45]. This was registered as gradual acidification of aqueous solutions. Thus, during the manually performed titration, after the addition of every new NaOH portion, and reaching the corresponding higher value, the pH of the solution starts drifting downward, and this drift continues if another portion of NaOH is not added. This acidification of GO solutions initiated by the addition of NaOH is the reason why the reverse titration curves (Figure 2.11) begin at lower pH values compared to the points where the direct titration curves end. The kinetics of the acidification reaction (Figure 2.11d) is typical of an ordinary chemical reaction. The process is intense in the first 60 minutes. The pH almost stabilizes in a few hours; however, acidification slowly continues during several days. This observation suggests that hydronium cations are generated in the course of the ordinary chemical reaction.

The direct titration curves are featureless (see also Figure 1.8). This observation contradicts the standing points of Boehm's titration protocol [48] (see section 1.4.5 for details). The protocol is based on the assumption that functional groups with different pK_a values should be neutralized sequentially with bases of different strength; this should inevitably lead to the existence of inflection points on titration curves. There were attempts to explain the lack of inflection points by overlapping of pK_a values of "numerous different acidic groups" [32, 49]. This is where the acidity-related studies come into conflict with existing structural models. According to the LK model, the only functional groups in GO having acidic character are carboxyls; and, as shown above, their content in GO is very low. The only other functional groups having some acidic character are phenols. Other types of functional groups possessing acidic character simply do not exist. If one assumes that both carboxyls and phenols indeed exist, then direct titration curves should contain two clear inflection points. To resolve this apparent contradiction between experimental facts and existing structural models, we suggested that acidic groups do not pre-exist in GO, but are generated by interaction with bases.

It was proposed that, in aqueous solutions, especially in basic conditions, vicinal diols undergo a chain of irreversible transformations that generate hydrogen cations. The proposed reaction mechanism is represented in Scheme 2.3.

The reaction between water and GO starts with the nucleophilic attack of a water molecule on tertiary alcohols, and ends up with C–C bond cleavage and the formation of ketones and enols on the newly formed edges. The proposed reaction mechanism is based on the

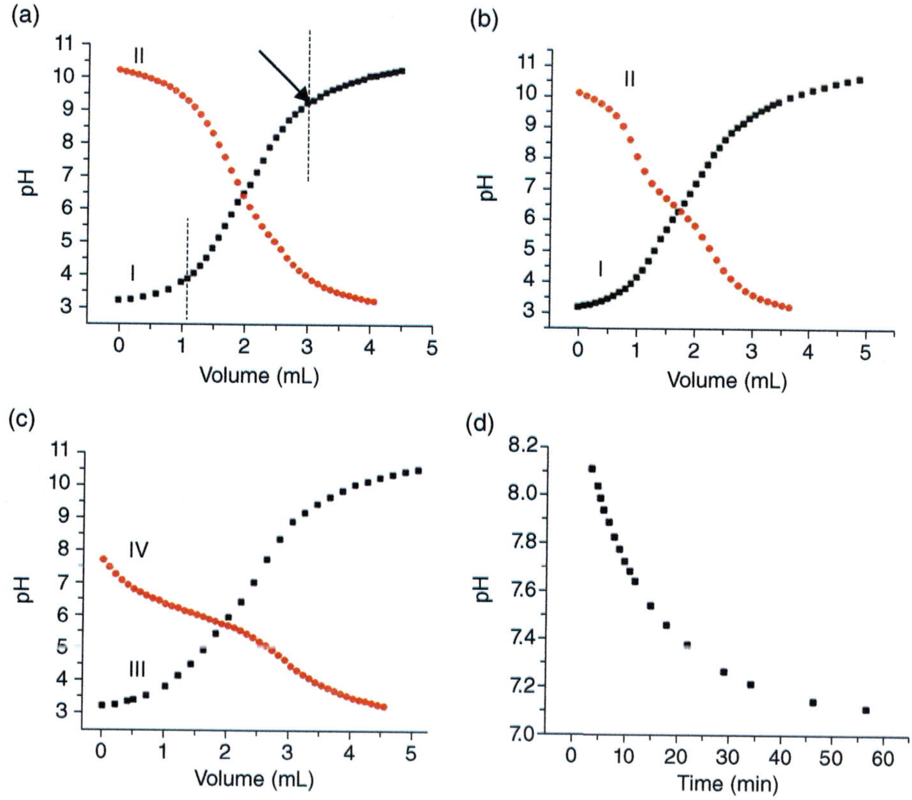

Figure 2.11 Acidic properties of GO solutions. (a–c) Forward (black squares) and reverse (red circles) titration curves. In the forward titration, the GO solution (1.0 mg ml^{-1}) was titrated with 0.100 M NaOH; and in the reverse titration, the conjugate base of GO was titrated with 0.100 M HCl. (a) The reverse titration was performed immediately after the forward titration. The reverse titration curve begins at almost the same pH. (b) Reverse titration was performed 6 h after the forward titration. The reverse titration curve begins at lower pH values due to the acidification during the 6 h time period. (c) The solution was heated at 60 °C for 15 h after the forward titration (III), and before performing the reverse titration (IV). The reverse titration begins at significantly lower pH values. (d) The change in pH of the GO solution with time after the addition of the 0.100 M NaOH solution. Zero time corresponds to the addition of NaOH. The increase in pH during the first 3 min 47 s after NaOH addition is not shown on the graph. All experiments are performed under nitrogen. Reproduced from [45] with permission of ACS

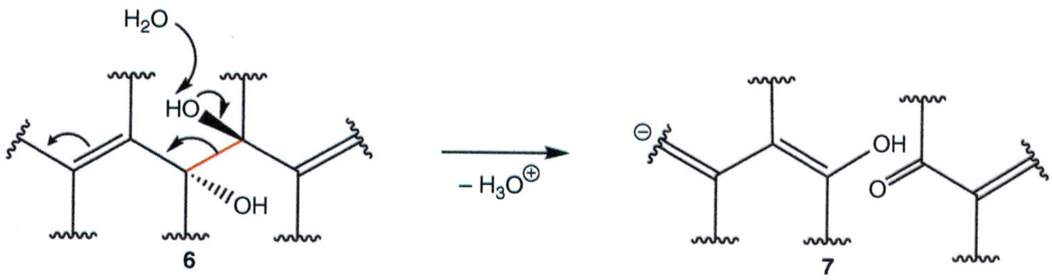

Scheme 2.3 Origin of GO acidity. The structure (**6**) is a GO fragment containing a vicinal diol. The C–C bond to be cleaved is shown in red. Nucleophilic attack of a water molecule on the hydrogen atom of the hydroxyl group results in C–C bond cleavage and in the formation of a ketone and an enol at newly formed edges (**7**). The reaction generates a hydronium cation; this explains the acidification of GO aqueous solutions. The hydrogen atom of the as-formed enol is highly acidic; this enol can further ionize, making the GO solution even more acidic

pseudo-acidity of tertiary alcohols that makes them prone to nucleophilic attack by water molecules. The acidity of the alcohol groups, in turn, is afforded by the stability of the conjugate base due to delocalization of a negative charge over the large area of a graphitic domain. The ability of GO to delocalize and accumulate negative charge is the necessary condition and the driving force for this reaction.

Another necessary condition is the aqueous solution. GO flakes can exist in the negatively charged state only when this charge is neutralized by the positive charge of counter-ions, building up an electrical double layer at the GO/water interface. This is possible only in aqueous solution. When GO reacts with bases, the attacking species, i.e. a hydroxide anion, is a stronger nucleophile compared to a water molecule. This makes a larger number of tertiary alcohols prone to nucleophilic attack. The metal cations do not form direct bonds with carboxylic groups; as discussed above, GO cannot contain so many carboxyl groups. Instead, metal cations are electrostatically attracted to the negatively charged GO planes. They partially or fully replace hydronium ions in the electrical double layer at the GO/water interface. This is the explanation for the high CEC values, when GO reacts with bases.

The formation of ketones and enols requires rupture of only one of the three C–C bonds of carbon atoms with neighbors; this makes a virtual enol-containing structure stronger, and thus more realistic compared to a carboxyl-containing structure. Enolic groups by themselves are not strongly acidic. However, on a GO platform, the acidity of enolic groups is increased by conjugation with two different entities. (i) In conjugation with ketones, enols form so-called vinylogous acids that have the same acidity as regular carboxylic acids. (ii) The conjugation of enols can extend to the entire graphitic domain, as shown for structure (**7**) in Scheme 2.3. This provides extremely high stability to enolate anions, thus rendering GO solutions highly acidic.

Based on the acidity-related studies, we proposed a new structural model of GO, which was referred to as the "dynamic structural model" (DSM) [45]. This model suggests that in aqueous solutions GO constantly interacts with water. Upon this reaction, tertiary alcohols slowly transform into enols and ketones by generating hydronium ions. While the essence of this model is represented by Scheme 2.3, the respective GO structure is shown in Figure 2.12. The DSM is sufficiently different from both the LK and SD models. While it accepts tertiary alcohols and epoxides as the main functional groups on the basal planes, it rejects carboxyls as more or less abundant species. Also, the DSM suggests the presence of enols and ketones on numerous points of the C–C bond cleavage. From this perspective, this model employs ideas proposed by Clauss, and incorporated more recently into the SD model. However, our explanation for enol formation is different from the keto–enol tautomerism proposed by Clauss. Our model derives enols directly from tertiary alcohols.

Figure 2.12 represents the simplified version of the DSM; it reflects only the structure of the basal planes. Note that the numerous points of C–C bond cleavage are not considered by the DSM as edges; they are structural features of the basal planes. Indirect confirmation of our model has been provided by HRTEM images. The image in Figure 2.13 clearly shows numerous points where the C–C bonds are broken. Also, one carbon atom seems to be missing.

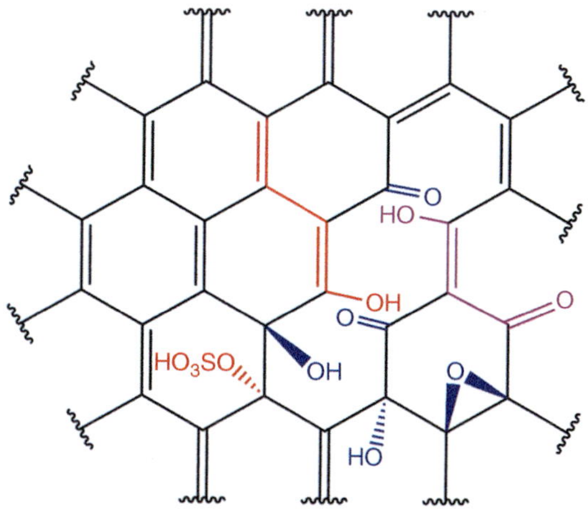

Figure 2.12 The simplified version of the GO dynamic structural model by Dimiev et al. The carbon framework is shown in black. Neutral oxygen functionalities are shown in blue. Functionalities with acidic properties are shown in red and purple. The enolic group shown in red is acidic due to conjugation with the graphitic domain. The enolic group shown in purple is acidic due to the formation of a vinylogous acid in conjugation with ketone. Organic sulfates are always adjacent to at least one tertiary alcohol. The density of organic sulfates depends on the washing procedures; for thoroughly washed GO samples, there is about one organic sulfate per 90 carbon atoms

Now, let us focus on the chemical structure of the edges of numerous holes in GO planes. Figure 2.14 represents the complete version of the Dimiev–Tour (DT) structural model, featuring all the functionalities proposed by the model. The GO fragment contains a hole situated on the border between the graphitic domain (lower right corner) and the oxidized domain (upper left corner). The main functional groups on the basal planes are epoxides and tertiary alcohols, in accordance with the LK model. However, the basal planes also contain numerous points of C–C bond cleavage as at **1** and **2** with ketones and enols at the newly formed edges.

In aqueous solutions, ketones undergo further transformation. One of the possible transformations is hydration and conversion into *gem*-diols (hydrates) such as **2** and **3** [31]. The *gem*-diols are stable in aqueous solutions, especially when carbonyls or double bonds are present in *α*-position to the *gem*-diol carbon atom, and the strain favors sp^3 hybridization [50, 51], as shown for the *gem*-diol **3**. There is another possible transformation, in that a *gem*-diol can be further converted to a hemiacetal such as **2**. Both *gem*-diols and hemiacetals contain an sp^3 carbon atom bonded to two oxygen atoms, which might be responsible for the 101 ppm signal in the ^{13}C SSNMR spectrum of GO. The DT structural model accounts for GO acidity. The enols formed at the point of C–C bond cleavage such as **1** are strongly acidic due to the conjugation over the large graphitic domain, and/or due to formation of a vinylogous acid in combination with ketones. The same is true for numerous vinylogous acids such as **4**, **5** and **6** situated on the edge of the hole. Thus the dynamic

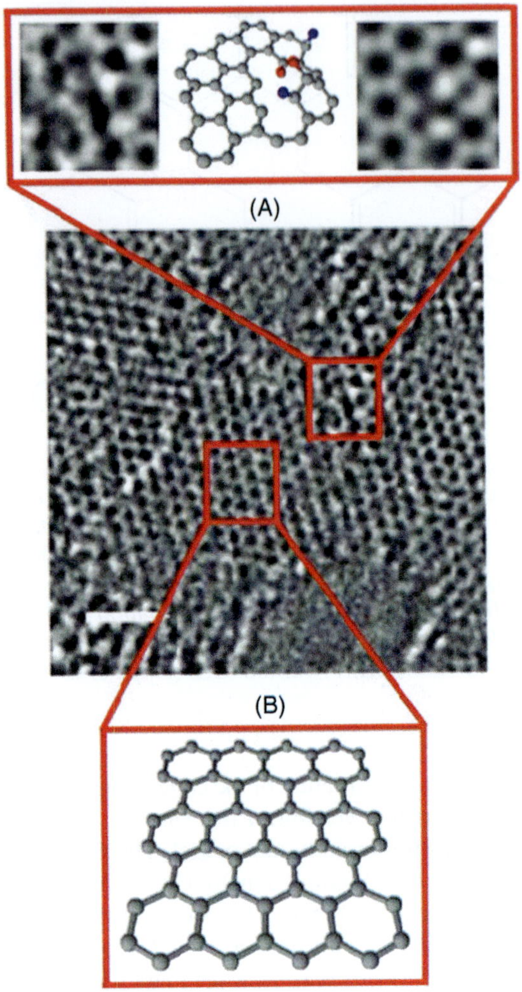

Figure 2.13 HRTEM images of GO. (A) Magnification of the oxidized domain area: left is the actual structure, right is the simulated model. (B) Magnification of the intact graphitic domain. The oxidized domain contains multiple points of C–C bond cleavage. Reproduced from [42] with permission of John Wiley & Sons

structural model logically and in the least controversial way explains many of the known GO properties, including the most important one: the acidity of aqueous solutions. This model describes the GO structure in aqueous solutions. We suggest that the fine chemical structure of GO in aqueous solutions is different from that in the dry solid state. Note that the presence of enols is not confirmed by ^{13}C SSNMR spectroscopy in solid graphite oxide samples. It is possible that enols exist only in aqueous solutions, and transform into other groups, such as ketones, when GO solutions are dried out, and graphite oxide is obtained in the bulk solid form.

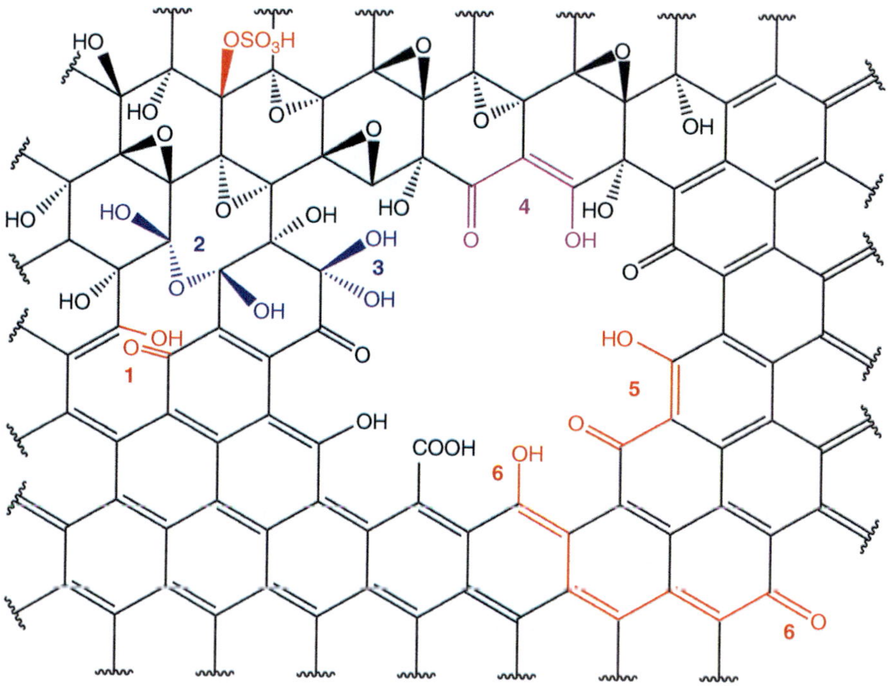

Figure 2.14 *The complete version of the Dimiev–Tour GO structural model (DT model), featuring all the proposed functionalities. A GO fragment contains a hole situated on the border between a graphenic domain (lower right corner) and an oxidized domain (upper left corner). Different structural features are represented by different numbers and colors.* **1** *Ketone and enol groups are formed at the point of C–C bond cleavage.* **2** *By hydration, ketones can turn into gem-diols, and further into hemiacetals.* **3** *Here a gem-diol is in α-position to a ketone; this favors the stability of gem-diols in aqueous solutions.* **4–6** *These are the vinylogous carboxylic acids. Conjugation in the carboxylic acid* **4** *is limited by two oxygen atoms. Conjugation of the vinylogous acids* **5** *and* **6** *extends to the entire graphitic domain; acids* **5** *and* **6** *are stronger acids than acid* **4**. *Organic sulfates are present in GO samples prepared in sulfuric acid medium*

Additional support for the dynamic structural model, with emphasis on the first word, came in a recent report by Kumar *et al.* [52]. It was demonstrated that, when a GO solution is processed hydrothermally at elevated temperatures, the as-synthesized metastable GO transforms into a different, relatively more stable state (Figure 2.15). This new quasi-stable state has very different spatial distribution of oxygen functional groups, and, as a consequence, exhibits different optical and electrical properties. With respect to our topic, this work demonstrates the mobility of oxygen functional groups on GO planes, and their ability to transform into each other when GO is in aqueous solutions. This process is intensified at higher temperatures, which help to overcome the activation energy barrier.

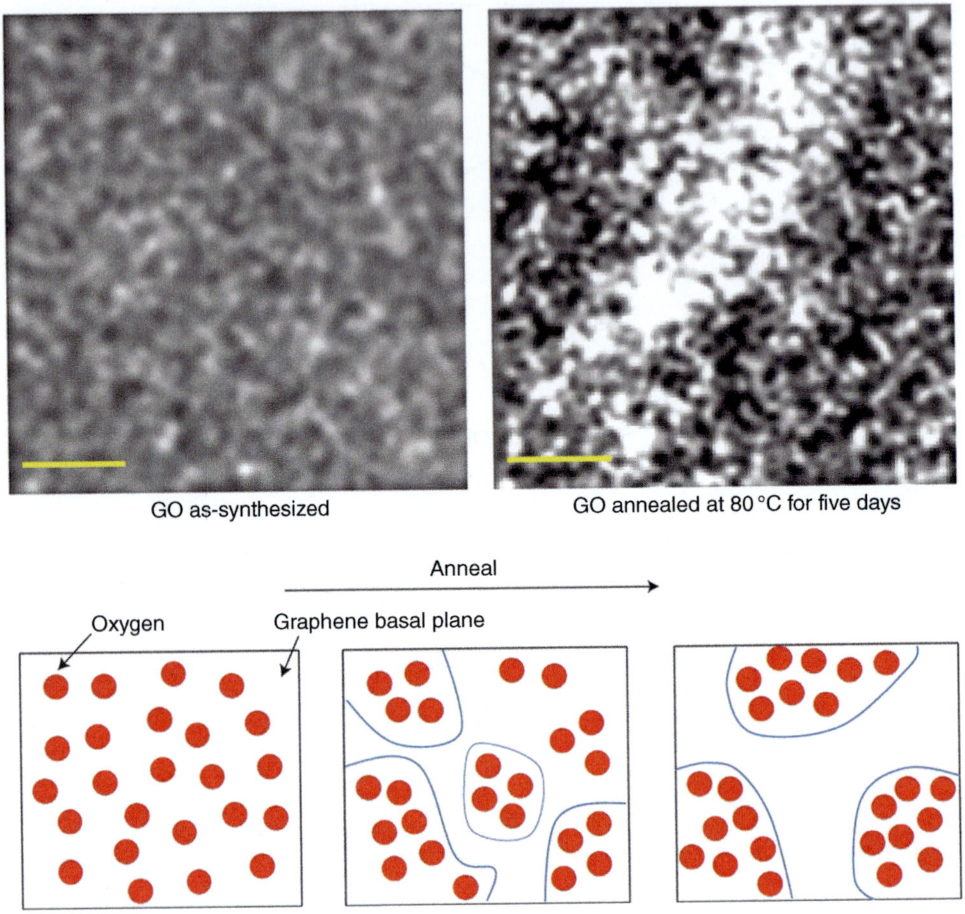

Figure 2.15 *Transformation of as-synthesized metastable GO into a different form upon hydrothermal treatment ("annealing" in the authors' terminology) in aqueous solution. The top two images are the Auger electron spectroscopy (AES) oxygen mapping of as-synthesized (left) and hydrothermally treated (right) GO films. The white spots indicate oxygen-rich regions, and the black spots indicate oxygen-deficient regions. The scale bar is 2 μm. On the bottom is the schematic depicting the proposed phase separation process. The as-prepared GO exhibits a uniform distribution of oxygen functionalities. This metastable GO has the potential to separate into two distinct oxidized and non-oxidized domains through diffusion of oxygen atoms on the graphene basal plane under the influence of an external stimulus, i.e. high temperature. Reproduced from [52] with permission of Nature Publishing Group*

2.7 Density of Defects and Introduction of Oxo-Functionalized Graphene

Let us finally discuss the origin and the density of the holes and the points of C–C bond cleavage on GO planes. The formation of such a defects is based on the divalent nature of oxygen. No permanent defects are introduced when graphene is functionalized with monovalent elements such as fluorine that can form single C–F bonds only. After hydrazine reduction of the fluorinated graphene, the original graphene plane can be almost fully restored [53]. Unlike fluorine, divalent oxygen tends to form double bonds with carbon atoms. After the epoxides and alcohols with the formal oxidation state of carbon +1 have been formed, oxygen tends to push the reaction even further to form ketones with formal oxidation state +2, and carboxyls with the formal oxidation state +3. As the extreme case, carbon dioxide with the oxidation state of carbon +4 is formed. In order to form C=O bonds, the C–C bonds need to be broken; this is the origin of the defects in GO.

It was reported that carbon dioxide is formed already in the course of GO production [8, 31]. The formation of CO_2, i.e. the removal of carbon, indicates the formation of holes. Defects are introduced in the various steps of GO production. Thus, significant evolution of CO_2 gas was detected during the second step of GO production, i.e. conversion of stage-1 GIC into PGO [31]. Additional defects are introduced during the third step of GO formation, i.e. washing of as-prepared PGO with water [45] according to the mechanism presented in Scheme 2.3. This is why, if one aims to obtain non-defective GO, one needs to conduct the oxidation reaction in such a way that allows the formation of monovalent functionalities, but avoids the formation of divalent carbonyls.

2.7.1 Oxo-Functionalized Graphene by Charpy–Hummers Approach

It was demonstrated back in 1909 by Charpy [8] that when graphite oxidation is conducted at elevated temperatures the loss of carbon increases: at room temperature almost no loss of carbon was registered, at 45 °C a small part of the graphite is lost, and at 100 °C already 50% of graphite is oxidized to CO_2. This observation has a simple explanation in terms of general chemistry: at higher temperatures, the activation energy barriers for many reactions are met, and oxygen realizes its full potential. At lower temperatures, oxidation is restricted to the formation of epoxides and alcohols, and the carbon grid is preserved. By keeping the reaction temperature below 10 °C, Eigler *et al.* [43] obtained minimally damaged GO, with an almost intact σ framework of C atoms. To stress the integrity of the honeycomb lattice, this GO can be termed “oxo-functionalized graphene” (oxo-G_1). The developed reaction sequence to oxo-G_1 is illustrated in Figure 2.16. It is critical that the temperature must be kept below 10 °C to avoid over-oxidation, especially during aqueous work-up.

The delamination of graphite oxide to oxo-G_1 proceeds in water for shaken, stirred or sonicated dispersions. The carbon lattice remains almost intact and Raman spectroscopy indicates a density of defects that is as low as 0.01% for the selected best quality of chemically reduced flakes (Figure 2.16A). It can be estimated that every second carbon atom of oxo-G_1 is sp^3-hybridized and functional groups are located on both sides of the basal plane. It is possible to conclude from analyses that functionalization at the edges of oxo-G_1 plays a minor role and oxo-G_1 is therefore a suitable precursor for wet-chemically synthesized graphene. In addition, the controlled chemistry of GO can be elaborated.

Flakes of oxo-G_1, dispersed in water, were deposited on Si/SiO_2 wafers by the Langmuir–Blodgett technique and subsequently chemically reduced by the vapor of hydriodic acid and trifluoroacetic acid. The graphene-like nature of the wet-chemically prepared graphene was determined by measuring magnetoresistance and Hall effect in magnetic fields up to 14 T, at a temperature of 1.6 K. Hole-like charge carriers with a charge carrier density $n = 1.6 \times 10^{12}\,cm^{-2}$ were found, a typical value for exfoliated graphene. The charge carriers

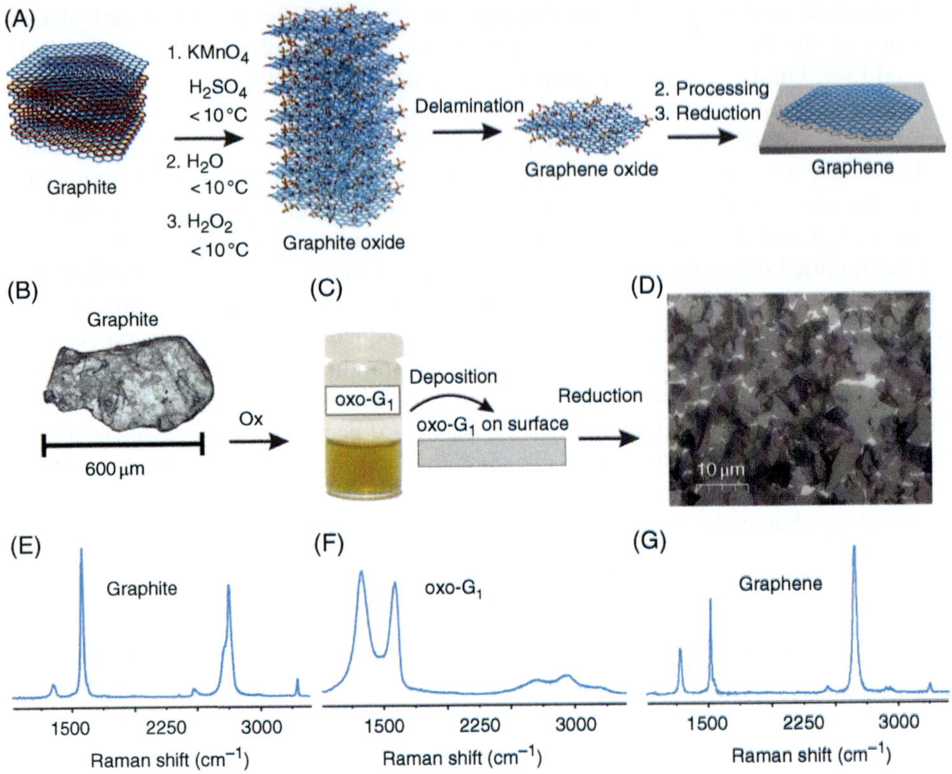

Figure 2.16 (A) Synthesis of GO with an almost intact carbon framework (oxo-functionalized graphene, oxo-G_1) starting from graphite in sulfuric acid with potassium permanganate as the oxidant. (B) Reflected light microscope image of natural graphite (flake size 600 μm). (C) Aqueous dispersion of oxo-G_1 at 0.1 mg ml^{-1}. (D) Scanning electron microscope image of graphene on SiO_2/Si. Raman spectra of (E) graphite, (F) oxo-G_1 and (G) a flake of graphene obtained from oxo-G_1 by chemical reduction. Reproduced from [43] with permission of Wiley-VCH Verlag GmbH & Co

responsible are attributed to the process of wet-chemical synthesis with measured mobility values exceeding 1000 cm^2V^{-1} s^{-1}. Furthermore, Shubnikov–de Haas (SdH) oscillations were observed for the first time in chemically synthesized graphene (Figure 2.16B). The analysis of Landau-level indices confirmed the presence of SdH oscillations, and they depend linearly on the inverse magnetic field as expected for SdH oscillations. The frequency of the oscillations plotted versus the charge carrier density revealed a linear dependence due to the linear density of states. This observation is only expected for 2D graphene (Figure 2.16C).

However, statistical Raman spectroscopy revealed heterogeneous properties of as-prepared oxo-G_1: different flakes exhibit different degrees of oxidation (Figure 2.17). The electronic transport properties of the flakes with an I_D/I_G ratio of 4 were also investigated (Figure 2.17D–F). A charge carrier density of 2.16×10^{12} cm^{-2} and a mobility of 270 cm^2V^{-1} s^{-1} were determined from the magneto-transport measurements. The sheet resistance R_{xx} is

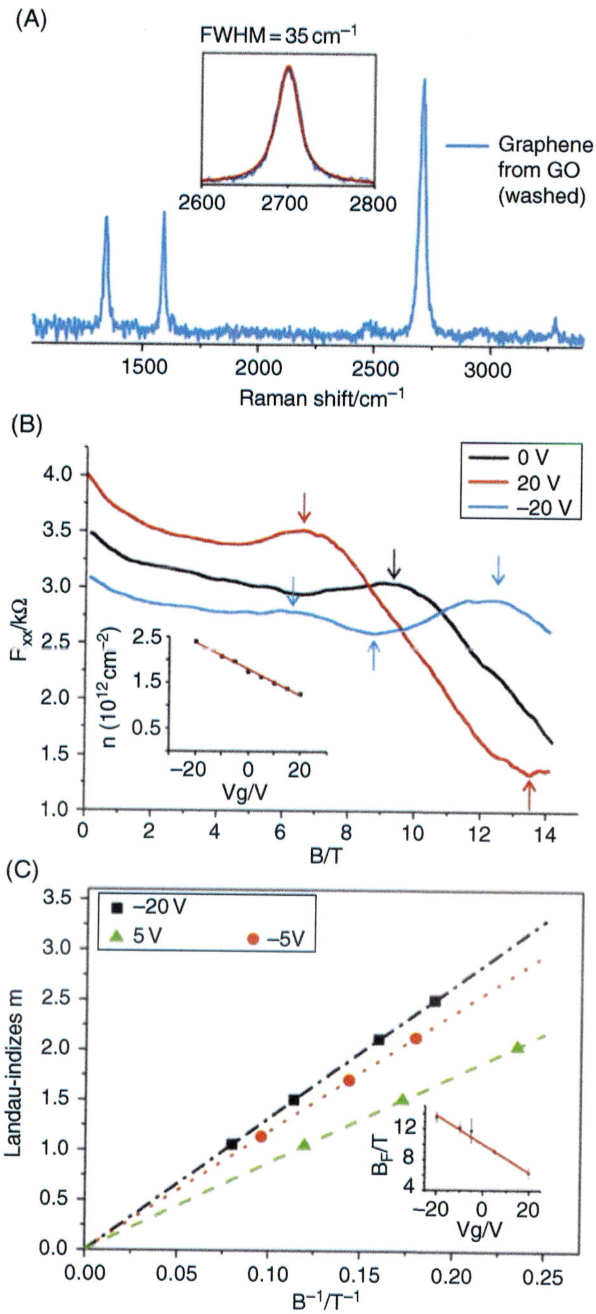

Figure 2.17 *Electrical properties of the wet-chemically prepared graphene derived from oxo-G_1. (A) Raman spectrum of a graphene flake ($\Gamma_{2D} = 35\,cm^{-1}$) used for the measurements shown in (B) and (C). (B) Four-point resistance R_{xx} at different gate voltages with visible Shubnikov–de Haas oscillations (maxima and minima indicated by arrows). The frequency of the oscillations varies with the charge carrier density. Inset: Linear dependence of the charge carrier density on the gate voltage V_G. The red line represents the fit $n \propto \alpha V_G$ with $\alpha = -3 \times 10^{10}\,cm^{-2}$. (C) Analysis of the Landau-level indices for the Shubnikov–de Haas oscillations for exemplary gate voltages. The Landau-level indices for a single gate voltage depend linearly on the inverse magnetic field. Inset: Dependence of the oscillation frequency B_F shows a linear dependence on the external electric field also. This is an important indication of a pure 2D material*

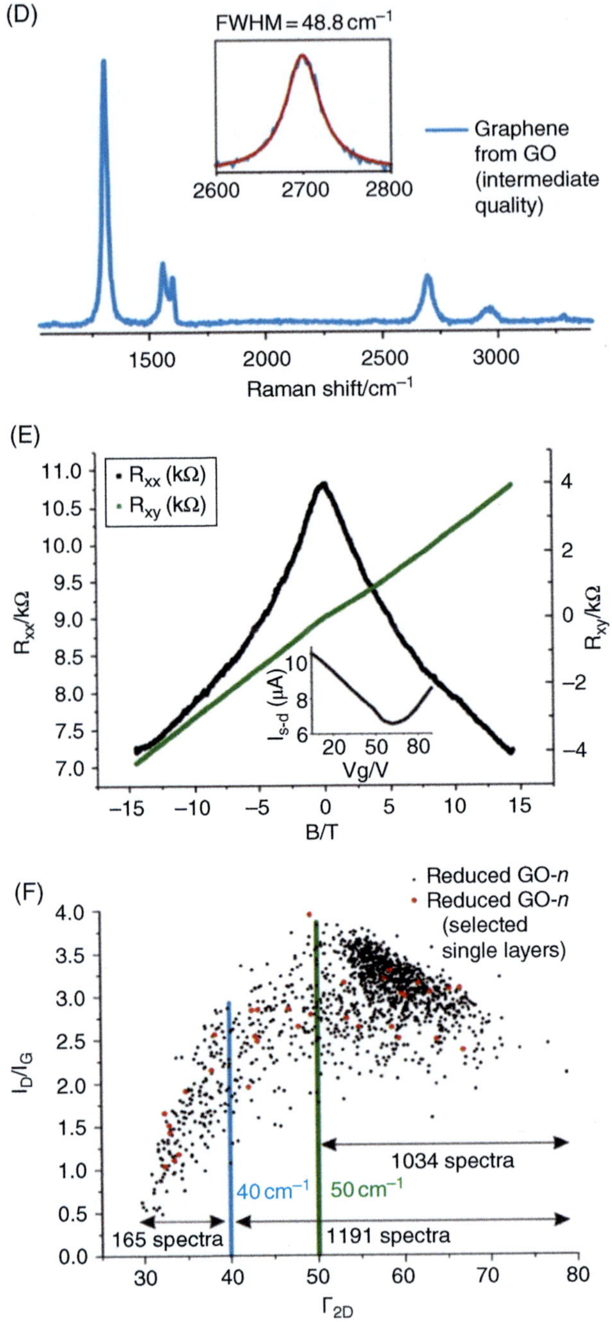

Figure 2.17 *(Continued) (D) Raman spectrum of graphene ($I_D/I_G = 4.0$). (E) Magnetotransport measurement of the sample characterized in (D). (F) Statistical Raman analysis of individual spectra of a film of flakes of graphene. Reproduced from [43] with permission of Wiley-VCH Verlag GmbH & Co*

dominated by a localization peak, and the field-effect measurement confirms the mobility of 250 cm^2V^{-1} s^{-1} for holes and 200 cm^2V^{-1} s^{-1} for electrons. An ambipolar field effect is observed, as expected for graphene samples (Figure 2.17E).

Oxo-G_1 is a suitable basis for the development of further functionalization reactions proceeding on the basal plane of graphene, while functional groups at the edges of the flakes (several micrometers) and defects play a minor role. Moreover, the performance limits observed for typical RGO are no longer limitations for graphene derived from oxo-G_1, as, for example, demonstrated by the charge carrier mobility values, which can exceed 1000 cm^2V^{-1} s^{-1}.

2.7.2 Oxo-Functionalized Graphene from Graphite Sulfate

As discussed in section 2.4, the stage-1 H_2SO_4-GIC, or graphite sulfate, is the first intermediate product in the course of GO production. Eigler *et al.* have demonstrated recently [54] that, upon quenching with water, graphite sulfate can be converted into oxo-functionalized material, as depicted in Figure 2.18A. The single-layer character and the oxidized nature of the as-obtained oxo-functionalized graphene was confirmed by atomic force microscopy and Raman spectroscopy (Figure 2.18B). A high degree of functionalization is evident from the presented Raman spectra: broad D, G and 2D peaks similar to the spectra of typical GO.

This type of oxo-G_1 exhibits a degree of functionalization of about 4%, and can be indicated as G_1-$(OH)_{4\%}$. Individual flakes of G_1-$(OH)_{4\%}$ with a diameter of several micrometers are identified by atomic force microscopy (Figure 2.18B). Flakes of graphene on a Si/SiO_2 substrate are obtained after reduction with the vapor of hydriodic acid and trifluoroacetic acid, a method that quantitatively removes oxo-functional groups from the carbon lattice (see section 1.6.4). The Raman analysis reveals that graphene with a residual density of defects of approximately 0.04% is obtained (Figure 2.18D). Therefore, the defect density was assumed to be only 0.04% in oxo-G_1 ($^{0.04\%}G_1$-$(OH)_{4\%}$). Such a low average density of defects was not reported for the wet-chemically prepared graphene from oxo-G_1 prepared by the Charpy–Hummers approach (section 2.7.1).

The formation of oxo-G1 by this approach is in sharp contrast with our earlier studies [23]: we did not observe sufficient oxidation of graphite upon intercalation–deintercalation cycle with the use of the same system; the oxidation degree of graphite was very low. The reason for this discrepancy is not yet completely clear. The possible explanation is that only one or a few outer graphene layers of the stage-1 H_2SO_4-GIC flake are being oxidized upon quenching it with water. The inner graphene layers most likely remain non-oxidized. Upon quenching, the graphite sulfate quickly deintercalates [24], graphite galleries close, and water does not have access to the inner graphene layers. Unlike the inner layers, the outer graphene layers are exposed to water while still in the charged (intercalated) state; this leads to the formation of covalent C–O bonds.

Such an oxidation sheds additional light on the mechanism of GO formation. Apparently, with this approach, the mechanism of oxo-G_1 formation is different from that discussed in section 2.4. Here, the second step, i.e. the formation of PGO, is completely missing, because no oxidation occurs in the concentrated persulfate–sulfuric acid system, used to prepare stage-1 H_2SO_4-GIC. There is no edge-to-center propagation of the oxidation similar to that for the $KMnO_4$–H_2SO_4 system. Oxidation occurs only upon quenching. Thus,

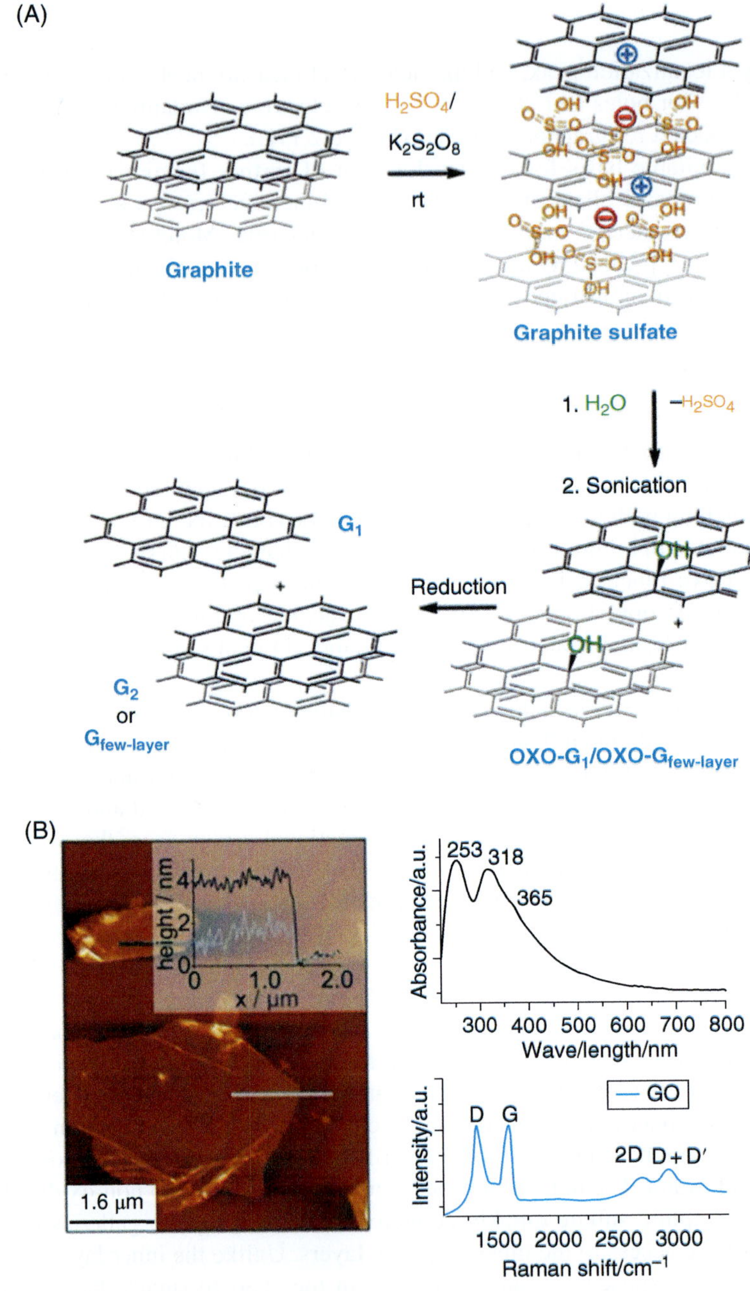

Figure 2.18 Oxo-functionalized graphene from graphite sulfate. (A) Synthesis of graphene and few-layer graphene from graphite sulfate, followed by the reaction with water to yield oxo-functionalized graphene and graphene after reduction. (B) Atomic force microscope image, ultraviolet–visible spectrum and Raman spectrum of oxo-functionalized graphene. (C) Thermogravimetric analysis of oxo-functionalized graphene coupled with mass spectrometry. (D) Statistical Raman analysis of a film of flakes of graphene. Inset: Histogram of Γ_{2D}. (E) Atomic force microscope image of graphene derived from G_1-$(OH)_{4\%}$. (F) Raman spectrum of graphene obtained from G_1-$(OH)_{4\%}$ after reduction, with $\Gamma_{D'}$, Γ_G and Γ_{2D} given in italic numbers. Reproduced from [54] with permission from RSC

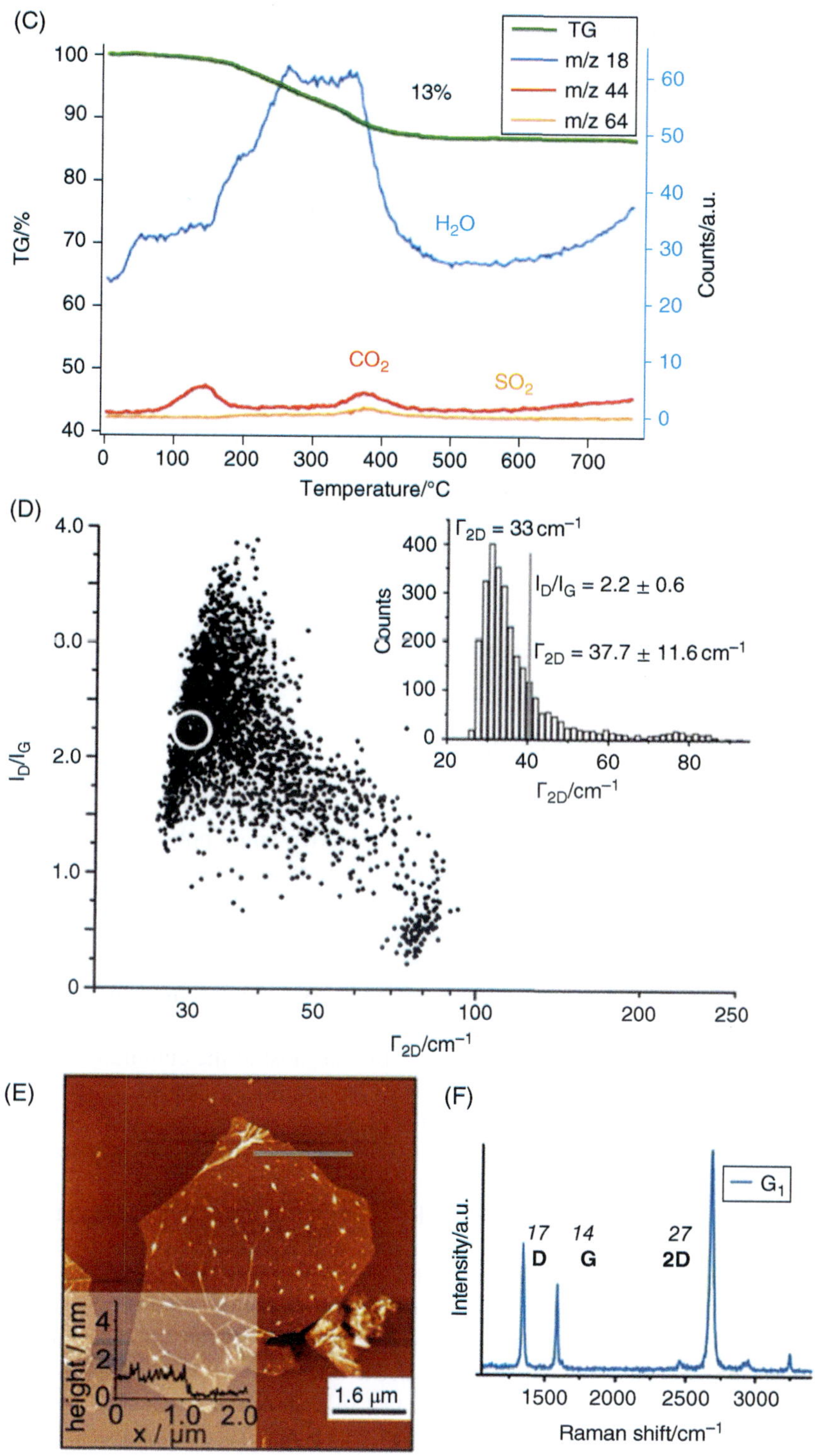

Figure 2.18 (Continued)

there is the possibility that C–O bonds form directly by contact of charged graphene with water molecules. However, it is more likely that the oxidizing agent here is some species formed by dilution of the persulfate–sulfuric acid mixture by water. The electrochemical potential of the persulfate–sulfuric acid system is lower than that for the permanganate–sulfuric acid system. Thus, with the former, oxygen cannot realize its full potential, C=O bonds are not formed, and oxidation stops on the step of C–O bond formation.

With the synthesis of oxo-functionalized derivatives of graphene and the minimization of defects within the carbon framework, the first step toward a controlled chemistry of oxo-G_1 is achieved. However, it is necessary to reliably characterize the carbon lattice of oxo-G_1 because of its heterogeneity. The chemistry of oxo-G_1 will be reviewed in more detail in Chapter 6.

Concluding this section, we need to stress that the minimally damaged oxo-functionalized graphene is a new material introduced recently by Siegfried Eigler. Typical GO is very different due to the higher density of defects. As was shown above, and will be additionally demonstrated in Chapter 6, these defects are responsible for the unique and rich chemistry of GO.

2.8 Addressing the Challenges of the Two-Component Structural Model

While discussing GO structure, one cannot avoid discussing the so-called two-component GO structural model, due to its strong impact on the scientific community. According to this model, proposed in 2011 by Rourke *et al.* [55], GO consists of two parts: (i) slightly oxidized, structurally intact, graphene-like sheets, and (ii) small polycyclic organic molecules, bearing most of the oxygen functional groups. The small molecules are adsorbed on the surface of graphene-like sheets. When an as-prepared GO (aGO) dispersion is heated with a strong base, such as NaOH, the small molecules can be separated from aGO, leaving behind clean "base-washed GO" (bwGO). In this section, the original terminology suggested by Rourke *et al.* [55] is preserved. Subsequently, these small molecules were named "oxidation debris" (OD) by analogy with earlier carbon-nanotube-related studies [56–58]. Being rich in oxygen groups, the OD is responsible for most of the cumulative ultraviolet–visible (UV-vis) absorption, IR absorption and photoluminescence of the integral aGO. The bwGO, in turn, contains fewer oxygen groups, and contributes significantly less to the integral properties mentioned above. Figure 2.19 represents this concept as seen by the model developers.

In the years following the original report, Rourke's group further developed this model, confirming and detailing the earlier-proposed concept [59, 60]. Up to the present day, the original report of the two-component model has amassed 150 citations, and several research groups have invoked this model to explain their findings [61–65]. The model was employed to describe such GO properties as surface adsorptive behavior [60, 61], fluorescence [59, 63, 65] and electrochemistry [64]. In the most recent GO-related review article, the model was discussed as being a matter of fact [3]. Thus, this model has evolved from a hypothesis proposed by one group into a theory that makes a notable impact on the entire GO field. The number of actual supporters of this model is probably even higher than the

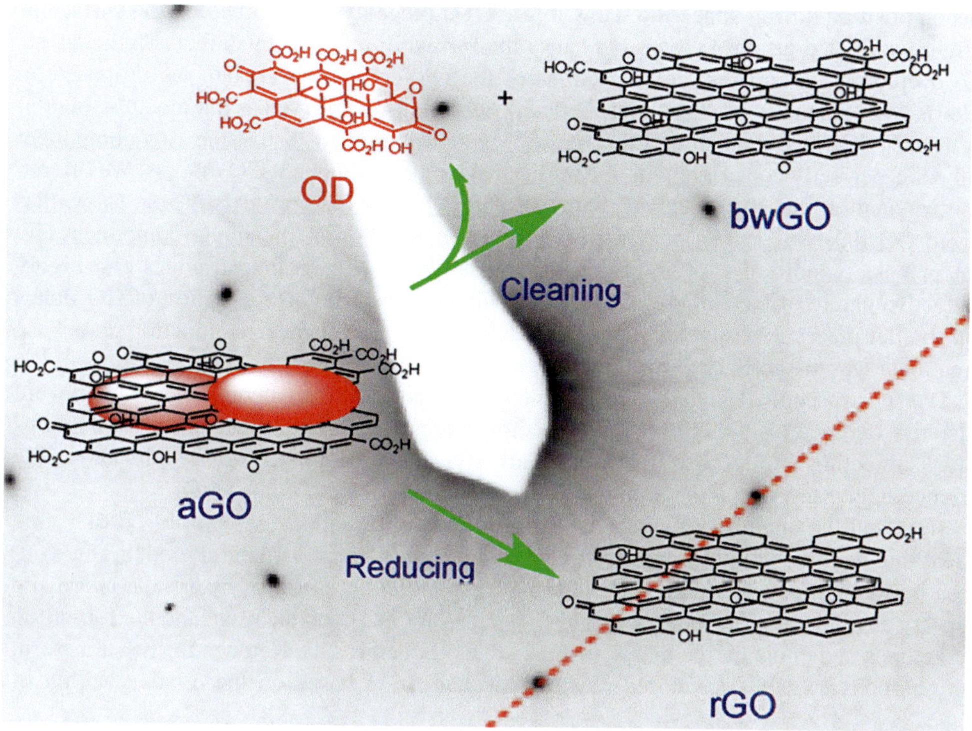

Figure 2.19 Schematic representation of the two-component GO structural model, as presented by the model developers. OD, which allegedly pre-exists in aGO samples, is an oxygen-rich complex polycyclic molecule strongly physisorbed on the surface of slightly oxidized bwGO. OD detaches from bwGO upon treatment in hot basic solutions. Reproduced from [59] with permission of ACS

number of published studies. In the author's informal correspondence with researchers working in the field, some of them admitted that they cannot fully rule it out. The attractiveness of this model lies in its apparent simplicity. This additionally illustrates that the modern research society likes simple solutions. The impact of the two-component model is so strong that some researchers, in the rush to fit their experimental observations into the model's framework, readily sacrifice not only fundamental principles of science, but sometimes even common sense. Thus, Guo *et al.* [65] suggested that intrinsically black bwGO appears bright yellow because OD is "strongly physisorbed" on its surface. When OD desorbs, the real black color of bwGO is revealed. Moreover, Rodriguez-Pastor *et al.* [66] concluded that the oxidation of graphite is carried out in such a way that some graphene layers remain almost intact (bwGO), while others are completely destroyed into small OD-like fragments.

In essence, the two-component model revises all the views on GO structure that were developed by generations of chemists. At closer look, the two-component model contains insurmountable contradictions that were discussed in our recent publication on this topic [67]. As an additional example, the two-component model fails to account for GO decomposition during annealing. How might OD, allegedly physisorbed on the surface of slightly oxidized graphene, possibly cause the formation of vacancy defects on the latter? To overcome the apparent contradictions of the two-component model, we proposed an alternative explanation to the experimental observations that served as the model's foundation [67]. It was demonstrated that all the evidence used to justify the two-component model is perfectly consistent with the traditional single-component GO models. We offered an explanation that the so-called oxidative debris, rather than pre-existing on as-synthesized GO flakes, is in fact generated by the decomposition of the single-component GO under basic conditions. As an argument, we demonstrated the jagged edges and highly defective nature of base-treated GO flakes which confirm the disintegration of GO flakes on smaller pieces in basic conditions (see Figure 3.22). The flake size reduction upon base treatment was also observed by Taniguchi *et al.* [68].

The spectroscopic data that were presented in support of the two-component model were also re-examined [67]. FTIR and XPS spectra suggest the formation of carboxylate groups after subjecting GO to base treatment. Carboxylate ion groups can exist only on newly formed edges, thus proving the highly defective nature of base-treated GO.

To reveal the chemistry behind the experimental observations, we employed the same ideas that were earlier placed in the foundation of our dynamic structural model. The reaction between strong base and GO starts with nucleophilic attack of hydroxide anions on tertiary alcohols on a GO platform, and ends up with C–C bond cleavage and the formation of ketones and enols on the newly formed edges (Scheme 2.4). Ketones further transform into carboxylates. The proposed reaction mechanism is based on the relative acidity of

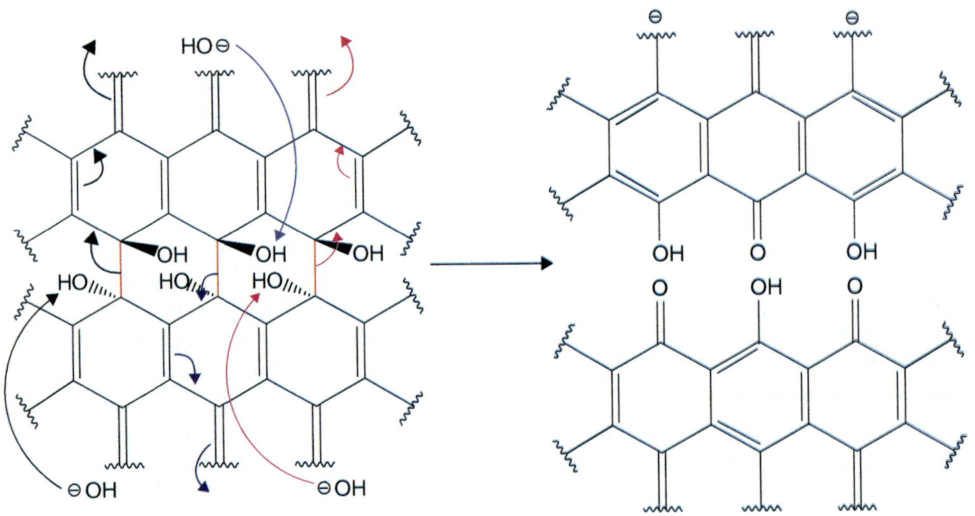

1 2

*Scheme 2.4 Disintegration of GO in basic conditions via C–C bond cleavage. The structure (**1**) is a GO fragment containing three vicinal diols. The C–C bonds to be cleaved are shown in red. The structure (**2**) shows the same GO fragment after the three C–C bonds have been cleaved. The three different transformations initiated by the attack of hydroxide ions on tertiary alcohols are represented by three separate sets of curly arrows shown in different colors* tertiary alcohols, which makes them prone to nucleophilic attack by hydroxide anions. The acidity of alcohol groups, in turn, is afforded by the stability of the conjugate base due to delocalization of negative charge over a large area of graphitic domains. Thus, the GO chemistry in basic conditions fits well into the earlier proposed dynamic structural model. The interaction between GO and strong bases can be considered as a deeper propagation of reactions running even in neutral aqueous reactions.

Figure 2.20 schematically represents the disintegration of GO flakes, and the formation of OD during base treatment. The GO fragment (**1**) represents a typical GO flake, drawn in accordance with the actual GO structure as recorded by HRTEM [42]. The oxidized areas of GO (shown in yellow) form a continuous network, while the intact graphitic domains (shown in blue) form isolated islands. In addition, GO contains structural defects, i.e. holes, shown in white. During the base treatment, numerous additional defects form on the GO

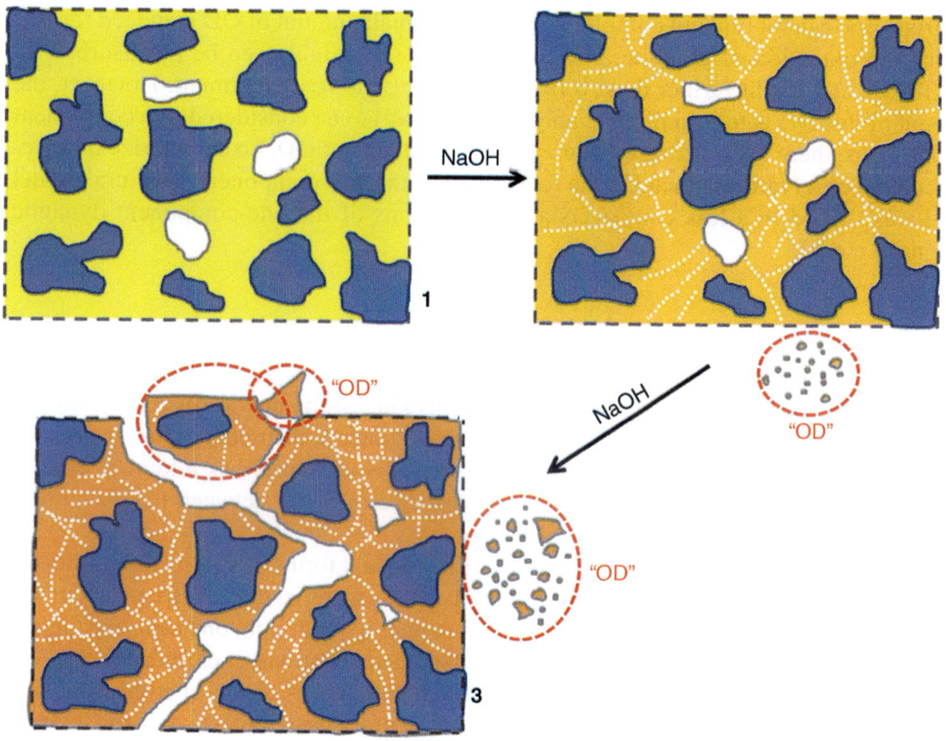

Figure 2.20 *Schematics representing the gradual disintegration of GO in basic conditions. Structure* **1** *is a fragment of a regular as-prepared GO. The blue-colored areas represent intact graphitic domains. The continuous yellow-colored network represents oxidized domains. White islands represent holes. Structures* **2** *and* **3** *represent progressive stages of GO disintegration. The gradual color change from yellow in* **1** *to orange in* **3** *represents the change in chemical structure of the oxidized domains. Dotted lines represent invisible cuts formed on GO as a result of the C–C bond cleavage caused by reaction. The small orange-colored shapes in dashed red ovals represent carbon fragments removed from the respective GO platform during base treatment. The size of the removed carbon fragments varies from subnanometer through several nanometers. Reproduced from [67] with permission of Elsevier*

platform (**2**). Some of these defects are simply points of C–C bond cleavage, where the carbon atoms remain on the GO platform. Other defects are tiny holes formed by the removal of small carbon fragments. These small defects are not detectable by scanning electron microscopy (SEM) and transmission electron microscopy (TEM), while the flake stays in one piece, as shown in the GO fragment (**2**). However, these defects serve as hidden pathways for future disintegration. They become pronounced during sonication and/or prolonged base treatment, leading to disintegration of the flakes into smaller pieces (**3**). The chemical structure of the oxidized domains changes during the base treatment; this is represented by the gradual color change from yellow to orange.

The entire confusion with the two-component model is probably rooted in some experimental mistakes made by Rourke's group during the work-up procedures, and in determining the yield of OD (see Ref. [67] for details). The high yields (25–41%) reported by Rourke's group have not been reproduced by other groups [65, 67]. To tell the truth, one cannot rule out completely that some small amount of OD is indeed formed during the oxidation of graphite. Furthermore, this OD might even be physisorbed on GO flakes, and might further desorb from GO during the base treatment. However, the amount of this OD is too small (probably <1%, if at all) to be considered as a component of GO. Thus, there is no strong evidence for the existence of OD. To summarize this section, all the observations placed in the foundation of the two-component structural model can be explained in a less controversial way in terms of the one-component dynamic structural model.

2.9 Structure of Bulk Graphite Oxide

Solid graphite oxide is the product formed by restacking of the previously exfoliated single-layer GO sheets. Due to the 2D nature of single GO sheets, their bulk accumulation, i.e. graphite oxide, has a lamellar paper-like structure. This substance has a loosely ordered structure in the *c*-axis direction, with a repeat distance of 6–9 Å between the constituent GO layers.

Due to its 2D nature, GO can acquire many different bulk forms. If simply dried from the gel or from aqueous paste, it forms dark-brown shapeless material that is difficult to cut and grind due to its lamellar structure. However, by an intelligent design, one can shape it into different forms, each of which has numerous potential applications. If extruded from solution, it can be spun into fiber (Figure 2.21a). The SEM images of the fiber cross-section (Figure 2.21b,c) demonstrate the ordered layered structure. The self-deposition of GO from solutions by drop-casting or by filtration leads to formation of very uniform paper-like materials (Figure 2.21d). The densely packed layered structure of these films is apparent from the SEM images of the film edges (Figure 2.21e,f). Clauss and Hofmann demonstrated back in 1956 that as-prepared GO films, when of a certain thickness, demonstrate permeability for water molecules, while they remain absolutely impermeable for other molecules and even atoms [69]. The same study with GO membranes reproduced in the modern graphene era [70] deserved publication in the journal *Science*. After annealing, the GO membranes become permeable for several species. They demonstrate selectivity toward different gases, and thus have great potential for separation of gases. This very important application of GO is reviewed in more detail in Chapter 9 by Ho Bum Park and

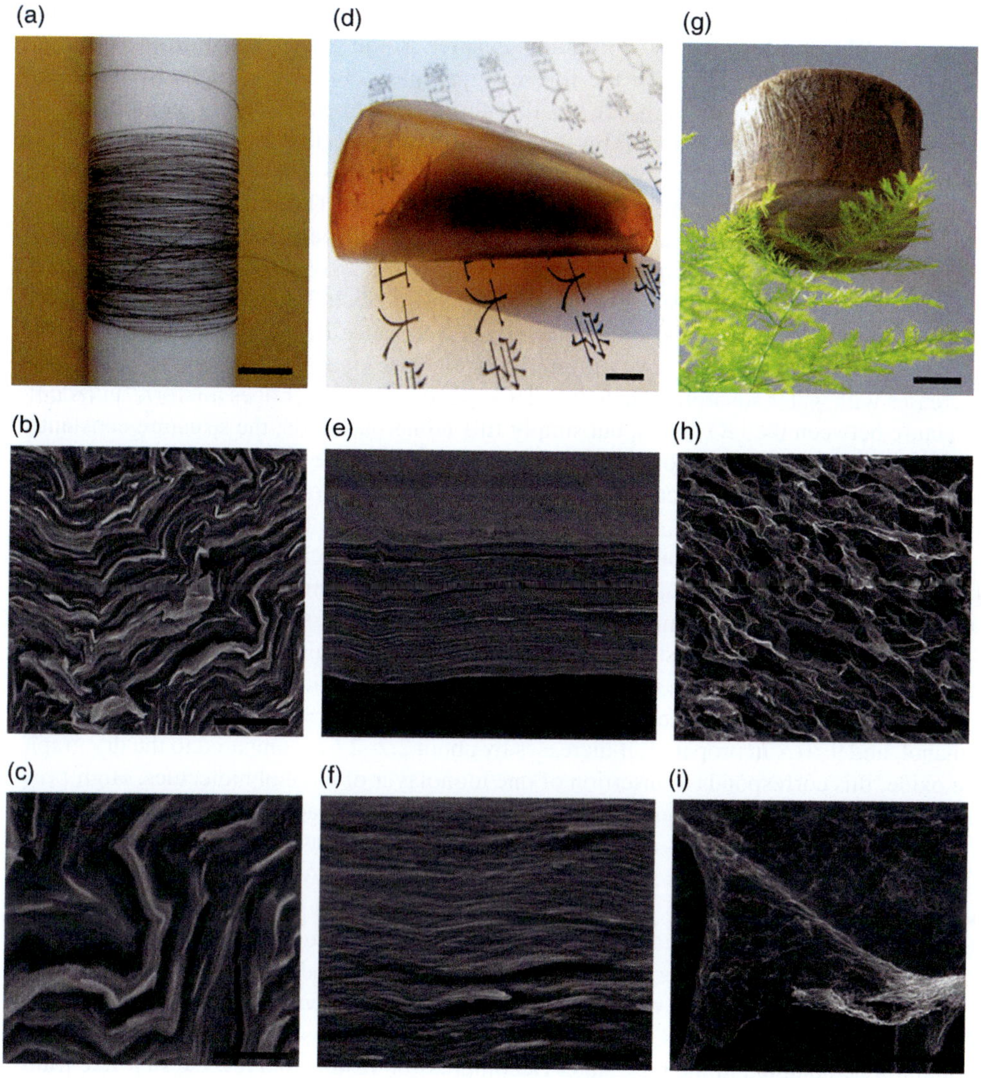

Figure 2.21 *Various macro-forms of GO. (a) A wet-spun 14 m long continuous fiber with diameter 10 μm and (b,c) SEM images of the cross-section of the fiber. (d) A paper-like film made by the filtration method and (e,f) SEM images of the paper edges. (g) Ultra-light-weight GO aerogel with a density of 2 mg cm^{-3} and (h,i) SEM images of the aerogel showing the voluminous morphology. Scale bars: 3 cm (a), 1 μm (b), 500 nm (c), 1 cm (d), 3 μm (e), 400 nm (f), 2 cm (g), 30 μm (h) and 2 μm (i). Reproduced from [14] with permission of Nature Publishing Group*

colleagues. When GO solutions are dried by freeze-drying, highly voluminous, sponge-like materials can be obtained (Figure 2.21 g). These materials, also referred to as aerogels or hydrogels, may find applications everywhere where high surface area is desired: supercapacitors, sorption materials and many more.

All the above-mentioned bulk forms of GO possess unique physical properties that we are simply unable to review within this chapter. Some properties of different graphite oxide forms are reviewed in the following chapters, in both Parts I and II. In this section we will only briefly touch on the ability of paper-like solid graphite oxide (Figure 2.21d) to be intercalated by several species, since this is interesting from the structural perspectives.

All solid graphite oxide samples contain intercalated water. The amount of water depends on the relative humidity and varies from 6–8% for carefully dried samples through 20–25% for samples in high-moisture atmospheric conditions. It is an open question whether water can be considered as an intercalant. The interlayer distance in bulk graphite oxide is not a fixed value, but depends on the amount of intercalated water, i.e. is dependent on humidity. Dry graphite oxide samples have an interlayer distance of about 5.3 Å, which gradually increases with water adsorption up to 9 Å. This suggests that water does not form an orderly structure between the GO layers, but simply fills numerous voids; the structure constantly changes when more water molecules are accommodated in the structure. Interestingly, the situation is different for intercalants other than water, where more or less discrete interlayer distances have been reported.

Several intercalated compounds of graphite oxide with intercalants other than water have been reported up to the present day. The reported intercalants include carbon dioxide [71], alcohols [72] and organic amines [73, 74]. We will now discuss some of these.

Talyzin *et al.* [72] demonstrated that primary alcohols could be intercalated into solid graphite oxide by immersion of a dry solid GO sample into liquid alcohols. The reported interlayer distance in the intercalated graphite oxide was: 8.85 Å in methanol, 9.23 Å in ethanol, and 9.50 Å in propanol. It increases by about 2.2–2.6 Å compared to the dry graphite oxide; this corresponds to insertion of one monolayer of alcohol molecules. High-pressure compression of the intercalated samples results in an additional structural change (Figure 2.22). For the GO–methanol sample, a sharp phase transformation starts at 0.37 GPa. A new phase with an interlayer distance of ~11.4 Å appears in addition to the previous phase. The patterns recorded at 0.37 and 0.43 GPa showed the coexistence of high- and low-pressure phases, while from 0.68 GPa and above the low-pressure phase has disappeared. This anomaly is explained by the insertion of additional methanol between the graphite oxide layers. A very similar phase transformation was also observed for the GO–ethanol system above 0.59 GPa. However, the high-pressure phase of GO–ethanol could not be obtained in the pure state. Decompression results in a reverse step-like transformation back to the low-pressure phase for the GO–methanol system, but in the case of GO–ethanol the high-pressure phase was observed to be unchanged even after full release of the pressure. At ambient pressure, the interlayer spacing of the high-pressure GO–ethanol phase reached a value of 13.4 Å, which is about 1 Å higher than its value at the pressure where this phase appeared during compression.

Intercalation of graphite oxide by four different aliphatic alkylamines ($C_nH_{2n}NH_2$ where $n=2, 4, 8, 12$) in water–ethanol solution was reported by Bourlinos *et al.* [73]. In all four cases, there was a systematic increment of the d_{001} value, as compared to the parent GO, indicating insertion of the alkylamines into the interlayer galleries of graphite oxide. Unlike alcohols, organic amines can chemically interact with GO. Bourlinos *et al.* suggested a nucleophilic substitution reaction of amines on epoxy groups of GO to account for the experimental observations. However, considering the aqueous medium, and in light of new

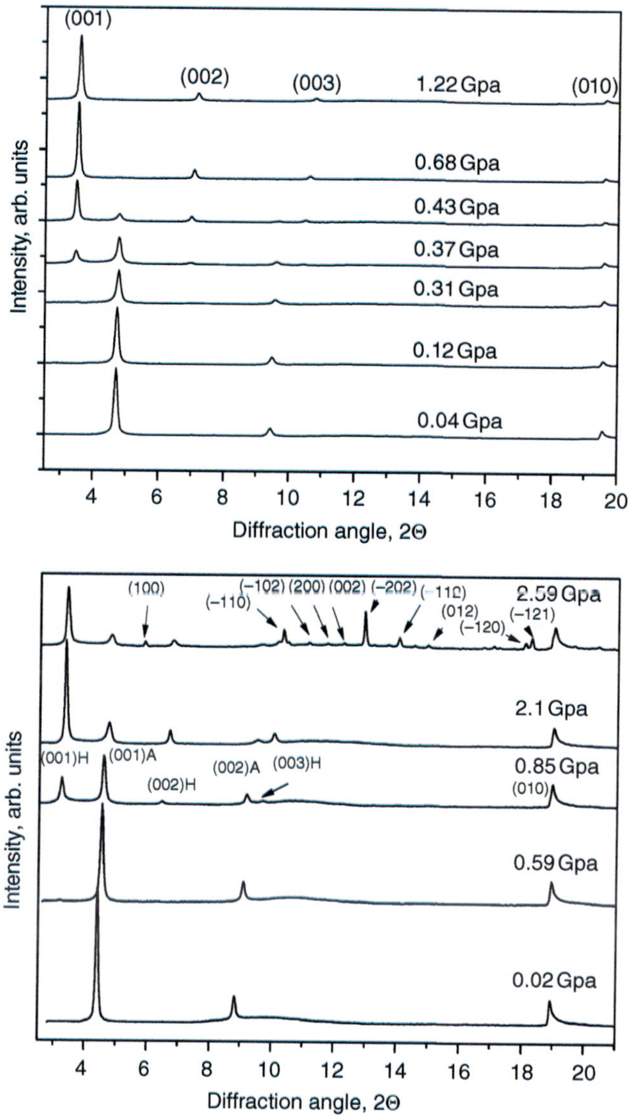

Figure 2.22 *XRD patterns (λ=0.7092 Å) recorded upon compression from a GO–methanol sample (top) and a GO–ethanol sample (bottom). Indexing for the GO–ethanol phases is given on the 0.85 GPa pattern: A, ambient-pressure phase; H, high-pressure phase. A solid ethanol phase is indexed on the 2.59 GPa pattern by a $P2_1/c$ structure with a=7.543 Å, b=4.738 Å, c=7.1904 Å and β=114.48°, in agreement with literature data. Reproduced from [72] with permission of ACS*

findings [75], this phenomenon can also be explained by electrostatic interaction between the negatively charged GO layers and positively charged amino groups.

Matsuo *et al.* [74] performed intercalation of the four different n-alkylamines ($C_nH_{2n}NH_2$; abbreviated as C_n, n=4, 8, 12, 16) simply by grinding the mixture of GO and C_n in a pestle

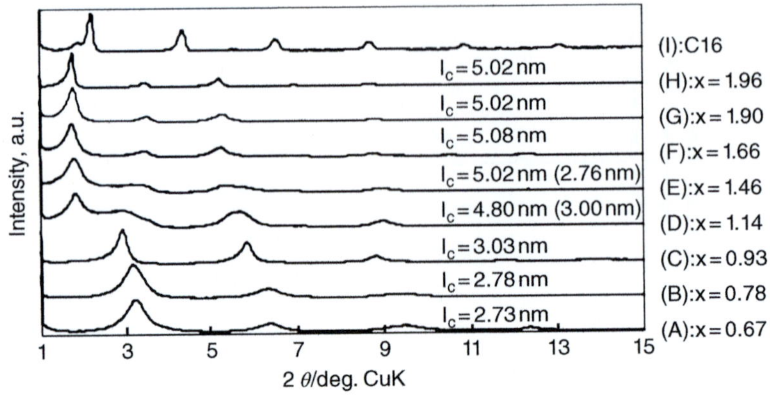

Figure 2.23 *X-ray diffraction patterns of the hexadecylamine–GO intercalation compound (C_{16}/GO) with various C_{16} contents, together with that of C_{16}. The values in parentheses are the interlayer spacing of a minor phase. The diffraction peaks at 2θ = 1.98°, 2.24°, 4.40°, 6.62°, 8.80°, 10.96° and 13.2° are due to the unreacted C_{16} phase. Reproduced from [74] with permission of Elsevier*

in the presence of a small amount of hexane. Based on elemental analysis of as-prepared intercalation compounds, Matsuo *et al.* claimed that water was removed from GO, and some amount of hexane co-intercalated GO along with alkylamines. With C_{16}, the C_{16}/GO ratio in the as-obtained intercalated product was 0.93–1.66. Within this ratio, only one phase, i.e. intercalation compound, was registered. Outside this region, the second phase, unreacted GO or C_{16}, was registered. Note that the varying intercalant content is specific for intercalation compounds of GO, but no for GICs. The C_{16}/GO samples possessed two different interlayer spacings (I_c) of 2.73–3.03 nm and 4.80–5.08 nm in which alkylamine molecules acquire interdigitated monolayer and bilayer orientations, respectively (Figure 2.23). Interestingly, for the shorter-chain alkylamines, the C_n/GO ratios decreased with decreasing chain length. The interlayer spacing of the samples with shorter alkyl chain lengths were very small, 2.48 nm and 0.90 nm for $(C_8)_{1.55}$GO and $(C_4)_{0.43}$GO, respectively. These I_c values were similar to those reported by Bourlinos *et al.* [73]; however, for the longer-chain alkylamines C_{12}/GO and C_{16}/GO the I_c values were larger than that reported by Bourlinos *et al.* This was explained by the competition between alkylamines and ethanol to intercalate GO and thus lower the intercalation degree in the work by Bourlinos *et al.* This observation demonstrates that intercalation into GO is accompanied and/or caused by chemical interaction of intercalant with GO, but not simply by charging graphene layers as in the case of GICs.

2.10 Concluding Remarks

Fundamental questions related to the mechanism of GO formation and the fine chemical structure of GO are still open. Concerning the mechanism of formation, the main avenues of future research should be revealing the nature of the actual oxidizing agents. The actual oxidizing agent might be the same for all the known preparation methods. When the actual oxidizing agent is identified, theoretical studies might help to describe the act of oxidation resulting in the formation of the C–O bond. However, real experimental conditions, i.e. positively charged state of graphene layers caused by sulfuric acid intercalant, should be taken into account. Concerning the fine chemical structure of the as-prepared GO, understanding the conditions for the transformation of the functional groups into each other should be developed. Despite the different preparation methods and levels of oxidation of as-produced GO, there is the same set of major functional groups for all the GO samples, with only little variation. This is confirmed by the very similar ^{13}C SSNMR, FTIR and XPS spectra acquired from different GO samples. It seems very likely that universal reaction mechanisms exist that account for the formation of the major functional groups and their transformation into each other. These mechanisms should be revealed and fully understood in the near future to pave the way to controlled GO production and functionalization.

References

[1] Loh, K.P.; Bao, Q.; Eda, G.; Chhowalla, M., Graphene oxide as a chemically tunable platform for optical applications. *Nature Chem.* **2010**, *2* (12), 1015–1024.
[2] Chua, C.K.; Pumera, M., Chemical reduction of graphene oxide; a synthetic chemistry viewpoint. *Chem. Soc. Rev.* **2014**, *43* (1), 291–312.
[3] Dreyer, D.R.; Todd, A.D.; Bielawski, C.W., Harnessing the chemistry of graphene oxide. *Chem. Soc. Rev.* **2014**, *43* (15), 5288–5301.
[4] Eigler, S.; Hirsch, A., Chemistry with graphene and graphene oxide – challenges for synthetic chemists. *Angew. Chem. Int. Ed.* **2014**, *53* (30), 7720–7738.
[5] Brodie, B., Note sur un nouveau procédé pour la purification et la désagrégation du graphite. *Ann. Chim. Phys.* **1855**, *45*, 351–353.
[6] Staudenmaier, L., Verfahren zur Darstellung der Graphitsäure. *Ber. Dtsch. Chem. Ges.* **1898**, *31* (2), 1481–1487.
[7] Hummers, W.S.; Offeman, R.E., Preparation of graphitic oxide. *J. Am. Chem. Soc.* **1958**, *80* (6), 1339.
[8] Charpy, G., Sur la formation de l'oxyde graphitique et la définition du graphite. *C. R. Hebd. Séances Acad. Sci.* **1909**, *148* (5), 920–923.
[9] Boehm, H.-P.; Scholz, W., Vergleich der Darstellungsverfahren für Graphitoxyd. *Liebigs Ann. Chem.* **1966**, *691* (1), 1–8.
[10] Poh, H.L.; Sanek, F.; Ambrosi, A.; *et al.*, Graphenes prepared by Staudenmaier, Hofmann and Hummers methods with consequent thermal exfoliation exhibit very different electrochemical properties. *Nanoscale* **2012**, *4* (11), 3515–3522.
[11] You, S.; Luzan, S.M.; Szabó, T.; Talyzin, A.V., Effect of synthesis method on solvation and exfoliation of graphite oxide. *Carbon* **2013**, *52*, 171–180.
[12] Kovtyukhova, N.I.; Ollivier, P.J.; Martin, B.R.; *et al.*, Layer-by-layer assembly of ultrathin composite films from micron-sized graphite oxide sheets and polycations. *Chem. Mater.* **1999**, *11* (3), 771–778.
[13] Marcano, D.C.; Kosynkin, D.V.; Berlin, J.M.; *et al.*, Improved synthesis of graphene oxide. *ACS Nano* **2010**, *4* (8), 4806–4814.
[14] Peng, L.; Xu, Z.; Liu, Z.; *et al.*, An iron-based green approach to 1-h production of single-layer graphene oxide. *Nature Commun.* **2015**, *6*, 5716–5724.
[15] Li, J.-L.; Kudin, K.N.; McAllister, M.J.; *et al.*, Oxygen-driven unzipping of graphitic materials. *Phys. Rev. Lett.* **2006**, *96* (17), 176101–176104.
[16] Sun, T.; Fabris, S., Mechanisms for oxidative unzipping and cutting of graphene. *Nano Lett.* **2012**, *12* (1), 17–21.
[17] Shao, G.; Lu, Y.; Wu, F.; *et al.*, Graphene oxide: the mechanism of oxidation and exfoliation. *J. Mater. Sci.* **2012**, *47* (10), 4400–4409.

[18] Boukhalov, D.W., Oxidation of graphite surface: the role of water. *J. Phys. Chem. C* **2014**, *118* (47), 27594–27598.
[19] Dimiev, A.M.; Tour, J.M., Mechanism of graphene oxide formation. *ACS Nano* **2014**, *8* (3), 3060–3068.
[20] Eklund, P.C.; Olk, C.H.; Holler, F.G.; *et al.*, Raman scattering study of the staging kinetics in the *c*-face skin of pyrolytic graphite-H_2SO_4. *J. Mater. Res.* **1986**, *1* (2), 361–367.
[21] Yosida, Y.; Tanuma, S., *In situ* observation of X-ray diffraction in a synthesis of H_2SO_4-GICS. *Synth. Met.* **1989**, *34* (1–3), 341–346.
[22] Inagaki, M.; Iwashita, N.; Kouno, E., Potential change with intercalation of sulfuric acid into graphite by chemical oxidation. *Carbon* **1990**, *28*, 49–55.
[23] Dimiev, A.M.; Bachilo, S.M.; Saito, R.; Tour, J.M., Reversible formation of ammonium persulfate/sulfuric acid graphite intercalation compounds and their peculiar Raman spectra. *ACS Nano* **2012**, *6* (9), 7842–7849.
[24] Dimiev, A.M.; Cierotti. G.; Behabtu, N.; *et al.*, Direct, real time monitoring of stage transitions in graphite intercalation compounds. *ACS Nano* **2013**, *7* (3), 2773–2780.
[25] Dreyer, D.R.; Park, S.; Bielawski, W.; Ruoff, R.S., The chemistry of graphene oxide. *Chem. Soc. Rev.* **2010**, *39* (1), 228–240.
[26] Glemser, O.; Schröder, H., Über Manganoxyde. II. Zur Kenntnis des Mangan(VII)-oxyds. *Z. Anorg. Allg. Chem.* **1953**, *271* (5–6), 293–304.
[27] Royer, D.J., Evidence for the existence of the permanganyl ion in sulfuric acid solutions of potassium permanganate. *J. Inorg. Nucl. Chem.* **1961**, *17* (1–2), 159–167.
[28] Dzhabiev, T.S.; Denisov, N.N.; Moiseev, D.N.; Shilov, A.E., Formation of ozone during the reduction of potassium permanganate in sulfuric acid solutions. *Russ. J. Phys. Chem.* **2005**, *79*, 1755–1760.
[29] Mermoux, M.; Chabre, Y., Formation of graphite oxide. *Synth. Met.* **1989**, *34* (1–3), 157–162.
[30] Sorokina, N.E.; Shornikova, O.N.; Avdeev, V.V., Stability limits of graphite intercalation compounds in the systems graphite–$HNO_3(H_2SO_4)$–H_2O–$KMnO_4$. *Inorg. Mater.* **2007**, *43* (8), 822–826.
[31] Dimiev, A.; Kosynkin, D.V.; Alemany, L.B.; *et al.*, Pristine graphite oxide. *J. Am. Chem. Soc.* **2012**, *134* (5), 2815–2822.
[32] Petit, C.; Seredych, M.; Bandosz, T.J., Revisiting the chemistry of graphite oxides and its effect on ammonia adsorption. *J. Mater. Chem.* **2009**, *19* (48), 9176–9185.
[33] Grasselli, J.G., *Atlas of Spectral Data and Physical Constants for Organic Compounds*. CRC Press, Cleveland, OH, **1973**.
[34] Kharasch, N. *Organic Sulfur Compounds*. Pergamon Press, New York, **1961**.
[35] Brimacombe, J.S., Foster, A.B., Hancock, E.B., *et al.*, Acidic and basic hydrolysis of some diol cyclic sulphates and related compounds. *J. Chem. Soc.* **1960**, 201–211.
[36] Eigler, S.; Dotzer, C.; Hof, F.; *et al.*, Sulfur species in graphene oxide. *Chem.: Eur. J.* **2013**, *19* (29), 9490–9496.
[37] Lerf, A.; He, H.; Forster, M.; Klinowski, J., Structure of graphite oxide revisited. *J. Phys. Chem. B* **1998**, *102* (23), 4477–4482.
[38] Szabó, T.; Berkesi, O.; Forgó, P.; *et al.*, Evolution of surface functional groups in a series of progressively oxidized graphite oxides. *Chem. Mater.* **2006**, *18* (11), 2740–2749.
[39] Clauss, A.; Plass, R.; Boehm, H.P.; Hofmann, U., Untersuchungen zur Struktur des Graphitoxyds. *Z. Anorg. Allg. Chem.* **1957**, *291* (5–6), 205–220.
[40] Scholz, W.; Boehm, H.P., Betrachtungen zur Struktur des Graphitoxids. *Z. Anorg. Allg. Chem.* **1969**, *369* (3–6), 327–340.
[41] Gomez-Navarro, C.; Meyer, J.C.; Sundaram, R.S.; *et al.*, Atomic structure of reduced graphene oxide. *Nano Lett.* **2010**, *10* (4), 1144–1148.
[42] Erickson, K.; Erni, R.; Lee, Z.; *et al.*, Determination of the local chemical structure of graphene oxide and reduced graphene oxide. *Adv. Mater.* **2010**, *22* (40), 4467–4472.
[43] Eigler, S.; Enzelberger-Heim, M.; Grimm, S.; *et al.*, Wet chemical synthesis of graphene. *Adv. Mater.* **2013**, *25* (26), 3583–3587.
[44] Gao, W.; Alemany, L.B.; Ci, L.; Ajayan, P., New insights into the structure and reduction of graphite oxide. *Nature Chem.* **2009**, *1* (5), 403–408.

[45] Dimiev, A.; Alemany, L.; Tour, J.M., Graphene oxide. Origin of acidity and dynamic structural model. *ACS Nano* **2012**, *7* (1), 576–588.
[46] Hofmann, U.; Holst, R., Über die Säurenatur und die Methylierung von Graphitoxyd. *Ber. Dtsch. Chem. Ges.* **1939**, *72* (4), 754–771.
[47] Hofmann, V.U.; König, E., Untersuchungen über Graphitoxyd. *Z. Anorg. Allg. Chem.* **1937**, *234* (4), 311–336.
[48] Boehm, H.P.; Heck, W.; Sappok, R.; Diehl, E., Surface oxides of carbon. *Angew. Chem. Int. Ed.* **1964**, *3* (10), 669–677.
[49] Szabó, T.; Tombacz, E.; Illes, E.; Dékány, I., Enhanced acidity and pH-dependent surface charge characterization of successfully oxidized graphite oxides. *Carbon* **2006**, *44*, 537–545.
[50] Shapiro, Y.M., Gem-polyols – a unique class of compound. *Russ. Chem. Rev.* **1991**, *60* (9), 1035–1049.
[51] Krois, D.; Langer, E.; Lehner, H., Selective conformational changes of cyclic systems – XII. Enhanced reactivity of ten membered ring ketones due to transannular steric compression: hydrates and hemiketals from oxo-[2,2]metacyclophanes. *Tetrahedron* **1980**, *36* (10), 1345–1351.
[52] Kumar, P.V.; Bardhan, N.M.; Tongay, S.; *et al.*, Scalable enhancement of graphene oxide properties by thermally driven phase transformation. *Nature Chem.* **2013**, *6* (2), 151–158.
[53] Robinson, J.T.; Burgess, J.S.; Junkermeier, C.E.; *et al.*, Properties of fluorinated graphene films. *Nano Lett.* **2010**, *10* (8), 3001–3005.
[54] Eigler, S., Graphite sulphate – a precursor to graphene. *Chem. Commun.* **2015**, *51* (15), 3162–3165.
[55] Rourke, J.P.; Pandey, P.A.; Moore, J.J.; *et al.*, The real graphene oxide revealed: stripping the oxidative debris from graphene-like sheets. *Angew. Chem. Int. Ed.* **2011**, *50* (14), 3173–3177.
[56] Salzmann, C.G.; Llewellyn, S.A.; Tobias, G.; *et al.*, The role of carboxylated carbonaceous fragments in the functionalization and spectroscopy of a single-walled carbon-nanotube material. *Adv. Mater.* **2007**, *19* (6), 883–887.
[57] Verdejo, R.; Lamoniere, S.; Cottam, B.; *et al.*, Removal of oxidation debris from multi-walled carbon nanotubes. *Chem. Commun.* **2007**, *2007* (5), 513–515.
[58] Fogden, S.; Verdejo, R.; Cottam, B.; Shaffer, M., Purification of single walled carbon nanotubes: the problem with oxidation debris. *Chem. Phys. Lett.* **2008**, *460* (1–3), 162–167.
[59] Thomas, H.R.; Vallés, C.; Young, R.J.; *et al.*, Identifying the fluorescence of graphene oxide. *J. Mater. Chem. C* **2013**, *1* (2), 338–342.
[60] Thomas, H.R.; Day, S.P.; Woodruff, W.E.; *et al.*, Deoxygenation of graphene oxide: reduction or cleaning? *Chem. Mater.* **2013**, *25*, 3580–3588.
[61] Faria, A.F.; Martinez, D.S.T.; Moraes, A.C.M.; *et al.*, Unveiling the role of oxidation debris on the surface chemistry of graphene through the anchoring of Ag nanoparticles. *Chem. Mater.* **2012**, *24* (21), 4080–4087.
[62] Coluci, V.R.; Martinez, D.S.T.; Honorio, J.G.; *et al.*, Noncovalent interaction with graphene oxide: the crucial role of oxidative debris. *J. Phys. Chem. C* **2014**, *118* (4), 2187–2193.
[63] Hu, C.; Liu, Y.; Yang, Y.; *et al.*, One-step preparation of nitrogen-doped graphene quantum dots from oxidized debris of graphene oxide. *J. Mater. Chem. B* **2013**, *1* (1), 39–42.
[64] Bonnani, A.; Ambrosi, A.; Chua, C.K.; Pumera, M., Oxidation debris in graphene oxide is responsible for its inherent electroactivity. *ACS Nano* **2014**, *8* (5), 4197–4204.
[65] Guo, Z.; Wang, S.; Wang, G.; *et al.*, Effect of oxidation debris on spectroscopic and macroscopic properties of graphene oxide. *Carbon* **2014**, *76*, 203–211.
[66] Rodriguez-Pastor, I.; Ramos-Fernandez, C.; Varela-Rizo, H.; *et al.*, Towards the understanding of the graphene oxide structure: how to control the formation of humic- and fulvic-like oxidized debries. *Carbon* **2015**, *84*, 299–309.
[67] Dimiev A.M.; Polson, T.A., Contesting the two-component structural model of graphene oxide and reexaminig the chemistry of graphene oxide in basic media. *Carbon* **2015**, *93*, 445–455.
[68] Taniguchi, T.; Kurihara, S.; Tateishi, H.; *et al.*, pH-driven, reversible epoxy ring opening/closing in graphene oxide. *Carbon* **2015**, *84*, 560–566.
[69] Clauss, A.; Hofmann, U., Graphitoxyd-Membranen zur Messung des Wasserdampf-Partialdruckes. *Angew. Chem.* **1956**, *68* (16), 522.

[70] Nair, R.R.; Wu, H.A.; Jayram, P.N.; *et al.*, Unimpeded permeation of water through helium-leak-tight graphene-based membranes. *Science* **2012**, *335* (6067), 442–444.
[71] Eigler, S.; Dotzer, C.; Hirsch, A.; *et al.*, Formation and decomposition of CO_2-intercalated graphite oxide. *Chem. Mater.* **2012**, *24* (7), 1276–1282.
[72] Talyzin, A.V.; Sundqvist, B.; Szabó, T.; *et al.*, Pressure-induced insertion of liquid alcohols into graphite oxide structure. *J. Am. Chem. Soc.* **2009**, *131* (51), 18445–18449.
[73] Bourlinos, B.; Gournis, D.; Petridis, D.; *et al.*, Graphite oxide: chemical reduction to graphite and surface modification with primary aliphatic amines and amino acids. *Langmuir* **2003**, *19* (15), 6050–6055.
[74] Matsuo, Y.; Miyabe, T.; Fukutsuka, T.; Sugie, Y., Preparation and characterization of alkylamine-intercalated graphite oxides. *Carbon* **2007**, *45*, 1005–1012.
[75] Feicht, P.; Kunz, D.A.; Lerf, A.; Breu, J., Facile and scalable one-step production of organically modified graphene oxide by a two-phase extraction. *Carbon* **2014**, *80*, 229–234.

3

Characterization Techniques

Siegfried Eigler and Ayrat M. Dimiev

3.1 Nuclear Magnetic Resonance Spectroscopy of Graphene Oxide

3.1.1 Nuclear Magnetic Resonance Spectroscopy in Solids

Nuclear magnetic resonance spectroscopy (NMR) revolutionized the structure determination of organic compounds and is now, next to X-ray crystal structure analysis and mass spectrometry, the method of choice to characterize biomolecules and organic compounds in general. With molecules of high molecular weight, the signal-to-noise ratio as well as the solubility of compounds may decrease. The molecular weight of a flake of graphite with a typical diameter of 5 μm is on the order of 10^{10} g mol^{-1}. Therefore, such molecules are more likely a material, rather than a molecule. However, NMR spectroscopy can also be conducted in solids, and solid-state nuclear magnetic resonance (SSNMR) spectroscopy has been developed and enhanced to a reliable method in recent years. Line shapes recorded by SSNMR are generally broader compared to NMR spectroscopy conducted on organic molecules; therefore, coupling constants cannot be determined. Nevertheless, a lot of information about the type of functional groups can be gained using SSNMR. A general overview about SSNMR as well as additional basic information is given by Kolodziejski and Klinowski [1].

In graphene oxide (GO), the carbon atoms dominate the structure, and therefore SSNMR spectroscopy has concentrated mostly on the analysis of the ^{13}C atoms of GO. Unfortunately, the ^{12}C nucleus has no spin and is therefore NMR-silent, but the spin of ^{13}C is ½, as it is for ^{1}H, and is therefore suitable for NMR spectroscopy. Other problems of ^{13}C NMR arise, because only part of the 1.1% ^{13}C is available in natural carbon and the magnetic moment is only about a quarter that of ^{1}H. Due to these facts, recording the ^{13}C NMR spectra of

Graphene Oxide: Fundamentals and Applications, First Edition. Edited by Ayrat M. Dimiev and Siegfried Eigler.

organic compounds requires much longer measurement times compared to ^{1}H spectra. Therefore, recording ^{13}C NMR spectra of materials in the solid state, such as GO, is time-consuming. Moreover, the anisotropy of the materials and the heterogeneous structure cause additional line broadening when recording ^{13}C directly. Moreover, ^{13}C is coupled to the ^{1}H nucleus, which would result in even broader lines. However, special techniques have been developed to realize short measurement times and sharp signals.

^{13}C spectra are normally combined with decoupling the ^{1}H nucleus, which is possible by a second pulse exciting the ^{1}H nucleus. Another technique that is combined is magic angle spinning (MAS). Line broadening is generally observed due to dipole and quadrupole interactions and the anisotropy of nuclei, because the matter in solids does not move as it does in the case of molecules in solution, where the anisotropy is reduced due to Brownian motion. The dipolar interaction and anisotropy of the chemical shift can be essentially reduced by rotating the probe at up to 70 kHz at the magic angle of 54.74° in relation to the external magnetic field. This MAS technique essentially reduces line broadening and makes the reliable recording of SSNMR spectra possible. Additional spinning sidebands are normally observed, but they can be identified by varying the spinning frequency, and no spinning sidebands are observed if spectra are recorded at very high spinning frequencies. However, direct excitation and recording of ^{13}C spectra is called a Bloch decay (BD) experiment and requires long measurement times. Moreover, for ^{13}C SSNMR spectroscopy, the spin–lattice relaxation is on the order of tens of seconds and thousands of scans are needed to reduce the signal-to-noise ratio.

Cross-polarization (CP) was developed as a method that enables fast measurement of ^{13}C SSNMR spectra (Figure 3.1) and the method is based on heteronuclear dipolar interactions. Another benefit of CP is due to its sensitivity to internuclear distances and the mobility of molecules or functional groups involved. Therefore CP is a method that gives additional information about dynamics and connectivity for structure determination. Although CP is a powerful technique, the optimization of the parameters is necessary and the contact time

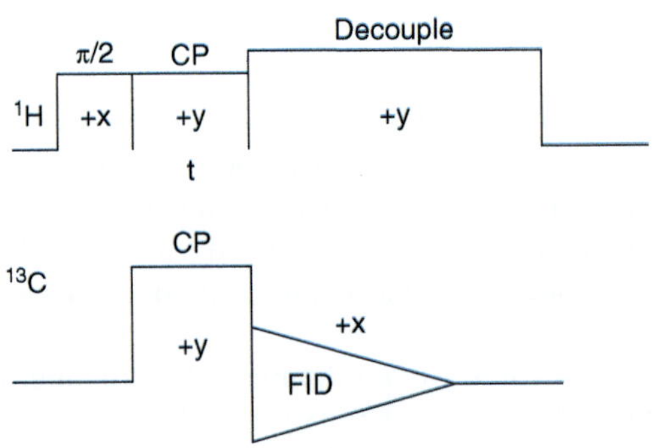

Figure 3.1 Example of a basic cross-polarization (CP) pulse sequence between ^{1}H nuclei and ^{13}C nuclei. First, a pulse is applied to polarize the sample in the magnetic field. Next, ^{1}H is excited by a pulse (π/2) along the x′-axis of the rotating frame. Then, the ^{1}H spin is locked by another pulse along the y′-axis accompanied by another pulse to ^{13}C nuclei along the y′-axis. The CP occurs during that time, and the time is called the contact time. Next, the radio frequency applied to the ^{13}C nuclei is switched off to record the free induction decay (FID) of the ^{13}C nuclei. Reproduced from [1] with permission of ACS

must be optimized. Moreover, dipolar dephasing is used as an additional technique to simplify NMR spectra. Signals originating from ^{13}C directly bound to ^{1}H can be identified and removed from the spectra, giving additional information. The technique can also be applied to, for example, ^{15}N.

SSNMR spectroscopy developed and became a useful method suitable as a chemical characterization technique for materials science. However, this technique is not yet used as a general characterization technique. The reason for that is due to the large amounts of substance needed for analysis, and this was true years ago. However, nowadays 10–20 mg may be enough, with measurement times of several hours only. But specialists are generally needed for that technique because pulse sequences need to be optimized to get reliable results and measurements still need much longer than for organic dissolved molecules, which can be characterized by routine measurements within minutes.

3.1.2 Nuclear Magnetic Resonance Spectroscopy of Graphene Oxide

The oxidation product of graphite is graphite oxide, as outlined in Chapter 2. Delamination of graphite oxide leads to single layers of GO. Restacking of GO leads then to randomly stacked graphite oxide. The two materials, graphite oxide and restacked GO, are chemically identical and therefore structure analysis by SSNMR can be conducted on both materials, graphite oxide or restacked GO. Thus, GO is used in this section as an abbreviation without distinguishing between GO and graphite oxide. In the following the SSNMR characterization and knowledge gained from SSNMR spectroscopy are summarized. The information gained from SSNMR led to the structure models of GO. The GO analyzed by SSNMR is generally oxidized by $KMnO_4$ in sulfuric acid. Therefore, the chemical structure of GO prepared by other techniques may differ. However, it was mentioned by Mermoux *et al.* [2] that the SSNMR spectra are the same for GO prepared by either the Staudenmaier or Hummers–Offeman protocols. Mermoux *et al.* published in 1991 a thorough ^{13}C SSNMR study [2]. Their main results are summarized here.

As depicted in Figure 3.2(A) three peaks are observed in the ^{13}C SSNMR spectrum using CP MAS conditions. In the original publication, these peaks were labeled as α resonating at 60.2 ppm, β at 71.2 ppm and γ at 132 ppm. The α and β peaks have dipolar dephasing characteristic times shorter than that of the γ peak (Figure 3.2B). This observation is expected for aliphatic carbon. Moreover, the spinning sidebands of the γ peak are typical for aromatic carbon. These observations allow the assignment of the peaks to ether (60.2 ppm), hydroxyl (71.2 ppm) and sp^2 carbon (132 ppm). The lower ^{13}C SSNMR spectrum of Figure 3.2(A) is ^{1}H-decoupled and allows the quantification of the relative amount of functional groups. The results were also correlated to chemical analysis, and a formula of $C_8(O)_{1.7}(OH)_{1.7}$ was determined. It is interesting to see that the ratio of the ether and hydroxyl groups is one. This ratio is also maintained for less oxidized samples. Moreover, no carboxylic groups could be detected, and it was concluded that the amount is below the detection limit. From dephasing experiments it was concluded that H atoms are too far away from sp^2 carbon to show a significant effect on the intensity of the sp^2 peak at 132 ppm and therefore enol groups and keto–enol tautomerism were excluded. In addition, the dephasing experiments show that the α and β peaks have similar dipolar relaxation times, which allows the conclusion that the H atoms are at an equal distance from the ether oxygen and hydroxyl groups. Thus, at short range, a certain order of the functional groups was

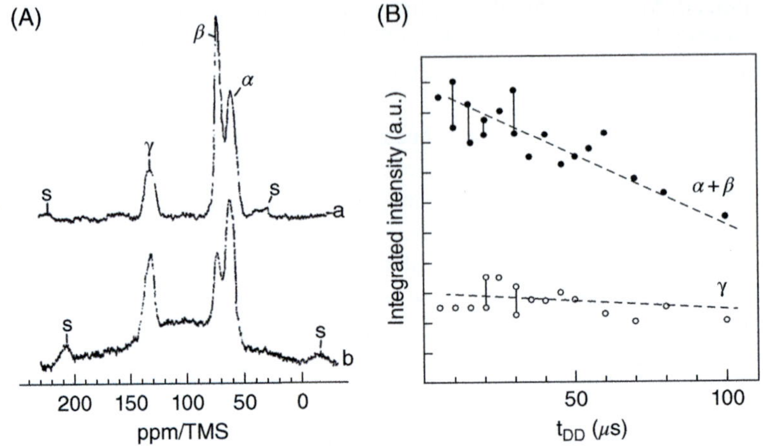

Figure 3.2 (A) CP MAS ^{13}C SSNMR spectrum of graphite oxide (top); proton-decoupled ^{13}C SSNMR spectrum measured under MAS conditions (bottom). (B) Development of the integrated intensity of the α, β and γ lines under a variable dipolar dephasing time ranging from a few microseconds to 100 μs. Reproduced from [2] with permission of Elsevier

suggested and an increased acidity of the tertiary alcohols was proposed. Regarding the structure, the authors [2] write:

> In this structure, the location of the two kinds of C–O bonds, with hydrogen bonding between the hydrogen of the hydroxyl groups and the oxygen of the epoxy groups, can explain the acidic properties of GO.

However, the model depicted in the paper shows 1,3-ethers, and 1,3-ethers have become a matter of debate in the years since.

Some years later, the structure of GO was revisited by Anton Lerf, Jacek Klinowski and coworkers Heyong He and Michael Forster [3]. Their investigations on the structure of graphite oxide were thoroughly studied in earlier publications [4–6]. The results and interpretation of the SSNMR data are summarized here. Their graphite oxide was synthesized by the protocol introduced by Hummers and Offeman, and the as-prepared graphite oxide was further washed with dilute HCl and subsequently purified by dialysis.

In the above-mentioned investigations, the ^{13}C NMR spectra of as-prepared GO consistently show three peaks, located at 60, 70 and 130 ppm (Figure 3.3). The peaks were assigned to 1,2-ethers (epoxide groups), tertiary alcohol groups and sp^2 carbon. The epoxy group assignment was further supported by analyzing the reaction products. Especially, the reaction of potassium iodide (KI) with GO was analyzed. It was well known that iodine is produced from HI and GO. However, KI reacts much slower and a reduced intensity of the peaks at 60 and 70 ppm was observed. In addition, a peak at 110 ppm was produced. The results suggest that GO was deoxygenated, forming another type of hydroxyl group that has similar chemical shifts as phenol and/or aromatic diols.

The observed deoxygenation reaction suggests that epoxy groups are the majority structural motifs, because 1,3-ethers should be stable upon treatment with iodide. In addition, no signals were detected at roughly 90 ppm and thus, C–OOH groups were excluded as

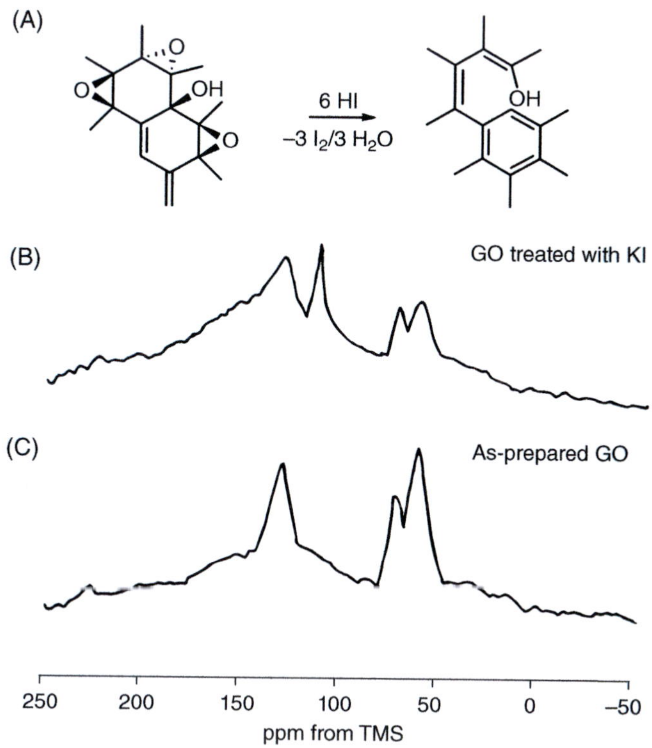

Figure 3.3 (A) Suggested chemical scheme for the deoxygenation of GO with HI, producing iodine and water, and forming phenol-like sp^2 carbon atoms. ^1H-decoupled ^{13}C MAS SSNMR spectra of (B) graphite oxide treated with KI and (C) as-prepared GO. Reproduced from [3] with permission from ACS

major species. Also thermally decomposed GO was analyzed and the peak at 110 ppm was assigned to phenolic sp^2 carbon (Figure 3.3). Moreover, GO was treated with several chemicals. Here, only the reaction with ethoxide will be mentioned. The reaction product contains ethyl groups that cannot be removed by washing procedures, suggesting that ethoxide was chemically bound by the reaction with epoxy groups.

Next, the role of water was further investigated. Dipolar dephasing CP spectra prove that all carbons are quaternary carbons. Short-contact-time CP spectra reveal that the protons are very close to the carbon atoms of epoxy groups and hydroxyl groups. However, dipolar dephasing experiments with delay times longer than 100 μs demonstrate that the protons are closer to the carbon of hydroxyl groups than to the carbon of epoxy groups.

In ^1H MAS NMR spectra (Figure 3.4), two peaks are found at 6.6 and 5.4 ppm. The water content is much too high to see tertiary alcohols. Therefore, two types of water can be distinguished. While the peak at 6.6 ppm decreases gradually between 20 and 140 °C, the peak at 5.4 ppm remains stable. Thus, the two types of water are assigned to mobile water and strongly adsorbed water.

The structure of GO was also studied in 2009 and the ^{13}C SSNMR spectrum is depicted in Figure 3.5 [7]. Well-resolved ^{13}C SSNMR spectra were obtained showing peaks not only

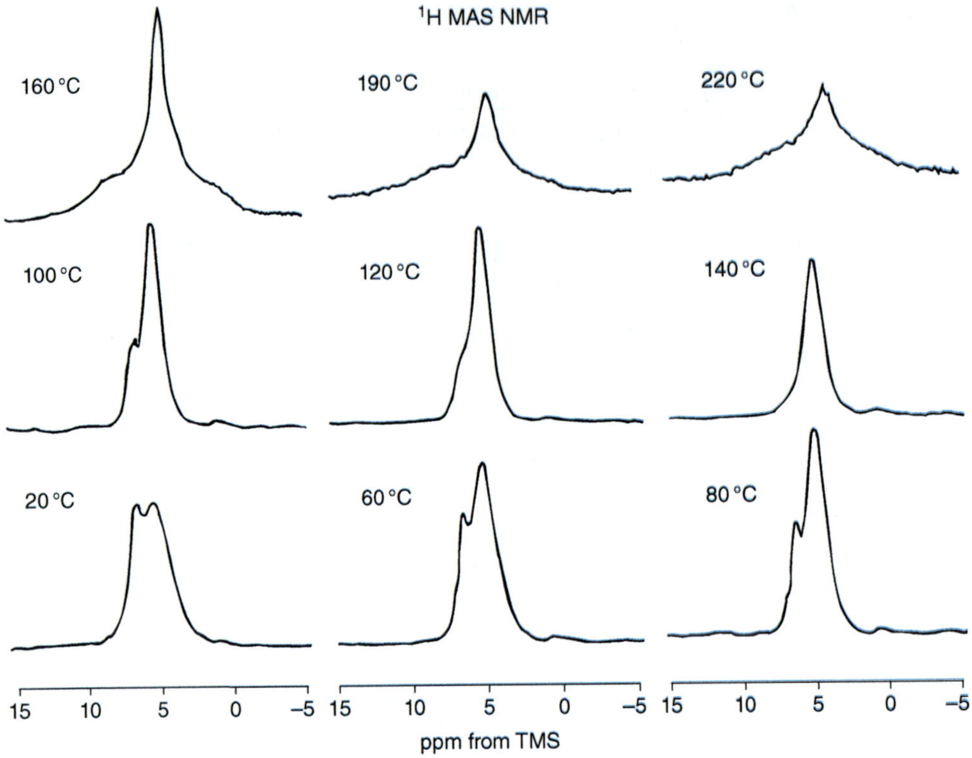

Figure 3.4 Different thermally treated ^1H MAS NMR spectra of graphite oxide in the temperature range between 20 and 220 °C to reduce the water content. Reproduced from [3] with permission from ACS

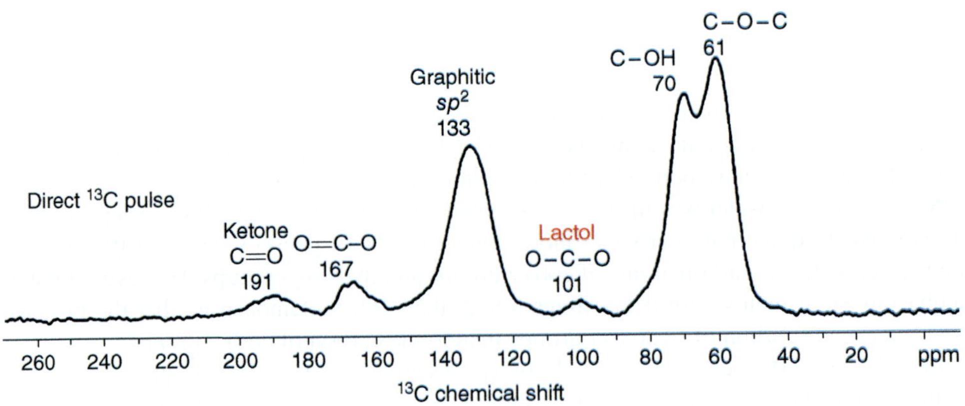

Figure 3.5 Direct ^{13}C pulsed MAS SSNMR spectrum of GO with chemical shifts of signals at 61, 70, 101, 133, 167 and 191 ppm. The minor groups have been assigned to ketones, carboxyl groups and lactols. The signal at 101 ppm may also be assigned to geminal diols (hydrates) or phenol-like moieties. However, ketones and carboxyl groups may not be directly bound to sp^2 carbon, suggesting that they are not part of the main structure [8]. Reproduced from [7] with permission of Nature Publishing Group

at 61 ppm (epoxy), 70 ppm (hydroxyl) and 133 ppm (sp^2) but also at 101, 167 and 191 ppm. The latter three signals were assigned in that study to lactol groups, carboxyl groups and ketones. Another explanation of the 101 ppm signal could be due to hydrates of ketone groups (geminal diols) [7].

As outlined in the introduction to this section, the percentage of ^{13}C carbon in naturally occurring carbon is 1.1%, and the measurements needed to get a low signal-to-noise ratio are time-consuming. Therefore, ^{13}C-labeled graphite was grown on Ni from ^{13}C-labeled methane [9]. The labeling of carbon atoms allows correlation spectra (^{13}C–^{13}C) to be recorded with measurement times that are about 160 000 times faster compared to conventional samples. The major peaks found are at about 60, 70 and 129 ppm and the assignments of the earlier publications were confirmed. Moreover, cross-peaks were found that confirm the connectivity between sp^2 C and ^{13}C–OH, as well as between sp^2 C and the ^{13}C of epoxides. Thus sp^2 carbon is close to sp^3 carbon. Moreover, also a cross-peak of the signal at 101 ppm to 60–70 ppm is found, but no cross-peaks are observed for the minor signals at 169 and 191 ppm. This observation suggests that these carbons are laterally separated from the majority of sp^2 carbons. It is speculative, but these signals may be related to oxidative debris, which would however need further investigations.

In another study on ^{13}C-labeled graphite oxide, the regiochemistry of functional groups is investigated further using two-dimensional (2D) correlation ^{13}C SSNMR spectroscopy [8]. The results of correlation spectroscopy were also simulated by *ab initio* methods, and two different types of arrangements of the functional groups were modeled. As shown in Figure 3.6, a structure with epoxy groups and hydroxyl groups conjugated to C–C double bonds is compared to a structure with 1,3-ethers and hydroxyl groups beneath aromatic units. The results of the modeling and the comparison with the experimental results strongly support the first model.

The ^{13}C labeling provides further benefits for solving the regiochemistry of the functional groups in GO. Special ^{13}C 2D correlation techniques allow the identification of networks of functional groups, such as the network of sp^2–sp^2–sp^2 motifs. As depicted in Figure 3.7 also other structural motifs are visible and sp^2 regions are terminated by epoxy or hydroxyl groups. Another epoxy or hydroxyl group is close to the epoxy group and hydroxyl group, respectively. In addition, structural motifs of hydroxyl–epoxy–epoxy and hydroxyl–hydroxyl–epoxy are found. Those results are modeled in Figure 3.6 and suggest that conjugated C–C double bonds are in close proximity to functional groups.

Unfortunately, the synthesis of ^{13}C-labeled graphite oxide is still a curiosity and not a method that is easily accomplished. Thus, the studies presented here are the only available ones. Note that generalization of the results of these experiments on all the GO samples needs to be done cautiously. The graphite used for oxidation in these studies was specifically grown by chemical vapor deposition (CVD) on a Ni substrate in the laboratory of the researchers. During the entire oxidation procedure, the graphite was still attached to the Ni substrate; the substrate was finally dissolved in diluted acid during the washing procedures. This can make the as-obtained GO substantially different from typical GO, due to the interaction with Ni salts. The role of defects could not be determined, and the size of aromatic units is not determined. Moreover, the stability of the functional groups and the change of the chemical structure with time were not revealed. Different degrees of oxidation and over-oxidation could give further insights into the diversity of the chemical structure of GO. It would also be helpful to analyze derivatives of GO by ^{13}C SSNMR spectroscopy. One benefit is the ability to gain information about the regiochemistry of those derivatives,

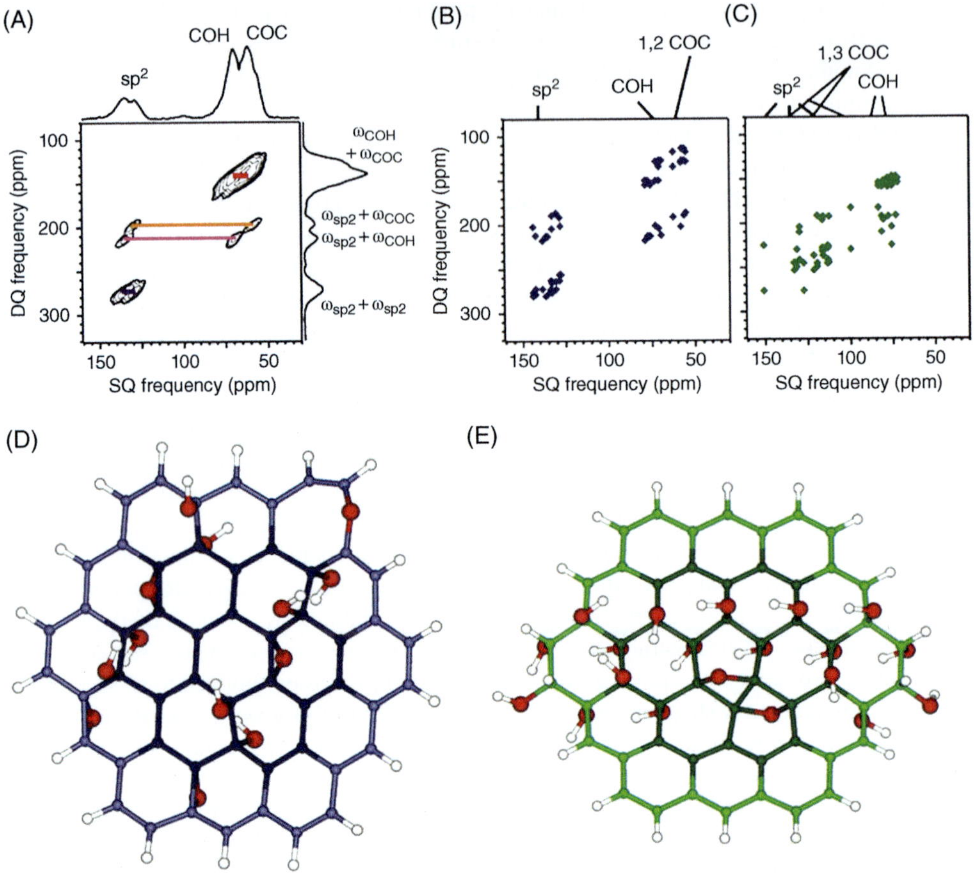

Figure 3.6 (A) The 2D ^{13}C double quantum/single quantum correlation SSNMR spectrum of ^{13}C-labeled GO, with (B) and (C) predicted correlation spectra assuming the chemical structures shown in (D) and (E), respectively. The results suggest epoxy groups and hydroxyl groups are adjacent to double bonds. Reproduced from [8] with permission of ACS

allowing for the development of a more controlled chemistry. Although it was suggested that edges are terminated by phenol-like functional groups, carboxylic acids or ketones, there is still no definitive proof.

3.1.3 Discussion

The experimental data accumulated through the years proves that SSNMR spectroscopy is a reliable tool to determine the structure of GO and potentially GO derivatives. Since the structure of GO depends on the preparation conditions, more SSNMR investigations would be highly appreciated in the future. It can be expected that systematic investigations will give insights into the results of reaction protocols applied to GO. Therefore it will become possible to elaborate reaction conditions to functionalize the carbon framework of graphene and GO, respectively. Especially, SSNMR spectroscopy has the potential to become

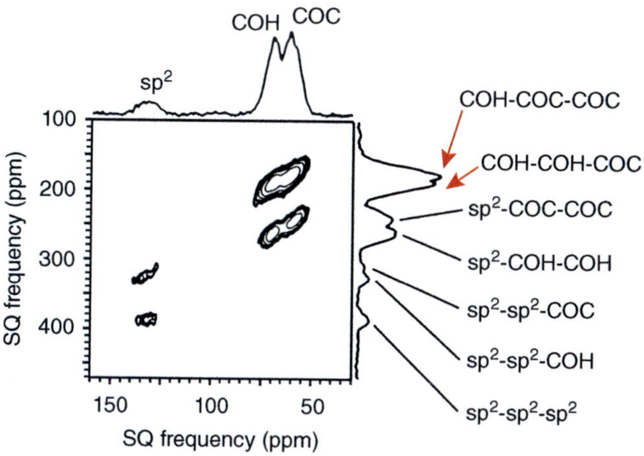

Figure 3.7 *Special correlation spectra of ^{13}C signals of GO allow the identification of motifs of functional groups and the determination of their regiochemistry. Thus, the major motifs are hydroxyl–epoxy–epoxy and hydroxyl–hydroxyl–epoxy, next to sp^2–epoxy–epoxy and sp^2–hydroxyl–hydroxyl. Consequently sp^2–sp^2–epoxy, sp^2–sp^2–hydroxyl and sp^2–sp^2–sp^2 motifs are found. Reproduced from [8] with permission of ACS*

a standard characterization tool for proving the success of chemical transformations in the future, as it is already for the characterization of organic molecular compounds. The elaboration of structure and function relationships will become possible, especially by combining with other analytical techniques. Thus, the development of applications with precision, related to the chemical structure, will become possible.

3.2 Infrared Spectroscopy

As reviewed in section 3.1, ^{13}C SSNMR was the primary method that for the first time allowed researchers to determine the chemical identity of GO functional groups. The presence of epoxides and tertiary alcohols on the GO platform was confirmed with high certainty based on ^{13}C SSNMR data.

While ^{13}C SSNMR is no doubt the most advanced and the most informative method for GO characterization, it also has some limitations. The most important disadvantage is a very long signal accumulation time. It takes several hours, possibly days, to collect a sufficiently strong signal even with advanced instrumentation. Another limitation is that ^{13}C SSNMR detects only carbon atoms, and thus it misses everything that is not directly bonded to the carbon grid. As the next downfall, the NMR spectra cannot be collected if ferromagnetic and/or paramagnetic elements such as Fe and Mn are present in the sample even in small quantities. Thus this method is not suitable for characterizing many composite materials of GO.

For these reasons, simple, quick and robust methods are very important for routine analysis of GO and especially GO-based composite materials. In the next two sections we will review two additional spectroscopic methods broadly used for GO characterization: Fourier-transform infrared spectroscopy (FTIR) and X-ray photoelectron spectroscopy (XPS). Against expectations, we do not review FTIR in the same section with Raman spectroscopy. The single chemical bond vibrations in GO are not Raman-active. The Raman spectroscopy in GO is sensitive to vibrations caused by the graphene lattice in-plane phonons rather than to the vibrations of single bonds, and it is reviewed in section 3.4.

FTIR is probably the most available and the fastest method suitable for GO characterization. The main disadvantage of FTIR spectroscopy is that it does not allow unambiguous assignment of absorption bands in the fingerprint region; only a few absorption bands can be assigned with a high level of confidence. Another disadvantage of FTIR is that it is almost purely qualitative. While some comparison of different samples against each other is possible, no solid quantitative data can be obtained from infrared (IR) spectra. At the same time, in some ways, FTIR is advantageous even over ^{13}C SSNMR. As will be shown below, in the authors' personal experience, FTIR has been able to provide very valuable information where ^{13}C SSNMR has been silent.

The principles of IR spectroscopy are well known, and thus they do not require a long introduction. IR spectroscopy is based on the fundamental properties of atoms joined by chemical bonds to absorb electromagnetic radiation in the 4000–400 cm^{-1} frequency region. As a platform bearing numerous functional groups, GO exhibits very rich and complex IR spectra. Theoretically, all the known FTIR sampling techniques can be used for GO analysis. However, two sampling techniques are used most commonly: transmission mode with a KBr pellet, and attenuated total reflection (ATR). In the traditional transmission mode, solid GO is pressed into a KBr pellet, and the radiation that passes through the pellet is measured. Indeed, several publications, including very recent ones, indicate that the transmission configuration was used to acquire the spectra. Note that grinding lamellar GO into a fine powder, in order to blend it with KBr, is difficult. Considering the uniform, lamellar, paper-like morphology of GO samples, the ATR sampling technique is probably the most convenient way to acquire FTIR spectra from GO samples. In this case, one needs only to firmly press soft GO paper against the crystal surface. The penetration depth, which is about 2 μm, provides sufficient information for the entire sample. With the ATR configuration, the entire analytical procedure, including sample preparation, takes only a few minutes. Most importantly, the sampling technique does not affect the GO spectra.

We begin the discussion here by warning the reader that at the present day the FTIR spectra of GO are commonly misinterpreted. Despite the existence of several high-quality studies, demonstrating the correct and well-argued interpretation, they are appreciated only by a small circle of experts, but remain largely ignored by the broader community. Most of the modern publications refer to works where the data had been previously misinterpreted; thus, misinterpretations are reproduced. Ironically, today, one can rarely find GO-related publications where the FTIR spectrum is correctly interpreted. Below, we show correctly interpreted spectra and highlight the most typical misinterpretations.

A typical FTIR spectrum of GO (Figure 3.8) can be arbitrarily divided into three characteristic regions: (i) an intense and very broad absorption band in the 3600–2400 cm^{-1} region, (ii) the two most recognizable absorption bands at 1723 and 1619 cm^{-1} in the middle of the spectrum, and (iii) a bunch of overlapping signals in the fingerprint region. The 3600–2400 cm^{-1} absorption band originates from the stretching modes of O–H bonds. Theoretically, this absorption can originated from two different species: tertiary alcohols

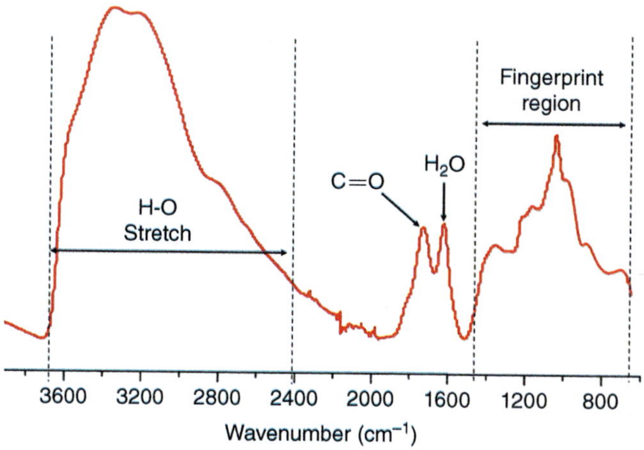

Figure 3.8 *A typical FTIR spectrum for a GO sample can be arbitrarily divided into three characteristic regions: (i) an intense and very broad absorption band in the 3600–2400 cm*$^{-1}$ *region, (ii) the two most recognizable absorption bands at 1723 and 1619 cm*$^{-1}$ *in the middle of the spectrum, and (iii) a bunch of overlapping signals in the fingerprint region*

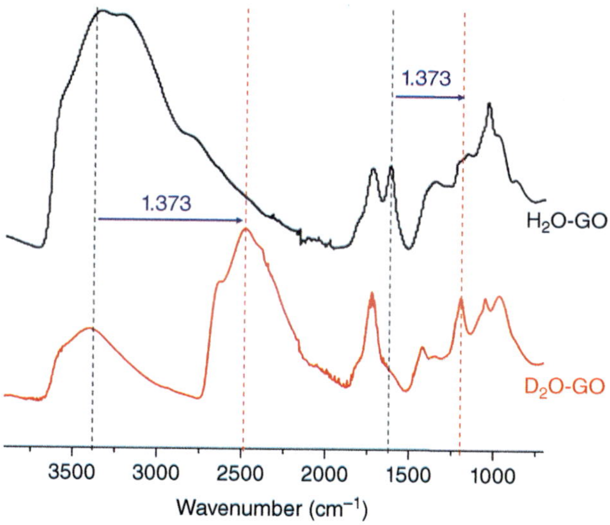

Figure 3.9 *FTIR spectra for* H_2O*-GO and* D_2O*-GO. The absorption bands associated with water molecules downshift upon replacing* H_2O *molecules by* D_2O *molecules. The shift corresponds to the factor 1.373 derived from the difference in masses of the H and D isotopes. Reproduced from [12] with permission of ACS*

that are an integral part of the GO platform, and water molecules incorporated into the GO structure. In fact, this absorption band originates mostly from water molecules. The hydroxyl groups of alcohols contribute significantly less. This has been demonstrated by experiments with deuterated water (Figure 3.9) [10–12]. In these experiments, normal air-dry GO was first dried under high vacuum over P_2O_5 to remove physisorbed H_2O

molecules. The as-prepared anhydrous GO was then equilibrated in D_2O vapor. This procedure affords rapid reinsertion of D_2O molecules into anhydrous GO, and results in the formation of D_2O-GO. Note that this procedure can only replace removable H_2O molecules by D_2O molecules, but not H in hydroxyl groups. After deuteration, the 3600–2400 cm^{-1} absorption band is shifted to the 2700–1900 cm^{-1} frequency region. This is in complete agreement with the frequency shift factor 1.373 that originates from the mass difference between the H and D isotopes. However, the 3600–2400 cm^{-1} absorption band is not eliminated completely. The remaining absorption in this region might originate from the hydroxyl groups of tertiary alcohols, and/or from remaining H_2O molecules that are strongly physisorbed with GO, and are not removable with the used drying procedures. The question about the origin of the remaining absorption in the 3600–2400 cm^{-1} region for D_2O-GO remains open up to the present day.

The pair of absorption bands in the middle of the spectrum at 1723 cm^{-1} and at 1619 cm^{-1} are the signature of GO FTIR spectra. These two bands are present in all the reported GO samples, and are easily recognizable due to the absence of other bands in the close vicinity. The exact positions of these bands vary slightly from work to work. The position of the first peak has been reported to be from 1719 cm^{-1} through 1734 cm^{-1}. The position of the second peak varies from 1615 cm^{-1} through 1626 cm^{-1}. The difference is most likely related to the instrumentation used rather than to the nature of the GO samples. The 1723 cm^{-1} band has been unambiguously assigned to the stretching mode of carbonyls. Often, it is attributed to carboxylic groups, in accordance with the Lerf–Klinowski GO structural model. However, it can originate from any carbonyl groups, including ketones and aldehydes. In any scenario, there is common agreement that this band originates from the stretching modes of C=O bonds. There is less agreement on assigning the 1619 cm^{-1} band. This is the most commonly misinterpreted absorption band in the modern literature. In the recent GO-related literature, it has been assigned to a variety of different species, with the most common assignment being to the stretching modes of C=C bonds. In fact, this absorption band originates from the bending modes of water molecules integrated into the GO structure. This has been unambiguously demonstrated in the deuterated water experiments, described above. In the FTIR spectrum of D_2O-GO, the 1619 cm^{-1} band is no longer present, and a new band centered at 1200 cm^{-1} appears in accordance with the frequency shift factor 1.373 based on the H/D mass ratio.

Assigning the bands from the fingerprint region is difficult due to overlapping of multiple bands from different functional groups. In contrast to the two bands discussed above, the reported positions and intensities of the bands in the fingerprint region vary from work to work. Thus, in our studies [12, 13] the most intense band in the fingerprint region was reported at 1039 cm^{-1}. Very similar spectra were reported elsewhere [14, 15]. In other works [10, 16–18] the most intense band in the fingerprint region was reported at 1051–1060 cm^{-1}. It is difficult to attribute such a large difference to instrumentation; it is possible that this difference is related to the difference in GO structure. Interestingly, this difference is irrelevant to the method used for GO preparation. This band is often assigned to epoxy rings. However, epoxy rings should absorb at ~1250 cm^{-1}, not at 1040–1060 cm^{-1}. Thus, the assignment of this band remains open. Other strong absorption bands reported in the fingerprint region are at 1368 and 1420 cm^{-1}. These bands have been assigned to the bending modes of tertiary C–OH groups [10, 11], and to the deformation modes of O–H in hydrogen bonds to heteroatoms of hydroxyl, epoxide or sulfate groups, respectively [15].

However, all these assignments are speculative, since they have never been confirmed in a way similar to the one with deuterated water experiments.

The only solid assignment of the adsorption bands in the fingerprint region was demonstrated in our experiments [13] with "pristine graphite oxide" samples. We found that graphite oxide samples purified by organic solvents (in contrast to commonly used water) exhibit two strong adsorption bands at 1221 and 1420 cm^{-1} (see Figure 2.6). These two bands were significantly weaker in the HCl-washed samples, and barely detectable in the water-washed GO samples. The signal intensity was in the opposite proportion to the nucleophilicity of the solvent used for GO purification, and in direct proportion to the amount of sulfur present in the GO samples. The two bands were assigned to the symmetric and asymmetric stretch of S=O bonds in organic sulfates, i.e. sulfuric acid esters of tertiary alcohols. This assignment allowed us to rule out the presence of sulfones, which were previously suggested to be present on the GO platform [15]. Thus, FTIR was the tool employed to propose the new functional group on the GO platform. This became possible due to the high sensitivity of IR spectroscopy to the S=O stretch vibrations. Even with a relatively low content of covalent sulfates in GO, the corresponding signal intensity was high. It is worth noting here that other analytical methods, including ^{13}C SSNMR, were not able to distinguish covalent sulfates from other functional groups. Thus, this was an example when FTIR spectroscopy was advantageous over NMR spectroscopy.

3.3 X-ray Photoelectron Spectroscopy

XPS is a unique and powerful method, and in general it is more advantageous compared to FTIR, and may even be so compared to NMR. The two main advantages are that it provides the elemental content of the sample and it is quantitative. Moreover, XPS not only reveals GO composition with respect to its constituent elements, but also provides the percentage of the different oxygen functional groups on the GO platform. The main drawback of XPS is the instrument cost: being expansive, it is not always available. Another drawback of XPS is that it is limited to surface analysis. However, for GO with its lamellar structure, XPS provides fairly reliable data for the entire sample. Considering that in many applications GO needs to be sampled as a film on a substrate surface, the latter obstacle can be considered not as a drawback, but as an advantage.

XPS is a surface chemical analysis technique. XPS spectra are obtained by irradiating a material with a beam of X-rays, and measuring the number and kinetic energy of the electrons that are released by the sample upon irradiation. The working principle of the method is based on the law of conservation of energy written as

$$E_{\mathrm{ph}} = E_{\mathrm{kin}} + \varphi + E_{\mathrm{bind}} \tag{3.1}$$

Here E_{ph} is the energy of the incident photons of X-ray radiation, E_{kin} is the kinetic energy of released electrons, φ is the work function and E_{bind} is the binding energy (ionization energy) measured in electronvolts (eV). From here, the binding energy, determined by the method, is represented by

$$E_{\mathrm{bind}} = E_{\mathrm{ph}} - E_{\mathrm{kin}} - \varphi \tag{3.2}$$

Depending on the tested material, the electrons are collected from the top 3–7 nm of the specimen. XPS is a very sensitive tool to determine the entity and quantity of the elements present in the sample. The detection limits for most of the elements on modern instruments are in the range of parts per thousand. The ability to detect heteroatoms makes XPS a very powerful tool in the characterization of functionalized GO samples and composites of GO with different materials. The elemental content is determined by running a survey spectrum that reveals all the elements present in the sample. XPS is also known as electron spectroscopy for chemical analysis (ESCA), to emphasize that the method also provides the chemical or electronic state of each element. With respect to GO, XPS provides not only the elemental content, but also the nature and the relative content of the functional groups. This information is obtained by running so-called elemental spectra for particular elements with lower incident radiation intensity that allows better resolution of the signals corresponding to different components. For GO, the most valuable information can be obtained from the C 1s spectrum, which reveals different states of carbon atoms in a sample.

Let us begin by reviewing the C 1s XPS spectrum for the graphite precursor, and then compare it with the spectrum for GO. The C 1s spectrum for graphite (Figure 3.10) is composed of one single component centered around 284.8 eV. The peak is not symmetrical. This asymmetry is due to the presence of the minor component situated at 285.6 eV that originates from imperfections, i.e. damage to the perfect sp^2 structure on the graphite surface [18, 19]. The low-intensity peak at 291.4 eV is due to the $\pi \rightarrow \pi^*$ shake-up interactions. This minor peak is the signature of any high-quality undisturbed sp^2 graphitic material.

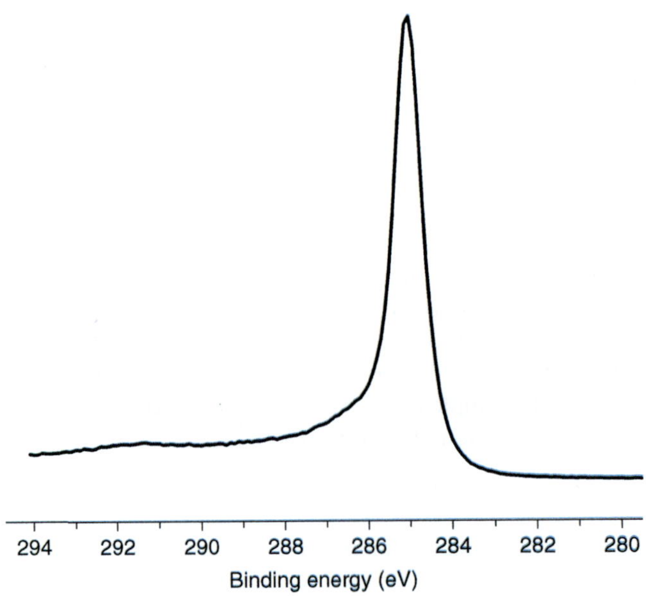

Figure 3.10 *The C 1s XPS spectrum for graphite. The spectrum consists of one non-symmetric peak centered at 284.8 eV and a low-intensity peak at 291.4 eV generated by the $\pi \rightarrow \pi^*$ shake-up interactions*

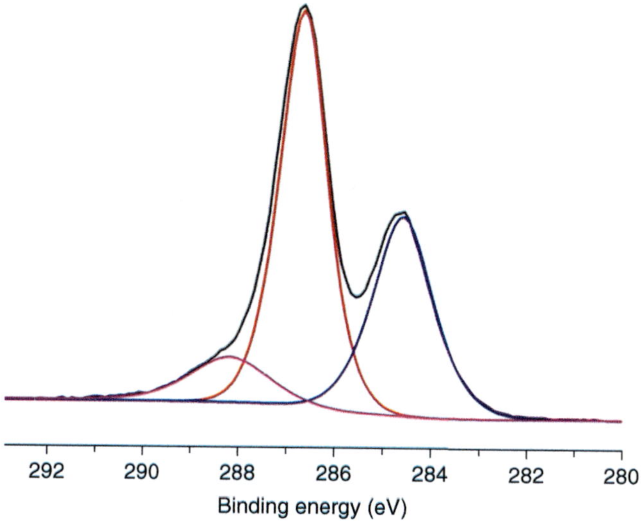

Figure 3.11 *The C 1s spectrum for GO. The black line is the actual experimental spectrum. The integral spectrum can be deconvoluted into three major components: at 284.8 eV (blue line), at 286.5 eV (red line) and a shoulder at 289.2 eV (pink line)*

It is present in the spectra of high-quality graphite and high-quality carbon nanotubes (CNTs). Generally, it is not present in the spectra of significantly damaged materials such as turbostratic graphite, functionalized graphene, functionalized CNTs, graphene nanoribbons and GO.

The C 1s spectrum for GO samples (Figure 3.11) normally consists of three apparent components. The peak at 284.5 eV is attributed to the carbon atoms of graphitic domains. Even if the carbon atom is not perfectly graphitic, it will generate a signal in this area as long as it is not chemically bound to oxygen atoms. The spectrum is dominated by the component centered at 286.5 eV. This peak is normally attributed to the carbon atoms of epoxides and tertiary alcohols; this is perfectly in line with information from databases for model small-molecule compounds. The integral spectrum contains a shoulder at 289.2 eV that is normally attributed to carboxylic acid groups. The still open question is where the signal from ketone groups should appear. Some researchers place the signal from ketones into the 286.5–287.5 eV interval together with epoxides and alcohols. Others place it into the 288.5–289.2 eV interval together with carboxyl groups. The actual position of the ketone carbon atoms is probably somewhere between. Most of the references in the XPS databases for known compounds report the ketone carbon signal between 287.7 and 288.1 eV (see e.g. http://www.lasurface.com/database/liaisonxps.php) [20, 21]. The databases for the known carboxyl and ester group carbon O–C=O provide values from 288.6 eV through 289.2 eV. Thus ketones bridge alcohols and epoxides on the right side of the spectrum with carboxyls on the left. Note that the component indicated by the pink line on Figure 3.11 is centered at 288.2 eV; this cannot be attributed either to ketones or to carboxyls, if based on the data for known compounds. Most likely, this component is the sum of the overlapping of the two peaks for carboxyls and ketones. Since the

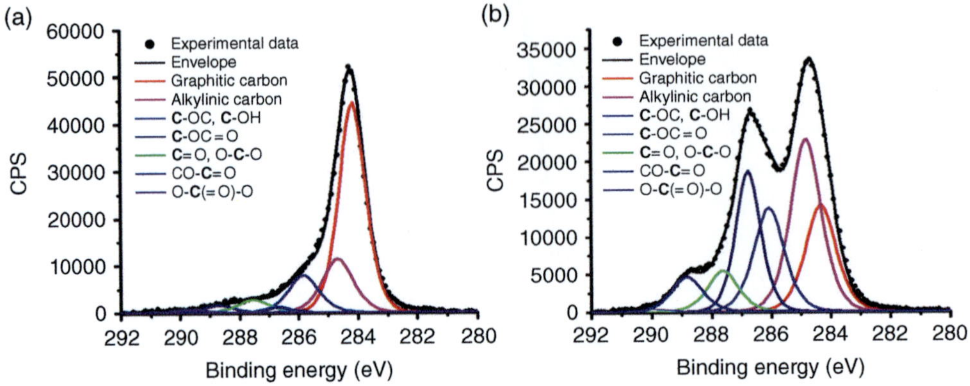

Figure 3.12 Deconvolution of the integral C 1s XPS spectra of two carbon-based materials that includes all the theoretical components. (a) The content of oxygen-free carbon is 79%. (b) The content of oxygen-free carbon is 49%. The authors disentangle the C–O–C and C–OH components, as well as graphitic carbon and alkylinic carbon. Reproduced from [22] with permission of ACS

question about the presence and content of ketones and carboxyls in GO remains open in principle, there is almost no way to unambiguously resolve the content of them both by XPS.

In the literature, one can find many different ways of deconvoluting the integral C 1s spectra. Some of these deconvolutions try to disentangle the emission of the C–OH carbons from the emission of the C–O–C atoms, and the emission of the C=O groups from that of the O–C=O groups [22, 23]. While these deconvolutions might indeed be correct, they are highly speculative (Figure 3.12). The number of possible ways to deconvolute the integral spectra is virtually unlimited. The question is how much one can trust these data. One thing can be stated with confidence: the C 1s XPS spectrum demonstrates well the overall oxidation level of GO and RGO. For this purpose, XPS is probably the most effective and the most precise method among all the spectroscopic methods. With its ability to quantitatively detect heteroatoms, this method is a very powerful tool in characterization of GO and its composites.

3.4 Raman Spectroscopy

Raman spectroscopy is a well-established characterization tool to probe the quality of graphene. A monochromatic laser excites a sample and the emitted light is detected. Among the excitation wavelength (Rayleigh scattering), other frequencies can also be detected. The interaction of matter and light results in a shift relative to the excitation wavelength that is specific for a material. Raman spectroscopy is based on the interaction of light with the polarizable electron density of molecules or materials [24]. Thus, polar solvents or polar impurities, such as water, do not essentially disturb the Raman process. Consequently, Raman spectroscopy is a characterization method suitable to probe GO (containing reasonable amounts of water), RGO and graphene, respectively.

3.4.1 Introduction

The Raman spectrum of graphene is well understood and readily used to characterize the quality of graphene in production processes as well as in research, because Raman spectroscopy is non-destructive [25]. Scanning Raman microscopy (SRM) can be performed with a high local resolution close to the diffraction limit [26]. In reality a resolution of about 250–300 nm can be reached using blue and green lasers, respectively. Tip-enhanced Raman spectroscopy (TERS) is also possible to further increase the resolution [27]. However, the tip of an atomic force microscope (AFM), operated in contact mode, must be coupled to a Raman spectrometer to get access to resolution on the nanometer scale. Even if TERS becomes more and more attractive, it is not yet a routine method.

In this section, Raman spectroscopy will be described in general and in particular for graphene. Then the Raman spectroscopy of different types of GO and RGO is described and discussed. While the Raman spectra of GO and RGO prepared by standard methods almost do not change, due to the manifold defects within the carbon lattice, various different Raman spectra are obtained for RGO if lattice defects can be avoided by the synthesis protocol. Raman spectroscopy can be used for the characterization of the carbon framework of RGO and it is possible to quantify the density of defects within the carbon framework of RGO, especially in the range between roughly 0.01% and 3%. It is even possible to make conclu sions about the density of defects within the carbon framework of GO itself. Also the term "defect" in graphene and GO is discussed and open questions are described.

3.4.2 Raman Spectroscopy on Molecules

Raman spectroscopy is a widely used technique to characterize molecules [24]. The method is complementary to infrared spectroscopy. The crucial difference is due to the selection rules and the excitation process. Without going into the details, Raman-active vibrations are possible in non-polar and easily polarizable bonds. Infrared-active modes are active for polar bonds, such as carbonyl C=O bonds or O–H bonds. Thus, polar water shows strong infrared bands, but is Raman-silent. Raman-active bonds are, for example, C=C bonds and aromatic subunits. The Raman process excites the molecule and this excitation can be either resonant or non-resonant (Figure 3.13). A resonant excitation requires the proper excitation wavelength to reach a real excited state. If the excitation is not resonant, a virtual energy state is reached. If the same frequency as the excitation frequency is emitted, the Rayleigh line is observed. However, the interaction of light and molecular vibrations or phonons results in a shift relative to the excitation. The emitted energy can be of either lower energy (Stokes scattering) or higher energy (anti-Stokes scattering). In most cases the matter is in the ground state and excitation into a virtual state occurs. The interaction of light and molecular vibrations leads to Stokes scattering, which is normally interpreted. On the contrary, anti-Stokes shifts are mostly of lower intensity and are often less studied.

3.4.3 Raman Spectroscopy on Graphene, GO and RGO

In 2006 the Raman spectra of graphene were measured [28]. Moreover, as well as graphene, also a few layers of graphene, RGO and graphite can be measured, and distinct differences in the spectra are observed [29]. Raman spectroscopy is applied to all types of

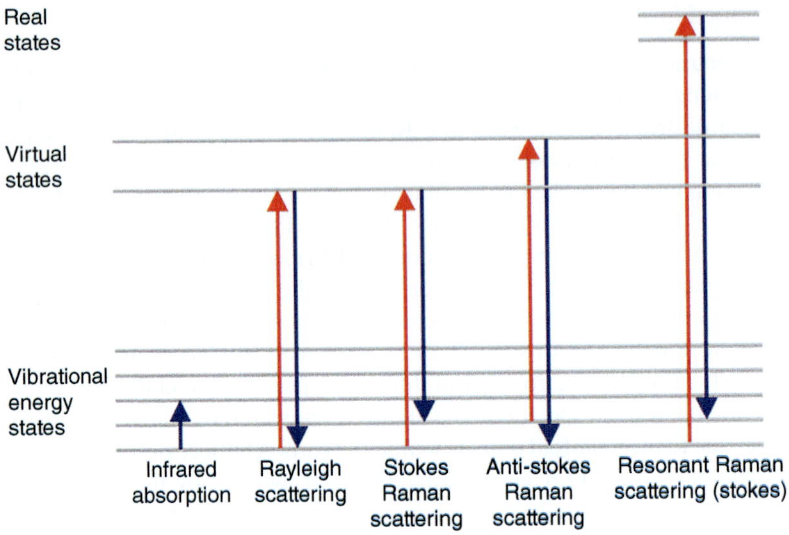

Figure 3.13 *Illustration of the Raman process in comparison to infrared excitation. In addition to Rayleigh scattering, Stokes shifts are observed in Raman spectroscopy, as well as anti-Stoke shifts. The excitation of molecules can occur into either virtual states or excited states, which is called resonant Raman spectroscopy*

carbon allotropes, such as graphite, fullerenes, carbon nanotubes, graphite, diamond or amorphous carbon, as well as polyconjugated molecules [30]. For graphene the Raman excitation process is always resonant. Graphene is a zero-bandgap semiconductor. Although, graphene bears no bandgap, there are energy states for a resonant excitation process using commonly available lasers for excitation, such as 532, 514, 473, 453 or 405 nm, as well as ultraviolet (UV) lasers [31]. Unwanted or unexpected reaction products, such as conjugated molecules, can be detected, as well as structural defects or functional groups, which makes Raman spectroscopy a powerful tool [32]. Currently, and in the past few years, knowledge about the Raman process as well as its interpretation have changed a lot. Thus, it is possible to determine stress and strain, or doping, the nature of edges and many more things. However, here we concentrate on the main features that are related to research on GO. We focus on the interpretation of spectra that relate to single layers. For the interpretation of spectra of a few layers, we refer to the literature [33–36].

The electronic structure is the basis for interpreting the Raman spectrum of graphene, as depicted in Figure 3.14(A) [37]. The electronic Brillouin zones of graphene are shown in Figure 3.14(C) and the first-phonon Brillouin zone is marked as the red rhombus. The Dirac cones illustrate the electronic dispersion. Phonon wavevectors connect electronic states in different valleys and are labelled in red. The Raman spectrum displays two major peaks at a Raman shift of about 1580 cm^{-1} (G peak) and 2700 cm^{-1} (2D peak). The G peak at about 1580 cm^{-1} is an allowed Γ-point phonon emission and corresponds to a high-frequency E_{2g} phonon at the Γ-point. The optical in-plane phonon dispersions of graphene in the energy and frequency range relevant for Raman scattering are depicted in Figure 3.14(D) (black). As illustrated in Figure 3.14(B) there is more than one option for

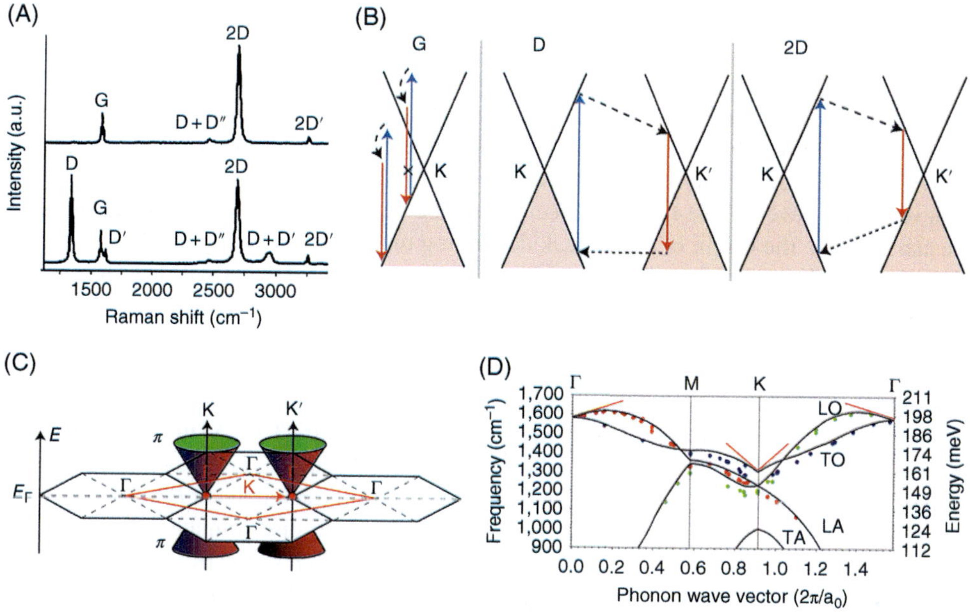

Figure 3.14 *(A) Raman spectra of graphene (top) and graphene with defects (bottom) with the main peaks labelled as D, G and 2D. (B) Illustrations of the excitation and emission in the Raman process responsible for the G, D and 2D peaks. (C) Electronic Brillouin zones of graphene and the first-phonon Brillouin zone marked in red. Electronic dispersion is visualized by Dirac cones. Phonon wavevectors that connect electronic states of different valleys are shown in red. (D) The black curves represent the dispersion of in-plane phonon modes of graphene in the frequency range (900–1700 cm^{-1}) relevant for Raman scattering and the interpretation of the Raman spectra of graphene. Reproduced from [37] with permission of ACS*

excitation that leads to the G peak. However, doping can eliminate one option (Figure 3.14B). The D peak at about 1340 cm^{-1} is a defect-activated peak and the 2D peak at about 2700 cm^{-1} is the overtone that is Raman-active without activation by a defect. For the latter, no defect is required because of momentum conservation due to two phonons with opposite wavevectors. A possible process is illustrated in Figure 3.14(B). The defect-activated D peak at about 1340 cm^{-1} is due to the breathing mode of a six-membered ring and it originates from TO phonons (Figure 3.14D) at the Brillouin zone corner K. The D peak is dispersive due to the interaction of phonons and electrons, and in practice the peak position varies with excitation wavelength between roughly 1310 cm^{-1} (red laser excitation) and 1400 cm^{-1} (ultraviolet laser excitation). More details about the origin of the peaks observed in the Raman spectra of graphene, G_2 or graphite are given in the specialized literature [33–37].

3.4.4 Defects in Graphene

The major peaks in graphene are the D, G and 2D peaks. As illustrated in Figure 3.16(A), the Raman spectra of graphene change their shape with the introduction of defects. The intensity of the D peak evolves on introducing defects [38]. These defects can be of structural nature, such as disorder within the carbon framework, e.g. misarranged carbon lattice atoms (other than six-membered rings), or missing atoms or even substituted atoms. Another possible evolution of the D peak is due to chemical functionalization of sp^2 carbon atoms of the lattice. With the introduction of addends, sp^3 carbon is formed and the Raman spectra evolve similarly to the depiction in Figure 3.16(A). Therefore, one cannot easily discriminate between these types of defects [26, 32, 39].

Nevertheless, even if the type of defect cannot be determined by Raman spectroscopy, one can estimate the origin of the D peak depending on the starting material. The example of Ar^+ ion bombardment on graphene that originally bears no defects will lead to a rearranged carbon lattice [38]. A D peak that evolves upon chemical reaction will be due to addends, and over-oxidation of the carbon lattice will lead to vacancies [26, 39, 40]. It is important to keep in mind that a D peak with, for example, the same intensity as the G peak does not necessarily mean that a high degree of functionalization was achieved or a high amount of defects was introduced [31, 38]. As depicted in Figure 3.15(A) the intensity ratio of the D peak and the G peak (I_D/I_G) follows a relation. A ratio $I_D/I_G=2$ can therefore be related either to a high degree of defects with a distance between defects $L_D=2$ nm or to a lower degree of defects with $L_D=6$ nm. However, it is easy to discriminate between these two values by taking the full width at half-maximum (FWHM) Γ of the peaks into account. Generally speaking, sharp peaks relate to a lower degree of defects and broad peaks to a higher degree of defects.

Examples of Raman spectra that show the important peaks, D, G and 2D, are shown in Figure 3.16(A) [31]. The L_D values that relate to these spectra are between 24 and 2 nm. At $L_D=24$ nm, the I_D/I_G ratio is 0.2, which increases to 2.9 for $L_D=5$ nm. For $L_D=5$ nm, one can expect that there are about 800 C atoms within the area between defects. As derived from the graphene lattice, one carbon atom fills an area of 0.026195 nm^2 and so the area between defects can be calculated from $(L_D)^2 \times (3)^{1/2}/2$ [26].

The illustration of the graphene structure in Figure 3.16(B) shows three red circles that mark the positions of possible defects [41]. For $L_D=10$ nm, the density of defects can be related to 0.03% and there are 3300 C atoms within this area. The Raman spectrum in Figure 3.16(C) (right) shows the corresponding I_D/I_G ratio of 1. The D peak is well resolved, but is much more narrow compared to the spectrum on the left of Figure 3.16(C). Such broad peaks relate to a density of defects of about 3% and thus there is almost no intact honeycomb lattice left. One could calculate that there are about 33 C atoms in place; however, the lattice might be highly distorted due to the large amount of defects.

3.4.5 Raman Spectra of GO and RGO

For interpreting the Raman spectra of RGO (and graphene), analysis of the exact FWHM Γ of the D, G and 2D peaks is necessary [39]. However, it should be clearly stated that the RGO mentioned here is not a typical RGO, as normally obtained by conventional preparation methods. RGO normally bears too many defects, so that Raman spectroscopy is not sensitive. Typical Raman spectra of GO and RGO are depicted in Figure 3.18 and described there. However, with RGO with less than roughly 1% of defects, it is possible to characterize this RGO by Raman spectroscopy. The interpretation is in line with the spectra obtained from graphene treated with ions, as outlined in section 3.4.4.

This analysis of typical Raman D, G and 2D peaks is the basis for characterizing RGO and the determination of the density of defects. A set of Raman spectra of RGO is depicted in Figure 3.17. The FWHM Γ values of the D, G and 2D peaks are printed in *italic* numbers. Values of Γ_D between 21 and 57 cm^{-1} are found in the I_D/I_G ratio range that goes from 1 up to about 4 and down to about 2 again. The Γ_G values are between 12 and 51 cm^{-1} and

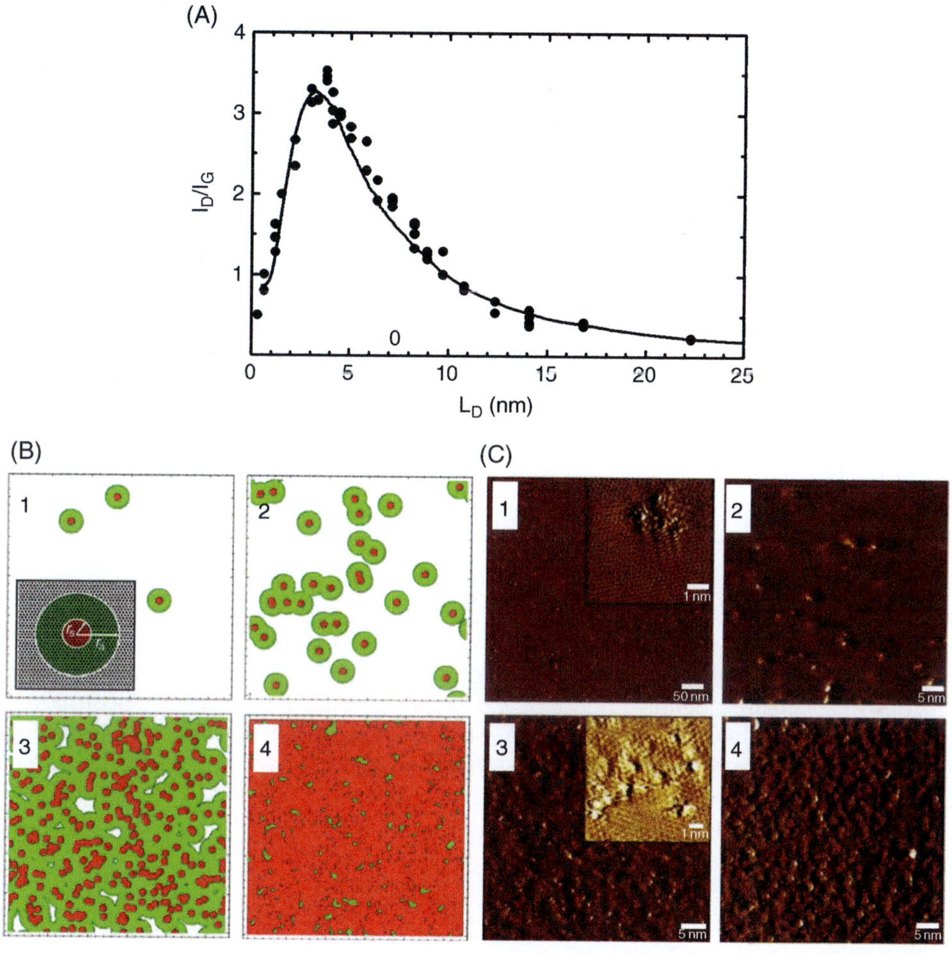

Figure 3.15 *(A) The I_D/I_G data points of different graphene samples as a function of the average distance L_D between defects. Defects generated by Ar^+ ion bombardment. The I_D/I_G ratio is shown as a function of L_D. (B) Activated regions (green) with radius r_a and structurally disordered regions (red) with radius r_s. The density of defects increases from panels 1 to 4 (as described in (C)). (C) STM images of the surface of a bulk highly oriented pyrolytic graphite (HOPG) sample subjected to 90 eV Ar^+ ion bombardment. The ion doses were varied between 10^{11} (1), 10^{12} (2), 10^{13} (3) and 10^{14} (4) Ar^+ ions per cm^2 and are the basis for the images shown in (B). The defective structure is depicted in the insets in panels 1 and 3. Reproduced from [38] with permission of Elsevier*

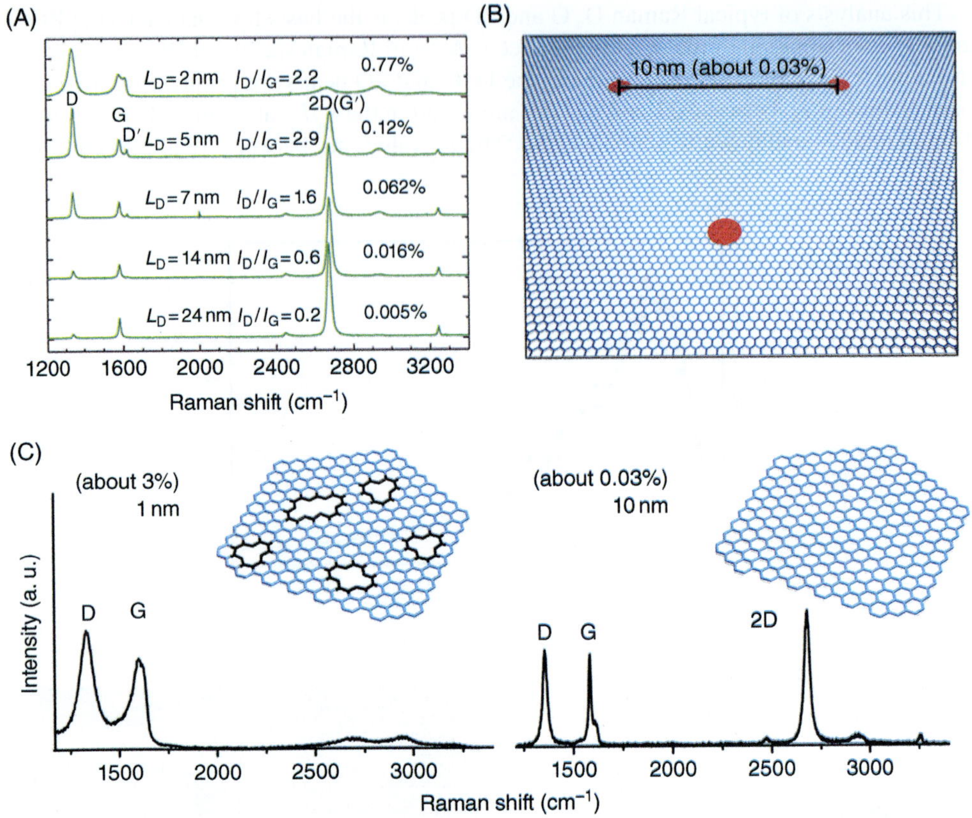

Figure 3.16 *(A) Raman spectra of graphene with a variable amount of defects between 0.005% and 0.77% based on the relation described and illustrated in Figure 3.15. The I_D/I_G ratio increases with increasing density of defects, and at about 0.3% the I_D/I_G ratio decreases again. In contrast, the $\Gamma_{D'}$, Γ_G and Γ_{2D} peaks increase with increasing density of defects. (B) Illustration of the carbon framework of graphene with symbolized defects marked as red circles. Here L_D = 10 nm and the expected I_D/I_G ratio would be 1. (C) Examples of Raman spectra of RGO with a density of defects of about 3% (left) and 0.03% (right). (A) Reprinted from Ref. [31] with permission from American Chemical Society. Copyright 2011. (B, C) Reproduced from [41] with permission of Wiley-VCH Verlag GmbH & Co*

the Γ_{2D} values are between 31 and 107 cm^{-1}. The density of defects estimated by the relation described above is between 0.03% and 1%.

For a high density of defects, the Γ value of the peaks broadens; however, it is difficult to make a reliable correlation. Hypothetically, RGO with a density of defects of about 5% would have the same spectral properties compared to RGO with a density of defects of 10%, or even 40%. As described above, chemical functionalization can also be a defect and thus a D peak is visible. On increasing the degree of functionalization to >3%, very broad D, G and 2D peaks are expected. In Figure 3.18(A) the Raman spectra of two different strongly oxidized layers of graphene are displayed. The characteristic peaks are all quite

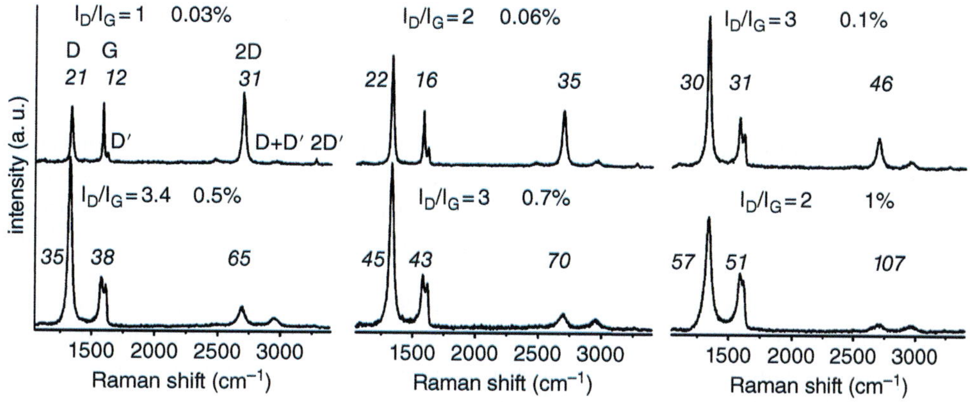

Figure 3.17 *Raman spectra of specially prepared RGO with a variable amount of defects. For the D, G and 2D peaks, the Γ values are given in* italic *numbers. For the displayed spectra, a density of defects between 1% and 0.03% can be estimated. Reproduced from [39] with permission of ACS*

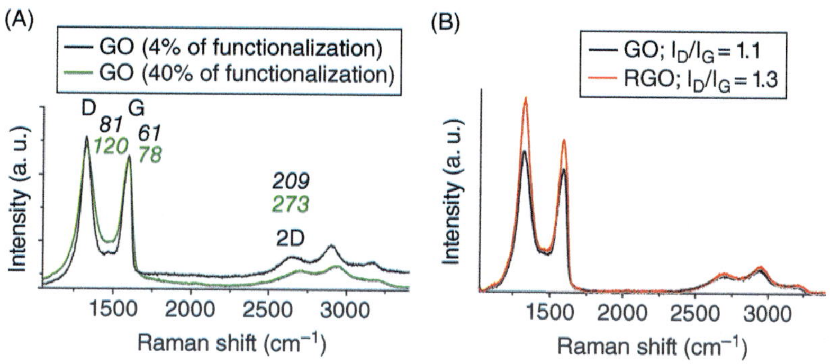

Figure 3.18 *(A) Raman spectra of GO with degree of functionalization of about 4% and about 40%, respectively (Γ values are given in* italic *numbers). A significant difference between the spectra and the relation to the degree of functionalization are not yet reliably established. (B) Spectra of typical GO and RGO prepared according to standard methods [43], outlined in Chapter 2. Spectral features almost do not change, because, although oxygen functional groups are removed, more than 3% permanent structural defects remain that also cause line broadening and a D peak. Only the I_D/I_G ratio increases slightly, which is related to a minimal increase of quality, as shown in Figure 3.15(A) for graphene with a density of defects of 0.77% and 0.12%, respectively. (B) Reproduced from [43] with permission of Elsevier*

broad with $\Gamma_D = 81\,cm^{-1}$, $\Gamma_G = 61\,cm^{-1}$ and $\Gamma_{2D} = 209\,cm^{-1}$ for a GO with a degree of oxidation roughly estimated to be 4% only (black line).

This type of oxo-functionalized graphene (GO) stems from water-quenched graphite sulfate, an intercalation compound of graphite, which can be delaminated to single layers [42]. Also shown in Figure 3.18(A) (green) is a spectrum of GO with about 40% of C atoms being sp^3 carbon, thus functionalized by oxygen groups, conventional GO [43]. The values of $\Gamma_D = 120\,cm^{-1}$, $\Gamma_G = 78\,cm^{-1}$ and $\Gamma_{2D} = 273\,cm^{-1}$ are about 30–40% broader compared to GO with a degree of functionalization of 4%. However, up to now, there is no reliable correlation between the degree of functionalization of GO with a degree of functionalization of more than 3%. In short, Raman spectroscopy on GO that is generally an oxo-functionalized derivative of graphene (degree of functionalization of about 40–50%) with a reasonable amount of additional structural defects can only be verified as GO and discriminated from unoxidized graphite. However, the degree of functionalization cannot be determined by Raman spectroscopy.

Another issue in Raman spectroscopy of GO and RGO is the change of spectra during chemical reduction. While the D peak in GO is mainly due to chemical functionalization and extended amounts of sp^3 C atoms, the D peak in RGO deviates from structural defects within the carbon framework. Functional groups have been removed by reduction from the π surface and structural defects, such as vacancies or rearranged structures of the carbon framework (five-membered or seven-membered rings), dominate [40]. However, as depicted in Figure 3.18(B) the Raman spectra of GO and RGO look very similar. In this example the I_D/I_G ratio slightly increases; however, the RGO structure is still far away from

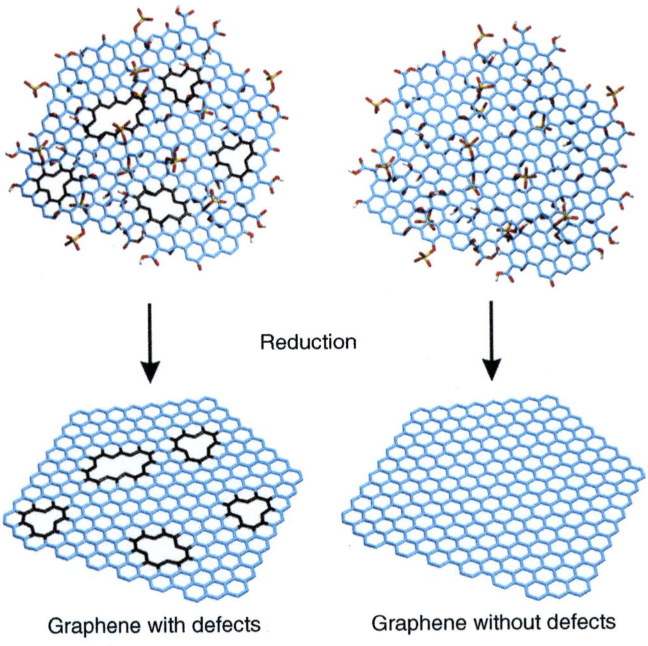

Figure 3.19 *Illustration of the chemical structure of GO. (left) GO with structural defects, such as vacancies, on the >3% scale. After chemical reduction, defects can be probed by Raman spectroscopy. Functional groups at defect sites are omitted for clarity. (right) GO with an intact carbon framework is shown, which can be chemically reduced to intact graphene (RGO without defects). Reproduced from [41] with permission of Wiley-VCH Verlag GmbH & Co*

being similar to graphene, and structural defects dominate the chemical structure and a density of defects of about 3% can be estimated.

Recent investigations revealed that chemical reduction of GO to RGO can result in RGO of different quality. Even graphene can be yielded that meets the same quality and shows the same Raman spectra as graphene prepared by other methods [44]. As described above, the D peak in graphene can be due to structural defects or functional addends as in GO. Thus, by removal of the addend, it should be possible to study the integrity of the carbon framework. This process is illustrated in Figure 3.19. For conventional GO that bears defects, the reduction removes functional groups, but with a density of structural defects of more than 3% the Raman spectra barely change, because Raman spectroscopy is very sensitive to scattering. Thus, the Raman spectra of RGO only change if the density of defects already present in GO is below at least about 1–3%, which can be seen as an experimentally determined upper limit for a suitable quantification of the density of structural defects [39, 44].

3.4.6 Statistical Raman Microscopy (SRM)

GO is a processable derivative of graphene; however, the quality of GO with respect to structural defects is difficult to determine, because conventionally prepared GO bears defects of more than 3% and Raman spectroscopy is not sensitive to variation of the density of defects on that scale. However, GO that bears less structural defects can be synthesized and oxo-functional addends are chemically bound on the surface of graphene. After chemical reduction and thus removal of functional groups, the quality of RGO can be probed. The D, G and 2D peaks can be fitted by Lorentz functions and parameters, such as the intensity of the peaks as well as the Γ values, can be determined [39].

Modern scanning Raman spectrometers are equipped with a microscope and a motorized x–y table. The table can be moved on the micrometer scale and even below. Raman spectra and the x–y coordinates can be recorded and large areas of a film of RGO can be scanned by this method. With such a dataset, the quality of RGO can be analyzed statistically. A histogram of I_G is shown in Figure 3.20(A). It is found that I_G is mainly dependent on the carbon atoms probed by the focused laser beam and thus it is possible to discriminate between substrate, edges of flakes, RGO and a few layers of RGO. A plot of I_G versus Γ_G allows determination of the I_G value that relates to single layers of RGO: Γ_G values below about 30 cm^{-1} are only obtained for single layers of RGO with a density of defects less than about 0.3% (Figure 3.20B). Moreover, the Γ_{2D} value is sensitive to the quality of the RGO and values below 30 cm^{-1} relate to single layers; and for a density of defects less than 0.3%, Γ_{2D} values are below 50 cm^{-1} (Figure 3.20C). A plot of the I_D/I_G ratio or A_D/A_G versus Γ_{2D} can visualize the quality of RGO (Figure 3.20D). The comparison of I_D/I_G ratio and A_D/A_G ratio reveals that qualitatively the same results are obtained. Because intensities are easier to determine, the I_D/I_G ratio is more often used.

It is also possible to further relate SRM results to other microscopic techniques, such as atomic force microscopy [39]. Probing the same position by other techniques allows proof of, for example, the assignment of spectral features made by SRM. Since the quality of graphene and RGO, respectively, is generally heterogeneous within a given sample, it is not sufficient to measure only one single-spot spectrum. Therefore, SRM was introduced, which allows for the determination of the quality within a complete sample using statistical

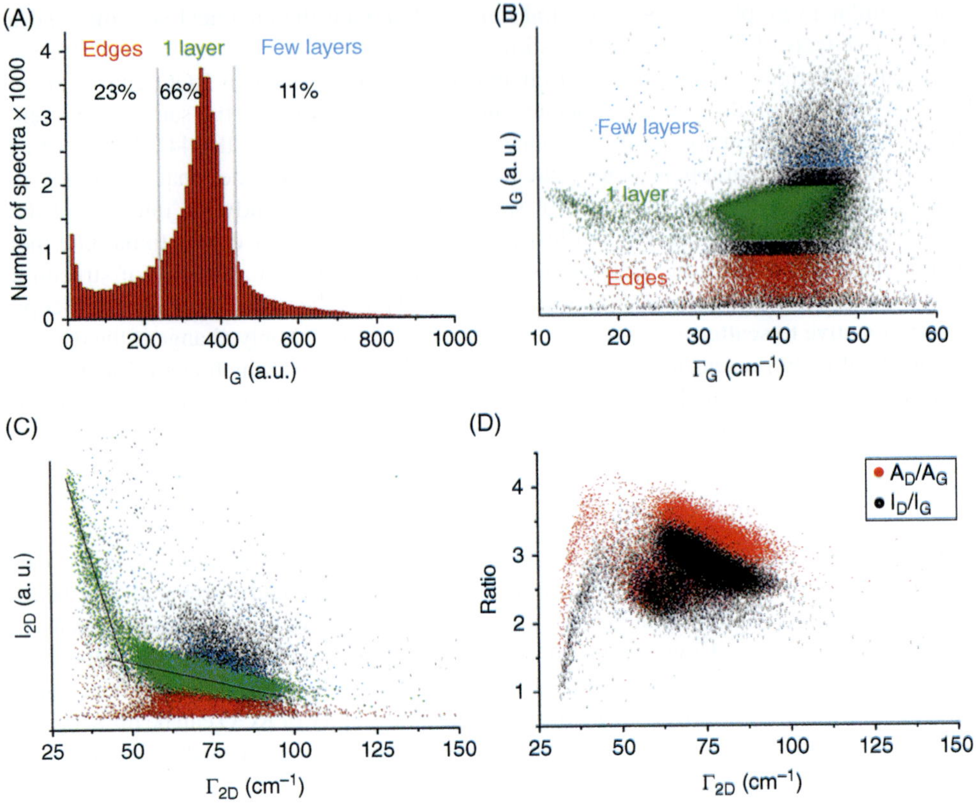

Figure 3.20 (A) Histogram of I_G and discrimination of RGO and few layers of RGO as well as edges of flakes of RGO. (B) Plot of I_G versus Γ_G extracted from a dataset of about 6×10^4 Raman spectra from RGO. The highest quality of RGO spectra is indicated by small Γ_G values. (C) Plot of I_{2D} versus Γ_{2D}; edges (red), monolayer graphene (green) and few-layer graphene (blue). (D) Plot of the I_D/I_G ratio and A_D/A_G versus Γ_{2D} that allows the quality of RGO to be illustrated. Reproduced from [39] with permission of ACS

methods. As outlined above, the *x–y* position and the corresponding Raman spectrum can be recorded. Furthermore, the spectral features can be extracted from the Raman spectra. With these data it is possible to build maps of films of RGO and it is possible to display spectral features. In Figure 3.21(A), the results obtained from a scanned film of flakes of RGO are shown. The single layers of RGO are color-coded in green and flakes of RGO with a value $\Gamma_{2D} < 45\,\text{cm}^{-1}$ are in orange. These flakes are RGO flakes that relate to $L_D > 3\,\text{nm}$ and they bear a density of defects less than about 0.3%.

3.4.7 Outlook

Raman spectroscopy is the most important technique to characterize graphene, as well as few-layer graphene. The borders between RGO and graphene start to vanish. However, Raman spectroscopy is only one tool for the characterization of GO. Although the integrity of the honeycomb lattice of GO can be quantitatively determined, there are several possible

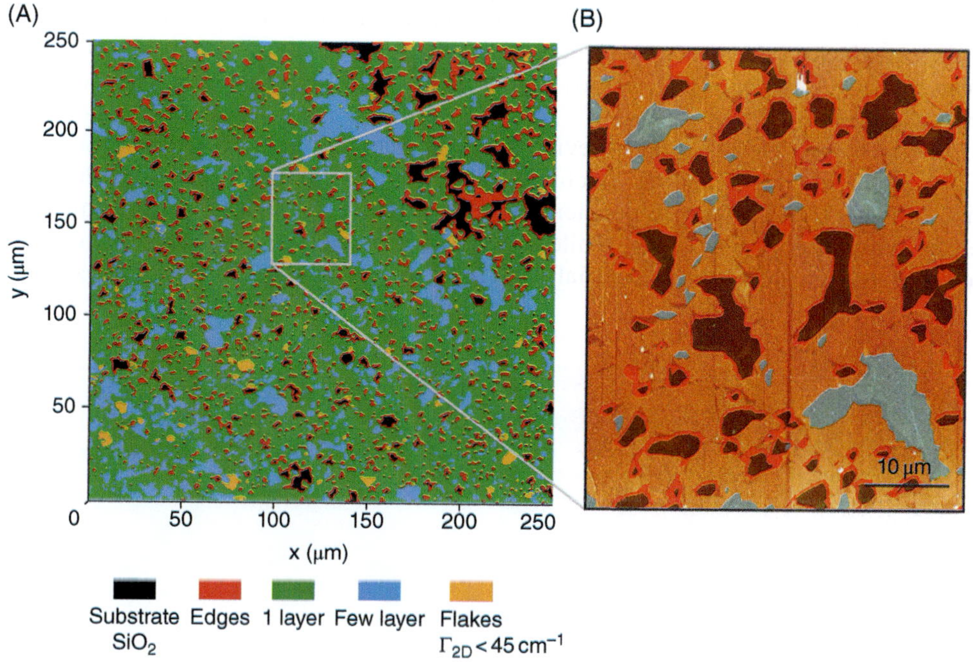

Figure 3.21 (A) SRM image of a film of 250×250 μm² of RGO. The I_G values are color-coded according to I_G and single layers of RGO are shown in green. (B) Magnification of the gray marked area of (A). Raman information about the thickness of the flakes is illustrated as an overlay of the corresponding AFM image. Reproduced from [39] with permission of ACS

origins for spectral features such as the D peak. Thus, other analytical techniques must be developed to thoroughly analyze GO. One recent development is statistical analysis of the quality of films of RGO. The quality can be visualized and quantified by SRM, a tool that gives information about the whole sample. In GO itself, up to now, it has been impossible to determine the density of structural defects beneath a high degree of functionalization of the carbon lattice by oxo-functional groups. Currently, mainly the D, G and 2D peaks of RGO are used to characterize samples. However, other peaks, such as D′, or a combination of peaks and especially their position may imply more information that needs to be revealed in future studies. A promising recent study on the relation of the size of defects and Raman spectra is only one example that may be used in the near future [45]. An open question about the reliability of the Γ value of single layers of GO with variable densities of defects is another field of research that must be conducted. It can be assumed that Raman spectroscopy of GO, RGO and graphene will remain an exciting topic in the future.

3.5 Microscopy Methods

Every GO flake is a two-dimensional nanoparticle. As such, a GO flake can be analyzed and characterized by microscopy techniques. Unlike spectroscopy methods that analyze bulk samples, microscopy methods provide full data about a single flake's morphology.

Most of the GO flakes are too small to be well resolved in an optical microscope. However, electron microscopy techniques generate images that reveal all the structural features of GO flakes. Thus, microscopy methods are a very important part of GO characterization.

In this section we will briefly review the four main methods most often used for GO-related studies: scanning electron microscopy (SEM), transmission electron microscopy (TEM), atomic force microscopy (AFM) and high-resolution transmission electron microscopy (HRTEM). While SEM can be broadly used for everyday routine analysis, the remaining three, especially HRTEM, are mostly used for specific research

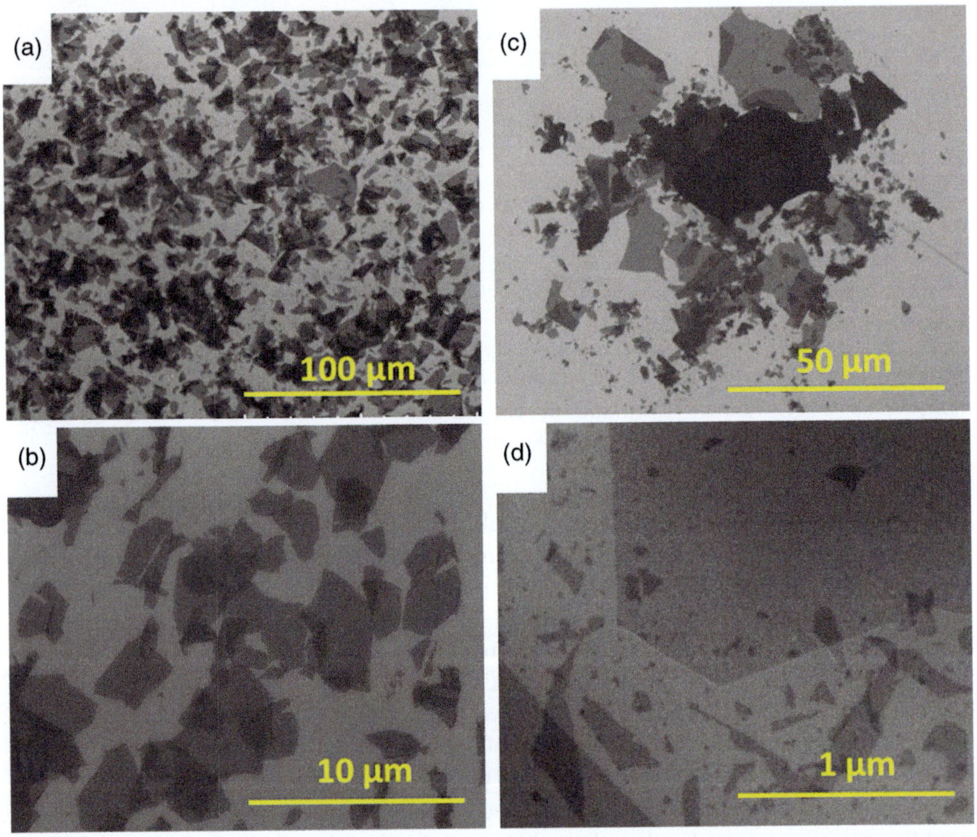

Figure 3.22 SEM images of different GO samples on Si/SiO_2 wafers. Images are acquired at different magnifications. (a) The image demonstrates that SEM allows simultaneous imaging of hundreds of flakes on the screen with a size 200 μm by 200 μm. (b) Typical monolayer GO flakes. The flake's size, edge roughness and other morphological features are easily visible. In the light-gray areas the flakes are of a single carbon layer. The double- and triple-layer areas, where flakes are folded or overlapped, are easily identifiable by flakes' opacity. (c) The image demonstrates GO flakes with different morphology; multi-layered flakes appear as black, non-transparent bodies. (d) A GO flake with smooth straight edges; the edges intersect at 120°

purposes. This is related to two factors: cost/availability, and the time needed to acquire the respective images.

3.5.1 Scanning Electron Microscopy

SEM images are generated by detecting and collecting the secondary electrons emitted by specimen atoms excited by the incident electron beam. SEM is probably the most robust microscopy method to analyze GO samples. It allows fast scanning of large surface areas of a specimen that includes tens and even hundreds of different GO flakes on the same image. High-quality GO images can be obtained with magnifications varying from 500 through 100 000. When GO flakes are deposited on a substrate (usually SiO_2-coated Si wafers), SEM provides complete information about the flake size distribution, number of carbon layers, flake morphology, etc. (Figure 3.22). For its simplicity and robustness, in principle, SEM can even be used as a routine quality control method when GO production comes to industrial scale. SEM was used as the main instrumental method in the study contesting the two-component structural model of GO (Figure 3.23) [46].

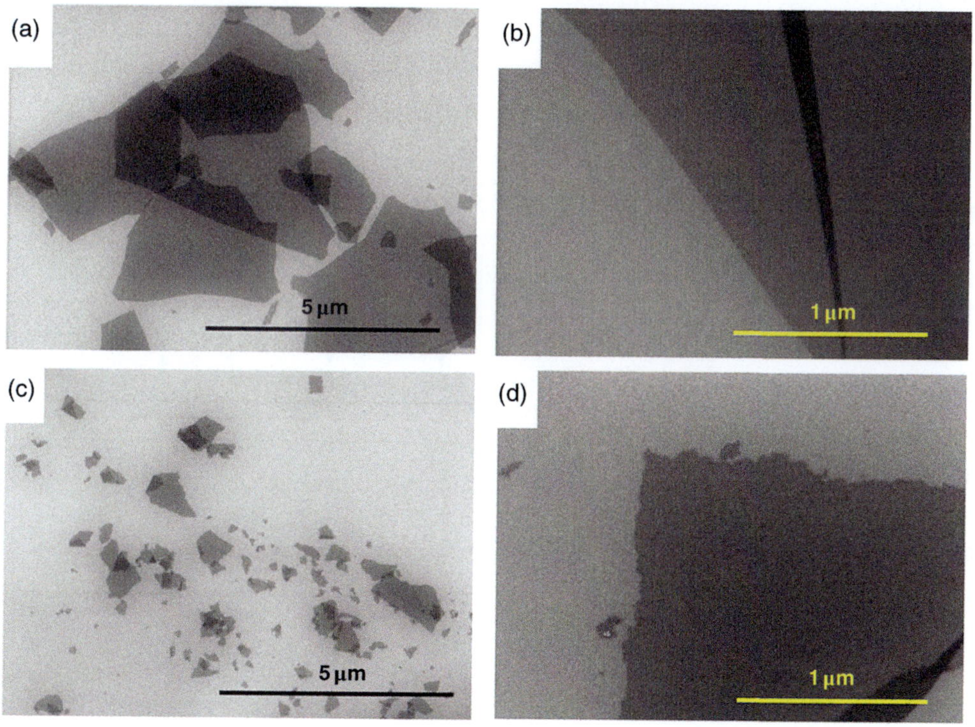

Figure 3.23 SEM images of GO flakes demonstrating the impact of base treatment on flakes' morphology. (a,b) Typical GO flakes retain their integrity and have smooth edges. (c,d) GO flakes damaged by base treatment and consecutive sonication for 20 s. These flakes exhibit smaller size and jagged edges. Reproduced from [46] with permission of Elsevier

3.5.2 Atomic Force Microscopy

AFM is based on continuous scanning of the specimen surface by the probing tip attached to the cantilever, while the cantilever is moving across the surface in the x and y directions. When the tip is brought into the proximity of a sample surface, forces between the tip and the sample lead to a deflection of the cantilever. Typically, the deflection is measured using a laser spot reflected from the top surface of the cantilever onto a photodiode. The AFM image is formed as a color-mapped signal obtained from the tip/cantilever as a function of the tip's x–y position. The lateral resolution of the AFM is related to the tip curvature diameter, which is currently on the order of 10 nm. In the z direction, the resolution is in the fractions of nanometers.

With respect to graphene and GO-related studies, AFM is normally employed to obtain the thickness of the flakes. When on a substrate surface, the thicknesses of monolayer GO flakes are normally within the 0.8–1.2 nm interval. These values are slightly higher than the 0.6–0.9 nm repeat distance obtained from the X-ray diffraction method. Unlike SEM,

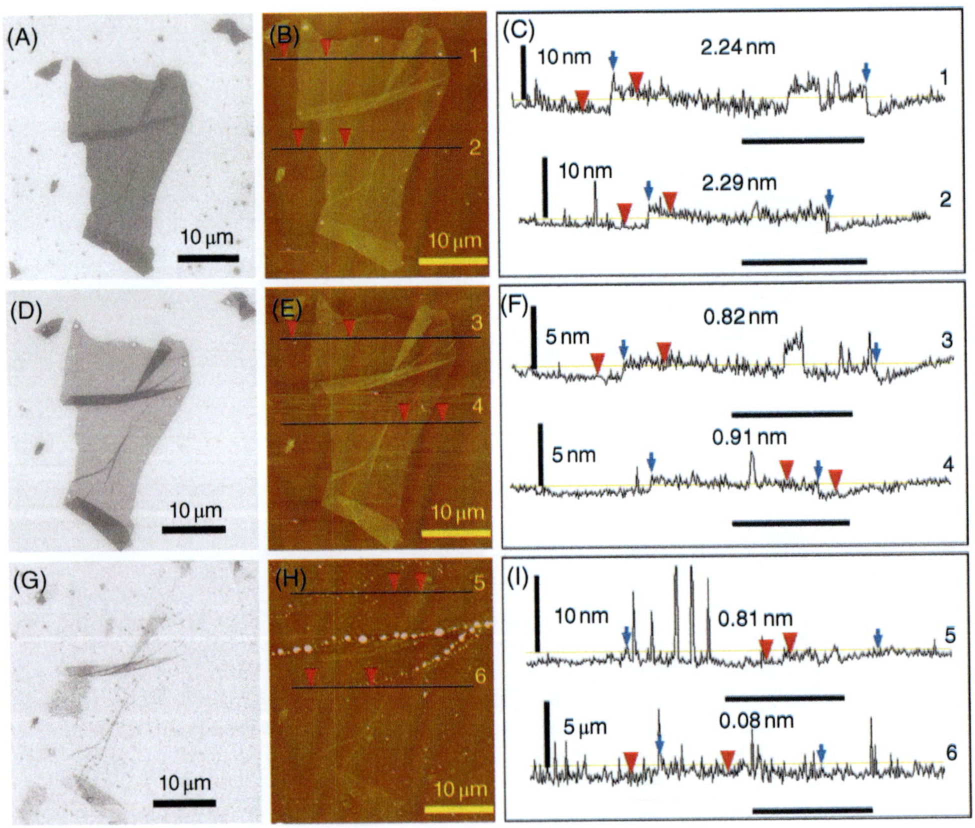

Figure 3.24 (A, D, G) SEM images and (B, E, H) corresponding AFM images of a GO flake. (C, F, I) Height profiles corresponding to the respective images (B, E, H). The image demonstrates removal of a single GO layer from an original bilayer flake. Reproduced from [47] with permission of AAA

acquisition of AFM images takes from several minutes through several tens of minutes. Also, most of the current atomic force microscopes require lengthy and complex alignment procedures. AFM also does not allow one to quickly move along the substrate to spot desired objects, as one can do with SEM. Nevertheless, AFM is the only method to obtain the exact flake thickness. Thus, AFM was used in combination with SEM to confirm the removal of a single layer of GO from multi-layered GO samples (Figure 3.24) [46]. Figure 3.24 demonstrates that, when GO is less than three layers, the number of layers can be obtained from the SEM images by the flake's opacity. In addition to the traditional height (used to obtain flake thickness) and phase modes, modern AFM instruments allow additional functional modes such as adhesion, surface potential, electrical conductivity, etc. The size of the oxidized and graphenic domains on GO is on the order of 2–4 nm. This is smaller than the tip diameter. This is why, unfortunately, AFM cannot resolve two types of domain on the GO platform.

3.5.3 Transmission Electron Microscopy

TEM is based on detecting the electrons transmitted through a thin specimen. It is a very convenient method for 2D materials such as graphene and GO, because the specimens can be easily prepared by suspending flakes on a TEM grid.

However, the use of TEM for GO-related studies is relatively limited. This is because of the relatively small viewing area in TEM, allowing one to view one single flake at a time. This is a disadvantage when compared to SEM, which allows one to image numerous flakes simultaneously, which is advantageous for comparison purposes and for statistical analysis. TEM might be beneficial for studying nanocomposite materials such as GO impregnated with nanoparticles. Since single-layer GO is completely transparent in the electron beam, good contrast between the GO background and nanoparticles can be obtained. See Figure 3.25 as an example.

3.5.4 High-Resolution Transmission Electron Microscopy

Due to the significantly higher magnification allowed by high-resolution transmission electron microscopy (HRTEM), this method affords numerous possibilities as compared to regular TEM. The level of resolution, allowing single atoms or groups of atoms to be seen, provides numerous details that are valuable for understanding GO structure. The first aberration-corrected HRTEM images for RGO (reduced under hydrogen plasma) were reported by Gómez-Navarro *et al.* [49] in 2010 (Figures 3.26 and 3.27). The reported images help to understand the actual chemical structure of GO. Below, we briefly discuss the reported images in relation to the fine chemical structure of GO discussed in Chapter 2.

It is apparent that the largest portion of the carbon layer comprises clean crystalline graphenic areas where the hexagonal lattice is clearly observed (light gray color in Figure 3.26). The average size of the visible well-crystallized areas is from 3 to 6 nm, and statistics reveal that they cover ~60% of the RGO plane. In part there are point defects visible within those areas of graphenic regions. In addition to graphenic areas, a significant amount of topologically defective areas that appear as quasi-amorphous single-layer carbon structures (marked in blue in Figure 3.26b) can also be observed. These extended defects incorporate a large number of carbon pentagons, heptagons and rotated hexagons in a limited nanometer-sized area. Nevertheless, all the carbon

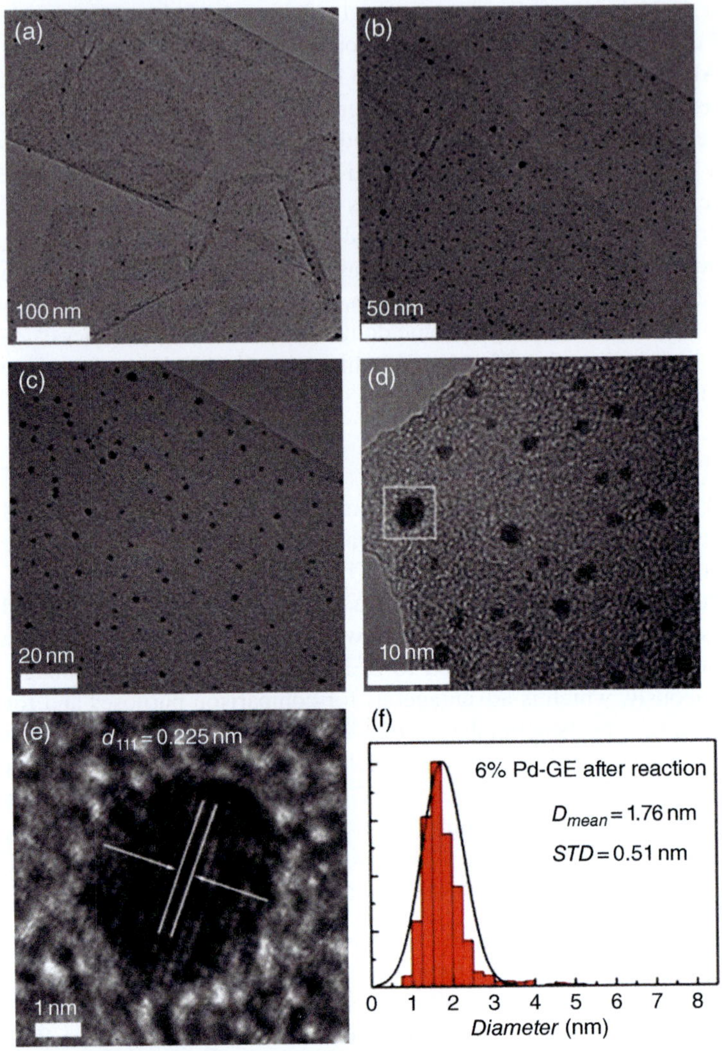

Figure 3.25 TEM images of GO–Pd nanoparticle (NP) composite. The tiny dark spots on panels (a)–(d) are Pd NPs. (e) The HRTEM image reveals the crystalline structure of a Pd NP. (f) Histogram of particle size distribution. Reproduced from [48] with permission of Elsevier

atoms in these areas are bonded to three neighbors in a planar sp^2 configuration. The extended topological defects cover ~5% of the surface and exhibit typical sizes of 1–2 nm in diameter.

Surprisingly, despite the presence of such a significant amount of topological defects, the long-range orientation of the graphene plane is maintained. This long-range order is partially preserved even within the areas color-coded as topological defects.

The defective areas are bordered by black curved lines. These are most likely the points of C–C bond cleavage (Figure 3.27d). The edges of the black lines have a higher density of

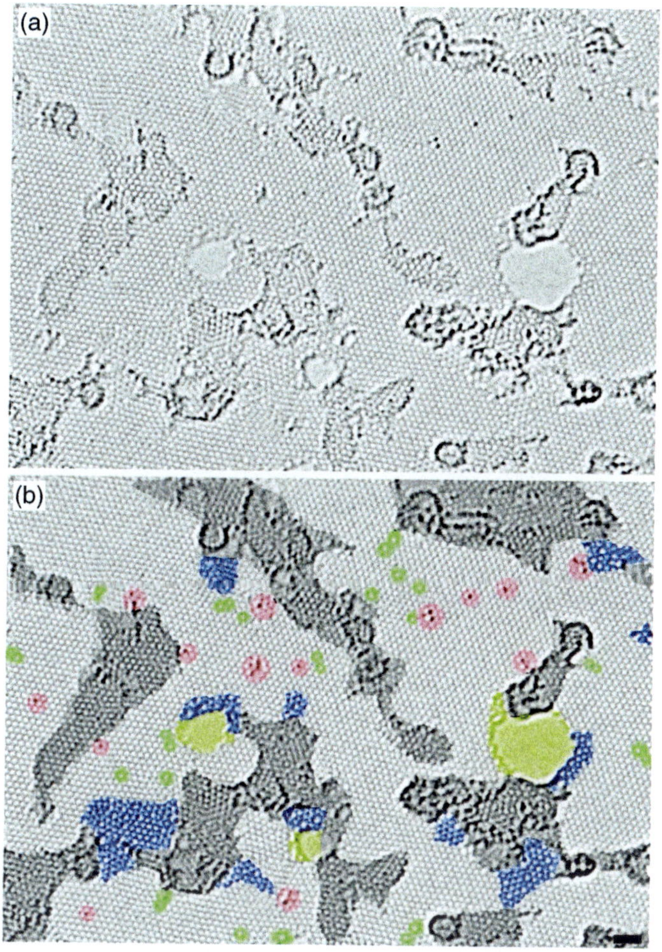

Figure 3.26 Atomic-resolution, aberration-corrected TEM image of a single-layer RGO. (a) Original image and (b) the same image color-coded to highlight different features. The defect-free crystalline graphene area is displayed in the original light-gray color. Defective areas are shaded in dark gray. Blue-colored regions are the disordered single-layer carbon networks, or extended topological defects, identified as remnants of the oxidation–reduction process. Red-colored areas highlight individual adatoms or substitutions. Green-colored areas indicate isolated topological defects, which are single bond rotations or dislocation cores. Holes and their edge reconstructions are colored in yellow. Scale bar (bottom right) is 1 nm. Reproduced from [49] with permission of ACS

white-colored clusters, presumably remaining oxygen atoms. Thus, the points of C–C bond cleavage are terminated by oxygen functional groups.

Note that the RGO sample analyzed in this study is notably less defective than that in the similar work by Erickson *et al.* [50] (see Figure 2.9). Thus, more HRTEM studies are needed to gain more knowledge about the actual structure of GO and RGO.

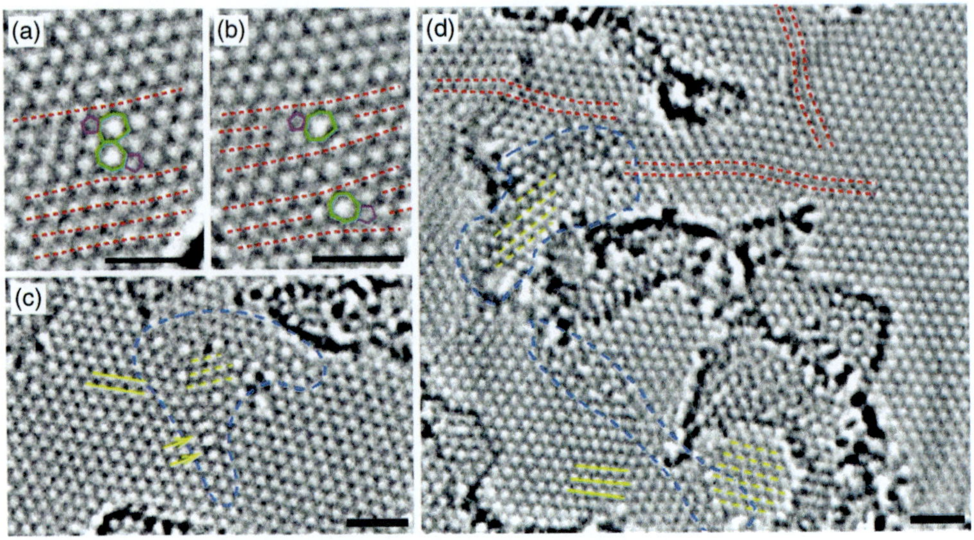

Figure 3.27 (a,b) Pentagon–heptagon dipole observed at two different times (2 min apart) in the TEM. The separation between the dislocation pairs has increased, indicating mobility of the carbon grid, and a significant amount of strain causing the separation. (c,d) Additional defect clusters, indicated by blue dashed lines. Yellow dashed lines indicate regions with a hexagonal lattice rotated to the dominant orientation (yellow solid lines for comparison). Red dashed lines indicate distortions in the hexagonal lattice. Yellow arrows in (c) indicate strongly elongated carbon polygons. All scale bars are 1 nm. Reproduced from [49] with permission of ACS

References

[1] Kolodziejski, W.; Klinowski, J., Kinetics of cross-polarization in solid-state NMR: a guide for chemists. *Chem. Rev.* **2002**, *102* (3), 613–628.
[2] Mermoux, M.; Chabre, Y.; Rousseau, A., FTIR and ^{13}C NMR study of graphite oxide. *Carbon* **1991**, *29* (3), 469–474.
[3] Lerf, A.; He, H.; Forster, M.; Klinowski, J., Structure of graphite oxide revisited. *J. Phys. Chem. B* **1998**, *102* (23), 4477–4482.
[4] He, H.; Riedl, T.; Lerf, A.; Klinowski, J., Solid-state NMR studies of the structure of graphite oxide. *J. Phys. Chem.* **1996**, *100*, 19954–19958.
[5] Lerf, A.; Heb, H.; Riedl, T.; *et al.*, ^{13}C and ^{1}H MAS NMR studies of graphite oxide and its chemically modified derivatives. *Solid State Ionics* **1997**, *101–103* (2), 857–862.
[6] He, H.; Klinowski, J.; Forster, M.; Lerf, A., A new structural model for graphite oxide. *Chem. Phys. Lett.* **1998**, *287*, 53–56.
[7] Gao, W.; Alemany, L.B.; Ci, L.; Ajayan, P.M., New insights into the structure and reduction of graphite oxide. *Nature Chem.* **2009**, *1* (5), 403–408.
[8] Casabianca, L.B.; Shaibat, M.A.; Cai, W.W.; *et al.*, NMR-based structural modeling of graphite oxide using multidimensional ^{13}C solid-state NMR and ab initio chemical shift calculations. *J. Am. Chem. Soc.* **2010**, *132* (16), 5672–5676.
[9] Cai, W.; Piner, R.D.; Stadermann, F.J.; *et al.*, Synthesis and solid-state NMR structural characterization of ^{13}C-labeled graphite oxide. *Science* **2008**, *321* (5897), 1815–1817.
[10] Szabó, T.; Berkesi, O.; Dékány, I., DRIFT study of deuterium-exchanged graphite oxide. *Carbon* **2005**, *43* (15), 3186–3189.

[11] Szabó, T.; Berkesi, O.; Forgó, P.; *et al.*, Evolution of surface functional groups in a series of progressively oxidized graphite oxides. *Chem. Mater.* **2006**, *18*, 2740–2749.

[12] Dimiev, A.M.; Alemany, L.B.; Tour, J.M., Graphene oxide. Origin of acidity, its instability in water, and a new dynamic structural model. *ACS Nano* **2013**, *7* (1), 576–588.

[13] Dimiev, A.; Kosynkin, D.V.; Alemany, L.B.; *et al.*, Pristine graphite oxide. *J. Am. Chem. Soc.* **2012**, *134* (5), 2815–2822.

[14] Pendolino, F.; Parisini, E.; Lo Russo, S., Time-dependent structure and solubilization kinetics of graphene oxide in methanol and water dispersions. *J. Phys. Chem. C* **2014**, *118* (48), 28162–28169.

[15] Seredych, M.; Bandosz, T.J., Combined role of water and surface chemistry in reactive adsorption of ammonia on graphite oxides. *Langmuir* **2010**, *26* (8), 5491–5498.

[16] Eigler, S.; Dotzer, C.; Hof, F.; *et al.*, Sulfur species in graphene oxide. *Chem. Eur. J.* **2013**, *19* (29), 9490–9496.

[17] Hontoria-Lucas, C.; López-Peinando, A.J.; López-González, J.D.D.; *et al.*, Study of oxygen-containing groups in a series of graphite oxides: physical and chemical characterization. *Carbon* **1995**, *33* (11), 1585–1592.

[18] Ambrosi, A.; Chua, C.K.; Bonanni, A.; Pumera, M., Lithium aluminum hydride as reducing agent for chemically reduced graphene oxides. *Chem. Mater.* **2012**, *24* (12), 2292–2298.

[19] Yang, D.Q.; Sacher, E., s–p hybridization in highly oriented pyrolytic graphite and its change on surface modification, as studied by X-ray photoelectron and Raman spectroscopies. *Surf. Sci.* **2002**, *504*, 125–137.

[20] Briggs, D., in *Handbook of X-ray Photoelectron Spectroscopy*, vol. 3, eds C.D. Wagner, W.M. Riggs, L.E. Davis, J.F. Moulder and G.E. Muilenberg. Perkin-Elmer Corp., Eden Prairie, MN, **1979**.

[21] Hofflund, G.B., Spectroscopic techniques: X-ray photoelectron spectroscopy (XPS), Auger electron spectroscopy (AES) and ion scattering spectroscopy (ISS). In *Handbook of Surface and Interface Analysis*. Marcel Dekker, New York, **1998**.

[22] Chiu, P.L.; Mastrogiovanni, D.D.; Wei, D.; *et al.*, Microwave- and nitronium ion-enabled rapid and direct production of highly conductive low-oxygen graphene. *J. Am. Chem. Soc.* **2012**, *134* (13), 5850–5856.

[23] Ganguly, A.; Sharma, S.; Papakonstantinou, P.; Hamilton, J., Probing the thermal deoxygenation of graphene oxide using high-resolution in situ X-ray-based spectroscopies. *J. Phys. Chem. C* **2011**, *115* (34), 17009–17019.

[24] Banwell, C.N.; McCash, E.M., *Fundamentals of Molecular Spectroscopy*. McGraw-Hill, London, **1983**.

[25] Bae, S.; Kim, H.; Lee, Y.; *et al.*, Roll-to-roll production of 30-inch graphene films for transparent electrodes. *Nature Nanotechnol.* **2010**, *5* (8), 574–578.

[26] Englert, J.M.; Vecera, P.; Knirsch, K.C.; *et al.*, Scanning-Raman-microscopy for the statistical analysis of covalently functionalized graphene. *ACS Nano* **2013**, *7* (6), 5472–5482.

[27] Stadler, J.; Schmid, T.; Zenobi, R., Nanoscale chemical imaging of single-layer graphene. *ACS Nano* **2011**, *5* (10), 8442–8448.

[28] Ferrari, A.C.; Meyer, J.C.; Scardaci, V.; *et al.*, Raman spectrum of graphene and graphene layers. *Phys. Rev. Lett.* **2006**, *97* (18), 187401.

[29] Graf, D.; Molitor, F.; Ensslin, K.; *et al.*, Spatially resolved Raman spectroscopy of single- and few-layer graphene. *Nano Lett.* **2007**, *7* (2), 238–242.

[30] Hoffmann, G.G.; de With, G.; Loos, J., Micro-Raman and tip-enhanced Raman spectroscopy of carbon allotropes. *Macromol. Symp.* **2008**, *265* (1), 1–11.

[31] Cançado, L.G.; Jorio, A.; Ferreira, E.H.M.; *et al.*, Quantifying defects in graphene via Raman spectroscopy at different excitation energies. *Nano Lett.* **2011**, *11* (8), 3190–3196.

[32] Koehler, F.M.; Jacobsen, A.; Ensslin, K.; *et al.*, Selective chemical modification of graphene surfaces: distinction between single- and bilayer graphene. *Small* **2010**, *6* (10), 1125–1130.

[33] Ni, Z.H.; Yu, T.; Lu, Y.H.; *et al.*, Uniaxial strain on graphene: Raman spectroscopy study and band-gap opening. *ACS Nano* **2008**, *2* (11), 2301–2305.

[34] Wang, Y.Y.; Ni, Z.H.; Yu, T.; *et al.*, Raman studies of monolayer graphene: the substrate effect. *J. Phys. Chem. C* **2008**, *112* (29), 10637–10640.

[35] Das, A.; Pisana, S.; Chakraborty, B.; *et al.*, Monitoring dopants by Raman scattering in an electrochemically top-gated graphene transistor. *Nature Nanotechnol.* **2008**, *3* (4), 210–215.
[36] Casiraghi, C.; Hartschuh, A.; Qian, H.; *et al.*, Raman spectroscopy of graphene edges. *Nano Lett.* **2009**, *9* (4), 1433–1441.
[37] Ferrari, A.C.; Basko, D.M., Raman spectroscopy as a versatile tool for studying the properties of graphene. *Nature Nanotechnol.* **2013**, *8* (4), 235–246.
[38] Lucchese, M.M.; Stavale, F.; Ferreira, E.H.M.; *et al.*, Quantifying ion-induced defects and Raman relaxation length in graphene. *Carbon* **2010**, *48* (5), 1592–1597.
[39] Eigler, S.; Hof, F.; Enzelberger-Heim, M.; *et al.*, Statistical-Raman-microscopy and atomic-force-microscopy on heterogeneous graphene obtained after reduction of graphene oxide. *J. Phys. Chem. C* **2014**, *118* (14), 7698–7704.
[40] Eigler, S., Mechanistic insights into the reduction of graphene oxide addressing its surfaces. *Phys. Chem. Chem. Phys.* **2014**, *16* (37), 19832–19835.
[41] Eigler, S.; Hirsch, A., Chemistry with graphene and graphene oxide – challenges for synthetic chemists. *Angew. Chem. Int. Ed.* **2014**, *53* (30), 7720–7738.
[42] Eigler, S., Graphite sulphate – a precursor to graphene. *Chem. Commun.* **2015**, *51* (15), 3162–3165.
[43] Eigler, S.; Dotzer, C.; Hirsch, A., Visualization of defect densities in reduced graphene oxide. *Carbon* **2012**, *50* (10), 3666–3673.
[44] Eigler, S.; Enzelberger-Heim, M.; Grimm, S.; *et al.*, Wet chemical synthesis of graphene. *Adv. Mater.* **2013**, *25* (26), 3583–3587.
[45] Pollard, A.J.; Brennan, B.; Stec, H.; *et al.*, Quantitative characterization of defect size in graphene using Raman spectroscopy. *Appl. Phys. Lett.* **2014**, *105* (25), 253107.
[46] Dimiev, A.M.; Polson, T.A., Contesting the two-component structural model of graphene oxide and reexamining the chemistry of graphene oxide in basic media. *Carbon* **2015**, *93*, 544–554.
[47] Dimiev, A.; Kosynkin, D.V.; Sinitskii, A.; *et al.*, Layer-by-layer removal of graphene for device patterning. *Science* **2011**, *331* (6021), 1168–1172.
[48] Li, Y.; Yu, Y.; Wang, J.-G.; *et al.*, CO oxidation over graphene supported palladium catalyst. *Appl. Catal. B: Environm.* **2012**, *125*, 189–196.
[49] Gómez-Navarro, C.; Meyer, J.C.; Sundaram, R.S.; *et al.*, Atomic structure of reduced graphene oxide. *Nano Lett.* **2010**, *10* (4), 1144–1148.
[50] Erickson, K.; Erni, R.; Lee, Z.; *et al.*, Determination of the local chemical structure of graphene oxide and reduced graphene oxide. *Adv. Mater.* **2010**, *22* (40), 4467–4472.

4

Rheology of Graphene Oxide Dispersions

Cristina Vallés

4.1 Liquid Crystalline Behaviour of Graphene Oxide Dispersions

4.1.1 Liquid Crystals and Onsager's Theory

Anisotropic molecules or particles in fluid states (i.e. in liquid media) have been demonstrated to be capable of forming highly ordered structures at the mesoscopic scale, which are known as liquid crystalline phases. Increasing contents of anisotropic particles in liquid media lead to an increasing restriction of the free rotation of the particles. As their free rotation becomes more and more difficult, the particles tend to organize themselves, forming ordered structures [1], which are known as colloidal liquid crystals (LCs). Colloidal liquid crystals attract a great deal of interest and attention due to their potential in the fabrication of highly ordered structures from molecular to higher levels [2] as well as their potential use in display devices [2–5], for which the processing of these well-organized materials is extremely important.

In 1940s Lars Onsager predicted that colloidal suspensions of one-dimensional (1D) rod-like particles and two-dimensional (2D) plate-like particles (hard platelets) could go from a disordered isotropic phase to an ordered nematic phase when the concentration of the 1D rod-like particles or, alternatively, the 2D hard platelets reached a critical concentration [6]. When the concentration of rod- or plate-like particles is low, there is no interaction between them and they can rotate freely. With increasing concentrations, the free rotation of the particles is more and more restricted due to the exclusion of free volume, and the rods or platelets are thus forced to orient and organize themselves going towards a nematic order [1, 6–8]. This is schematically represented for 2D platelets in Figure 4.1.

Graphene Oxide: Fundamentals and Applications, First Edition. Edited by Ayrat M. Dimiev and Siegfried Eigler.

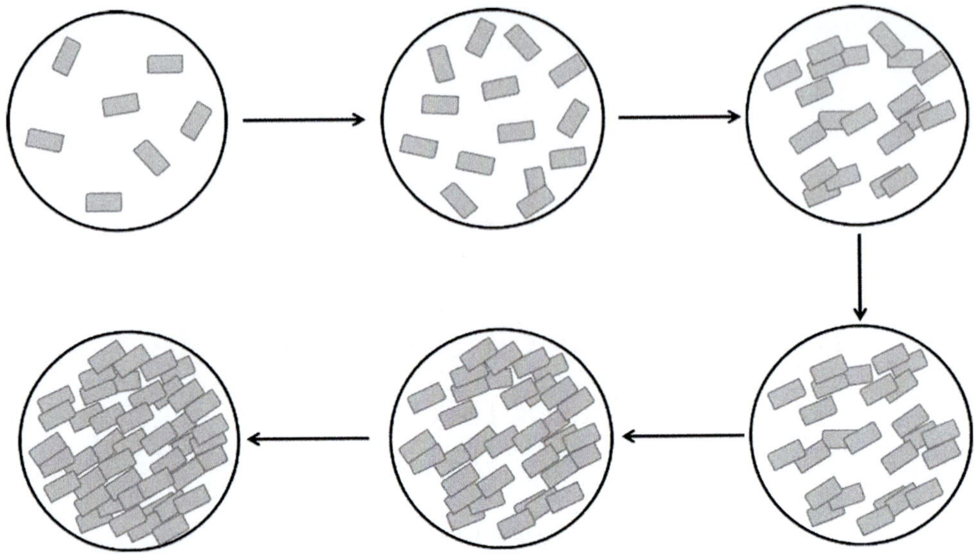

Figure 4.1 Schematic representation of the formation of highly ordered structures of 2D anisotropic platelets in liquid media when the particle concentration is increased. Increasing contents of anisotropic particles in liquid media leads to an increasing restriction of their free rotation. As a consequence of this restricted free rotation, the particles tend to organize themselves, forming highly ordered structures, which are known as colloidal liquid crystals (LCs)

Onsager's theory also predicted a critical concentration in colloidal dispersions of rod-like or plate-like particles in liquid media for the transition from isotropic to nematic phase, in which both phases coexist (called a biphase). This critical concentration strongly depends on the aspect ratio of the rods (length/diameter) or platelets (diameter/thickness) [6–11].

4.1.2 Nematic Phases in Carbon Nanomaterials

In particular, the formation of lyotropic LC phases (nematic phases) has been observed for a wide range of 1D and 2D anisotropic nanoparticles in the concentrated regime, such as metal nanorods [13], clays [14], protein fibres [15], carbon nanotubes (CNTs) [16, 17], inorganic nanosheets [18] and graphene [7–10, 19, 20], resulting in highly organized (aligned) nanostructures. These highly ordered carbonaceous mesophases raise a particular research interest as they represent effective routes to functional carbon-based materials [21–26].

Among these carbonaceous nanomaterials, graphene oxide (GO) is particularly interesting, as it offers some important advantages with respect to other carbon nanomaterials. Graphite oxide is a promising precursor for the bulk production of graphene-based materials. It is a highly oxidized layered material that exfoliates into ultra-thin layers of GO when it is in contact with water [7–10, 19, 20] and some organic solvents [27–29]. GO sheets are not electrically conductive due to the high amount of defects and oxygen-containing functional groups attached to the surface of the flakes. They can, however, be converted back to highly conductive graphene through chemical reduction [30] and thermal annealing [31], or alternatively through a simple base washing, which has been demonstrated to remove a high amount of oxidative debris attached to the surface to the graphene flakes [32]. In addition, GO sheets have an extremely high aspect ratio with monoatomic thickness and a lateral size of tens of micrometres [7, 8, 33], which is essential for the formation of highly ordered LCs. Aspect ratios as high as 10 000 have been reported for GO sheets [8], which are higher than those previously observed for CNTs [21, 23], putting GO materials in a better position to form nematic phases and highly ordered carbonaceous structures relative to CNTs.

For the formation of LC phases from dispersions, good colloidal stabilities of high concentrations of highly anisotropic particles are a strong requirement [10, 34]. GO can reach very good colloidal stabilities in water as a consequence of the GO sheets being negatively charged (highly oxidized graphene sheets), which can be explained using the classic theory of Derjaguin, Landau, Verwey and Overbeek (DLVO) [34, 35]. The existence of negative charges on the surface of the GO sheets generates large repulsion forces between the negatively charged colloids in water, which work to avoid the aggregation of colloidal particles through strong π–π and van der Waals interactions. Recently, lyotropic LC phases have been identified in dispersions of GO and reduced GO in water [7–10, 20]. Furthermore, Jalili *et al.* [28] and Xu *et al.* [29] reported the formation of nematic LC phases from dispersions of GO in organic solvents, with the transition from isotropic to nematic phase occurring at a very low critical concentration of GO sheets (~0.1 wt.%), strongly depending on the aspect ratio of the GO sheets (diameter/thickness ratio) in agreement with Onsager's predictions. The formation of these lyotropic LC phases is evidenced by the appearance of a birefringence under polarized light, which means that the light is refracted in two different directions (as a consequence of the anisotropy of the LC phases) and it can be easily seen using a polarized optical microscopy. As an example, Figure 4.2 shows that, for very low concentrations of GO sheets, no birefringence is observed due to a random

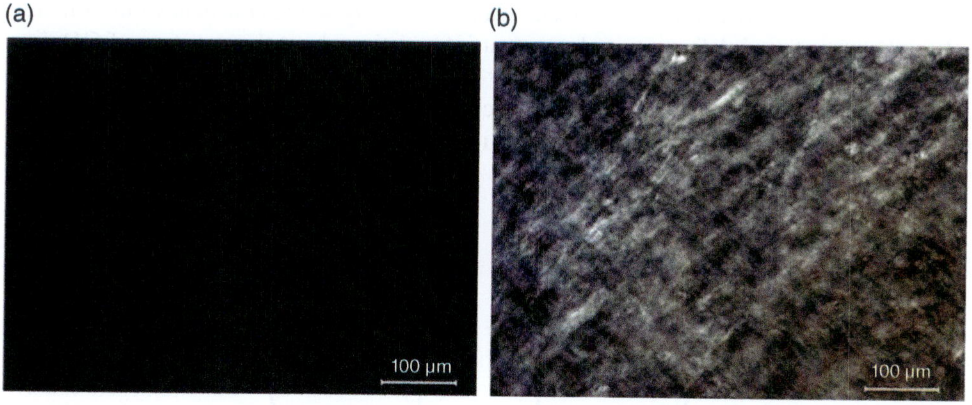

Figure 4.2 Polarized light optical micrographs of (a) GO sheets randomly oriented in a very dilute dispersion, and (b) GO sheets forming an LC phase when the concentration is increased, showing the appearance of birefringence under polarized light when, at a higher concentration of particles, they organize themselves, forming a nematic LC phase. Reproduced from [12] with permission of RSC

orientation of the sheets. When the concentration of GO sheets is high enough, a nematic LC phase is formed, which is evidenced by the appearance of birefringence under polarized light. This finding represents the possibility to design and manufacture aligned graphene-based macrostructures, such as aligned fibres or layered composites [7–10, 20, 36–40].

The exceptional mechanical properties (stiffness and strength), high functionality and ultra-high aspect ratio, combined with the recently found capacity to form LC phases not only in water but also in some organic solvents leading to perfectly aligned structures, makes GO a very promising material to be used in the fabrication of highly ordered, strong and conductive graphene-based structures [34, 41].

4.2 Rheological Behaviour of Aqueous Dispersions of LC-GO

In order to understand the potential of GO materials to form graphene-based highly ordered structures and become capable to exploit their industrial applications, their rheological properties need to be evaluated. Due to its great importance, the rheological behaviour of aqueous dispersions of GO in both the diluted and the concentrated regime has been recently investigated and reported [12, 42, 43]. They are studied as a model system to understand and predict the influence of graphene on the rheological properties and processing of more complex systems (with different and more complex interactions between the graphene and either solvent or polymeric matrix). It is also of great importance to understand how GO dispersions can be processed in order to develop fabrication protocols for GO dispersions using different processing techniques. The advantage of using water as the dispersive medium is that it is a Newtonian fluid, which makes the effect of the graphene obvious, eliminating any other additional effects, and that aqueous GO dispersions are highly stable up to high concentrations.

The following subsections describe the typical rheological behaviour of GO dispersions both under dynamic shear and under steady shear. Rheology studies the behaviour of materials under an external deformation. Important information about the structure of GO dispersions can be obtained by studying how they behave under dynamic and steady shear (i.e. sliding deformation). Figure 4.3 describes schematically the dynamic and steady shear deformation applied to the GO dispersions to investigate their structure.

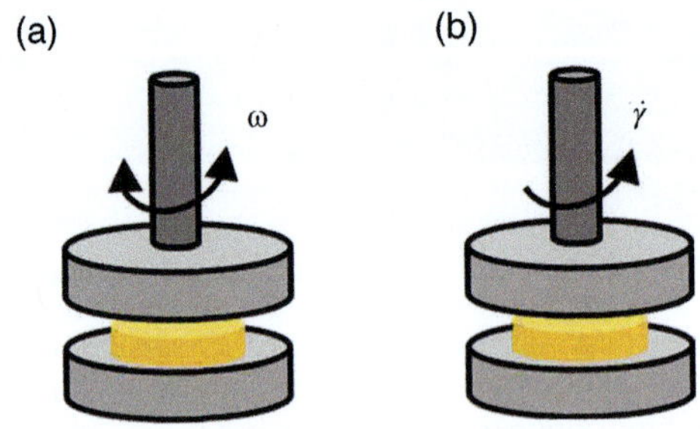

Figure 4.3 *Schematic representation of (a) dynamic and (b) steady shear measurements, showing how the samples are deformed (sliding deformation)*

4.2.1 Dynamic Shear Properties

A typical dynamic shear experiment consists on the application of an increasing strain (or, alternatively, stress) on the sample while keeping constant the frequency (ω), as shown in Figure 4.3(a). It allows one to follow the variation of the storage modulus (G') and loss modulus (G'') with increasing strain/stress, which gives information about the viscoelastic behaviour of the GO network in our case, with G' and G'' representing the elastic and the viscous part of the network, respectively.

Aqueous GO dispersions have been found to show the typical behaviour for space-filling networks of particles. Figure 4.4 shows the typical dynamic strain sweep measurements performed on dispersions of GO at different concentrations from 0.1 up to 8 vol.% conducted at a constant frequency of 1 rad s^{-1} and increasing strain amplitudes. Due to the existence of electrostatic interactions between them, the GO flakes form an elastic network in the dispersive medium, which shows viscoelastic behaviour under dynamic shear up to a critical strain. Within this region, which we refer to as the 'linear viscoelastic region' (LVR), the values for the storage (G') and loss (G'') moduli are independent of the applied strain, with G' being higher than G'' (Figure 4.4).

The LVR ends at a critical strain, above which the network of GO flakes breaks, G' and G'' cross and become highly strain-sensitive (rapidly decreasing with increasing strains). At this critical strain the dispersion loses its viscoelastic properties and it represents the transition from solid-like behaviour to liquid-like behaviour. As shown in Figure 4.4, the critical strain varies between 0.1% and 10% strain depending on the concentration of GO flakes [42]. The solid–liquid transition region, which is entered when G' and G'' cross, tends to occur typically at higher strains for higher GO concentrations. For the high concentrations,

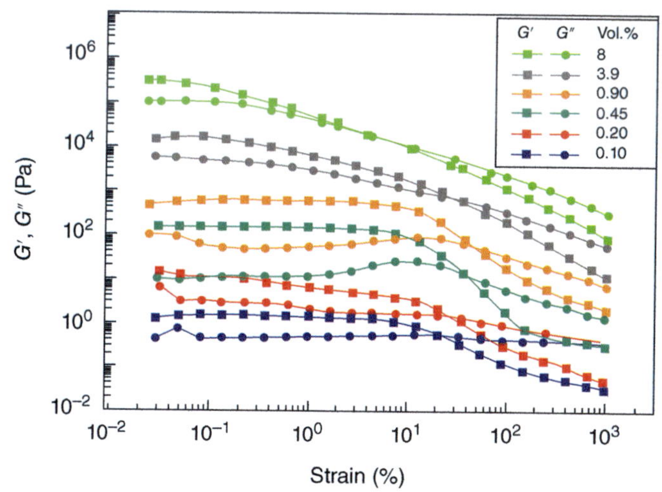

Figure 4.4 Typical dynamic strain sweeps on GO dispersions at different concentrations conducted at a constant frequency (1 rad s^{-1}) and increasing strain amplitude [42], showing the presence of a LVR (with G′ and G″ independent of strain and G′ > G″) up to a critical strain, which strongly depends on the concentration. Above the critical strain, a liquid-like behaviour region is entered. Reproduced from [42] with permission of Springer

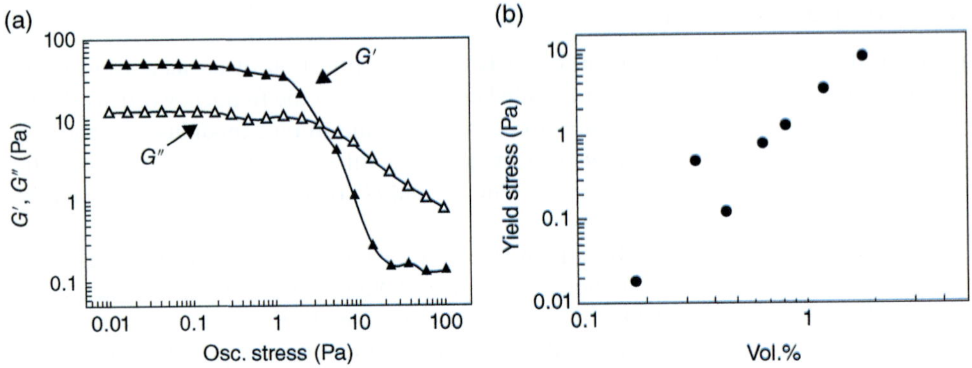

Figure 4.5 (a) Typical dynamic stress sweeps for a 0.83 vol.% GO dispersion, showing the LVR up to a critical stress, above which the two moduli cross and the liquid-like behaviour region is entered. (b) Variation of the yield stress with concentration of GO, showing a minimum at 0.45 vol.%, which represents an orientation of the sheets [43]. Reproduced from [43] with permission of Elsevier

the crossover between the two moduli occurred at a lower strain than expected, which is a consequence of the sample forming a gel (orientation, alignment of the flakes).

Similar results have been found under dynamic shear when a stress is applied to the dispersions instead of a strain (stress sweep measurements, shown in Figure 4.5a). Both G' and G'' moduli are constant with stress up to a critical stress of 1.27 Pa for a liquid crystalline dispersion of GO flakes with a concentration of 1.2 vol.% [43]. Similarly to what happens above the critical strain, above the critical stress both G' and G'' moduli become strongly stress-dependent. The G' modulus begins to decrease with increasing stress due to the rupture of the viscoelastic GO network and eventually crosses G'', leaving the solid-like and entering the liquid-like behaviour region. Similarly to the critical strain, the critical stress strongly depends on the concentration of GO. Figure 4.5(b) shows the variation of the critical stress with concentration of GO. It can be seen that, with increasing concentration, the yield stress first moves to high values (concentration = 0.33 vol.%), then decreases at a concentration of 0.45 vol.% and finally increases again to higher values [43].

This minimum observed at 0.45 vol.% corresponds to an alignment or orientation of the GO sheets, considerably reducing their entanglement. This variation of the yield stress with the concentration of GO flakes is similar to those previously reported for discotic laponite gels and concentrated yield stress fluids [44]. Dynamic frequency sweeps conducted at a constant strain within the LVR give information about the structure of the dispersions of GO before the rupture of the GO elastic network, as they allow one to monitor the variation of both G' and G'' moduli with increasing frequency. Typical dynamic frequency sweeps for different concentrations of GO dispersions are represented in Figure 4.6. Within the LVR the behaviour of the GO aqueous dispersions under dynamic frequency sweeps strongly depends on the concentration:

i. For very low concentrations of GO (~0.03 vol.%), the dispersions behave like a typical fluid, in which the particles do not interact between themselves. The moduli can be observed to increase with increasing frequency.

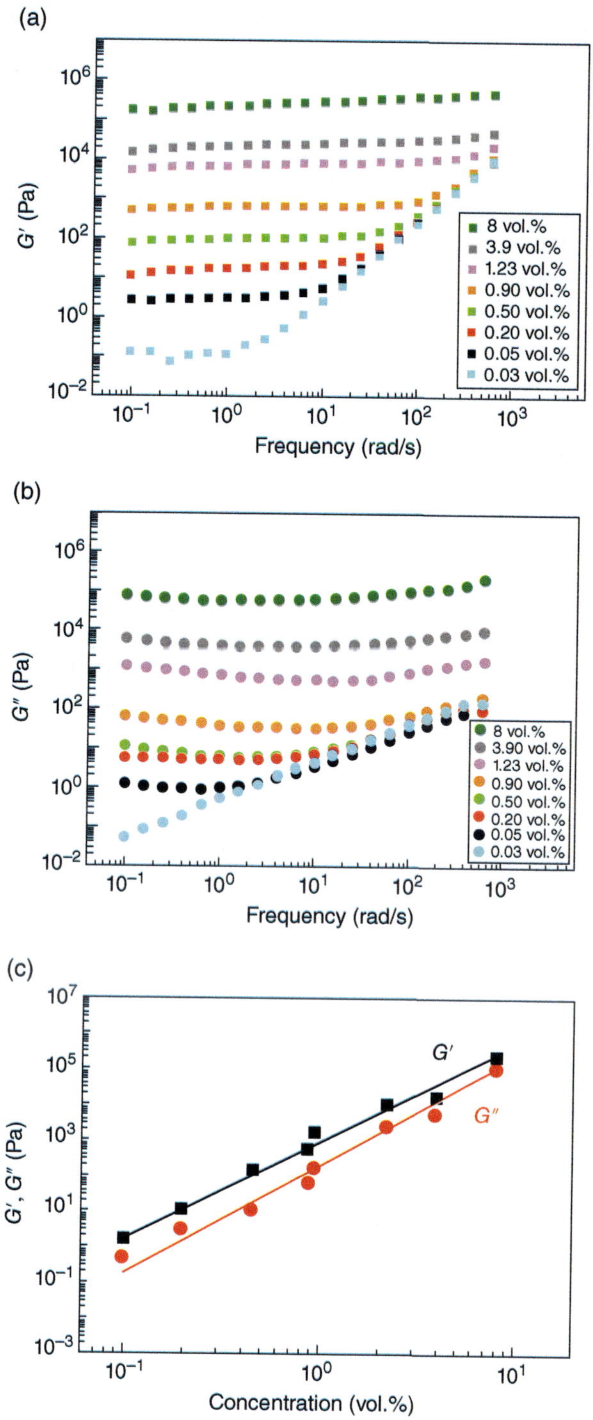

Figure 4.6 *Dynamic frequency sweep on aqueous GO dispersions conducted within the LVR (at a constant strain amplitude of 0.1%). Variation of (a) G′ and (b) G″ with increasing frequencies, showing different behaviours depending on the GO concentration. (c) Values of the moduli G′ and G″ within the LVR at a constant frequency of 0.1 rad* s^{-1} *as a function of the concentration. Reproduced from [42] with permission of Springer*

ii. At medium concentrations of GO (0.03–1.2 vol.%), an intermediate response between fluid-like and gel-like behaviour is found. At low frequencies, G' and G'' are independent of the frequency, with G' being larger than G'', which corresponds to the GO typical solid-like behaviour of an elastic structure. At higher frequencies though, G' and G'' become frequency-dependent, suggesting the rupture of the GO structure and the entry into a liquid-like behaviour region.
iii. At concentrations above 1.2 vol.%, the GO system behaves like a classic strong-gel spectrum, with G' and G'' being independent of the applied frequency in all the frequency range.

Sometimes, a viscoelastic soft solid behaviour can be additionally defined between liquid behaviour and gel behaviour. The dynamic shear behaviour of GO 'ultra-large' flakes (aspect ratio = 45 000) has recently been described as being composed of four different regions depending on the concentration: a viscoelastic liquid, a transition state from viscoelastic liquid to viscoelastic soft solid, a viscoelastic soft solid (behaving like a solid below the yield stress while flowing readily above the yield stress [45]), and a viscoelastic gel [12]. The viscoelastic gel behaviour is observed at concentrations above 0.83 vol.% of 'ultra-large' GO flakes (aspect ratio = 45 000), which is slightly lower than that observed for smaller GO flakes (aspect ratio = 800) at a concentration of ~1.2 vol.%. Independently of the aspect ratio, GO dispersions typically behave like a percolated network, with the low-frequency moduli values (G' and G'') depending on the concentration (c) following a power-law relationship [42] as shown in Figure 4.6(c):

$$G', G''(\text{Pa}) = x \times c(\text{vol.}\%)^{y} \tag{4.1}$$

Exponents (y) of 2.7 and 3 have been found for G' and G'', respectively, in aqueous dispersions of GO flakes with very good quality of the fits ($R^2 > 0.90$) [42]. Similar power-law relationships between moduli and concentration have been found experimentally in other percolated networks with index values between 2 and 3.5, which are commonly attributed to random three-dimensional (3D) networks [46]. This suggests the presence of similar 3D networks (i.e. liquid crystalline gels) of GO in aqueous dispersions.

Two different types of force between the sheets are competing in the formation of GO-LC nematic phases: repulsive flake–flake forces due to the negatively charged functional groups (oxygen functional groups) present on their surfaces, and attractive forces through π–π and van der Waals interactions among the graphitic regions of the GO flakes. An equilibrium between the static repulsions and the bonding interactions between the flakes, in combination with the large aspect ratio and high flexibility of the 2D GO flakes, is responsible for the formation of GO-LC gel or 3D GO gel network. Increasing bonding forces (i.e. decreasing repulsion forces) between the sheets lead to stronger networks. Thus, at high GO concentrations, the strength of attractive forces will dominate due to small inter-sheet separation, which will result in the formation of strong interlinked GO gel-like phase.

4.2.2 Steady Shear Properties

A typical steady shear experiment (Figure 4.3b) consists on the application of an increasing shear rate ($\dot{\gamma}$) on the sample while monitoring the variation of its viscosity. In systems composed of anisotropic colloidal particles, polymers or worm-like micelles, the application of an external shear can originate some alignment and orientational molecular order. Steady-state shear flow curves help us to understand how the GO flakes dispersed in a liquid medium can orient and align under shear.

The typical steady shear behaviour observed for aqueous GO dispersions is shown in Figure 4.7. GO dispersions exhibit a Newtonian or non-Newtonian behaviour under steady shear flow depending on the concentration (Figure 4.7a). At low concentrations (~0.08 vol.%), the shear viscosity of GO dispersions shows Newtonian behaviour at low and high shear rates. With increasing concentrations, typical shear-thinning behaviours (i.e. viscosities decreasing very rapidly with increasing shear rates) are found, which are normally associated with the existence of some orientation of the GO flakes forming a nematic phase [43].

The relationship between the viscosities obtained studying the flow behaviour (steady shear) and the applied shear rate can be approximated using a power-law relationship (i.e. a power-law shear-thinning behaviour) [43]:

$$\eta \approx k\dot{\gamma}^{x-1} \tag{4.2}$$

where η is the viscosity, k is a constant, $\dot{\gamma}$ is the shear rate and x is the power-law exponent [47]. An exponent with $x=1$ corresponds to Newtonian behaviour, whereas $x<1$ is related to shear-thinning fluids. For low concentrations of GO (0.08 vol.%), x has been found to be 0.9 [43], which fits well with the exponent expected for an isotropic dispersion of low concentrated GO flakes showing a Newtonian behaviour. For higher concentrations (1.8 vol.%), the power-law exponent has been found to be 0.29, which corresponds well to the values typically attributed to weak gels or concentrated dispersions.

The shear-thinning behaviour typically found from low to high shear rates at high concentrations of GO can also be described by the simple Carreau equation [43, 48, 49].

Alternatively, as shown in Figure 4.7, two different regions have also been considered by some authors to describe the steady shear behaviour typically observed for aqueous GO dispersions, depending on the applied shear rate [42]:

i. a first region is found at low shear rates, where the dispersions significantly shear-thin (i.e. the viscosity decreases very rapidly with increasing shear rates, while the stress remains approximately constant);
ii. a second region is found at higher shear rates, where the viscosities remain approximately constant with increasing shear rates (while the stress increases).

There is a pseudo-yield stress between these two regions. As shown in Figure 4.7(b), this pseudo-yield stress can be scaled up using the Péclet number (*Pe*), which is a number that gives the relative time scales for Brownian motion of, and for the hydrodynamic forces on, the particles. If we approximate the GO flakes to circular hard discs, the Péclet number can be defined using the following equation [46]:

$$Pe = \frac{\text{time scale for Brownian motion}}{\text{time scale for convection motion}} = \frac{\dot{\gamma}}{D_r} = \dot{\gamma}\,\frac{32\eta_s b^3}{3kT} \tag{4.3}$$

where D_r is the rotary Brownian diffusion coefficient, k is Boltzmann's constant, T is absolute temperature, b is the radius of the disc and η_s is the viscosity of the solvent [46, 50].

In this kind of systems the Brownian forces (working to attract flakes to each other) are competing with the hydrodynamic forces (working to set the flakes apart). In the first

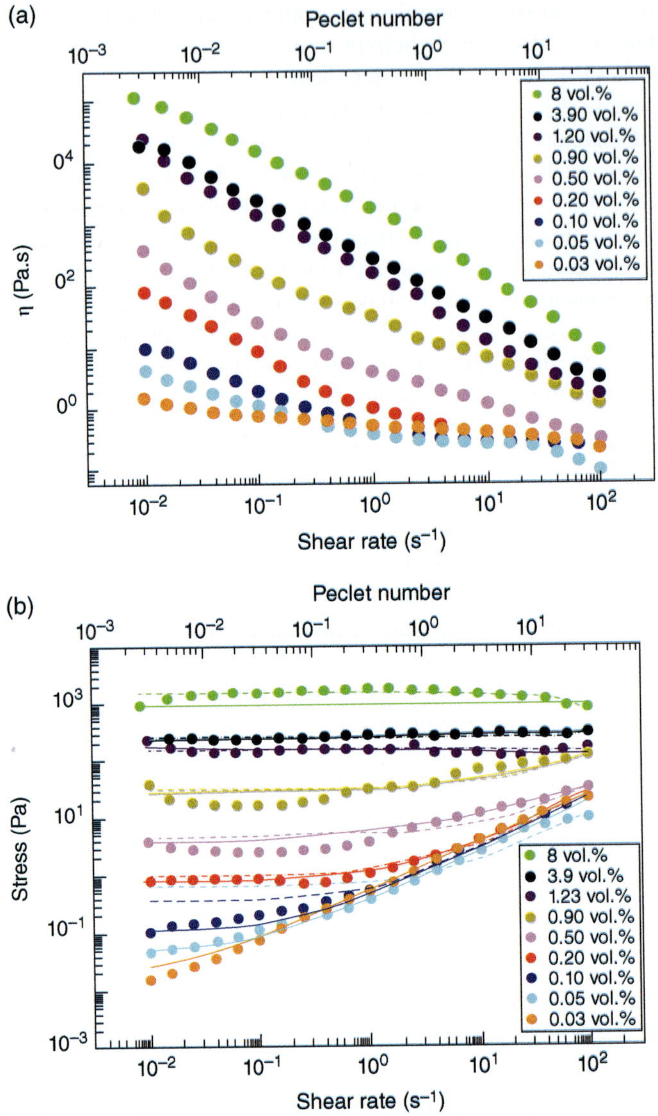

Figure 4.7 *(a) Steady shear rate flow behaviour of the GO dispersions at different concentrations, showing Newtonian or non-Newtonian behaviour depending on the concentration. (b) The steady shear flow data can be fitted to either the Bingham model (dashed lines) defined by (4.4) or the Herschel–Bulkley model (solid lines) defined by (4.5), as both models incorporate a yield stress. Reproduced from [42] with permission of Springer*

region (thinning behaviour, before the yield stress), the network re-forms as it is sheared, as the Brownian motion of the particles dominates over the hydrodynamic forces, with $Pe < 0.3$. In the second region (with constant viscosities and increasing stress, observed at higher shear rates after the yield stress), the network breaks because the hydrodynamic forces are dominating over the Brownian motion. In this second region the shear stress can be described by a power-law relationship with $n \sim 1$ at some loadings. The transition between these two regions occur at a certain *Pe* (≥0.3) depending on the concentration of GO and indicates the moment from which the hydrodynamic forces on the particles start to dominate over their Brownian motion. At high concentrations of GO (≥ 1.2 vol.%) the first region extends until very high shear rates and the dispersions do not enter the second region, which indicates the formation of a gel (nematic phase).

The Bingham and Herschel–Bulkley models have been widely used in the industrial field for pastes, slurries and suspensions [51] and they can also fit well the steady shear flow data for aqueous GO dispersions, as they incorporate a yield stress. The Bingham model has a zero strain rate at stresses beneath the yield stress whereas the fluid acts as a Newtonian one above the yield stress (as expressed in (4.4)); and the Herschel–Bulkley model is an Ostwald–de Waele (or power-law) model with a yield stress added (as expressed in (4.5)):

$$\text{Bingham model} \qquad \tau = \tau_y + \eta_\infty \dot{\gamma} \tag{4.4}$$

$$\text{Herschel – Bulkley model} \qquad \tau = \tau_y + K\dot{\gamma}^n \tag{4.5}$$

where τ is the shear stress, η_∞ is the infinite-shear-rate viscosity, n and K are constants and τ_y is the yield stress.

Although these two models fit the data quite well at concentrations of GO below 1.2 vol.%, they tend to fail to fit the experimental data at higher concentrations where the samples form a gel and never entered the viscous flow region.

The rheological behaviour of the dispersions of GO sheets at high concentrations is similar to that of a flocculated network. We have seen so far that, when the dispersions are at rest, as a consequence of the electrostatic forces between them, the GO flakes form a space-filling elastic network. At very low strain rates ($Pe \ll 1$), the majority of the network of GO flakes remains intact (within the LVR). Above the critical strain (outside the LVR), the structure of the GO network breaks down, which causes a very rapid decrease of the viscosity and the moduli with increasing rates. The microstructure of the system is represented in Figure 4.8. Some authors use the flocculated networks proposed by Barnes [52] to explain this rheological behaviour, as this model can explain the apparent yield stress observed. According to Barnes, a flocculated network under shear breaks down into flocs, which decrease in size with increasing shear rates. There is a Brownian motion in the system (forming the flocs), coexisting and competing with the shear forces (breaking the flocs apart). The size of the flocs in the system depends strongly on the competition between the Brownian motion and the shear forces, which is given by the Péclet number and leads to an equilibrium floc size for each applied shear rate, as mentioned previously.

The increased stress observed at high shear rates can also be explained using this theory. With increasing shear rates, the existing flocs break into progressively smaller flocs, which require less stress to flow, thus helping to maintain the stress virtually constant even with increasing shear rates (Figure 4.7b). Eventually (at a sufficiently high shear rate) the flocs break up into fully dispersed GO flakes, which are responsible for the increase observed in the stress at high rates.

Figure 4.9 shows the variation of the viscosity with concentration of GO. This variation is non-monotonic because the behaviour of the viscosity of liquid crystalline dispersions depends not only on the GO concentration but also on the particle arrangements. At low

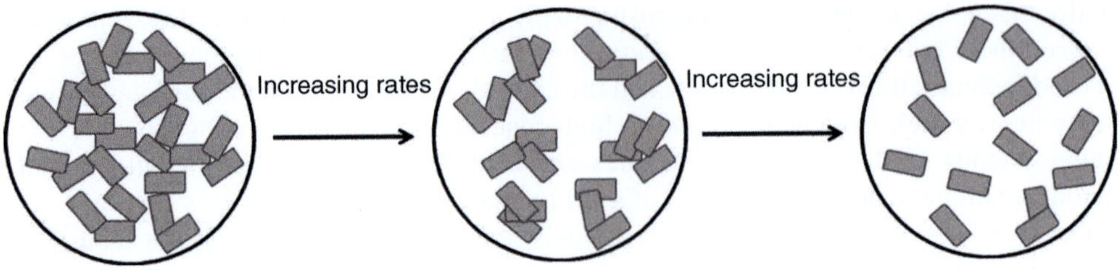

Figure 4.8 Schematic representation of the elastic network of GO flakes existing within the LVR, which breaks down into flocs by applying increasing rates leaving the LVR. The size of the formed flocs decreases as the shear rates increase

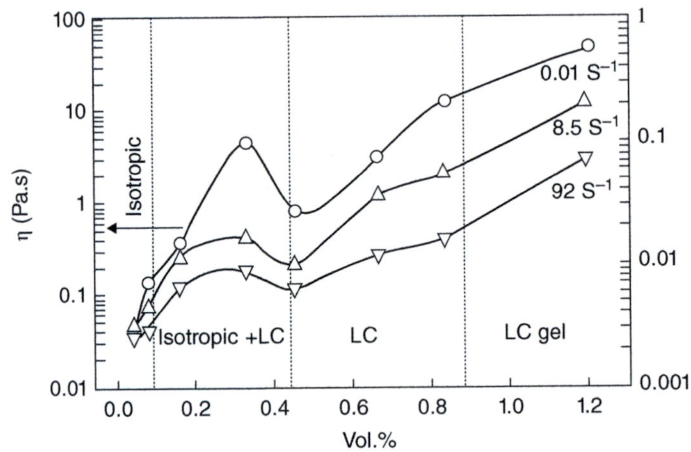

Figure 4.9 Variation of the shear viscosity of aqueous GO dispersions with GO composition, showing a non-monotonic relationship between them, which demonstrates the isotropic–nematic phase transition. Reproduced from [43] with permission of Elsevier

concentrations of GO (~0.08 vol.%) there is an isotropic phase. Then the viscosities increase with increasing concentrations up to a critical volume fraction (~0.33%), reaching a maximum value. After the maximum, with further increased concentrations, the viscosities are reduced due to the formation of a low-viscosity nematic liquid crystalline phase, reaching a viscosity minimum value, which corresponds approximately to the formation of the entire nematic phase. After this minimum, the values for the viscosities start increasing again (Figure 4.9).

The transition from isotropic to nematic phase can thus be followed under steady shear, and typically occurs at a critical GO composition of ~0.33%, which is close to the theoretical value predicted for polydispersed infinitely thin platelets (0.25%) [11], confirming a typical colloidal isotropic–nematic phase transition.

It should be noted at this point that the values found for the viscosities observed for GO dispersions with very low concentrations (where the particles do not interact with each other) are normally higher than those predicted by the classic Einstein–Stokes equation [53, 54] and the classic Brenner prediction for dilute spheres [55], both of which are employed to describe or predict the behaviour of dilute spheres. These underestimations of the viscosities have been related to the high aspect ratio of flexible and negatively charged GO sheets, as these theoretical classic models were assuming spherical hard, rigid particles, with no charges and no attractive forces between them.

4.2.3 Recovery of the Structure

We have seen above that there is an elastic GO network formed in a dispersive medium, which is broken under dynamic and steady shear. It is of great importance to evaluate whether the GO dispersions can recover their initial structure after their rupture, as well as to establish how long the recovery takes. We refer to the dispersions as 'shear-thinning' or 'thixotropic' depending on whether the structure recovers instantly or after a period of time, respectively.

The recovery of the structure of the GO can be evaluated by conducting a series of dynamic strain sweeps on a GO dispersion and varying the rest time between the sweeps. To evaluate the recovery of the structure after steady shear sweeps, a dynamic stress is applied after the dispersions are sheared in order to follow the recovery of the structure. A sufficiently high dynamic stress must be selected so that the recovery can actually be measured, but it does not have to be so high that it is actually impeding a complete recovery.

GO dispersions have been found to be 'thixotropic', as they can recover their original structure with resting after being broken under dynamic strain or steady shear rate sweeps.

After breaking the GO network by applying dynamic strain sweeps, it takes typically between 30 and 60 minutes for both the G' and G'' moduli to recover their initial values. We assume the recovery of the moduli values as evidence of the recovery of the flakes network structure. As the recovery of the structure is independent of the concentration of the dispersion, similar recovery processes are expected with different contents of GO. On some occasions, the recovery of the structure may not be complete, which is indicated by higher viscous structures (slightly higher G'' values) for the recovered structure in comparison to the initial one [42].

The structure of the samples is broken down even more severely by applying steady shear rate sweeps in comparison to the damage caused by the dynamic strain sweeps. By steady shearing a GO dispersion, the structure breaks severely, which is shown by a dramatic decrease of the values of both G' and G'' moduli. Although the structure recovers upon resting, the recovery process is not linear. Initially, both moduli increase very slowly with time, followed by a steeper increase, reaching values that are close to the initial ones after about 6 hours, although they never reach the original values. This means that, even if it gets close, the initial structure of the GO dispersion is never fully recovered after steady shear.

The LC-GO dispersions exhibit thixotropic fluid-like features coming from the breakdown of the interconnected elastic network of GO flakes, which is followed by an enhanced orientational ordering of the GO flakes in the medium.

4.2.4 Tuning the Rheology of GO Dispersions to Enable Fabrication

Throughout this chapter we have seen that the rheological properties of GO dispersions depend strongly on the concentration, which means that they can be tuned and controlled by adjusting the concentration of GO particles. The establishment of a correlation between the rheological properties and GO concentrations represents an important point for the use of various manufacturing techniques, which will enable us to process and fabricate macroscopic GO structures via a wide range of industrial techniques.

Several regions with different rheological behaviours are distinguished depending on the GO concentration, in which the dispersions should be processed using different processing techniques, as represented in Figure 4.10.

i. Although, at very low concentrations, the GO dispersions were thought to be completely isotropic and to show only liquid-like behaviour (only viscous phase), for aqueous dispersions of 'ultra-large' GO flakes, considerable elastic components have been found recently even in the very low concentration regime (0.25 mg ml^{-1}), which makes them behave like a viscoelastic liquid even at very low concentrations (consequence of a high aspect ratio). The transition from solid-like to liquid-like behaviour (yield stress) in these ultra-large GO dispersions occurs at a very low critical concentration (approximately three orders of magnitude lower than the theoretical value for colloidal suspensions) [12], due to an extremely large aspect ratio (~45 000). The existence of an important elastic

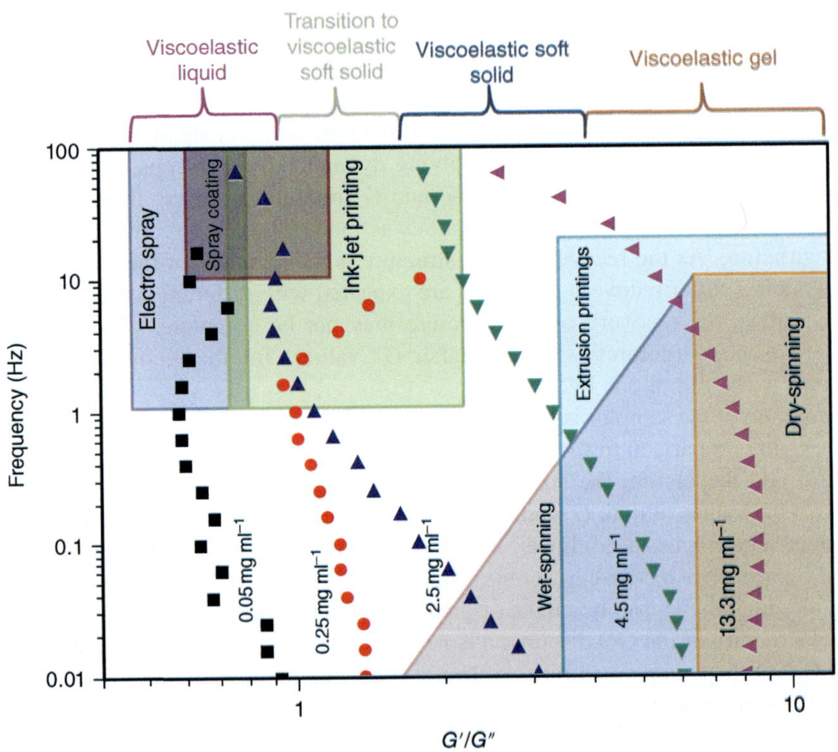

Figure 4.10 Variation of the ratio of elastic and storage moduli (G′/G″) with frequency for GO dispersions at different concentrations, showing which processing techniques are suitable to allow the use of GO dispersions for industrial fabrication. For example, when the viscous component (G″) dominates, high-rate processing methods where the dispersion must spread on a substrate are suitable; whereas when the elastic component (G′) dominates, extrusion printing and fibre spinning are appropriate fabrication methods. Reproduced from [12] with permission of RSC

component in addition to the viscous component at such low loadings enables highly controlled electro-spray and spray coating at extremely low concentrations. This makes possible the fabrication of ultra-transparent GO thin films without the typical perturbations found when viscous fluids with no elastic components are sprayed. Spray coating of ultra-large GO dispersions has great potential for a wide range of industrial processes, such as electronics, painting, microencapsulation, electroemulsification, fine powder production, and micro- and nano-thin-film deposition [56].

ii. At a GO concentration of ~0.25–0.75 mg ml^{-1}, GO dispersions show a biphasic behaviour, in which both isotropic and nematic phases coexist. In this range of concentrations, because 'long-range' rearrangements tend to occur very slowly, the GO dispersions behave like a viscoelastic soft solid over long time scales (with $G' > G''$). In an intermediate time scale, however, a dominant viscous behaviour (liquid-like behaviour) is found (with $G'' > G'$) because the 'short-range' rearrangements occur fast. The fluid properties of these GO dispersions thus seem ideal for inkjet printing. (For good ink formulation, the original viscosity has to be recovered quickly (at short time scales) while preserving the shapes and structures at long time scales after printing [57]). This is of great importance to formulate stable inks from pure dispersions of GO with no need for binders or additives to impart the dominant elastic components required for inkjet printing. Recently GO has been successfully printed without any clogging side effects as a consequence of the highly flexible nature of GO sheets [12], which enables the fabrication of electrode materials to be used in electronic applications, such as organic field-effect transistors (OFETs).

iii. At higher concentrations, in the range of 0.75–2.5 mg ml^{-1}, the GO dispersions form single-phase nematic liquid crystals. Despite the low concentration, a network of GO flakes is formed in the dispersive medium with a dominant elastic part (i.e. solid-like behaviour). At a concentration of 2.5 mg ml^{-1} the GO dispersions tend to behave quite similarly to cells, soft glassy materials (SGMs), liquid crystals or weak gels [58–60], being slightly different from how polymer networks behave (in a purely elastic polymer network, G' shows a frequency-independent plateau even at low frequencies [59, 61], whereas in the typical behaviour of cells G' would increase slowly with frequency [58]). This rheological behaviour at this concentration offers the possibility to use processing techniques for GO dispersions, which have been typically used for the processing of weak gels, such as wet spinning [62–66].

iv. As the concentration increases up to 4.5 mg ml^{-1}, the viscous part (G'') dominates only at very short time scales. With increasing concentrations, the network of GO flakes becomes more and more robust, showing a gel-like behaviour (viscoelastic gel), with the elastic part (G') becoming more and more dominant. The crossover between the two moduli ($G' = G''$) thus shifts to shorter time scales and, eventually, at a concentration of GO as high as 13.35 mg ml^{-1}, G' and G'' do not cross at any frequency, which suggests gel-like behaviour. These viscoelastic liquid crystal gels show very high elastic modulus (350–490 Pa), which is considerably higher than those calculated for suspensions of single-walled carbon nanotubes at a similar concentration (~60 Pa) [67]. The viscoelastic liquid crystal gels combine the typical anisotropy of an LC network with the typical exceptional uniformity of a network structure. This range of high concentrations for the GO dispersions make possible the use of some industrial processes such as gel extrusion printing and dry spinning for GO dispersions for the first time.

The rheological behaviour (viscoelastic behaviour) of GO dispersions is different from those observed normally for other viscoelastic materials, such as polymers. This is very interesting both from fundamental research perspectives as well as for applications, potentially providing a wide range of processing techniques.

4.2.5 Electro-Optical Switching of LC-GO with an Extremely Large Kerr Coefficient

The self-assembling properties into highly ordered structures in combination with the outstanding properties of GO materials (mechanical, thermal and electrical after reduction) make GO (and reduced graphene oxide) very promising materials for industrial applications [41, 68]. In order to maximize the exploitation of graphene-based materials, controllable alignments of GO-LCs are strongly required [19, 23, 69–71].

The fluidic liquid nature of LCs allows one to control the alignment of their molecules (particles) by applying an electric field [72, 73]. Although electric fields have been proved to induce order in LCs in general, the microscopic ordering and the macroscopic alignment of GO-LCs can only be controlled by applying low electric fields (~5 V mm^{-1}, which is three orders of magnitude weaker than the electric field typically required for molecular LC switching) for very low concentrations of GO (i.e. very weak inter-flake interactions). Induced GO-LC switching can thus only be observed in the low-concentration nematic phase, the biphase and the isotropic phase) [74]. Under different conditions (higher electric fields and higher GO concentrations), the application of an electric field only leads to electrophoretic drift, reduction of the GO and electrolysis of water [9], but no induced orientation of the GO. This selective induced GO-LC switching is evidenced by the induced birefringence observed under polarized light, as shown in Figure 4.11.

This intrinsic selective sensitivity of GO to high electric fields must be related to the viscous rheology, similar to what happens with polymeric liquid crystals [9, 23]. Although the removal of the electric field typically leads to the disappearance of the nematic ordering of the LCs (including LC-GO), some weak nematic ordering still remains after the electric field is removed. The sensitivity to weak external fields disappears totally however with increasing concentration (i.e. decreasing inter-flake distance).

GO is a Kerr material, due to its capacity to change its refractive index in response to an applied electric field. A maximum Kerr coefficient (used to quantify the Kerr effect of any material) of approximately 1.8×10^{-5} m V^{-2} has recently been determined for GO-LCs [74], which is extremely large in comparison to other known systems (for example, the Kerr coefficients of nitrobenzene and aqueous 2D gibbsite platelet suspensions are of the order of 10^{-12} and 10^{-9} m V^{-2} respectively [75]).

The interactions between the flakes, which are responsible for the nematic ordering (related to Onsager's excluded-volume effect, described in section 4.1), also contribute enormously to the selective sensitivity of GO materials to external fields and their extremely large Kerr coefficient below a certain concentration. When the distance between the flakes decreases, the π–π and van der Waals attractive forces between the GO flakes became dominant [76], which increased the frictional effect, eliminating the extremely high sensitivity of the GO dispersion to external electric fields and preventing the Kerr effect.

The Kerr effect of GO dispersions can also be reduced or even totally prevented (desensitization of GO to external electric fields) by adding a salt (e.g. NaCl), as its

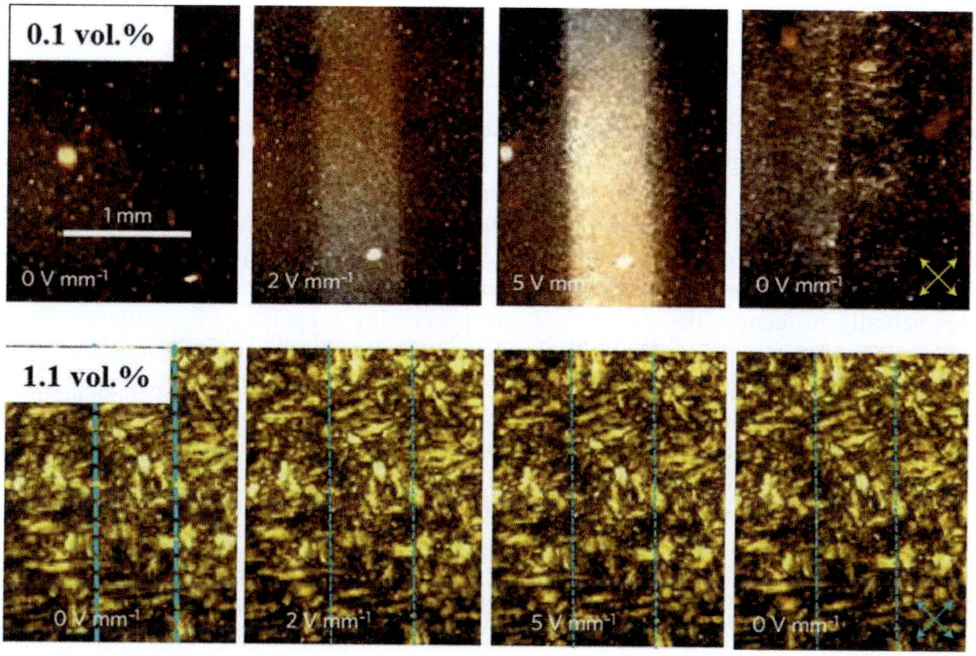

Figure 4.11 Field-induced birefringence generated on an aqueous GO dispersion with concentrations of 0.1 and 1.1 vol.% by applying an electric field of 10 Hz, showing the appearance of birefringence only for the 0.1 vol.% dispersion. The disappearance of the birefringence when the electric field is removed can also be seen [74]. Reproduced from [74] with permission of Nature Publishing Group

addition increases the conductivity of the electrolyte and enhances the interactions between the flakes.

The performance of an electro-optic device has recently demonstrated the significance of the extremely large Kerr coefficient found for GO materials [74]. Similarly to the already known and commonly used interdigitated electrodes in liquid crystal displays (LCDs) including blue-phase LCDs, a simple GO electro-optic device has recently been achieved using two simple wire electrodes. Wide areas of electrically aligned, highly concentrated nematic GO were successfully prepared by slowly evaporating the dispersive water from an isotropic 0.056 vol.% aqueous GO dispersion between two electrodes on a substrate while applying an electric field. Once the water was completely dried, the obtained uniform GO-LC alignment on the substrate was maintained even after the fields were switched off. In addition, by changing the direction of the applied electric fields, the director of the field-induced GO could be easily rotated. These GO electro-optic devices have been found to work very well, showing optical switching behaviours that are comparable to those found for typical nematic LC devices. The performance of these electro-optic devices from GO dispersions shows that GO-LC alignment can be achieved easily by designing an appropriate electric field and controlling the GO concentration.

4.3 Comparison with Other Systems

4.3.1 Comparison of Aqueous and Polymer Matrix Systems

By adding nano-sized particles, such as graphene materials, into a polymer matrix, a particulate graphene–polymer system is created. An understanding of the rheological behaviour of aqueous dispersions of GO flakes considered as a model system (previously described) is of great importance to understand and predict the rheological and processing properties of more complex systems, for example, graphene–polymer composites. This model system allows us to understand how the incorporation of nanoparticles actually influences the rheological and processing properties of a polymer. The rheological properties of particulate suspensions are sensitive to the structure, particle size, shape and surface modification of the particles. The rheological properties change significantly with favourable particle–matrix interactions compared to non-interacting systems or strong particle–particle attractions. When GO flakes are added to a polymer melt or any other liquid phase, changes in the rheological behaviour are expected. The van der Waals forces and π–π interactions between the flakes tend to form agglomerates, which increase the viscosity of the dispersive medium. The rheological properties of a system with well-dispersed flakes are very different from those of a system with the flakes forming agglomerates. Particle–particle and particle–matrix interactions (i.e. surface chemistry) thus play an important role in the rheological behaviour of particulate–polymer systems.

Figure 4.12 shows the typical rheological behaviour of GO–PMMA (poly(methyl methacrylate)) melts as a representative GO–polymer system. Figures 4.12(a) and (b) show the typical rheological behaviour of GO–PMMA melts with different GO contents under dynamic shear. When GO flakes are incorporated into a polymer matrix, both G' and G'' moduli increase gradually with increasing GO content (Figures 4.12a,b), which indicates a reinforcement of the polymer melt by adding GO. A gradual development of a percolating particulate network of GO in the matrix occurs as the GO content is increased. The existence of this elastic GO network at rest at very low frequencies is evidenced by the low-frequency plateau found for G' (where G' is independent of the frequency).

Figure 4.12(c) shows the variation of the viscosities of GO–PMMA composites with different contents of GO at increasing shear rates (determined from steady shear experiments). The increase observed in the value of the viscosity of the host polymer matrix with increasing content of GO suggests again a reinforcement of the polymer. Typically, the viscosities of the GO–polymer composites present an initial Newtonian plateau, in which they are independent of the applied shear rate, which corresponds to the particulate network of GO flakes formed in the matrix. The shear-thinning region (viscosities decreasing very rapidly with increasing shear rates) found following the Newtonian plateau evidences the network breakdown.

The addition of GO flakes to a polymer matrix leads to the formation of an interconnected network of GO flakes in the matrix, which remains intact at rest and low shear rates. The presence of this network of GO flakes in the polymer matrix is responsible for the increases in the viscosity of the host polymer (viscosities increase with increasing contents of GO), which provides a more solid-like behaviour. We refer to the

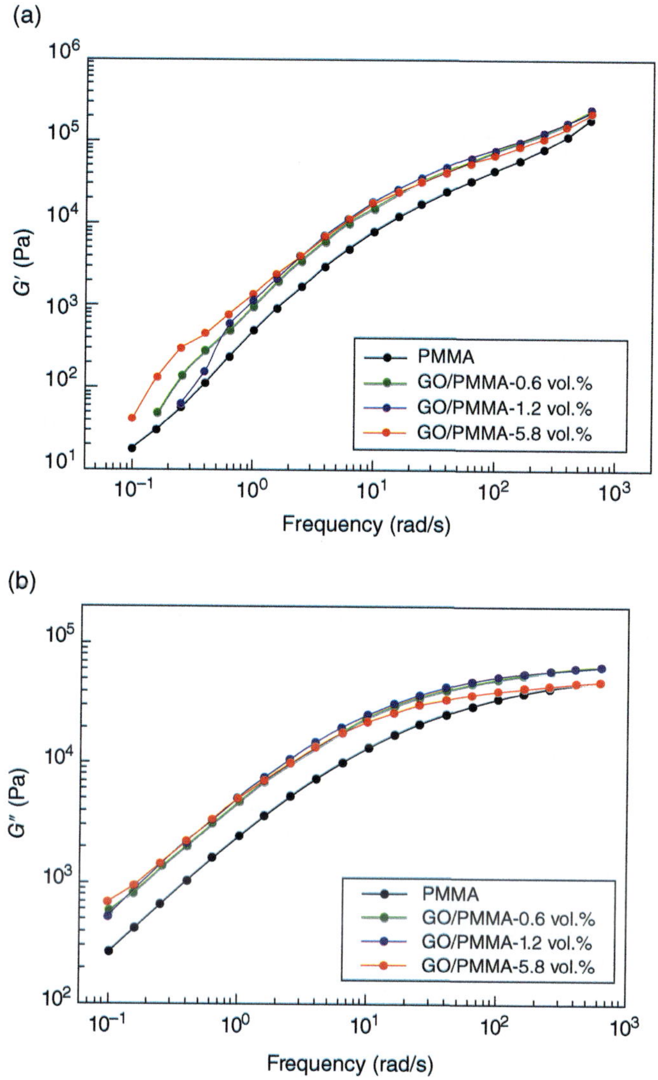

Figure 4.12 *Rheological properties of GO–PMMA melts with different GO contents (performed at the polymer melting temperature, i.e. 230 °C). Dependence of (a) G′ and (b) G″ with frequency (dynamic frequency behaviour at a constant strain amplitude of 1%), showing the gradual development of a GO percolated network in the matrix with increasing GO content. (c) Dependence of the viscosity with increasing shear rates (steady shear behaviour), showing an initial Newtonian plateau, which corresponds to the presence of the particulate network of GO flakes in the matrix, followed by a shear-thinning region, which indicates the breakdown of the GO network. Reproduced from [42] with permission of Springer*

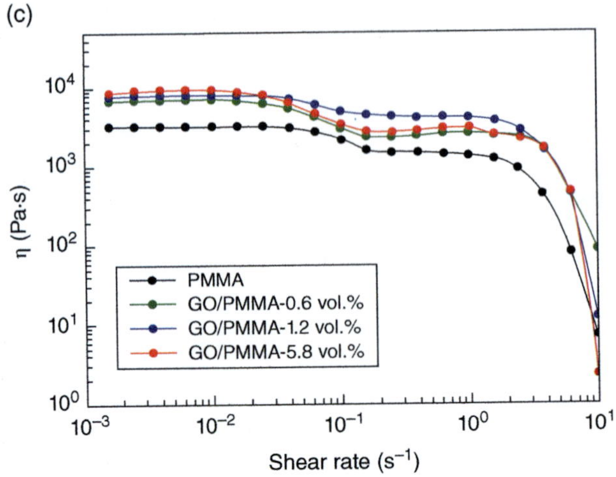

Figure 4.12 (Continued)

formation of a long-range connectivity in graphene–polymer systems as the percolation threshold and it corresponds to their optimal (rheological, mechanical or electrical) properties. In this graphene–PMMA system the percolation threshold is ~1.2 vol.% content of GO [42].

The viscosities and rheological percolation thresholds in graphene–polymer systems strongly depend on the dispersion of the flakes in the matrix, as well as on the interactions between graphene and the polymer (graphene/polymer interface). It is thus possible to modify the interactions between graphene and the polymer in order to be able to tune the viscosities and percolation thresholds of any graphene–polymer systems through modifying either the surface chemistry of the flakes or the nature of the polymer. For example, the oxygen content at the surface of the graphene flakes strongly influences the rheological properties of PMMA composites, due to both a matching polarity between the GO flakes and the polymer matrix and the quality of the dispersion of the GO sheets in the matrix. Graphene flakes with higher C/O ratios provide a more homogeneous dispersion in PMMA, which helps to provide a more effective impact on the viscoelasticity of the polymer (i.e. lower rheological percolation threshold and higher storage moduli and viscosities) [77].

As an alternative to modifying the surface chemistry of the graphene or GO flakes, the rheological properties of graphene–polymer systems can also be tuned by choosing a different matrix, a matrix with which the interactions are different. For example, for chemically reduced GO–PP (polypropylene) composites, larger increases in viscosity and a transition from liquid- to solid-like behaviour occurring at lower concentrations (0.2–0.4 vol.% of graphene) have been found [78] in comparison to those found for a GO–PMMA system (~1.2 vol.%). Different degrees of interaction between GO or graphene materials and polymers are responsible for the different rheological behaviours.

The formation and rheological strength of flocculated networks and hence the viscosities can be easily tuned through controlling the degree of interaction between GO sheets and the polymer, which is of great importance for practical applications.

4.3.2 Comparison Between Aqueous Dispersions of GO and Oxidized Carbon Nanotubes: Role of Dimensionality

Graphene materials are very often compared to its 'older brother', carbon nanotubes (CNTs), as they present many similarities in morphology, composition, reactivity, properties and applications. When the rheological behaviour of aqueous dispersions of GO flakes are investigated, it seems natural to compare it with those of aqueous dispersions of oxidized CNTs. The rheological behaviour of aqueous dispersions of oxidized CNTs in the concentrated regime has already been studied [79].

By refluxing CNTs in concentrated acid, their surfaces are successfully oxidized, which leads to the break-up of the CNT aggregates into individual tubes, as well as to an important improvement of their dispersivity in water, other polar dispersive media and polymers [80]. Typically the oxygen content in the GO flakes (~33 at.% oxygen) is slightly higher than that present in the surface of the CNTs (<20 at.% oxygen). Due to the existence of electrostatic interactions between the activated (oxidized) functional groups on both types of carbon nanomaterials, the aqueous dispersions of both GO and oxidized CNTs are highly stable. Besides, GO and oxidized CNTs are comparable in terms of dimensions, with the length of the CNTs being similar to the diameter of the GO flakes. Aqueous dispersions of oxidized CNTs can thus be easily compared to aqueous GO dispersions in terms of rheological properties.

Many similarities can actually be found between the rheological behaviours of these two systems. To start with, both GO and CNT dispersions act as reversible concentrated flocculated networks showing an elastic behaviour (LVR, with G' and G'' being independent of the strain or stress applied, and the viscosity being independent of the shear rate applied) up to a critical strain (or stress), above which G' and G'' become highly strain/stress-sensitive and the viscosity becomes highly shear-dependent. For GO dispersions, the values found for G', G'' and viscosities within the LVR are slightly higher than those obtained for oxidized CNT dispersions with similar concentrations (G', $G'' \sim 10^5$ Pa and $\eta^* \sim 10^6$ Pa have been reported for aqueous dispersions of 8 vol.% of GO and 9.2 vol.% of CNTs).

Under dynamic shear, the LVR extended to strains one order or magnitude higher for GO (up to ~10% strain) than for CNTs (up to ~1% strain). Similarly, the transition from solid-like to liquid-like behaviour occurs at higher strains for dispersions of GO than for oxidized CNTs at similar concentrations. In addition, the transition from liquid-like behaviour to strong gel spectrum (shown by dynamic frequency sweeps conducted within the LVR) occurs at lower concentrations for GO than for CNTs (1.2 vol.% for GO, 4 vol.% for CNTs). In order to break the structure of the GO network formed in aqueous dispersions, higher strains (or, alternatively, stress) are thus required relative to those required to break the network of oxidized CNTs at similar concentrations. These small differences observed in the rheological behaviour of the two systems under dynamic shear can be perfectly explained in terms of their charge concentration. The higher oxygen contents present in the GO flakes (~33%) relative to the CNTs nanotubes (<20%) must provide different degrees of inter-particle interactions for similar concentrations, which are responsible for the slightly different behaviours found for these two systems under dynamic sweeps. The behaviours of the two systems are thus comparable.

Under steady shear, the equilibrium floc sizes for GO are smaller than those for CNTs for identical shear rates and concentrations, which is evidenced by the apparent yield stress found at a much lower *Pe* for GO ($Pe < 1$) than for CNTs ($Pe \sim 1$–10). Lower shear rates are required to reduce the GO flocs to the primary particles (i.e. fully dispersed flakes) relative to those required to reduce the CNT flocs to individual tubes. Despite having comparable lengths, the 1D tubes tend to retain an entangled network up to higher shear rates, whereas 2D flakes are more easily dispersed from each other. Similarly, although both systems are thixotropic, the different dimensionality (and aspect ratio) between 1D CNT rods and 2D GO flakes is responsible for the slightly quicker recovery of the structure of a CNT network with respect to a GO network after shearing.

4.4 Summary and Perspectives

Anisotropic molecules or particles with high aspect ratio in liquid media are capable of forming highly ordered structures at the mesoscopic scale, known as liquid crystalline (LC) phases. Increasing concentrations of anisotropic particles (i.e. rod-like or plate-like particles) restrict the free rotation of the particles due to the exclusion of free volume. As Onsager's theory predicted, the rods or platelets are thus forced to orient and organize themselves, forming highly ordered structures. There is a critical concentration for the transition from isotropic to nematic phase with both phases coexisting (biphase), which depends strongly on the aspect ratio of the particles. The fabrication of highly organized LC structures from molecular to higher levels attracts great interest due to their potential in display devices.

The recently found capacity of GO materials to form LC phases in water and organic solvents, leading to perfectly aligned structures, in combination with their exceptional mechanical properties (stiffness and strength), high functionality and ultra-high aspect ratio make GO a very promising material for the fabrication of highly ordered, strong and conductive graphene-based structures. Investigating the rheological behaviour of GO dispersions is of great importance to understand how GO dispersions can be processed in order to develop fabrication protocols using different processing techniques, as well as to understand and predict the influence of graphene materials on the rheological properties and processing of more complex systems (systems with more complex interactions between graphene and solvent or polymeric matrix).

The rheological (viscoelastic) behaviour of GO dispersions is different from that typically found for other viscoelastic materials, i.e. polymers, which presents GO as a new type of soft material. This is very interesting from the points of view of both fundamental research on 2D materials and applications (wide range of processing techniques).

Throughout this chapter, we have seen that aqueous GO dispersions show the typical behaviour for space-filling networks of particles. Due to the existence of electrostatic interactions between the GO flakes, an elastic GO network is formed in dispersive media, which shows viscoelastic behaviour under dynamic shear up to a critical strain. Within this LVR, the values for the storage (G') and loss (G'') moduli are independent of the applied strain, with G' being higher than G''. The critical strain represents the breakdown of the elastic GO network and thus the transition from solid-like behaviour to liquid-like behaviour, with G' and G'' crossing and starting to decrease rapidly with increasing strains. The liquid-like region is entered when G' and G'' cross, and tends to occur at higher strains for higher GO concentrations, except for the very high concentrations in which the sample forms a gel (orientation, alignment of the flakes).

Under steady shear flow, GO dispersions exhibit a Newtonian or non-Newtonian behaviour depending on the concentration. Typically, at very low concentrations the shear viscosity of GO dispersions shows Newtonian behaviour, whereas at higher concentrations they show a typical shear-thinning behaviour (i.e. viscosities decreasing very rapidly with increasing shear rates), which is normally associated with the existence of some orientation of the GO flakes forming a nematic phase. For ultra-large GO sheets, however, due to the high aspect ratio (~45 000), an elastic component is already found with very low GO content, showing the strong dependence of the rheological properties on the aspect ratio of the particles.

Two different types of force between the sheets are competing in the formation of GO liquid crystalline nematic phases: repulsive flake–flake forces due to the negatively charged functional groups (oxygenated functional groups) present on their surfaces, and attractive forces through π–π and van der Waals interactions among the graphitic regions of the GO flakes. The combination of an equilibrium between the two types of interaction between the 2D GO flakes, their large aspect ratio and their high flexibility is responsible for the formation of GO liquid crystalline gels or 3D GO gel networks. At high GO concentrations, there are increasing bonding forces due to smaller inter-sheet separation, which will lead to stronger networks (stronger interlinked GO gel-like phases).

This strong dependence of the rheological properties of GO dispersions on the concentration offers the possibility to tune and control their rheological behaviour by adjusting the concentration of GO particles. A number of regions with different rheological properties are distinguished depending on the GO content. In each of these regions, different processing techniques should be used for the GO dispersions, such as spray coating, inkjet printing, wet spinning, gel extrusion printing or dry spinning, which are of great importance for industrial applications.

In addition, the alignment of GO-LCs can be controlled using external stimuli. The microscopic ordering and the macroscopic alignment of GO liquid crystals (LCs) can be controlled by applying low electric fields (of 5 V mm^{-1}, which is three orders of magnitude weaker than the electric field typically required for molecular LC switching) when the inter-flake interactions are very weak, i.e. for very low concentrations of GO. The sensitive response of the nematic GO phase to external stimuli and the extremely large Kerr coefficient found for GO maximize the exploitation of graphene-based materials, allowing, for example, the fabrication of electro-optic devices.

The rheological properties of aqueous GO dispersions are compared to other systems, such as more complex graphene–polymer systems and aqueous dispersions of oxidized CNTs.

References

[1] Hurt, R.H.; Chen, Z.Y., Liquid crystals and carbon materials. *Phys. Today* **2000**, *53* (3), 39–44.

[2] Bisoyi, H.K.; Kumar, S., Liquid-crystal nanoscience: an emerging avenue of soft self-assembly. *Chem. Soc. Rev.* **2011**, *40* (1), 306–319.

[3] Lagerwall, J.P.F.; Scalia, G., A new era for liquid crystal research: applications of liquid crystals in soft matter nano-, bio- and microtechnology. *Curr. Appl. Phys.* **2012**, *12* (6), 1387–1412.

[4] Kaafarani, B.R., Discotic liquid crystals for opto-electronic applications. *Chem. Mater.* **2011**, *23* (3), 378–396.

[5] Yan, J.; Rao, L.; Jiao, M.; *et al.*, Polymer-stabilized optically isotropic liquid crystals for next-generation display and photonics applications. *J. Mater. Chem.* **2011**, *21* (22), 7870–7877.

[6] Onsager, L., The effects of shape on the interaction of colloidal particles. *Ann. NY Acad. Sci.* **1949**, *51* (4), 627–659.
[7] Aboutalebi, S.H.; Gudarzi, M.M.; Zheng, Q.B.; Kim, J.K., Spontaneous formation of liquid crystals in ultralarge graphene oxide dispersions. *Adv. Funct. Mater.* **2011**, *21* (15), 2978–2988.
[8] Dan, B.; Behabtu, N.; Martinez, A.; *et al.*, Liquid crystals of aqueous, giant graphene oxide flakes. *Soft Matter* **2011**, *7* (23), 11154–11159.
[9] Kim, J.E.; Han, T.H.; Lee, S.H.; *et al.*, Graphene oxide liquid crystals. *Angew. Chem. Int. Ed.* **2011**, *50* (13), 3043–3047.
[10] Xu, Z.; Gao, C., Aqueous liquid crystals of graphene oxide. *ACS Nano* **2011**, *5* (4), 2908–2915.
[11] Bates, M.A.; Frenkel, D., Nematic–isotropic transition in polydisperse systems of infinitely thin hard platelets. *J. Chem. Phys.* **1999**, *110* (13), 6553–6559.
[12] Naficy, S.; Jalili, R.; Aboutalebi, S.H.; *et al.*, Graphene oxide dispersions: tuning rheology to enable fabrication. *Mater. Horiz.* **2014**, *1* (6), 326–331.
[13] Li, L.S.; Walda, J.; Manna, L.; Alivisatos, A.P., Semiconductor nanorod liquid crystals. *Nano Lett.* **2002**, *2* (6), 557–560.
[14] Van Der Kooij, F.M.; Lekkerkerker, H.N.W., Formation of nematic liquid crystals in suspensions of hard colloidal platelets. *J. Phys. Chem. B* **1998**, *102* (40), 7829–7832.
[15] Jung, J.M.; Mezzenga, R., Liquid crystalline phase behavior of protein fibers in water: experiments versus theory. *Langmuir* **2010**, *26* (1), 504–514.
[16] Davis, V.A.; Parra-Vasquez, A.N.G.; Green, M.J.; *et al.*, True solutions of single-walled carbon nanotubes for assembly into macroscopic materials. *Nature Nanotechnol.* **2009**, *4* (12), 830–834.
[17] Behabtu, N.; Young, C.C.; Tsentalovich, D.E.; *et al.*, Strong, light, multifunctional fibers of carbon nanotubes with ultrahigh conductivity. *Science* **2013**, *339* (6116), 182–186.
[18] Miyamoto, N.; Nakato, T., Liquid crystalline nanosheet colloids with controlled particle size obtained by exfoliating single crystal of layered niobate $K_4Nb_6O_{17}$. *J. Phys. Chem. B* **2004**, *108* (20), 6152–6159.
[19] Behabtu, N.; Lomeda, J.R.; Green, M.J.; *et al.*, Spontaneous high-concentration dispersions and liquid crystals of graphene. *Nature Nanotechnol.* **2010**, *5* (6), 406–411.
[20] Guo, F.; Kim, F.; Han, T.H.; *et al.*, Hydration-responsive folding and unfolding in graphene oxide liquid crystal phases. *ACS Nano* **2011**, *5* (10), 8019–8025.
[21] Zhang, S.; Kumar, S., Carbon nanotubes as liquid crystals. *Small* **2008**, *4* (9), 1270–1283.
[22] Ko, H.; Tsukruk, V.V., Liquid-crystalline processing of highly oriented carbon nanotube arrays for thin-film transistors. *Nano Lett.* **2006**, *6* (7), 1443–1448.
[23] Zakri, C.; Blanc, C.; Grelet, E.; *et al.*, Liquid crystals of carbon nanotubes and graphene. *Phil. Trans. R. Soc. A* **2013**, *371* (1988), 20120499.
[24] Zakri, C., Carbon nanotubes and liquid crystalline phases. *Liq. Cryst. Today* **2007**, *16* (1), 1–11.
[25] Lee, S.H.; Lee, D.H.; Lee, W.J.; Kim, S.O., Tailored assembly of carbon nanotubes and graphene. *Adv. Funct. Mater.* **2011**, *21* (8), 1338–1354.
[26] Maiti, U.N.; Lee, W.J.; Lee, J.M.; *et al.*, Chemically modified/doped carbon nanotubes & graphene for optimized nanostructures & nanodevices. *Adv. Mater.* **2014**, *26* (1), 40–67.
[27] Gudarzi, M.M.; Moghadam, M.H.M.; Sharif, F., Spontaneous exfoliation of graphite oxide in polar aprotic solvents as the route to produce graphene oxide–organic solvents liquid crystals. *Carbon* **2013**, *64*, 403–415.
[28] Jalili, R.; Aboutalebi, S.H.; Esrafilzadeh, D.; *et al.*, Organic solvent-based graphene oxide liquid crystals: a facile route toward the next generation of self-assembled layer-by-layer multifunctional 3D architectures. *ACS Nano* **2013**, *7* (5), 3981–3990.
[29] Xu, Z.; Liu, Z.; Sun, H.; Gao, C., Highly electrically conductive Ag-doped graphene fibers as stretchable conductors. *Adv. Mater.* **2013**, *25* (23), 3249–3253.
[30] Stankovich, S.; Dikin, D.A.; Dommett, G.H.B.; *et al.*, Graphene-based composite materials. *Nature* **2006**, *442* (7100), 282–286.
[31] Steurer, P.; Wissert, R.; Thomann, R.; Mülhaupt, R., Functionalized graphenes and thermoplastic nanocomposites based upon expanded graphite oxide. *Macromol. Rapid Commun.* **2009**, *30* (4–5), 316–327.
[32] Rourke, J.P.; Pandey, P.A.; Moore, J.J.; *et al.*, The real graphene oxide revealed: stripping the oxidative debris from the graphene-like sheets. *Angew. Chem. Int. Ed.* **2011**, *50* (14), 3173–3177.

[33] Dimiev, A.; Kosynkin, D.V.; Alemany, L.B.; *et al.*, Pristine graphite oxide. *J. Am. Chem. Soc.* **2012**, *134* (5), 2815–2822.
[34] Li, D.; Müller, M.B.; Gilje, S.; *et al.*, Processable aqueous dispersions of graphene nanosheets. *Nature Nanotechnol.* **2008**, *3* (2), 101–105.
[35] Cote, L.J.; Kim, F.; Huang, J., Langmuir–Blodgett assembly of graphite oxide single layers. *J. Am. Chem. Soc.* **2009**, *131* (3), 1043–1049.
[36] Xu, Z.; Gao, C., Graphene chiral liquid crystals and macroscopic assembled fibres. *Nature Commun.* **2011**, *2* (1) 571.
[37] Hu, X.; Xu, Z.; Gao, C., Multifunctional, supramolecular, continuous artificial nacre fibres. *Sci. Rep.* **2012**, *2*, 767.
[38] Cong, H.P.; Ren, X.C.; Wang, P.; Yu, S.H., Wet-spinning assembly of continuous, neat, and macroscopic graphene fibers. *Sci. Rep.* **2012**, *2*, 613.
[39] Xu, Z.; Sun, H.; Zhao, X.; Gao, C., Ultrastrong fibers assembled from giant graphene oxide sheets. *Adv. Mater.* **2013**, *25* (2), 188–193.
[40] Yousefi, N.; Gudarzi, M.M.; Zheng, Q.; *et al.*, Self-alignment and high electrical conductivity of ultralarge graphene oxide-polyurethane nanocomposites. *J. Mater. Chem.* **2012**, *22* (25), 12709–12717.
[41] Dreyer, D.R.; Park, S.; Bielawski, C.W.; Ruoff, R.S., The chemistry of graphene oxide. *Chem. Soc. Rev.* **2010**, *39* (1), 228–240.
[42] Vallés, C.; Young, R.J.; Lomax, D.J.; Kinloch, I.A., The rheological behaviour of concentrated dispersions of graphene oxide. *J. Mater. Sci.* **2014**, *49* (18), 6311–6320.
[43] Kumar, P.; Maiti, U.N.; Lee, K.E.; Kim, S.O., Rheological properties of graphene oxide liquid crystal. *Carbon* **2014**, *80*, 453–461.
[44] King Jr, H.E.; Milner, S.T.; Lin, M.Y.; *et al.*, Structure and rheology of organoclay suspensions. *Phys. Rev. E* **2007**, *75* (2) 021403.
[45] Seth, J.R.; Mohan, L.; Locatelli-Champagne, C.; *et al.*, A micromechanical model to predict the flow of soft particle glasses. *Nature Mater.* **2011**, *10* (11), 838–843.
[46] Macosko, C.W., *Rheology: Principles, Measurements and Applications.* Wiley-VCH, Weinheim, **1994**.
[47] Bird, R.B.; Armstrong, R.C.; Hassager, O., *Dynamics of Polymeric Liquids*, vol. *1*, *Fluid Mechanics*. John Wiley & Sons, Inc., New York, **1987**.
[48] Douglas, J.F., 'Shift' in polymer blend phase-separation temperature in shear flow. *Macromolecules* **1992**, *25* (5), 1468–1474.
[49] Tanner, R.I., *Engineering Rheology*. Oxford University Press, New York, **2000**.
[50] Russel, W.B.; Saville, D.A.; Schowalter, W.R., *Colloidal Dispersions*. Cambridge University Press, Cambridge, **1991**.
[51] Lapasin, R.; Pricl, S., *Rheology of Industrial Polysaccharides: Theory and Applications*. Chapman & Hall, London, **1995**.
[52] Barnes, H.A., Thixotropy – a review. *J. Non-Newtonian Fluid Mech.* **1997**, *70* (1–2), 1–33.
[53] Einstein, A., Eine neue Bestimmung der Moleküldimensionen. *Ann. Phys.* **1906**, *19*, 289–306.
[54] Einstein, A., Berichtigung zu meiner Arbeit: "Eine neue Bestimmung der Moleküldimensionen". *Ann. Phys.* **1911**, *34*, 591–592.
[55] Brenner, H., Rheology of a dilute suspension of dipolar spherical particles in an external field. *J. Colloid Interface Sci.* **1970**, *32*, 141–158.
[56] Jaworek, A.; Sobczyk, A.T., Electrospraying route to nanotechnology: an overview. *J. Electrostat.* **2008**, *66* (3–4), 197–219.
[57] Derby, B., Inkjet printing of functional and structural materials: fluid property requirements, feature stability, and resolution. *Annu. Rev. Mater. Res.* **2010**, *40*, 395–414.
[58] Chen, D.T.N.; Wen, Q.; Janmey, P.A.; *et al.*, Rheology of soft materials. *Annu. Rev. Condens. Matter Phys.* **2010**, *1*, 301–322.
[59] Sollich, P.; Lequeux, F.; Hébraud, P.; Gates, M.E., Rheology of soft glassy materials. *Phys. Rev. Lett.* **1997**, *78* (10), 2020–2023.
[60] Kroon, M.; Vos, W.L.; Wegdam, G.H., Structure and formation of a gel of colloidal disks. *Int. J. Thermophys.* **1998**, *19* (3), 887–894.

[61] Ovarlez, G.; Barral, Q.; Coussot, P., Three-dimensional jamming and flows of soft glassy materials. *Nature Mater.* **2010**, *9* (2), 115–119.
[62] Jalili, R.; Aboutalebi, S.H.; Esrafilzadeh, D.; *et al.*, Scalable one-step wet-spinning of graphene fibers and yarns from liquid crystalline dispersions of graphene oxide: towards multifunctional textiles. *Adv. Funct. Mater.* **2013**, *23* (43), 5345–5354.
[63] Jalili, R.; Razal, J.M.; Wallace, G.G., Exploiting high quality PEDOT:PSS–SWNT composite formulations for wet-spinning multifunctional fibers. *J. Mater. Chem.* **2012**, *22* (48), 25174–25182.
[64] Jalili, R.; Razal, J.M.; Innis, P.C.; Wallace, G.G., One-step wet-spinning process of poly(3,4-ethylenedioxythiophene):poly(styrenesulfonate) fibers and the origin of higher electrical conductivity. *Adv. Funct. Mater.* **2011**, *21* (17), 3363–3370.
[65] Jalili, R.; Razal, J.M.; Wallace, G.G., Wet-spinning of PEDOT:PSS/functionalized-SWNTs composite: a facile route toward production of strong and highly conducting multifunctional fibers. *Sci. Rep.* **2013**, *3*, 3438.
[66] Esrafilzadeh, D.; Razal, J.M.; Moulton, S.E.; *et al.*, Multifunctional conducting fibres with electrically controlled release of ciprofloxacin. *J. Control. Release* **2013**, *169* (3), 313–320.
[67] Hough, L.A.; Islam, M.F.; Janmey, P.A.; Yodh, A.G., Viscoelasticity of single wall carbon nanotube suspensions. *Phys. Rev. Lett.* **2004**, *93* (16), 168102.
[68] Williams, G.; Seger, B.; Kamat, P.V., TiO_2–graphene nanocomposites. UV-assisted photocatalytic reduction of graphene oxide. *ACS Nano* **2008**, *2* (7), 1487–1491.
[69] Hernandez, Y.; Nicolosi, V.; Lotya, M.; *et al.*, High-yield production of graphene by liquid-phase exfoliation of graphite. *Nature Nanotechnol.* **2008**, *3* (9), 563–568.
[70] Eda, G.; Fanchini, G.; Chhowalla, M., Large-area ultrathin films of reduced graphene oxide as a transparent and flexible electronic material. *Nature Nanotechnol.* **2008**, *3* (5), 270–274.
[71] Loh, K.P.; Bao, Q.; Eda, G.; Chhowalla, M., Graphene oxide as a chemically tunable platform for optical applications. *Nature Chem.* **2010**, *2* (12), 1015–1024.
[72] Hisakado, Y.; Kikuchi, H.; Nagamura, T.; Kajiyama, T., Large electro-optic Kerr effect in polymer-stabilized liquid-crystalline blue phases. *Adv. Mater.* **2005**, *17* (1), 96–98.
[73] Demus, D.; Goodby, J.; Gray, G.W.; Spiess, H.-W.; Vill, V., *Handbook of Liquid Crystals*, vol. *1*, *Fundamentals*. Wiley-VCH, Weinheim, **1998**.
[74] Shen, T.-Z.; Hong, S.-H.; Song, J.-K., Electro-optical switching of graphene oxide liquid crystals with an extremely large Kerr coefficient. *Nature Mater.* **2014**, *13* (4), 394–399.
[75] Jiménez, M.L.; Fornasari, L.; Mantegazza, F.; *et al.*, Electric birefringence of dispersions of platelets. *Langmuir* **2012**, *28* (1), 251–258.
[76] Dimiev, A.M.; Alemany, L.B.; Tour, J.M., Graphene oxide. Origin of acidity, its instability in water, and a new dynamic structural model. *ACS Nano* **2013**, *7* (1), 576–588.
[77] Zhang, H.B.; Zheng, W.G.; Yan, Q.; *et al.*, The effect of surface chemistry of graphene on rheological and electrical properties of polymethylmethacrylate composites. *Carbon* **2012**, *50* (14), 5117–5125.
[78] El Achaby, M.; Arrakhiz, F.E.; Vaudreuil, S.; *et al.*, Mechanical, thermal, and rheological properties of graphene-based polypropylene nanocomposites prepared by melt mixing. *Polym. Compos.* **2012**, *33* (5), 733–744.
[79] Kinloch, I.A.; Roberts, S.A.; Windle, A.H., A rheological study of concentrated aqueous nanotube dispersions. *Polymer* **2002**, *43* (26), 7483–7491.
[80] Esumi, K.; Ishigami, M.; Nakajima, A.; *et al.*, Chemical treatment of carbon nanotubes. *Carbon* **1996**, *34* (2), 279–281.

5

Optical Properties of Graphene Oxide

Anton V. Naumov

5.1 Introduction

Since its Nobel Prize-winning discovery [1], graphene has been among the frontier topics of scientific research in optics, electronics, materials science and biotechnology. Its high conductivity and remarkable optical properties have attracted attention toward a number of potential applications of graphene in touchscreens, liquid crystal displays (LCDs) and light-emitting diodes [2]. Due to its 97.7% transparency in the visible [3], in these applications graphene is envisioned to act in the form of transparent conductive electrodes. Other optical properties of graphene, including its Raman spectra that vary with the number of graphene layers [4], its infrared transitions [5] and optical phonon–electron coupling [6] also give rise to important potential applications in electronics. However, despite the benefits of using graphene in optical and electronic devices, in the majority of the above-mentioned applications, graphene can serve only as a passive non-emissive element. Such a limitation is dictated by its structure: because of two-dimensional (2D) symmetry, graphene electronically is a zero-bandgap semiconductor, in which electrons, when excited to the higher energy levels, will relax down non-radiatively.

In order to obtain a detectable emission from graphene, a sizable optical bandgap should be opened. This task could be accomplished in principle by decreasing the dimensionality of graphene, which leads to a bandgap opening due to quantum confinement. Thus, lower-dimensional structures such as graphene nanoribbons and quantum dots are expected to show optical emission. Recently a number of such fluorescent structures have been produced by electrochemical exfoliation [7], hydrothermal scission of graphene sheets [8], acidic treatment of carbon fibers [9] and microwave heating [10]. However, a limited

Graphene Oxide: Fundamentals and Applications, First Edition. Edited by Ayrat M. Dimiev and Siegfried Eigler.

particle size and complexity in production often hinders potential applications of these novel nanostructures in larger-scale optoelectronics. Thus, there is a need for a more applicable graphene-based structure that could be mass-produced with a specified set of optical properties. Functionalized graphene and, specifically, a more common form of it, graphene oxide (GO), appear to be more suitable for that role.

Functionalization by oxygen-containing addends results in a gap between the valence and the conduction bands in GO, turning it into a semiconductor, or even an insulator, depending on the degree of derivatization [11–13]. Such variability gives GO a significant advantage, since its structure can be tailored for a specific microelectronics application by additional functionalization or removal of oxygen functionalities through reduction [11, 12]. Simple and cost-effective production of GO from graphite [14, 15] together with large size variability and modifiable electronic properties make GO an attractive candidate for modern optoelectronics. Modification of the bandgap in GO leads to the change in its optical properties, including absorption, photoluminescence and Raman spectra.

5.2 Absorption

The absorption spectra of GO are dictated by its electronic transitions in the ultraviolet (UV), with their tail extending into the visible. The most prominent absorption feature in all of the GO samples is a $\pi \rightarrow \pi^*$ electronic transition, which is common for aromatic C=C bonds. The p orbitals of carbon can be combined either in-phase (Figure 5.1A) or out of phase (Figure 5.1B), producing bonding and antibonding combinations. This gives rise to π and π^* orbitals, where the π orbital is lower in energy than the π^*, allowing for a photon-induced transition between π and π^*. For GO this transition dominates the absorption spectrum and usually occurs at ~230 nm [16–20].

In addition to the $\pi \rightarrow \pi^*$ transition, a typical absorption spectrum of the GO prepared by the modified Hummers method [14] appears to have a shoulder at ~300 nm (Figure 5.2) that is attributed to the $n \rightarrow \pi^*$ transition in the C=O bonds of oxygen-containing functional groups [16–20]. This transition involves promoting an electron from a lone-pair non-bonding n orbital to the antibonding π^* orbital via photon absorption. Since, on the energy scale, the

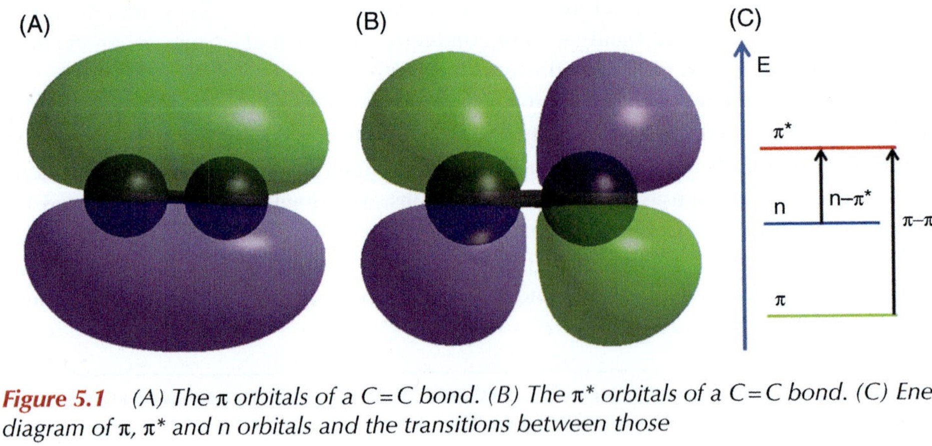

Figure 5.1 (A) The π orbitals of a C=C bond. (B) The π orbitals of a C=C bond. (C) Energy diagram of π, π* and n orbitals and the transitions between those*

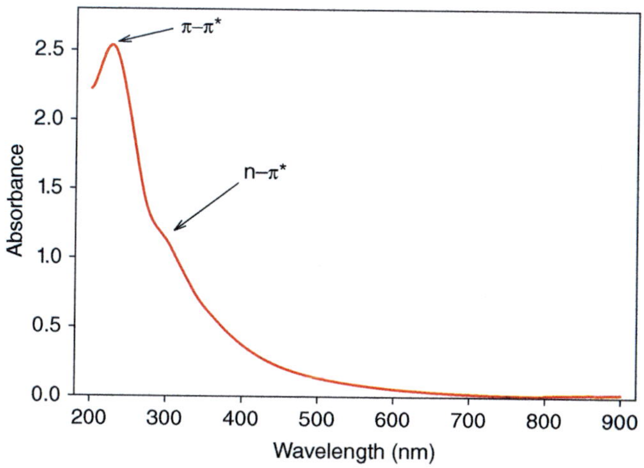

Figure 5.2 *Absorption spectrum of GO produced by the modified Hummers method*

non-bonding orbital is between π and π* (Figure 5.1C), the n → π* transition is less energetic and somewhat red-shifted from the π → π* absorption feature. The ratio in peak intensities between the two features can be explained by a lower molar absorptivity for the n → π* transition as opposed to π → π* [21].

Thus, absorption at ~300 nm is characteristic of the presence of carbonyl groups in GO. This is also supported by the X-ray photoelectron spectroscopy (XPS) data, according to which there are four main oxygen-containing functionalities of carbon present in a variety of GO samples. The dominant addend groups are hydroxyl (–OH) and epoxy, with a smaller contribution from carbonyl (C=O) and carboxyl (COOH) groups [16, 18, 22, 23]. The presence of these functional groups in GO is also verified by Fourier-transform infrared (FTIR) spectra [23] (Figure 5.3). Vibrations associated with hydroxyl, epoxy and carboxyl groups are present below 1500 cm^{-1}, whereas above 1500 cm^{-1} at ~1700 cm^{-1} there is a detectable band for the C=O stretch of the carbonyl group [19, 24, 25].

Some reports show GO with a greater content of carbonyl groups [26] that become depleted upon controllable reduction under a Xe lamp. Absorption spectra could be employed to monitor such changes in GO composition. For example, upon chemical reduction via acid-mediated dehydration of GO flakes, the absorption feature at ~300 nm corresponding to the n → π* transition in C=O exhibited a gradual decrease in intensity [27]. This was followed by spectrum broadening, yielding increased absorption in the visible. The main absorption peak experienced a non-monotonic change in intensity. This was explained as the result of two competing processes: decrease in oxygen content, yielding suppression of the 300 nm absorption feature; and restoration of the sp^2 carbon network, resulting in the absorption broadening throughout the visible and an increase in the intensity of the π → π* absorption peak. A different reduction treatment of GO with either KOH [23] or hydrazine [28, 29] also caused the loss of the 300 nm absorption feature combined with broadening into the visible, that was associated with the removal of oxygen-containing groups. A reverse process involving controlled ozone-induced oxidation of reduced graphene oxide (RGO) into GO produced completely opposite changes: a blue shift of absorption spectra and significant reduction of

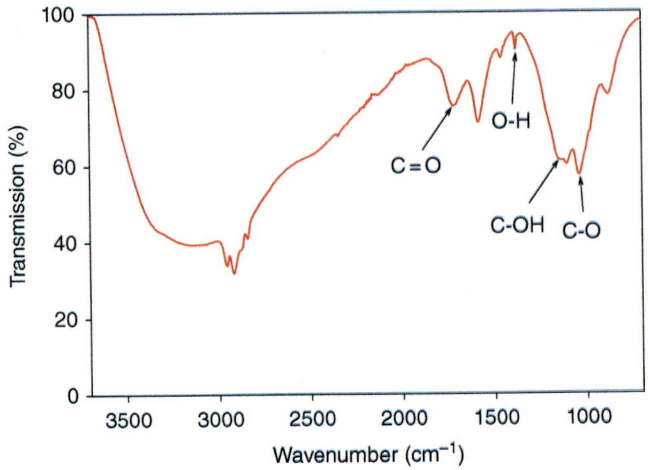

Figure 5.3 *Infrared (IR) spectrum of GO prepared by the modified Hummers method. Labeled features correspond to vibrational bands in functional groups*

absorption background in the visible with oxidation time (Figure 5.4A). That was indicative of the decrease in the sp^2 graphitic structure, which is highly absorbing in the visible, due to the formation of oxidized sp^3 regions. Such descriptive spectral changes suggest visible absorption as a possible characterization technique indicating the variation in the amount of oxygen-containing functional groups in GO due to oxidation or reduction.

Unlike oxidation or reduction, a change in pH of the surrounding medium appeared to introduce only slight variations to the UV–visible absorption spectra of GO, which did not affect the absolute intensities of the peaks [17, 24] or their positions. This signifies that, even though pH change might cause protonation/deprotonation of the oxygen-containing addends [24], the functional groups would still remain intact. Upon passivation of the functional groups with alkylamines [30] again, no change in peak position or loss of absorption features was observed; moreover, they became more pronounced due to modification but retention of the oxygen functionalities.

Despite the lack of significant change in UV–visible absorption with pH (Figure 5.4B), IR absorption showed substantial variation indicative of protonation/deprotonation of the functional groups, resulting in animation of different vibrational transitions [23]. Figure 5.5 illustrates the IR spectra of GO in acidic and basic conditions, showing significant differences between transition energies, and pointing to potential differences in oxidation state of the functional groups in either acidic or basic environments. A substantial spectral change in the IR is also observed in the case of alkylamine passivation of the functional groups [30]. Thus, IR absorption serves as an indicator of modification of the functional groups in GO. In addition to the single-photon absorption described above, ultrafast pump–probe spectroscopy in GO has shown that, in more heavily functionalized GO materials with larger sp^3 regions, two-photon absorption is possible at high excitation intensities [31]. This effect was attributed to the presence of a larger bandgap in sp^3 domains, which allows for the absorption of two

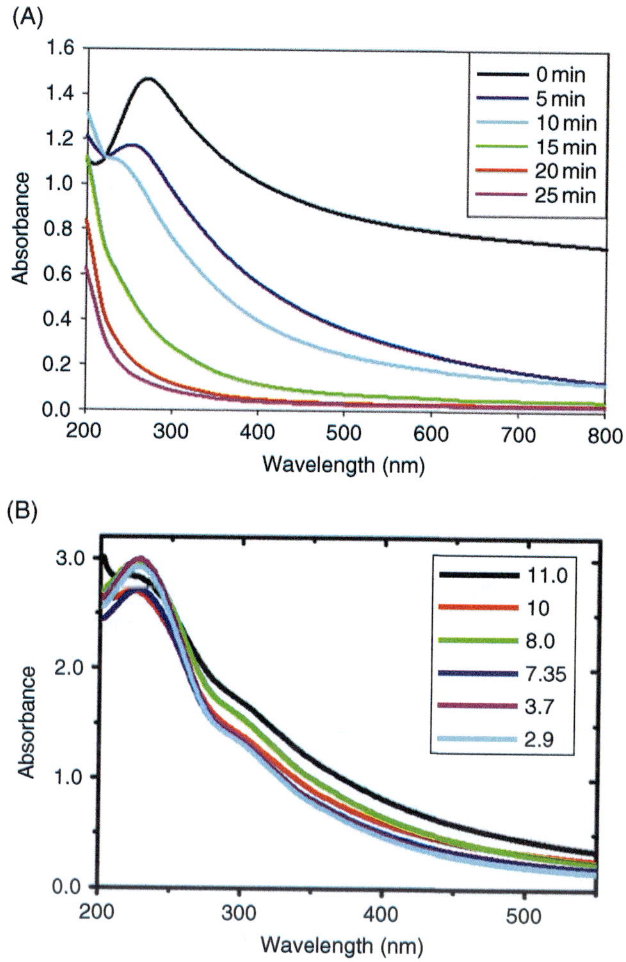

Figure 5.4 *(A) Absorbance spectra of RGO oxidized via controlled ozone treatment for 0 to 25 minutes. (B) Absorbance spectra of GO at different pH values [24]. Adapted from [24] with permission of Macmillan Publishers Limited*

photons. In contrast, the sp^2 domains in GO with smaller predicted bandgaps were expected to experience only a single-photon absorption. Thus, varying the ratio of sp^2 to sp^3 domains in GO through controlled oxidation/reduction can allow the adjustment of its nonlinear optical properties, which is highly desirable for applications in second harmonic generation.

Oxidation/reduction treatment altering sp^2/sp^3 structure could be sufficient to introduce substantial changes, which would be reflected in the transient absorption [32, 33]. This was probed by experiments involving pump–probe transmission spectroscopy: the pump beam was used to excite electrons, altering the electronic configuration in GO, which was further monitored with a probe beam. Such spectroscopic experiment through the absorption or transmission of the probe beam after the pump event allowed the relaxation kinetics of the

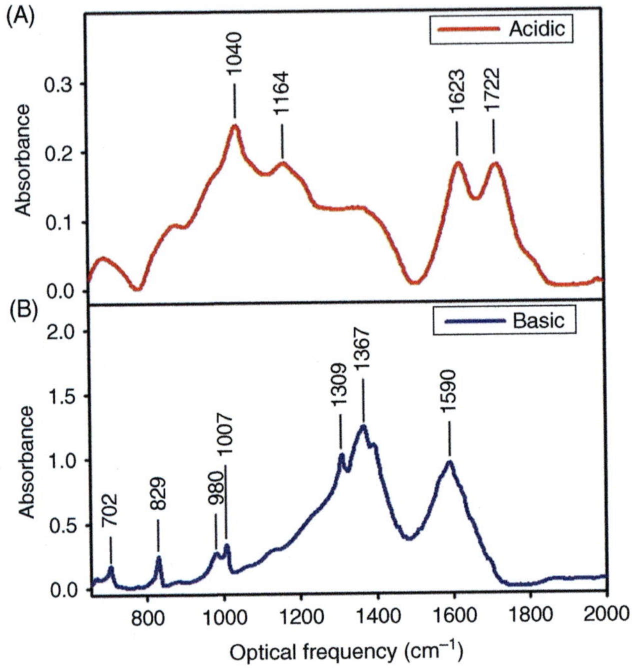

Figure 5.5 *Attenuated total reflectance Fourier-transform infrared (ATR-FTIR) spectra of solid GO samples isolated from (A) acidic and (B) basic aqueous suspensions. Major bands and proposed assignments: (A) 1040 cm^{-1} (C–O stretch), 1164 cm^{-1} (C–OH stretch), 1623 cm^{-1} (adsorbed water and skeletal vibrations of unoxidized graphitic domains), 1722 cm^{-1} (C=O stretch); and (B) 829 cm^{-1} (C–H out-of-plane wag), 980 cm^{-1} (possibly epoxide stretch), 1007 cm^{-1} (C–H in-plane bend), 1309 cm^{-1} (C–O stretch), 1367 cm^{-1} (COO– symmetric stretch), 1590 cm^{-1} (COO– antisymmetric stretch) [24]. Adapted from [24] with permission of Macmillan Publishers Limited*

excited charge carriers to be studied. It was observed that, upon reduction, transient absorption experiences substantial changes: the relaxation rates of photoinduced charge carriers appeared to be significantly faster for the RGO [32, 33]. In order to properly describe that effect, transient absorption spectra were studied in femtosecond spectroscopy experiments for both GO and RGO with excitation at 400 nm [32]. Transient decay kinetics recorded at 670 nm were fitted by exponential components to recover charge relaxation rates. The kinetics of RGO were 90% dominated by fast lifetime components (0.26 and 2.5 ps) attributed to relaxation due to electron–phonon interactions. Such a fast relaxation mechanism was also present in pristine photoexcited graphene, which makes the RGO carrier dynamics similar to that of graphene [34]. A slow >400 ps component associated with electron traps contributed only 4% of RGO relaxation kinetics, indicating low levels of defects and charge traps. Relaxation times in GO had a greater share of the slow >400 ps component, suggesting that oxygen-related traps and defect states control the relaxation dynamics in GO. Such behavior is consistent with previous studies on carbon nanotubes that have undergone acid purification, indicating that carboxyl and epoxy groups can act as charge traps [35]. An ultrafast pump–probe experiment at 790 nm [33] showed a similar correlation: GO experienced slower decay rates than RGO, but only fast decay components (below 5 ps) were detected for both. The time constants for these components associated with interband carrier–carrier scattering and carrier–phonon scattering were determined to be greater for GO due to its lower carrier density as opposed to RGO. As a result, such ultrafast pump–probe absorption experiments allow for optical sampling of GO electronic structure, indicating slower photoexcited carrier relaxation in GO as opposed to RGO or graphene. They complement steady-state visible and IR absorption spectroscopy that provides characterization of GO structure and the state of its functional groups.

5.3 Raman Scattering

Raman spectroscopy of GO can also serve as an effective characterization method elucidating structure-specific vibrational phonon modes. The Raman spectra of graphite and graphite-based materials such as GO have several distinctive features, including D (~1385 cm^{-1}), G (~1580 cm^{-1}) and G′ or 2D (~2700 cm^{-1}) bands [19, 32, 36–39]. The D band originates from a second-order effect involving a phonon and a defect in the sp^2 graphitic structure; thus it is often attributed to disorder in sp^2 carbon material. The G band comes from first-order Raman scattering in graphitic materials involving a single phonon from the E_{2g} mode involving doubly degenerate in-plane optical vibration in sp^2 carbon. The G′, often called 2D, band originates from the two-phonon process involving phonons with opposite momenta and depends strongly (inversely) on the number of graphene layers.

Other less prominent features such as D′, D+G, D+D′, G+D′, 2D′, etc. are also detectable in some cases [40]; however, since the G band describes the sp^2 graphitic structure and the D band is associated with defects in that structure, in the current discussion we will be mostly concerned with the G and D bands. The ratio between their intensities, I_D/I_G, is thus a good measure of the relative defect content in the sp^2 carbon lattice. These bands in GO (Figure 5.6) are somewhat different from those in graphite and graphene: the D band is broad and intense in GO, and the G band is somewhat suppressed, broadened substantially [41] and sometimes slightly shifted to higher frequencies [27, 42]. This shift can possibly be associated with doping effects [43] in accordance with the doping-induced G-band shifting behavior observed in as-synthesized GO [44, 45]. Another explanation is based on theoretical calculations [41] suggesting that an alternating pattern of single- and double-bonded carbons in GO would result in the blue shift in the Raman G band corresponding to those double bonds.

The I_D/I_G ratio for GO is larger than that for graphene or graphite, where the D band is often indistinguishable from the background and I_D/I_G is on the order of 0.01. In order to use this ratio for quantitative assessment of disorder, it is important to take into account two defect regimes [47]. As disorder in the graphene lattice begins to increase at low defect densities, the I_D/I_G ratio first also experiences an increase. With further increase in the defect densities ultimately leading to the loss of the carbon lattice, the I_D/I_G ratio starts to decrease due to the nonlinear attenuation of all the Raman peaks when the material starts turning into amorphous carbon [47]. In the low-defect-density regime, I_D/I_G can be

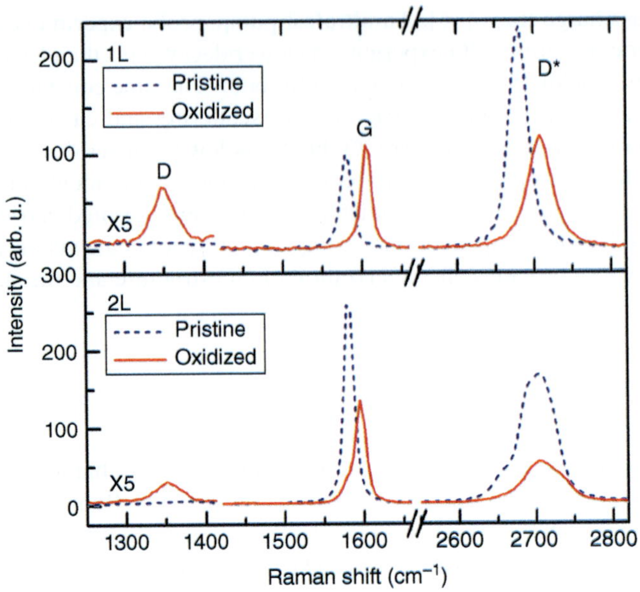

Figure 5.6 *Raman spectra of single-layer (top) and double-layer (bottom) pristine and oxidized graphene. Reproduced from [46] with permission of American Chemical Society*

approximated by an empirical formula relating the ratio to the distance between two adjacent defects, L_a:

$$\frac{I_D}{I_G} = \frac{C(\lambda)}{L_a^2} \tag{5.1}$$

Here λ is the Raman excitation wavelength and $C(\lambda)$ is an empirical parameter: $C(\lambda)=102\ \text{nm}^2$ for excitation wavelength $\lambda=514$ nm [47]. For GO, L_a determines the size of the graphitic domains encircled by oxygen-containing functional groups. For the edge defects, another form of (5.1), the Tuinstra–Koenig relation [48], is more appropriate:

$$\frac{I_D}{I_G} = \frac{C(\lambda)}{L_a} \tag{5.2}$$

Here $C(\lambda)=(2.4\times10^{-10}\ \text{nm}^{-3})\times\lambda^4$ [49] and λ is the Raman excitation wavelength. As opposed to the low-defect-density, in the high-defect-density regime the I_D/I_G ratio depends directly on the graphitic cluster size [50, 51]:

$$\frac{I_D}{I_G} = C'(\lambda)L_a \tag{5.3}$$

Here $C'(514\ \text{nm})=0.55\ \text{nm}^{-2}$ [50, 51]. As a result, the ratio between Raman peak intensities can serve as an estimate for the density of defects in GO and also the size L_a of the graphitic sp^2-hybridized carbon clusters surrounded by sp^3-hybridized functionalized sites.

Considering this dependence of I_D/I_G on the density of oxygen-containing functional groups, we expect significant changes in the Raman spectra depending on the oxidation state of graphene. This was observed in oxidative etching experiments on single- and multilayer graphitic substrates [46]: the I_D/I_G ratio increased with oxidation, indicating the decrease in size of the graphitic carbon domains. In addition to that, as discussed above, a blue shift of the Raman spectra was observed and quantified. This shift exhibited an increase with oxidation treatment and could again be explained by electron transfer doping in oxidation etching experiments [46]. A stronger oxidation via oxygen plasma treatment [52] resulted in a non-monotonic change in I_D/I_G ratio that first experienced a substantial decrease and then levelled off at ~0.7 with even a slight increase. This behavior could be accounted for by considering both low- and high-defect-density scenarios. Due to significant oxidation, at a certain time a number of defects in the graphitic structure may have reached the point at which due to the degradation of six-membered rings the intensity of the Raman D band started to decrease. At such high defect densities the I_D/I_G ratio would decrease with decreasing graphitic cluster sizes according to (5.3). Combining the dependences of the I_D/I_G ratio on the size of the sp^2-hybridized regions for low- and high-defect-density regimes provided a suitable theoretical explanation for the observed behavior and allowed to estimate that the size of the graphitic regions L_a decreased down to ~1 nm upon oxidation. Other works report sp^2 domain sizes in GO thin films and bulk GO samples obtained by Hummers method to be 2.5–6 nm [53, 54].

The reverse process, reduction of GO, has consistently shown red shifts of the G band [55, 56], which agrees with previously doping-attributed blue shifts acquired with oxidation. However, the I_D/I_G ratio changes differently for various works with different reduction procedures and endpoints. Some studies of GO chemical reduction showed no difference in the I_D/I_G ratio [53, 54], implying that, even though oxygen-containing groups were removed by reduction, the defects left in the place of those groups were not cured. Other studies indicate a decrease of the I_D/I_G ratio, suggesting the removal of defects through reduction [57–59]. There are also cases where an increase in D : G intensity ratio was observed [60–62]. That could arise from bringing more disorder into a system and increasing the D-band intensity due to random nucleation of small sp^2 domains. A non-monotonic trend in I_D/I_G ratio was also observed for experiments with gradual chemical reduction: at first D : G intensity ratio was decreased and then increased [63]. This was attributed to growth of sp^2 regions upon initial reduction causing the decrease of I_D/I_G ratio followed by random formation of smaller sp^2 regions inside sp^3 clusters, which increased the disorder in the lattice and the I_D/I_G ratio. Combination of several reduction/oxidation treatments could also result in such non-monotonic behavior [64, 65]. In addition to oxidation and reduction, chemical functionalization of GO also affects Raman spectra. Functionalization treatment of GO with anthryl moieties has shown a red shift of the G band, possibly indicating a change in electronic doping, and an increase in I_D/I_G ratio, indicating additional disorder [66]. Thus Raman spectroscopy of GO can provide important information about the state of lattice disorder, the size of sp^2 carbon regions and electronic doping from chemical addends during oxidation, reduction or functionalization procedures.

5.4 Photoluminescence

Due to functionalization-induced opening of the energy gap, GO is expected to emit photoluminescence across its optical bandgap. The bandgap in graphitic oxide materials was first studied by Hummers [14] in 1958; however, the optical properties of a similar material obtained from graphite by chemical oxidation were studied even earlier, in the mid-nineteenth century by Brodie [67]. Currently, due to the emergence of a large number of potential applications of GO in optoelectronics, its optical properties, including photoluminescence, have been actively studied. At first, two distinct types of emission were observed from aqueous suspensions of GO by a number of scientific groups. Photoluminescence in the blue (350–450 nm) was observed from few-layer GO that had been purified by centrifugation [28]. That emission was attributed to electron–hole recombination in microscopic sp^2 graphitic regions carved out by the oxidation-induced sp^3 structure in GO. At the same time, a green to infrared (500–800 nm) emission from chemically oxidized GO was reported by a number of groups [24, 52, 68, 69]. Such emission (Figure 5.7) was predicted

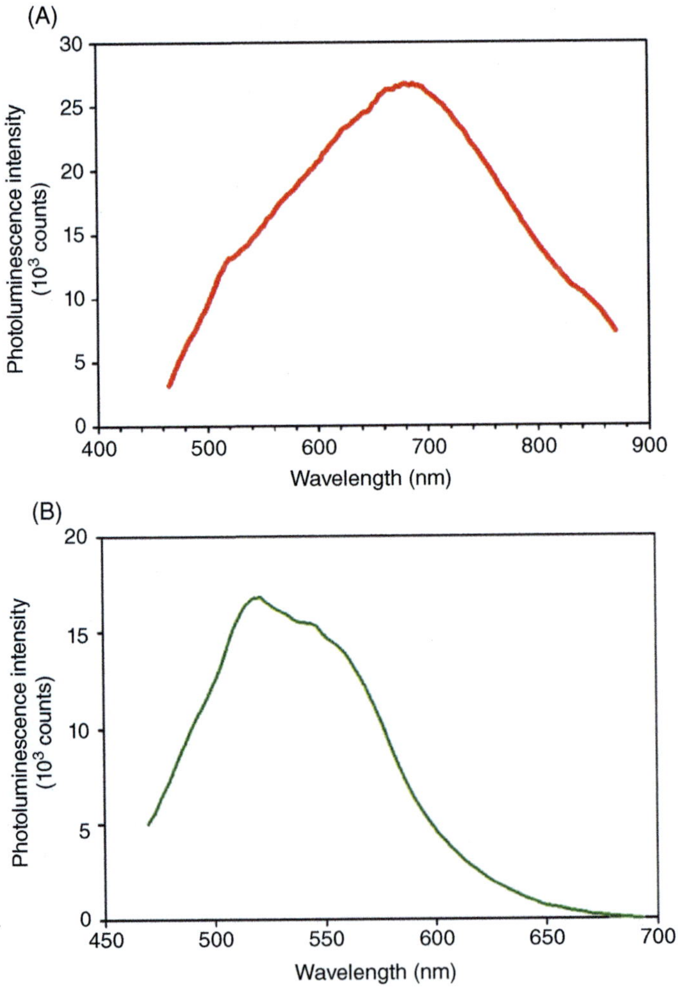

Figure 5.7 *(A) Photoluminescence spectrum of GO produced from graphite by the modified Hummers method [14, 70], 440 nm excitation. (B) Photoluminescence of GO produced by controlled oxidation from RGO; 440 nm excitation*

to arise from localized electronic structure, surrounding functional groups [24, 52], sp^2 carbon regions encircled by repulsive sp^3 carbon hard wall barriers, or bond alteration inducing inter-valley scattering effects [68]. Eda *et al.* [28] suggested that red GO emission is a result of a red shift from interlayer relaxation processes in multi-layer graphene flakes.

These initial works were further followed by a number of predictions and experimental studies of the origin of GO photoluminescence. In order to understand the emission mechanism, a number of factors needs to be considered. Photoluminescence itself can be envisioned as an excitation/emission process over the bandgap opened in graphene due to functionalization. The schematics of such process is depicted in Figure 5.8(A).

Due to the absorption of an incoming photon, an electron is promoted over the bandgap to the higher orbital, leaving a positively charged hole behind. Such an electron–hole pair created through the absorption of a photon is called an exciton. The creation of the exciton is then followed by non-radiative decay of the electron to the lowest unoccupied molecular orbital (LUMO) and of the hole to the highest occupied molecular orbital (HOMO). This process is further concluded by radiative recombination of the exciton. As a result, a photon is emitted with the energy lower than that of the excitation light. The efficiency of this process is described by the fluorescence quantum yield expressed as a ratio of absorbed to emitted photons. Quantum yield is usually below 100% (only 0.02–0.5% reported for GO) [20, 23], due to the electron-hole recombination through a number of non-radiative pathways, including charge defects/traps or phonon-assisted relaxation. Thus, the quantum yield could be expressed as the ratio of the decay rates:

$$Q = \frac{k_r}{k_r + k_{nr}} \tag{5.4}$$

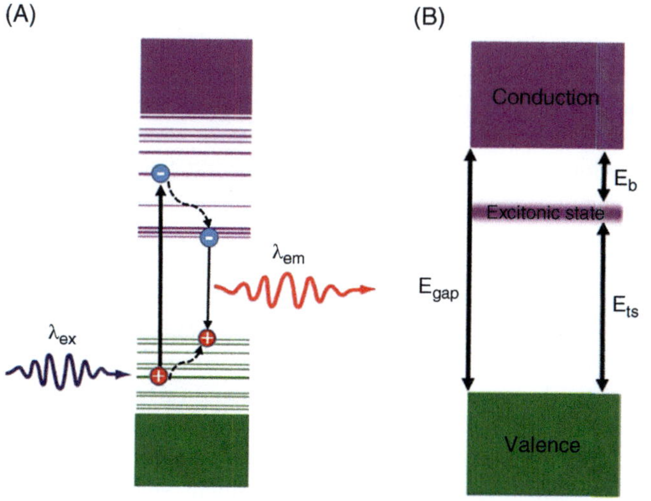

Figure 5.8 *(A) General schematics of the bandgap photoluminescence process. (B) Simplistic schematics of electronic transitions considering excitonic effects, where E_{gap} is a one-electron bandgap energy, E_b is the binding energy of the exciton and E_{ts} is the first allowed optical transition energy*

where k_r is the radiative decay rate and k_{nr} is the non-radiative decay rate. In this way the fluorescence quantum yield is literally the ratio of the rate of radiative emission to the total rate of depopulation of the excited state (i.e. the ratio of the number of emitted photons to the total number of photons in the excited state). Thus, the time that an electron spends in the excited state (fluorescence lifetime, τ) for such a fluorescence process will be the inverse of the total (radiative plus non-radiative) decay rate:

$$\tau = \frac{1}{k_r + k_{nr}} \tag{5.5}$$

Fluorescence lifetimes of the conventional fluorophores are usually in the nanosecond range; however, picosecond lifetimes are not uncommon for nanomaterials [71]. Fluorescence lifetimes of several picoseconds to nanoseconds have been reported in different studies for GO [16, 17, 28, 72]. Chen *et al.* [20] reported different fluorescence lifetimes of the blue and the red emission features with τ for red emission in the picosecond range and for the blue-in the nanosecond range.

There are several models describing photoluminescence in GO. A $\pi \rightarrow \pi^*$ bandgap photoluminescence mechanism envisioned by Eda *et al.* [28] implies that GO is split into smaller and larger clusters of graphitic sp^2 carbon separated by oxidized sp^3 regions. These clusters experience strong localization by surrounding sp^3-hybridized atoms acting as potential barrier walls. Since in graphene electrons act as massless particles, the confinement should induce energy quantization and, therefore, a bandgap [52] on the order of

$$E_{gap} \approx \frac{v_F h}{2d} \approx \frac{2\ \text{eV nm}}{d} \tag{5.6}$$

Here v_F is the Fermi velocity of the electron and d is the diameter of the confined sp^2 region. In their work Eda *et al.* [28] assumed that the photoluminescence process occurs in the small sp^2 carbon clusters that have a larger bandgap on the order of 3.2 eV and enclose roughly about a dozen aromatic rings according to the density functional theory (DFT) calculation presented therein. These clusters were localized by surrounding sp^3 oxidized carbon and, possibly, a few larger sp^2 clusters with smaller bandgaps. Such a model of GO emission is supported by previous theoretical predictions of photoluminescence in functionalized carbon materials [73, 74] that attribute emission to bandgaps in sp^2 graphitic domains. The bandgap values calculated in those works by tight-binding molecular dynamics simulation varied from 0.3 to 2.7 eV depending on the degree of functionalization. GO bandgap calculations were further performed using the DFT method [75]. This rigorous approach yielded $\pi \rightarrow \pi^*$ gaps on the order of 0.3–4 eV, although it did not consider electronic contributions from surrounding sp^3 sites. These results were supported by another model [76] accounting for the effects of functional groups and suggesting that, due to the specific arrangement of those, a bandgap of up to 4 eV could arise in GO from localized stripes of graphitic carbon with zigzag edges. Even such calculations, though, provide only a general estimate, since they were restricted only to epoxide and hydroxyl functional groups.

As opposed to the predictions by Eda *et al.* [28], the size of the graphitic sp^2 islands in GO sheets was estimated by high-resolution transmission electron microscopy (HRTEM) to be somewhat larger, on the order of 1–2 nm [77]. This would correspond to a smaller bandgap and thus a red-shifted emission in the green, red (Figure 5.7) or even near-infrared (NIR) parts of the spectra. Such NIR emission was observed in aqueous suspensions of GO and described using theoretical modeling [78]. In that scenario, the GO structure was represented by circular islands of sp^2 carbon several nanometers in diameter surrounded by sp^3 functionalized regions acting as potential barriers. The single-electron bandgap of such graphene nanodisks calculated using first-principles DFT-based GW approximation and Bethe–Salpeter approaches was shown to decrease with the size of the nanodisk. However, unlike the previously mentioned works, attributing photoluminescence to the emission across the electronic gap, the one by Kozawa *et al.* [78] suggested also taking into account the binding energy of the exciton. The emission energy was found to be less than the bandgap by an amount equal to the exciton binding energy (Figure 5.8B) similarly to the exciton-related fluorescence emission in carbon nanotubes [79]:

$$E_{ts} = E^{GW} - E_b \tag{5.7}$$

Here E^{GW} is the single-electron bandgap calculated by the GW approach, E_b is the exciton binding energy and E_{ts} is the optical transition energy (emission energy) [78]. Consistent with this model, the exciton binding energy in functionalized graphene was earlier predicted to constitute a large portion of the bandgap (up to 40%) [80]. The exciton binding energy in graphene nanodisks was found to decrease (Figure 5.9A) with their size, [78] similarly to the diameter dependence of the exciton binding energy in carbon nanotubes. Further calculations of the emission energy (E_{ts}) showed an inverse dependence on the nanodisk diameter, resulting in a nearly linear 1/D trend (Figure 5.9B). These results allow correlation of GO photoluminescence emission energy with the size of the sp^2 graphitic circular islands and highlight the importance of the excitonic effects in GO.

Based on the preparation of GO, one would expect a wide distribution of sizes of graphitic sp^2 regions. This is in part confirmed by the shift in the emission spectra observed in a number of studies [16, 72, 78] upon the change in excitation the wavelength (Figure 5.10A). When excited in resonance, graphitic regions of particular sizes would be responsible for photoluminescence emission. However, when excitation is shifted, for example, to the red, the larger sp^2 regions would then appear in resonance and exhibit red-shifted emission. Due to the width of the emission spectra, it is likely, though, for photoluminescence to originate from different sizes of graphene nanodisks, resulting in a wider distribution of emission energies according to Figure 5.9(B). This can be possible with just a single excitation by the means of energy transfer between adjacent nanodisks of different size. Such transfer between sp^2 graphitic regions is suggested by the inverse dependence of photoluminescence decay lifetimes on emission energy [78]. The longest decay times were shown by emission at low energies, indicating an energy build-up behavior at large nanodisks with low transition energy. That can be explained by cascade energy transfer from smaller nanodisks with higher emission energies to the larger-diameter ones. Thus in the process of minimizing the energy of the excited electron, excitation could be transferred from small nanodisks with large gap to larger nanodisks with lower energetic states where the emission would happen.

This relaxation process involves multiple energy transfer steps and therefore results in longer lifetimes for less energetic emission [78]. That also suggests that due to the energy

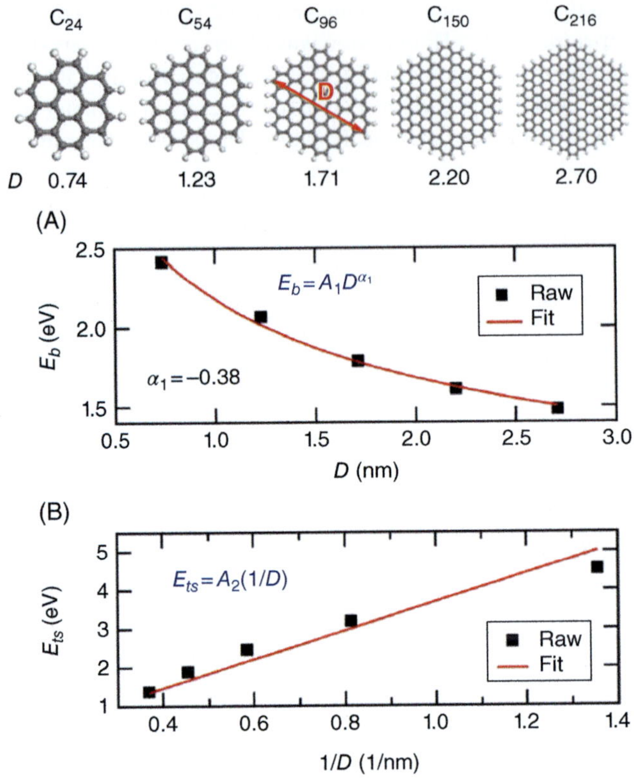

Figure 5.9 Structural models of graphene nanodisks and the scaling characteristics of optical properties. (A) Scaling relations for the exciton binding energy (E_b) with the size of the graphene nanodisks (D). (B) Transition energy (E_{ts}) as a function of the inverse size (1/D). Reproduced from [78] with permission of American Chemical Society

transfer between nanodisks, GO photoluminescence would be more likely to be observed in the red or NIR, region.

Thus, the position of the photoluminescence peak could be described by a combination of two contributing factors: the relative abundance of sp^2 graphitic regions of a particular size and the efficiency of energy transfer to the regions of the larger size. The first is governed by the degree of oxidation in GO and, possibly, the type of functionalization process. The second factor influencing the energy transfer is given in part by the proximity of graphitic regions in the sp^3 matrix, which, again, relies on the state of GO functionalization and the distribution of sp^2 clusters. The dependence of these factors on the degree of oxidation is, however, hard to predict: stronger oxidation could result in the decrease of graphitic regions in size, or, possibly, the disappearance of smaller sp^2 domains, or a combination thereof. These effects could be studied experimentally through either additional oxidation or reduction of GO.

Reduction experiments have been carried out by a number of scientific groups [18, 26, 28, 56, 63, 68, 81]. Even though some of these studies report partly contradicting observations, most of them agree on the decrease in GO photoluminescence intensity upon

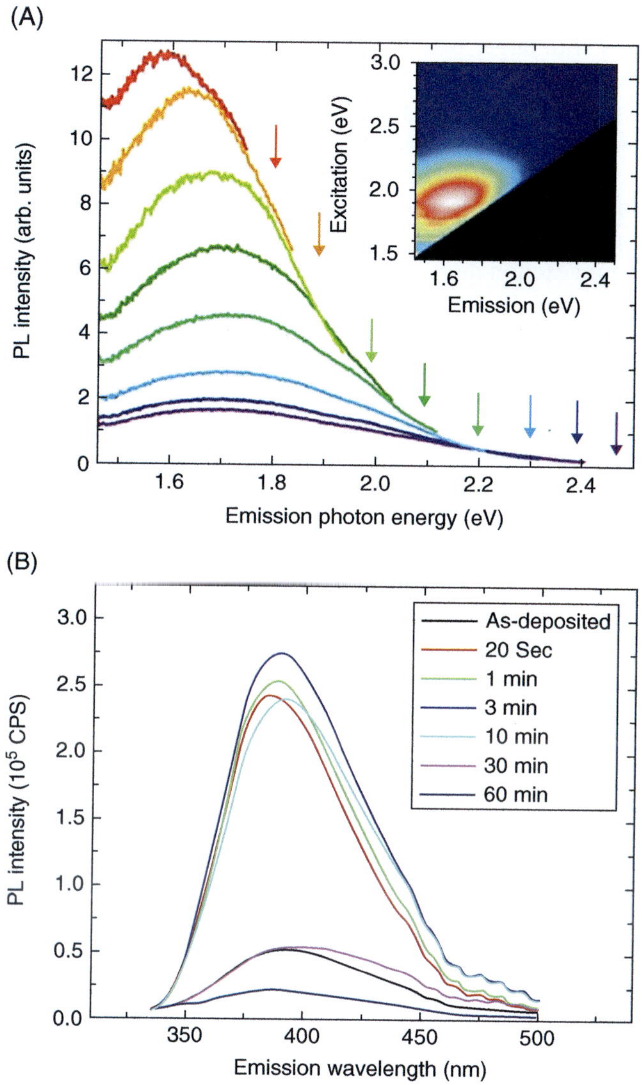

Figure 5.10 (A) PL spectra of aqueous GO suspensions at excitation energies from 1.8 to 2.5 eV indicated by arrows. Inset: A photoluminescence emission intensity map of GO. (B) Change of photoluminescence spectra of GO with hydrazine reduction time. (A) Reproduced from [78] with permission of American Chemical Society. (B) Adapted from [28] with permission of John Wiley & Sons

reduction (Figure 5.10B). In addition to decrease in the emission intensity, a number of works also report a blue shift [18, 82] that contradicts the theoretical prediction of bandgap shrinkage upon reduction of GO [75]. Such unexpected behavior could be explained by reduction-induced formation of small sp^2 islands with larger optical bandgaps: it had been suggested that reduction at first does not induce the increase of sp^2 cluster size in GO, but rather stimulates nucleation of smaller-sized sp^2 domains [26]. According to that, newly created sp^2 domains with larger bandgaps would indeed shift the emission into the blue. This hypothesis is also supported by observation of slightly non-monotonic behavior of emission intensity upon chemical or thermal reduction of GO [28, 56]: in those experiments photoluminescence at first experienced an increase followed by a steady decrease with further GO reduction. This may point to the formation of new sp^2 regions responsible for the increase in photoluminescence at the initial stages of the process. Further reduction could cause photoluminescence quenching due to percolation between sp^2 regions allowing for increased exciton migration eventually leading those to non-radiative recombination sites [28].

GO optical behavior has also shown a dependence on the type of reduction process. Thermal reduction resulted in a less abrupt change in GO photoluminescence than chemical reduction [81] and appeared to yield less blue-shifted photoluminescence [22] possibly due to a weaker excitonic confinement (larger sp^2 clusters). Gradual reduction under the steady-state Xe lamp irradiation (500 W) [26] has shown the complex behavior of GO photoluminescence: an initial emission feature in the red has decreased significantly with concurrent increase of a sharper emission peak at ~450 nm. The authors attribute this to a dual nature of photoluminescence in GO. The red predominant emission feature of as-produced GO was suggested to arise from optical transitions in functionalization disorder-induced localized states (Figure 5.11A) that lie deep within the $\pi \rightarrow \pi^*$ gap. The feature in the blue arising as a result of reduction was attributed to the emission from confined sp^2 regions appearing during the initial reduction of GO (Figure 5.11B) as suggested previously [28].

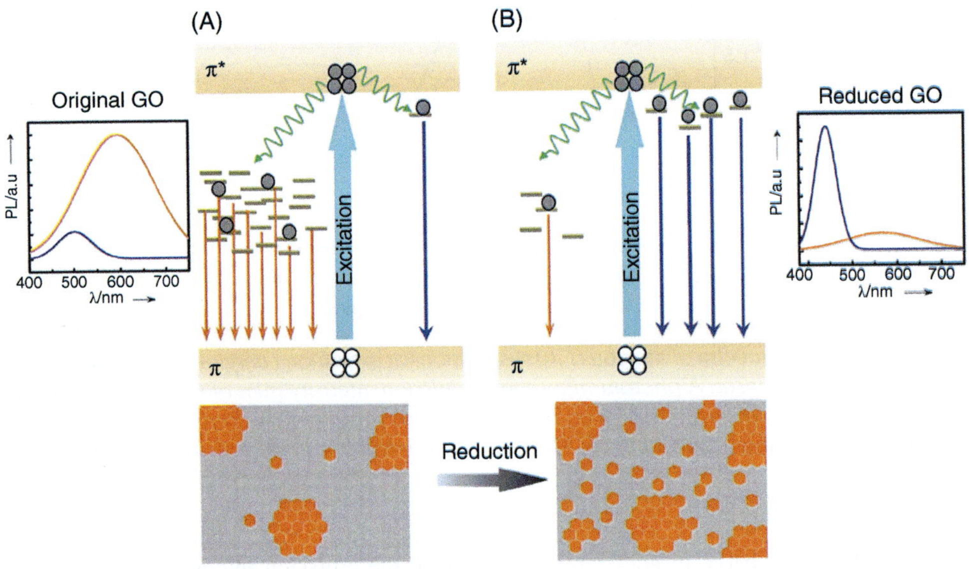

Figure 5.11 Schematics of the dual nature of photoluminescence in reduced GO. (A) The predominant emission in GO from disorder-induced localized states. (B) The predominant emission in thermally reduced GO from confined cluster states. Adapted from [26] with permission of John Wiley & Sons

These findings confirm the sp^2 cluster-based model of GO emission in the blue and suggest an alternative explanation for the commonly observed GO photoluminescence in the red.

A light chemical reduction was observed to cause a red shift in GO photoluminescence spectra [63]. This finding contradicts the reduction-induced blue shifts observed by others [18, 81] and suggests that initially upon a minor reduction some growth of sp^2 regions in GO could happen. However, based on experimental evidence, such growth was suggested to be limited by the domain size of 2 nm, after which the nucleation of new sp^2 domains would take place [63]. At a greater reduction, as proposed previously [28], photoluminescence is quenched because of exciton recombination at the defect sites due to their increased migration allowed by reduction-facilitated percolation of sp^2 regions. This model of GO reduction is supported by the above-mentioned Raman spectral changes [63]: the I_D/I_G ratio decreases at a mild reduction corresponding to the initial growth of sp^2 domains and, therefore, decrease in disorder. At a stronger reduction, the I_D/I_G ratio starts to rise due to the chaotic appearance of multiple newly nucleated sp^2 clusters which increase the disorder. Different approaches to reduction of GO might yield different behaviors; however, when analyzed, they provide a better picture of the complex and variable lattice morphologies arising during this process and affecting the optical properties of GO.

In order to achieve a better understanding of the origins of GO photoluminescence, it is also critical to study the changes introduced into emission by the reverse process, an oxidation. Different oxidation routines involving plasma treatment of graphene and acid-assisted oxidation of graphite have all resulted in bandgap opening [52, 82, 83]. That gap in GO could be chemically controlled through timed oxidation treatment with nitric acid [83]. As a result, bandgaps of 1.7 to 2.4 eV were controllably produced, with stronger oxidation causing larger bandgap. That is consistent with oxidation-induced decrease of the size of graphitic islands down to small ones with higher gap energies. Another procedure allowing controllable oxidation of graphitic materials involves timed ozone treatment of non-emissive reduced graphene oxide (RGO) obtained from GO by hydrazine reduction. Upon oxidation, initially non-emissive RGO has exhibited bright photoluminescence in the visible, with its intensity dependent on the time of ozone treatment (Figure 5.12). In that process, a comparatively small shift in the emission energies was observed, possibly because RGO has already inherited lattice defects that were not removed upon reduction. As suggested by Dong *et al.* [25], confined sp^2 domains could still have been present in RGO from the previous oxidation. These defects would have predetermined the initial size and distribution of oxidation-induced sp^2 islands and thus the emission wavelength range. The other possibility involves a different photoluminescence mechanism, with emission arising from functional groups, which dictates somewhat weaker dependence of emission energy on the degree of oxidation. Such a mechanism, proposed in multiple works studying photoluminescence of GO [22, 24, 52], involves emission from localized states at the functional groups. For example, Cuong *et al.* [22], while observing the strong carrier localization in thermally processed GO (tpGO), suggested that photoluminescence in thermally treated GO occurs due to exciton recombination in localized electronic states rather than from band-edge transitions. The important role of functional groups was also supported by calculation of molecular orbitals in GO [16]. It was shown that photoluminescence is governed by a HOMO–LUMO transition mainly occurring over carbon atoms situated next to oxygen-containing functional groups. According to this work all three kinds of oxygen functionalities, C–O, C=O and

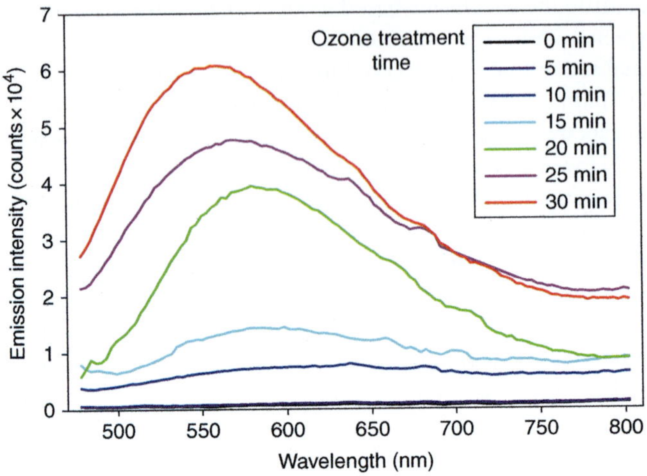

Figure 5.12 *Photoluminescence spectra of ozone-treated RGO*

O=C–OH, were related to the emission. A similar conclusion was achieved while fingerprinting GO photoluminescence to electronic transitions in functional groups [23]. In that study, GO was treated with KOH in order to remove hydroxyl groups and convert epoxy groups to hydroxyl, or with HNO_3 in order to further oxidize GO and produce additional carboxyl groups. Features apparent in the photoluminescence spectra of these variations of GO were attributed to particular electronic transitions (Figure 5.13): $\pi^* \rightarrow \pi$ transition of the C=C bond, $\pi^* \rightarrow n$ transition of the C=O groups and $\sigma^* \rightarrow n$ transition of C–OH groups. As seen from Figure 5.13, the $\pi^* \rightarrow n$ transition was amplified for HNO_3-treated GO and the $\sigma^* \rightarrow n$ transition-for KOH-treated GO as expected due to corresponding excess in C=O and C–OH groups in those. Based on that analysis, photoluminescence transitions in GO were fingerprinted primarily to oxygen-containing functional groups.

The width of GO emission spectra was also discussed: short photoluminescence lifetimes suggest that emission proceeds before charge relaxation to the band edge, therefore bypassing band-edge transitions [23]. Due to the wide spectrum of vibrational states in GO, there is a variety of possible transitions yielding a broad emission spectrum. Similarly to reports discussed above, this work also indicates a dependence of emission wavelength on excitation. However, the Stokes shift between those two appeared to be constant for a particular electronic transition and, therefore, a particular functional group. Such strong correspondence of optical effects in GO with oxygen-containing functional groups together with the evidence provided by photoluminescence spectral mapping to transitions in those groups sets a solid background for the functional-group-related emission model in GO.

Another indication of the functional-group-attributed emission is presented in a joint work by Novoselov, Geim, Ferrari, Hartschuh and coworkers [52]. In that study, the authors confront the emission model based on sp^2 cluster confinement on the grounds of spectral hole burning in oxygen-plasma-oxidized graphene. They predict that, within the framework of the sp^2 domain model, upon intense excitation at a particular wavelength a subset of sp^2 carbon-cluster-based emitters of a certain size should be bleached, leaving a spectral

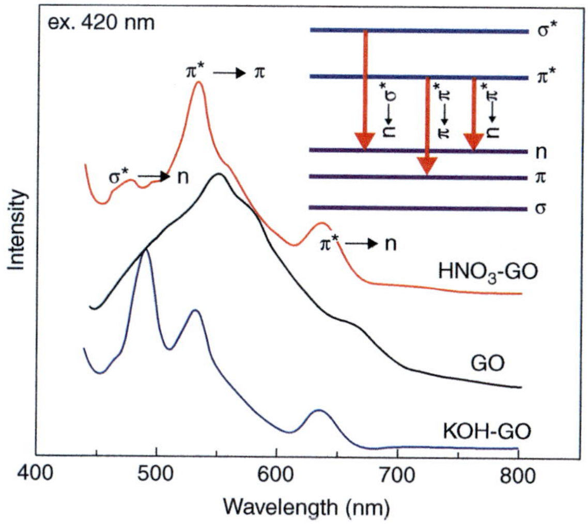

Figure 5.13 Photoluminescence spectra of GO, KOH- and HNO_3-treated GO sheets in deionized water (100 mg ml^{-1}) under excitation of 420 nm. Inset: A schematic energy level diagram depicting the multiple photoluminescence emissions due to electron transitions between different molecular orbitals. Reproduced from [23] with permission of The Royal Society of Chemistry

hole in the piece of plasma-oxidized graphene. However, experimentally, that did not happen; instead, the emission across all wavelengths was quenched in that region. Thus, photoluminescence was traced to localized electronic states at oxidation sites; and the emission spectral width was attributed to homogeneous broadening of the energy states at oxygen-containing emissive species. Such localized electronic states were modeled by Galande *et al.* [24] showing a region of negative electrostatic potential around oxygen-containing addends in GO (Figure 5.14A,B). In that work, the authors for the first time observe a strong pH dependence of GO photoluminescence: upon pH change from acidic to basic a broad GO emission in the red has been quenched and a sharp blue-shifted molecular-looking photoluminescence peak has emerged (Figure 5.14C).

That transition happened sharply at ~ pH 8 and resembled a spectrophotometric titration of molecular fluorophores. From similar excitation spectral features of basic and acidic species, the authors inferred a similar ground state of those emitters, indicating that the pH-induced change had to happen in the excited state. This change was reversible and therefore was attributed to the excited-state protonation of oxygen-containing functional groups. The assignment of the functional groups responsible was made on the basis of comparing excited-state protonation pK_a of GO to the values found previously for organic aromatic compounds with oxygen functionalities. The best match was found with aromatic carboxylic acids having their $pK_a^* \sim 8$ and exhibiting fluorescence in the visible. A model of the graphene sheet with a carboxylic group on it (Figure 5.14A) has shown that in the basic conditions there is a region of electrostatic potential encircling a carboxylic group. That region carves out about 34 carbon atoms from the graphene sheet (Figure 5.14B), which is similar to the size of aromatic carboxylic acids that have $pK_a^* \sim 8$. As a result, pH-induced variations in GO emission

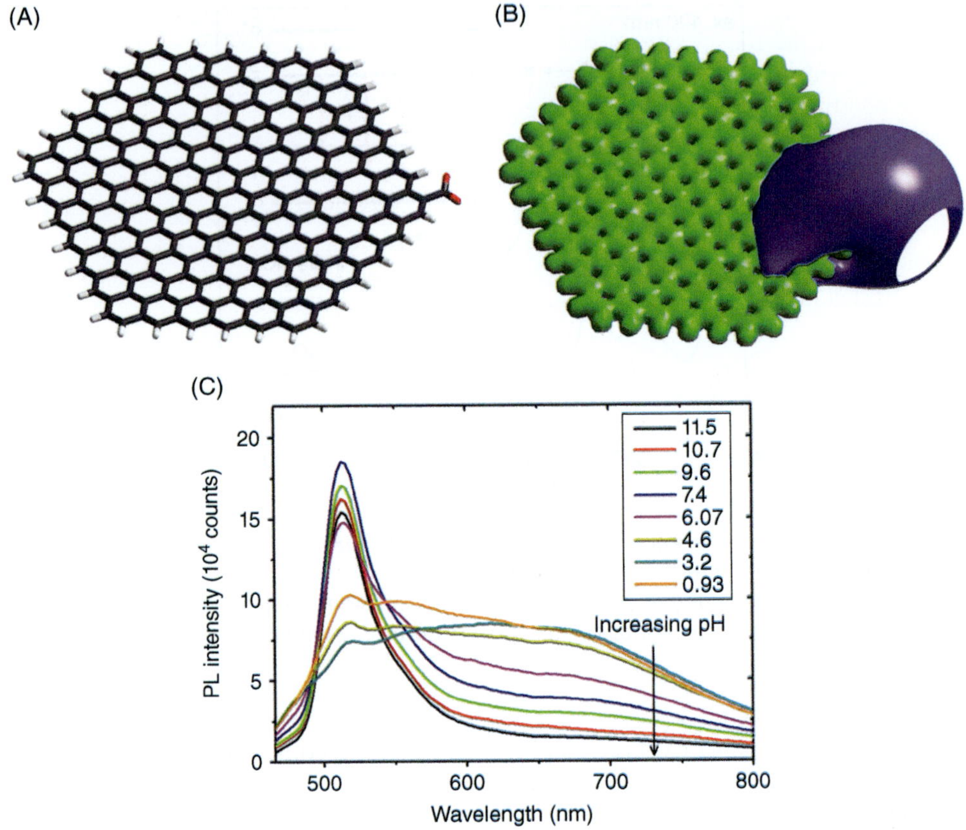

Figure 5.14 *(A) Model of graphene sheet fragment with a COO⁻ functional group at the edge. (B) Computation of electrostatic potential isosurface for a graphene sheet fragment with a COO⁻ functional group at the edge. (C) pH-dependent photoluminescence spectra of GO [24]. Adapted from [24] with permission of Macmillan Publishers Limited*

were attributed to quasi-molecular fluorophores arising from localized electronic structure at the carboxylic functional groups in GO. Further studies also confirm the contribution of carboxylic functional groups to pH-induced changes in GO photoluminescence [17, 19, 72, 84]. A blue basic and a red acidic feature enhanced by deprotonation and protonation in their corresponding pH ranges were shown by Zhang *et al.* [17] to exhibit different fluorescence lifetimes. Thus, such dual pH-dependent emission was attributed not to the band-edge transitions but rather to oxygen-group-based quasi-molecular fluorophores in GO. Other reports of pH-dependent effects have shown a blue emission shift upon the increase in pH, [85] that was associated with the change in GO electronic structure. Some researchers also mentioned a non-monotonic response from two emission peaks in the ultraviolet and blue attributed to graphitic carbon fragments bound to one or several functional groups [72], but most report the decrease of emission intensity of the red emission feature with pH [72, 84]. Such prominent effects of surrounding pH on the optical properties of GO fall in good agreement with the model of functional-group-related photoluminescence in GO.

As a result, we point out two major theories explaining the origins of photoluminescence in GO within the framework of confined sp^2 cluster-related the emission and emission associated with oxygen-containing functional groups. However, there are also scientific works reporting combined optical effects. For example, Chien *et al.* [26] observe two emission peaks in GO with a blue feature attributed to quantum confined graphitic domains and the red peak arising from functional-group-induced disorder states. Reports of such dual fluorescence are not uncommon; however, many of them present different explanations of the nature of the blue and red emission features in the work of Chen *et al.* those emission peaks were proposed to have different luminescence origins since they exhibit fluorescence lifetimes that differ by nearly three orders of magnitude [20]. Exarhos *et al.* [86] suggest that blue and red emission are tied through spectral migration. In their time-resolved photoluminescence measurements, GO and GO reduced via Xe lamp irradiation have also shown notably different lifetimes. Unreduced GO exhibited significantly longer fluorescence lifetimes for its emission in the red versus the blue RGO emission. However, when measured right after the excitation pulse, the emissions of both RGO and GO were identical and detected in the blue region. Such initial similarity suggests that, with time, due to non-radiative spectral relaxation in GO, the emission shifts to the red. This process was explained best by the disordered gap model (Figure 5.15), where at certain spatial regions in GO the gap energies are in the blue range and in some regions they are in the red. Therefore, upon excitation, the exciton in GO could migrate from larger (blue) gap regions to smaller (red) gap regions, where it would experience radiative recombination. In RGO, due to short carrier lifetimes dictated by photoluminescence quenching through non-radiative decay in newly formed zero-gap sp^2 regions, spectral relaxation and shifting would not have enough time to occur. As a result, long-lived excitons that in GO experience spectral migration to the red do not survive in RGO, which yields the short-lived emission in the blue.

Thomas *et al.* [87] even suggested that the blue-shifted feature in GO photoluminescence arises from oxidative debris, heavily oxidized small organic fragments produced

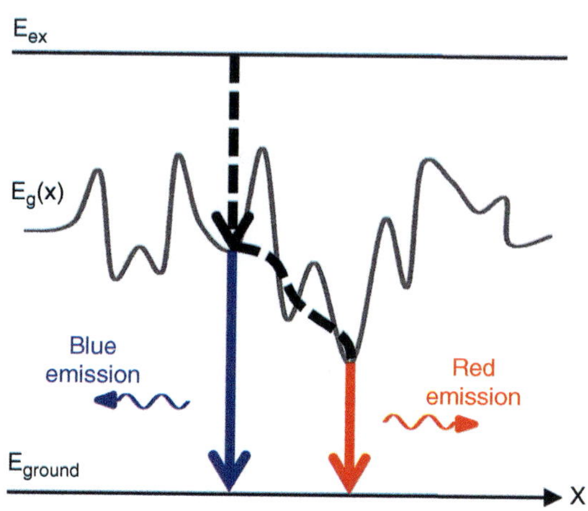

Figure 5.15 *Implied relaxation of excited carriers within the context of a disordered band model of GO. Reproduced from [54] with permission of American Chemical Society*

during GO synthesis. Their work indicates that, when washed with a weak base, the oxidative debris could be separated from the slightly oxidized sheets of GO that exhibit weak to negligible fluorescence in the red. This effect was suggested to explain the appearance of the sharp blue-shifted fluorescence feature at high pH reported by Galande *et al.* [24] in terms of recovery of the fluorescence of oxidative debris through separation from GO in basic conditions. When combined with weakly oxidized GO, due to interaction with the large GO matrix, oxidative debris was described to have highly quenched and red-shifted emission that is commonly observed in GO suspensions. Such an unconventional hypothesis provides yet another possible mechanism for the observed photoluminescence in GO; however, it does not explain the photoluminescence from oxidized graphene films [52].

Studies of the oxidation and reduction of GO not only elucidate possible mechanisms of fluorescence emission in GO but also provide important routes to modifying its optical properties. Together with oxidation and reduction, a significant modulation of GO emission has been reported upon surface derivatization with a variety of functional groups [30, 66, 88, 89]. A substantial increase in fluorescence quantum yield of GO up to 13% was observed upon functionalization of the epoxy and carboxyl groups in GO by alkylamines [30]. Such an increase was explained within the framework of localized graphitic domain-related GO photoluminescence theory as passivation of epoxy and carboxyl groups that were proposed to act as quenching centers, while still retaining the sp^2 domain structure in GO that, according to the theory, is responsible for the emission. Highly enhanced photoluminescence was reported also in anthryl-functionalized GO [66] and Mn^{2+}-bonded RGO [89], where the increase in emission was assigned to fluorescence resonance energy transfer from Mn ions to emissive sp^2 clusters in graphene. Due to the energy level structure in Mn, resonance energy transfer could significantly enhance the emission of RGO, strongly increasing the photoluminescence intensity at ~550 nm. Such photoluminescence enhancements suggest that chemical functionalization could serve as an efficient route to improve the fluorescence quantum yield of GO.

In summary, photoluminescence from GO was detected in the blue and red, with some reports indicating green ultraviolet and near-infrared emission. The origin of the photoluminescence has been proposed by several theories, including emission from confined domains of sp^2 graphitic carbon, emission from localized electronic environments surrounding oxygen-containing functional groups, combinations of these two scenarios, or even emission from oxidation-produced graphitic debris. Controlled reduction, oxidation and functionalization of GO allow for the tailoring of its optical properties for particular applications.

5.5 Graphene Oxide Quantum Dots

In addition to functionalization, cutting GO into quantum dots bears the promise of modifying its optical properties. Decreasing the dimensionality and the size of GO sheets down to several nanometers in diameter could result in quantum-confined structures, whose optical properties would be determined by their size [90]. This opens a range of possibilities of controllably adjusting the optical response from quantum dots through governing their size distribution. However, there could be a limiting size of GO quantum dots, above which their properties would strongly resemble those of GO. For example, recently, GO quantum dots obtained by hydrothermal treatment of GO have shown excitation- and pH-dependent photoluminescence in the blue spectral region [8], similar to GO, attributed to free zigzag sites, with pH dependence explained by protonation/deprotonation of zigzag edges. A similar explanation was proposed for emission of quantum dots prepared from GO by ultrasonication combined with thermal treatment [91]: their emission in the blue was assigned to zigzag edge states, whereas green photoluminescence was proposed to arise from disordered energy traps. The emission wavelength in those quantum dots could be controlled by varying the degree of oxidation just like for the GO [83]. Optical properties of such quantum dots appear to be similar to those of GO, therefore suggesting that the optical effects in both could originate at a size scale that is smaller than the size of quantum dots, i.e. small confined islands of sp^2 carbon or functionalization-induced defect states at the oxygen-containing groups. Considering the quantum dot size distribution of 5–13 nm [83], it is plausible that their blue and green photoluminescence would arise from smaller confined features. For example, Dong *et al.* [25] specifically attributed emission in quantum dots synthesized from citric acid to small confined sp^2 clusters within 15 nm GO dots.

However, for smaller (~5 nm) quantum dot sizes [92] studies suggest that such optical effects as excitation-dependent photoluminescence could be attributed not to the wide distribution of sizes of sp^2 graphitic regions as in GO, but rather to a distribution of sizes of the quantum dots themselves [92]. These GO quantum dots also exhibited strong solvent- and pH-dependent emission, with a quantum yield of 11.4%, which is significantly higher than the values reported for GO [20, 23]. In general, photoluminescence of graphene quantum dots is often attributed to the effects of oxygen functionalities present in these low-dimensional structures [93]. Similarly to GO they can exhibit excitation wavelength-dependent emission explained by a combination of factors, including surface defects, confinement effects and edge shapes [94]. Thus, some optical phenomena in GO could be focused upon in order to explain the optical properties of graphene quantum dots. Conversely, since GO quantum dots can be envisioned as fragments of the GO structure, studies of their photoluminescence have a great promise to provide additional information regarding the mechanism of emission in GO. Finally, the production of GO quantum dots of a particular size may allow for a significant improvement of their quantum yield as compared to GO together with retaining the general optical properties characteristic of GO.

5.6 Applications

As a result of the recent findings, there is already a large market of applications of the optical properties of GO in optoelectronics, sensing and biological imaging. The discovery of GO photoluminescence suggested a number of its potential uses in photonic devices, including electroluminescent cells and photodetectors [42]. Due to its electronic bandgap, GO has also been successfully employed as an acceptor material in organic bulk heterojunction (BHJ) photovoltaic devices, yielding a power conversion efficiency of 1.1% [95]. Environment-dependent effects in GO emission have been utilized in multiple sensing applications: based on its photoluminescence quenching due to the attachment of biological molecules and viruses, GO was applied as an optical biosensor [20, 96]. The strong pH dependence of GO emission also positions it as a nanoscale sensor for extracellular biological pH [24, 84]. Potential environmental applications of GO include metal ion sensing by N-doped GO due to the selective sensitivity of its photoluminescence to normal and transition metal ions [88]. In addition to sensing, GO and GO quantum dots have been successfully used as vehicles for drug/gene delivery and biological imaging [69, 90, 97–99]. GO is water-soluble and provides a large platform for the attachment of multiple drug molecules on both sides of the graphene sheets [97]. It has been reported as a high-capacity carrier for anticancer drugs, genes and targeting agents [69, 97]. In addition to that, the intrinsic photoluminescence of GO [69, 98] or GO quantum dots [90] could be used for the imaging of drug/gene delivery pathways without the need for additional fluorophores. Due to the transparency of biological tissue in the near-infrared, GO photoluminescence in that spectral region [69] makes it highly attractive for both *in vitro* and *in vivo* fluorescence imaging. As a result, GO poses a significant advantage to existing delivery/imaging modalities due to its multi-functionality and large transport capacity.

References

[1] Novoselov, K.S.; Geim, A.K.; Morozov, S.V.; *et al.*, Electric field effect in atomically thin carbon films. *Science* **2004**, *306* (5696), 666–669.

[2] Bonaccorso, F.; Sun, Z.; Hasan, T.; Ferrari, A.C., Graphene photonics and optoelectronics. *Nature Photon.* **2010**, *4* (9), 611–622.

[3] Nair, R.R.; Blake, P.; Grigorenko, A.N.; *et al.*, Fine structure constant defines visual transparency of graphene. *Science* **2008**, *320* (5881) 1308.

[4] Ferrari, A.C.; Meyer, J.C.; Scardaci, V.; *et al.*, Raman spectrum of graphene and graphene layers. *Phys. Rev. Lett.* **2006**, *97*, 187401.

[5] Jiang, Z.; Henriksen, E.A.; Tung L.C.; *et al.*, Infrared spectroscopy of Landau levels of graphene. *Phys. Rev. Lett.* **2007**, *98*, 197403.

[6] Yan, J.; Zhang, Y.; Kim, P.; Pinczuk, A., Electric field effect tuning of electron–phonon coupling in graphene. *Phys. Rev. Lett.* **2007**, *98*, 166802.

[7] Lu, J.; Yang, J.; Wang, J.; *et al.*, One-pot synthesis of fluorescent carbon nanoribbons, nanoparticles, and graphene by the exfoliation of graphite in ionic liquids. *ACS Nano* **2009**, *3* (8), 2367–2375.

[8] Pan, D.; Zhang, J.; Li, Z.; Wu, M., Hydrothermal route for cutting graphene sheets into blue-luminescent graphene quantum dots. *Adv. Mater.* **2010**, *22* (6), 734–738.

[9] Peng, J.; Gao, W.; Gupta, B.K.; *et al.*, Graphene quantum dots derived from carbon fibers. *Nano Lett.* **2012**, *12* (2), 844–849.

[10] Tang, L.; Ji, R.; Cao, X.; *et al.*, Deep ultraviolet photoluminescence of water-soluble self-passivated graphene quantum dots. *ACS Nano* **2012**, *6* (6), 5102–5110.

[11] Eda, G.; Mattevi, C.; Yamaguchi, H.; *et al.*, insulator to semimetal transition in graphene oxide. *J. Phys. Chem. C* **2009**, *113* (35), 15768–15771.

[12] Loh, K.P.; Bao, Q.; Eda, G.; Chhowalla, M., Graphene oxide as a chemically tunable platform for optical applications. *Nature Chem.* **2010**, *2* (12), 1015–1024.

[13] Li, S.-S.; Tu, K.-H.; Lin, C.-C.; *et al.*, Solution-processable graphene oxide as an efficient hole transport layer in polymer solar cells. *ACS Nano* **2010**, *4* (6), 3169–3174.

[14] Hummers, W.S.; Offeman, R.E., Preparation of graphitic oxide. *J. Am. Chem. Soc.* **1958**, *80* (6), 1339.

[15] Kovtyukhova, N.I.; Ollivier, P.J.; Martin, B.R.; *et al.*, Layer-by-layer assembly of ultrathin composite films from micron-sized graphite oxide sheets and polycations. *Chem. Mater.* **1999**, *11* (3), 771–778.

[16] Shang, J.; Ma, L.; Li, J.; *et al.*, The origin of fluorescence from graphene oxide. *Sci. Rep.* **2012**, *2*, 792.

[17] Zhang, X.-F.; Shao, X.; Liu, S., Dual fluorescence from graphene oxide: a time-resolved study. *J. Phys. Chem. A* **2012**, *116* (27), 7308–7313.

[18] Chen, J.-L.; Yan, X.-P., A dehydration and stabilizer-free approach to production of stable water dispersions of graphene nanosheets. *J. Mater. Chem.* **2010**, *20*, 4328–4332.
[19] Kochmann, S.; Hirsch, T.; Wolfbeis, O.S., The pH dependence of the total fluorescence of graphite oxide. *J. Fluoresc.* **2012**, *22* (3), 849–855.
[20] Chen, J.-L.; Yan, X.-P.; Meng, K.; Wang, S.-F., Graphene oxide based photoinduced charge transfer label-free near-infrared fluorescent biosensor for dopamine. *Anal. Chem.* **2011**, *83* (22), 8787–8793.
[21] Ingle, J.D.; Crouch, S.R., *Spectrochemical Analysis*. Prentice-Hall, Englewood Cliffs, NJ, **1988**.
[22] Cuong, T.V.; Pham, V.H.; Shin, E.W.; *et al.*, Temperature-dependent photoluminescence from chemically and thermally reduced graphene. *Appl. Phys. Lett.* **2011**, *99*, 041905.
[23] Li, M.; Cushing, S.K.; Zhou, X.; *et al.*, Fingerprinting photoluminescence of functional groups in graphene oxide. *J. Mater. Chem.* **2012**, *22* (44), 23374–23379.
[24] Galande, C.; Mohite, A.D.; Naumov, A.V.; *et al.*, Quasi-molecular fluorescence from graphene oxide. *Sci. Rep.* **2011**, *1*, 85.
[25] Dong, Y.; Shao, J.; Chen, C.; *et al.*, Blue luminescent graphene quantum dots and graphene oxide prepared by tuning the carbonization degree of citric acid. *Carbon* **2012**, *50* (12), 4738–4743.
[26] Chien, C.-T.; Li, S.-S.; Lai, W.-J.; *et al.*, Tunable photoluminescence from graphene oxide. *Angew. Chem. Int. Ed.* **2012**, *51* (27), 6662–6666.
[27] Chen, J.-L.; Yan, X.-P., A dehydration and stabilizer-free approach to production of stable water dispersions of graphene nanosheets. *J. Mater. Chem.* **2010**, *20* (21), 4328–4332.
[28] Eda, G.; Lin, Y.-Y.; Mattevi, C.; *et al.*, Blue photoluminescence from chemically derived graphene oxide. *Adv. Mater.* **2010**, *22* (4), 505–509.
[29] Li, D.; Müller, M.B.; Gilje, S.; *et al.*, Processable aqueous dispersions of graphene nanosheets. *Nature Nanotechnol.* **2008**, *3* (2), 101–105.
[30] Mei, Q.; Zhang, K.; Guan, G.; *et al.*, Highly efficient photoluminescent graphene oxide with tunable surface properties. *Chem. Commun.* **2010**, *46* (39), 7319–7321.
[31] Liu, Z.-B.; Zhao, X.; Zhang, X.-L.; *et al.*, Ultrafast dynamics and nonlinear optical responses from sp^2- and sp^3-hybridized domains in graphene oxide. *J. Phys. Chem. Lett.* **2011**, *2* (16), 1972–1977.
[32] Kaniyankandy, S.; Achary, S.N.; Rawalekar, S.; Ghosh, H.N., Ultrafast relaxation dynamics in graphene oxide: evidence of electron trapping. *J. Phys. Chem. C* **2011**, *115* (39), 19110–19116.
[33] Kumar, S.; Anija, M.; Kamaraju, N.; *et al.*, Femtosecond carrier dynamics and saturable absorption in graphene suspensions. *Appl. Phys. Lett.* **2009**, *95*, 191911.
[34] Ruzicka, B.A.; Werake, L.K.; Zhao, H.; *et al.*, Femtosecond pump–probe studies of reduced graphene oxide thin films. *Appl. Phys. Lett.* **2010**, *96*, 173106.
[35] Zhang, Y.; Shi, Z.; Gu, Z.; Iijima, S., Structure modification of single-wall carbon nanotubes. *Carbon* **2000**, *38* (15), 2055–2059.
[36] Ferrari, A.C.; Meyer, J.C.; Scardaci, V.; *et al.*, Raman spectrum of graphene and graphene layers. *Phys. Rev. Lett.* **2006**, *97*, 187401.
[37] Malard, L.M.; Pimenta, M.A.; Dresselhaus, G.; Dresselhaus, M.S., Raman spectroscopy in graphene. *Phys. Rep.* **2009**, *473* (5–6), 51–87.
[38] Reich, S.; Thomsen, C., Raman spectroscopy of graphite. *Phil. Trans. R. Soc. Lond. A* **2004**, *362* (1824), 2271–2288.
[39] Ferrari, A.C., Raman spectroscopy of graphene and graphite: disorder, electron–phonon coupling, doping and nonadiabatic effects. *Solid State Commun.* **2007**, *143* (1–2), 47–57.
[40] Kaniyoor, A.; Ramaprabhu, S., A Raman spectroscopic investigation of graphite oxide derived graphene. *AIP Advances* **2012**, *2*, 032183.
[41] Kudin, K.N.; Ozbas, B.; Schniepp, H.C.; *et al.*, Raman spectra of graphite oxide and functionalized graphene sheets. *Nano Lett.* **2008**, *8* (1), 36–41.
[42] Eda, G.; Chhowalla, M., Chemically derived graphene oxide: towards large area thin-film electronics and optoelectronics. *Adv. Mater.* **2010**, *22* (22), 2392–2415.
[43] Das, A.; Pisana, S.; Chakraborty, B.; *et al.*, Monitoring dopants by Raman scattering in an electrochemically top-gated graphene transistor. *Nature Nanotechnol.* **2008**, *3* (4), 210–215.

[44] Ang, P.K.; Wang, S.; Bao, Q.; *et al.*, High-throughput synthesis of graphene by intercalation–exfoliation of graphite oxide and study of ionic screening in graphene transistor. *ACS Nano* **2009**, *3* (11), 3587–3594.

[45] Voggu, R.; Das, B.; Rout, C.S.; Rao, C.N.R., Effects of the charge-transfer interactions of graphene with electron-donor and -acceptor molecules examined by Raman spectroscopy and cognate techniques. *J. Phys. Condens. Matter* **2008**, *20*, 472204.

[46] Liu, L.; Ryu, S.; Tomasik, M.R.; *et al.*, Graphene oxidation: thickness-dependent etching and strong chemical doping. *Nano Lett.* **2008**, *8* (7), 1965–1970.

[47] Lucchese, M.M.; Stavale, F.; Ferreira, E. H.; *et al.*, Quantifying ion-induced defects and Raman relaxation length in graphene, *Carbon* **2010**, *48* (5), 1592–1597.

[48] Tuinstra, F.; Koenig, L., Raman spectrum of graphite. *J. Chem. Phys.* **1970**, *53* (3), 1126–1130.

[49] Cançado, L.G.; Takai, K.; Enoki, T.; *et al.*, General equation for the determination of the crystallite size L_a of nanographite by Raman spectroscopy. *Appl. Phys. Lett.* **2006**, *88*, 163106.

[50] Novoselov, K.S.; Geim, A.K.; Morozov, S.V.; *et al.*, Two dimensional gas of massless Dirac fermions in graphene. *Nature* **2005**, *438* (7065), 197–200.

[51] Ferrari, A.C.; Robertson, J., Interpretation of Raman spectra of disordered and amorphous carbon. *Phys. Rev. B* **2000**, *61*, 14095.

[52] Gokus, T.; Nair, R.R.; Bonetti, A.; *et al.*, Making graphene luminescent by oxygen plasma treatment. *ACS Nano* **2009**, *3* (12), 3963–3968.

[53] Mattevi, C.; Eda, G.; Agnoli, S.; *et al.*, Evolution of electrical, chemical, and structural properties of transparent and conducting chemically derived graphene thin films. *Adv. Funct. Mater.* **2009**, *19* (16), 2577–2583.

[54] Gómez-Navarro, C.; Weitz, R.T.; Bittner, A.M.; *et al.*, Electronic transport properties of individual chemically reduced graphene oxide sheets. *Nano Lett.* **2007**, *7* (11), 3499–3503.

[55] Krishnamoorthy, K.; Veerapandian, M.; Mohan, R.; Kim, S.-J., Investigation of Raman and photoluminescece studies of reduced graphene oxide sheets. *Appl. Phys. A* **2012**, *106* (3), 501–506.

[56] Cuong, T.V.; Pham, V.H.; Tran, Q.T.; *et al.*, Optoelectronic properties of graphene thin films prepared by thermal reduction of graphene oxide. *Mater. Lett.* **2010**, *64* (6), 765–767.

[57] Mohanty, N.; Nagaraja, A.; Armesto, J.; Berry, V., High-throughput, ultrafast synthesis of solution-dispersed graphene via a facile hydride chemistry. *Small* **2010**, *6* (2), 226–231.

[58] Lee, V.; Whittaker, L.; Jaye, C.; *et al.*, Large-area chemically modified graphene films: electrophoretic deposition and characterization by soft X-ray absorption spectroscopy. *Chem. Mater.* **2009**, *21* (16), 3905–3916.

[59] Wang, S.; Ang, P.K.; Wang, Z.; *et al.*, High mobility, printable, and solution-processed graphene electronics. *Nano Lett.* **2010**, *10* (1), 92–98.

[60] Stankovich, S.; Dikin, D.A.; Piner, R.D.; *et al.*, Synthesis of graphene-based nanosheets via chemical reduction of exfoliated graphite oxide. *Carbon* **2007**, *45* (7), 1558–1565.

[61] Tung, V.C.; Allen, M.J.; Yang, Y.; Kaner, R.B., Throughput solution processing of large-scale graphene. *Nature Nanotechnol.* **2009**, *4* (1), 25–29.

[62] Paredes, J.I.; Villar-Rodil, S.; Solís-Fernández, P.; *et al.*, Atomic force and scanning tunneling microscopy imaging of graphene nanosheets derived from graphite oxide. *Langmuir* **2009**, *25* (10), 5957–5968.

[63] Xin, G.; Meng, Y.; Ma, Y.; *et al.*, Tunable photoluminescence from graphene oxide from near-ultraviolet to blue. *Mater. Lett.* **2012**, *74*, 71–73.

[64] Gao, W.; Alemany, L.B.; Ajayan, L.C.M., New insights into the structure and reduction of graphite oxide. *Nature Chem.* **2009**, *1* (5), 403–408.

[65] Wang, H.; Robinson, J.T.; Li, X.; Dai, H., Solvothermal reduction of chemically exfoliated graphene sheets. *J. Am. Chem. Soc.* **2009**, *131* (29), 9910–9911.

[66] Lu, Y.; Jiang, Y.; Wei, W.; *et al.*, Novel blue light emitting graphene oxide nanosheets fabricated by surface functionalization. *J. Mater. Chem.* **2012**, *22* (7), 2929–2934.

[67] Brodie, B.C., On the atomic weight of graphite. *Phil. Trans. R. Soc. London* **1859**, *149*, 249–259.

[68] Luo, Z.; Vora, P.M.; Mele, E.J.; *et al.*, Photoluminescence and band gap modulation in graphene oxide. *Appl. Phys. Lett.* **2009**, *94*, 111909.

[69] Sun, X.; Liu, Z.; Welsher, K.; *et al.*, Nano-graphene oxide for cellular imaging and drug delivery. *Nano Res.* **2008**, *1* (3), 203–212.

[70] Kovtyukhova, N.I.; Ollivier, P.J.; Martin, B.R.; *et al.*, Layer-by-layer assembly of ultrathin composite films from micron-sized graphite oxide sheets and polycations. *Chem. Mater.* **1999**, *11* (3), 771–778.

[71] Gokus, T.; Hartschuh, A.; Harutyunyan, H.; *et al.*, Exciton decay dynamics in individual carbon nanotubes at room temperature. *Appl. Phys. Lett.* **2008**, *92*, 153116.

[72] Kozawa, D.; Miyauchi, Y.; Mouri, S.; Matsuda, K., Exploring the origin of blue and ultraviolet fluorescence in graphene oxide. *J. Phys. Chem. Lett.* **2013**, *4* (12), 2035–2040.

[73] Robertson, J., Photoluminescence mechanism in amorphous hydrogenated carbon. *Diamond Relat. Mater.* **1996**, *5* (3–5), 457–460.

[74] Mathioudakis, C.; Kopidakis, G.; Kelires, P.C.; *et al.*, Electronic and optical properties of a-C from tight-binding molecular dynamics simulations. *Thin Solid Films* **2005**, *482*, (1–2), 151–155.

[75] Johari, P.; Shenoy, V.B., Modulating optical properties of graphene oxide: role of prominent functional groups. *ACS Nano* **2011**, *5* (9), 7640–7647.

[76] Yan, J.-A.; Xian, L.; Chou, M.Y., Structural and electronic properties of oxidized graphene. *Phys. Rev. Lett.* **2009**, *103*, 086802.

[77] Erickson, K.; Erni, R.; Lee, Z.; *et al.*, Determination of the local chemical structure of graphene oxide and reduced graphene oxide. *Adv. Mater.* **2010**, *22* (40), 4467–4472.

[78] Kozawa, D.; Zhu, X.; Miyauchi, Y.; *et al.*, Excitonic photoluminescence from nanodisc states in graphene oxides. *J. Phys. Chem. Lett.* **2014**, *5* (10), 1754–1759.

[79] Spataru, C.D.; Ismail-Beigi, S.; Benedict, L.X.; Louie, S.G., Excitonic effects and optical spectra of single-walled carbon nanotubes. *Phys. Rev. Lett.* **2004**, *92*, 077402.

[80] Pedersen, T.G.; Jauho, A.-P.; Pedersen, K., Optical response and excitons in gapped graphene. *Phys. Rev. B* **2009**, *79*, 113406.

[81] Cuong, T.V.; Pham, V.H.; Tran, Q.T.; *et al.*, Photoluminescence and Raman studies of graphene thin films prepared by reduction of graphene oxide. *Mater. Lett.* **2010**, *64* (3), 399–401.

[82] Nourbakhsh, A.; Cantoro, M.; Vosch, T.; *et al.*, Bandgap opening in oxygen plasma-treated graphene. *Nanotechnology* **2010**, *21* (43), 435203.

[83] Jeong, H.K.; Jin, M.H.; So, K.P.; *et al.*, Tailoring the characteristics of graphite oxides by different oxidation times. *J. Phys. D: Appl. Phys.* **2009**, *42*, 065418.

[84] Chen, J.-L.; Yan, X.-P., Ionic strength and pH reversible response of visible and near-infrared fluorescence of graphene oxide nanosheets for monitoring the extracellular pH. *Chem. Commun.* **2011**, *47* (11), 3135–3137.

[85] Gómez-Navarro, C.; Weitz, R.T.; Bittner, A.M.; *et al.*, Electronic transport properties of individual chemically reduced graphene oxide sheets. *Nano Lett.* **2007**, *7* (11), 3499–3503.

[86] Exarhos, A.L.; Turk, M.E.; Kikkawa, J.M., ultrafast spectral migration of photoluminescence in graphene oxide. *Nano Lett.* **2013**, *13* (2), 344–349.

[87] Thomas, H.R.; Vallés, C.; Young, R.J.; *et al.*, Identifying fluorescence of graphene oxide. *J. Mater. Chem. C* **2013**, *1* (2), 338–342.

[88] Qian, Z.; Zhou, J.; Chen, J.; *et al.*, Nanosized N-doped graphene oxide with visible fluorescence in water for metal ion sensing. *J. Mater. Chem.* **2011**, *21* (44), 17635–17637.

[89] Gan, Z.X.; Xiong, S.J.; Wu, X.L.; *et al.*, Mn^{2+}-bonded reduced graphene oxide with strong radiative recombination in broad visible range caused by resonant energy transfer. *Nano Lett.* **2011**, *11* (9), 3951–3956.

[90] Yan, X.; Li, B.; Cui, X.; *et al.*, Independent tuning of the band gap and redox potential of graphene quantum dots. *J. Phys. Chem. Lett.* **2011**, *2* (10), 1119–1124.

[91] Zhu, S.; Zhang, J.; Liu, X.; *et al.*, Graphene quantum dots with controllable surface oxidation, tunable fluorescence and up-conversion emission. *RSC Advances* **2012**, *2* (7), 2717–2720.

[92] Zhu, S.; Zhang, J.; Qiao, C.; *et al.*, Strongly-green-photoluminescent graphene quantum dots for bioimaging applications. *Chem. Commun.* **2011**, *47* (24), 6858–6860.

[93] Lee, J.; Kim, K.; Park, W.I.; *et al.*, Uniform graphene quantum dots patterned from self-assembled silica nanodots. *Nano Lett.* **2012**, *12* (12), 6078–6083.

[94] Habiba, K.; Makarov, V.I.; Avalos, J.; *et al.*, Luminescent graphene quantum dots fabricated by pulsed laser synthesis. *Carbon* **2013**, *64*, 341–350.

[95] Liu, Q.; Liu, Z.; Zhang, X.; *et al.*, Polymer photovoltaic cells based on solution-processable graphene and P3HT. *Adv. Funct. Mater.* **2009**, *19* (6), 894–904.
[96] Jung, J.H.; Cheon, D.S.; Liu, F.; *et al.*, A graphene oxide based immuno-biosensor for pathogen detection. *Angew. Chem. Int. Ed.* **2010**, *49* (33), 5708–5711.
[97] Liu, Z.; Robinson, J.T.; Sun, X.; Dai, H., PEGylated nanographene oxide for delivery of water-insoluble cancer drugs. *J. Am. Chem. Soc.* **2008**, *130* (33), 10876–10877.
[98] Kim, H.; Namgung, R.; Singha, K.; *et al.*, Graphene oxide–polyethylenimine nanoconstruct as a gene delivery vector and bioimaging tool. *Bioconjug. Chem.* **2011**, *22* (12), 2558–2567.
[99] Wang, Y.; Li, Z.; Wang, J.; *et al.*, Graphene and graphene oxide: biofunctionalization and applications in biotechnology. *Trends Biotechnol.* **2011**, *29* (5), 205–212.

6

Functionalization and Reduction of Graphene Oxide

Siegfried Eigler and Ayrat M. Dimiev

6.1 Introduction

The chemical properties of graphene oxide (GO) are the most complex and the most controversial topic in the entire GO field. Chemical reactions must be interpreted in accordance with the chemical structure of the original GO. Since the exact chemical structure is still elusive, a clear understanding of chemical transformations is not possible. Another reason is GO's multi-functionality. In GO, different functional groups are situated in close proximity to each other, allowing conjugation and cross-self-protection. This is why classical organic chemistry reactions do not run, or run differently, with GO functional groups. For the same reason, reactions involving one selective type of functional group, while preserving others, are probably not possible. The situation becomes even more complicated due to the heterogeneous nature of GO. It is clear that GO and its derivatives will never form monodisperse compounds, especially because every sp^3 carbon atom is a chiral center. Many theoretically possible reactions are sterically hindered due to the fixed position of functional groups on the GO platform. All the obstacles listed above require extra caution especially when designing and interpreting results of covalent functionalization reactions.

The chemistry of GO, which was developed in recent years, will be introduced in this chapter. First, the diverse chemical structure will be reviewed one more time with special focus on the degradation mechanism leading to defective GO (see also Chapter 2). Defects generally determine the reactivity of GO, and therefore the structure of GO that can explain its chemical reactivity is reviewed here briefly. The non-covalent chemistry is briefly discussed, and more attention is given to the covalent chemistry. For covalent chemistry,

Graphene Oxide: Fundamentals and Applications, First Edition. Edited by Ayrat M. Dimiev and Siegfried Eigler.

general considerations of the reaction principles are summarized followed by a description of chemical approaches. Theoretically, the reaction protocols known from organic chemistry should allow one to address the different types of functional groups. However, for GO, it was observed that not only one specific functional group is always selectively reacting, and therefore the chemistry and reactivity of GO must be critically discussed.

The main functional groups in GO are those situated on basal planes: hydroxyl and epoxy groups. The influence of functional groups located at the edges of defects and flakes will also be discussed. Another type of functional group is the residual sp^2 system that can also be utilized for chemical reactions, eventually forming C–C bonds. The portion of C–C bonds in GO can be increased by chemical reduction, and thus reduction methods to produce reduced graphene oxide (RGO) will be introduced. Moreover, the discussion of the chemical structure of RGO is briefly compared to that of graphene with the aim of understanding the chemical reactivity. Several reaction principles are available to functionalized RGO, leading to soluble RGO, also often called soluble graphene.

Due to the heterogeneity of GO and RGO, the control of their chemistry is challenging. The chemical structure of GO depends in part on the preparation conditions. However, keeping in mind that the structure of GO has been a matter of discussion for more than a century, it is clear that it is still difficult to conduct an analysis of the reaction products after chemical modifications. Moreover, it can be expected that functional groups located at edges are more reactive than functional groups on the basal plane. Anyhow, it is difficult to distinguish between those.

Nevertheless, there are efforts to introduce a controlled chemistry of graphene, based on GO. Here, a short overview about the requirements for conducting and developing this controlled chemistry will be given. The synthesis of this type of GO, termed oxo-functionalized graphene (oxo-G_1) is given in Chapter 2. Finally, a discussion about the current results of chemical modification and prospects and applications will complete this chapter.

6.2 Diverse Structure of Graphene Oxide

The structure of GO is highly variable and the absolute structure depends to some extent on the preparation conditions. Thus, no molecular formula can be given and no precise structure can be formulated. Many different types of GO can be imagined and the type of functional groups will therefore vary. Having in mind that every sp^3 carbon introduced into graphene is a stereo center, it becomes clear that no pure derivative of GO can exist. However, to establish a controlled chemistry of GO, it is necessary to know as much as possible about the chemical structure of GO. The development of the dynamic structural model by Tour and Dimiev can explain many experimental results and therefore the reader should also study Chapter 2 [1].

Ideally, graphite of high quality with a hexagonal σ framework of carbon atoms is the starting material. This carbon framework can become ruptured during the oxidation procedure, due to over-oxidation. The most accepted structural model for GO is the Lerf–Klinowski model that was refined using nuclear magnetic resonance (NMR) spectroscopy [2–5]. However, the dynamic structural model (DSM) provides additional details that can also explain the degradation and the acidity of GO [1]. Generally speaking, layers of graphite become oxidized by the oxidant and a series of oxo-functional groups are introduced, while some sp^2 clusters remain. The actual oxidizing species remains unclear. Potentially,

Mn_2O_7 or MnO_3^+ are the reactive species if potassium permanganate is used. Among epoxy and hydroxyl groups, decorating both sides of the basal plane, also ketones and carboxyl groups are introduced. However, the latter functional groups require the breaking of C–C bonds and are therefore located at the edges of flakes and defects. Note that, according to the dynamic structural model, ketones are also located at the points of C–C bond cleavage where carbon atoms are not removed; these points are not considered by the model as edges, but as features of the basal planes. The toxic and explosive ClO_2 is formed when potassium chlorate is used as the oxidant and it is important to keep the temperature below 45 °C to avoid detonations. Other oxidants, such as bichromate, have also been used, but over-oxidation is reported: carbon atoms can be fully oxidized to CO_2. In the sense of GO production, this means over-oxidation. Ideally, carbon loss is prevented at all. Otherwise the six-membered basal plane of C atoms becomes ruptured, and ketones, carboxyl or phenolic structures can be formed or even be major species. It is proposed that various oxo-functional groups, such as ketones, carboxyl, lactols or phenols, functionalize the edges of defects [6–8]. The unoxidized graphene-like patches are part of the GO structure, but generally approximately 50% of C atoms remain sp^2 carbon. The size of sp^2 carbon patches increases with time. Therefore, the structure of GO was described as metastable [9]. Also, unpaired spins have been reported as part of the structure [10, 11]. Other interesting and in part plausible structural models have been proposed by Dékány *et al.* [12], Nakajima and Matsuo [13], Hofmann *et al.* [14], Ruess [15], Scholz and Boehm [16] and others [6] (Figure 6.1).

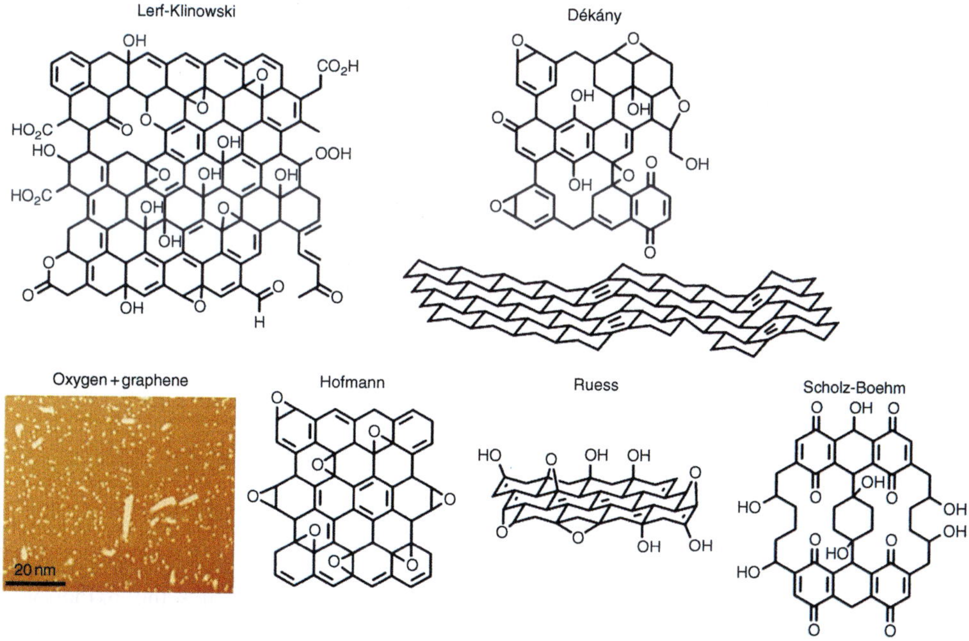

Figure 6.1 Different proposed structural models of GO and graphite oxide. The Hofmann model can be applied to the structure of graphene epitaxially grown from SiC and exposed to oxygen atoms (scanning tunneling microscopy image displays individual oxygen atoms covalently bound to graphene). Reproduced from [17, 18] with permission of ACS and Nature Publishing Group

Defects within the carbon framework have been proposed as a structural motif in the Clauss, Dékány, Scholz–Boehm and most recently Dimiev–Tour models. The edges of defects are functionalized by arbitrary oxo-groups (Figure 6.1). Also quinoid structures are plausible and stable motifs. Flakes of GO that were investigated by high-resolution transmission electron microscopy display preserved regions of ordered structures with lateral dimensions of about 1 nm among about 80% of disordered regions that were correlated to functionalized regions [19]. Furthermore, holes with diameters of about 1 nm could be identified. With respect to the structure of GO, it can be concluded that the term GO implies a partially amorphous structure with respect to the carbon framework.

Several authors realized more than 100 years ago that GO is an amorphous material and highly defective with properties related to humic acid. It seems likely that the structure models depicted in Figure 6.1 that bear a single carbon framework are all valid to describe GO. However, the formed structure motifs depend on the amount of defects and the preparation conditions.

It can be assumed that some applications may work best for GO with a carbon framework with quinoid-like motifs, and other applications require an intact carbon framework. The problem is however that it is difficult to determine the degree of defects and the type and concentration of functional groups and structure motifs. The synthetic protocol and the grade of graphite has a strong influence on the kinetics of the oxidation and the structure. Thus, reliable analysis of the product is required to enable a controlled synthesis [20]. Residual metal impurities in GO were determined by inductively coupled plasma mass spectrometry (ICP-MS) and iron, cobalt, nickel and manganese were found, as a consequence of impurities from the natural graphite source or from reagents used during the synthesis of GO. While >1000 ppm of iron was found, other metals were detected in concentrations of approximately 5–20 ppm. Although this concentration seems to be low, the metals can dominate catalytic properties and strongly influence the oxygen reduction reaction that was also claimed to work in the absence of metals [21]. The isolation of GO from the reaction mixture is normally done by filtration or centrifugation and redispersion in water. Recently, an alternative work-up procedure was demonstrated for GO that involves the addition of dodecylamine to graphite oxide [22]. In the latter example an electrostatic interaction of dodecylammonium is exploited in acidic media to isolate individual layers of GO. However, that GO is different from normally isolated GO because a composite is already formed by the ammonium compound.

6.3 Stability of Graphene Oxide

6.3.1 Thermal Stability of Graphene Oxide

The thermal decomposition of GO was intensively studied by Boehm and Scholz and large variations were found [23]. Generally, impurities and metal salts, such as added $FeCl_3$, can decrease the thermal stability. The main weight loss for GO prepared with permanganate is found at about 180–200 °C. Other investigations revealed that a certain potassium content can dramatically decrease the stability of GO and explosive decompositions were observed [24–26]. Thermogravimetric analysis (TGA), coupled with mass spectrometry(MS), gives a first hint toward the decomposition process of GO. Adsorbed water is released up to 120 °C followed by CO_2 formation, with the main decomposition temperature at about 180 °C (section 6.8.2). However, GO that bears defects on the few percent scale (>2–3%) already decomposes at a temperature of about 45 °C, and is significantly less stable than oxo-G_1 that bears an almost intact carbon framework. Films of flakes of GO, placed on ZnSe, can be analyzed by infrared spectroscopy. The spectra measured in transmission mode on ZnSe between 25 and 150 °C reveal a new absorption at 2336 cm^{-1}. The maximum intensity of this new absorption band is found at 120 °C. Also nanometer-sized blisters could be visualized in films of GO during annealing. The infrared signal at 2336 cm^{-1} can be related to the stretching vibration of CO_2, intercalated between sheets of GO. The presence of the CO_2 signal indicates that the carbon lattice of GO starts to decompose at temperatures as low as 50 °C, although the trapping of CO_2 prevents its detection by mass spectrometry in the thermogravimetric experiment [27]. The quantification of the CO_2 concentration indicates that one CO_2 molecule is formed on every 2 nm. Moreover, the role of water in the decarboxylation process of GO was investigated. Experiments with ^{18}O-labeled water indicate the exchange of ^{16}O water with ^{18}O-labeled water. TGA-MS analysis detects CO_2 at *m/z* 44 ($C^{16}O_2$) and *m/z* 46 ($C^{16}O^{18}O$) for $^{18}OH_2$-treated GO. Thus, water facilitates the formation of CO_2 (Figure 6.2A). This experimental evidence can be explained by carbonyl chemistry, such as the formation of hydrates from ketones or carboxylic acids, and a mechanism for the rearrangement of α,β-epoxy ketones to α-hydroxy carboxylic acids is proposed in Figure 6.2(B). The carbonyl chemistry, as proposed in Figure 6.2(B), is also supported by this experiment.

6.3.2 Stability and Chemistry of Graphene Oxide in Aqueous Solution

Pristine graphite oxide, a type of graphite oxide that was treated with alcohols instead of water during work-up, was described as bearing cyclic organosulfate groups that hydrolyze during water impact [28]. Moreover, it was determined that there is approximately one carbonyl group on 10–12 carbon atoms. This high amount of carbonyl groups can stem only from functional groups at the edges of defects and not only from functional groups at the edges of flakes. It was further determined that about one CO_2 molecule is formed on 35–55 carbon atoms during the synthesis of GO. The formation of CO_2 and the formation of carbonyl groups within the carbon lattice require breaking of C–C bonds. In relation to recent publications of Dimiev *et al.*, a formation mechanism of CO_2 from the carbon lattice of GO is proposed and illustrated in Figure 6.3(D), starting from vicinal diols (**1**). The reaction with hydroxide can lead to C–C bond cleavage (**2**, **3**) and further base treatment may form carboxylic acids and facilitate CO_2 formation (**8–10**). This explanation also agrees with the observations of TGA-MS measurements that demonstrated the incorporation of ^{18}O in CO_2 from $^{18}OH_2$. Subsequent reactions can further degrade the carbon lattice. Moreover, the degradation of GO can lead to acidic functional groups as illustrated by the reaction of **1** to **2** and to **7**. Therefore, a new dynamic structure model of GO was proposed by Dimiev *et al.* that describes the origin of acidity of GO that is accompanied by the degradation of the carbon lattice towards humic acid [1]. This instability of GO towards base treatment was further demonstrated to lead to oxidative debris that was formed upon etching of flakes of GO [28]. The disintegration of flakes upon base treatment is depicted in Figure 6.3.

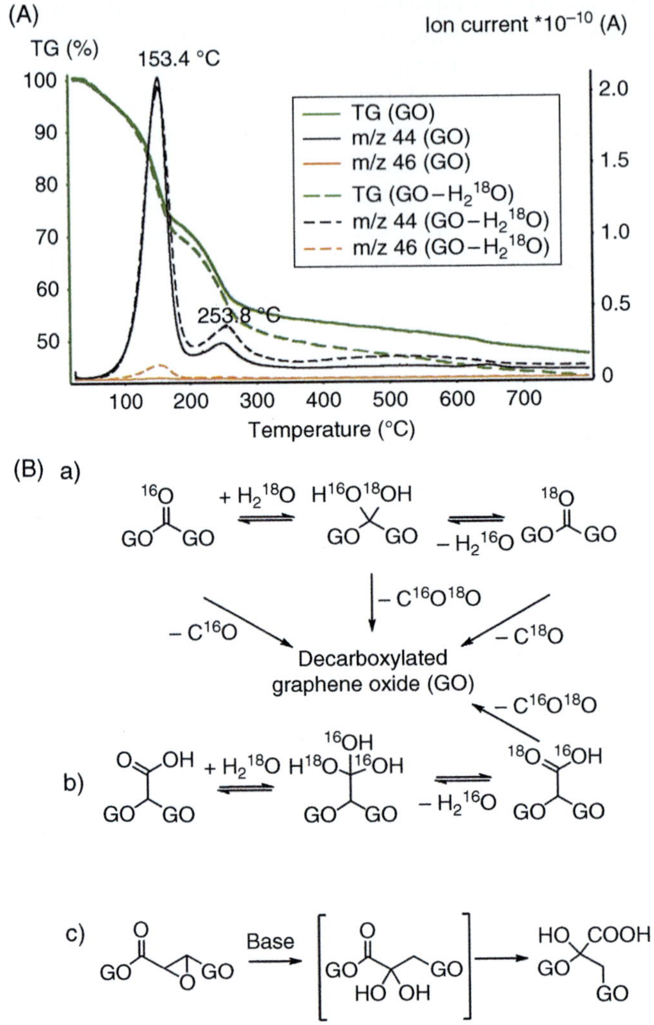

Figure 6.2 *(A) Thermogravimetric analysis coupled with mass spectrometry of GO and GO treated with ^{18}O-labeled water ($^{16}OH_2$ exchanged by $^{18}OH_2$). $C^{18}O^{16}O$ can be detected, evidence for the formation of hydrate species generated from carbonyl or carboxyl groups. (B) Suggested reaction pathways of decarboxylation processes in GO. Reproduced from [27] with permission of ACS*

The lability of functional groups in GO is problematic for developing a controlled chemistry of GO, because many reactions that could in principle be developed need thermal activation. Moreover it is plausible that batches of GO may have different contributions of functional groups and concentrations of defects, depending on the synthetic protocol and especially the degree of degradation. It was observed that GO with almost no defects within the carbon framework (oxo-G_1) is thermally more stable, as outlined in section 6.3.3.

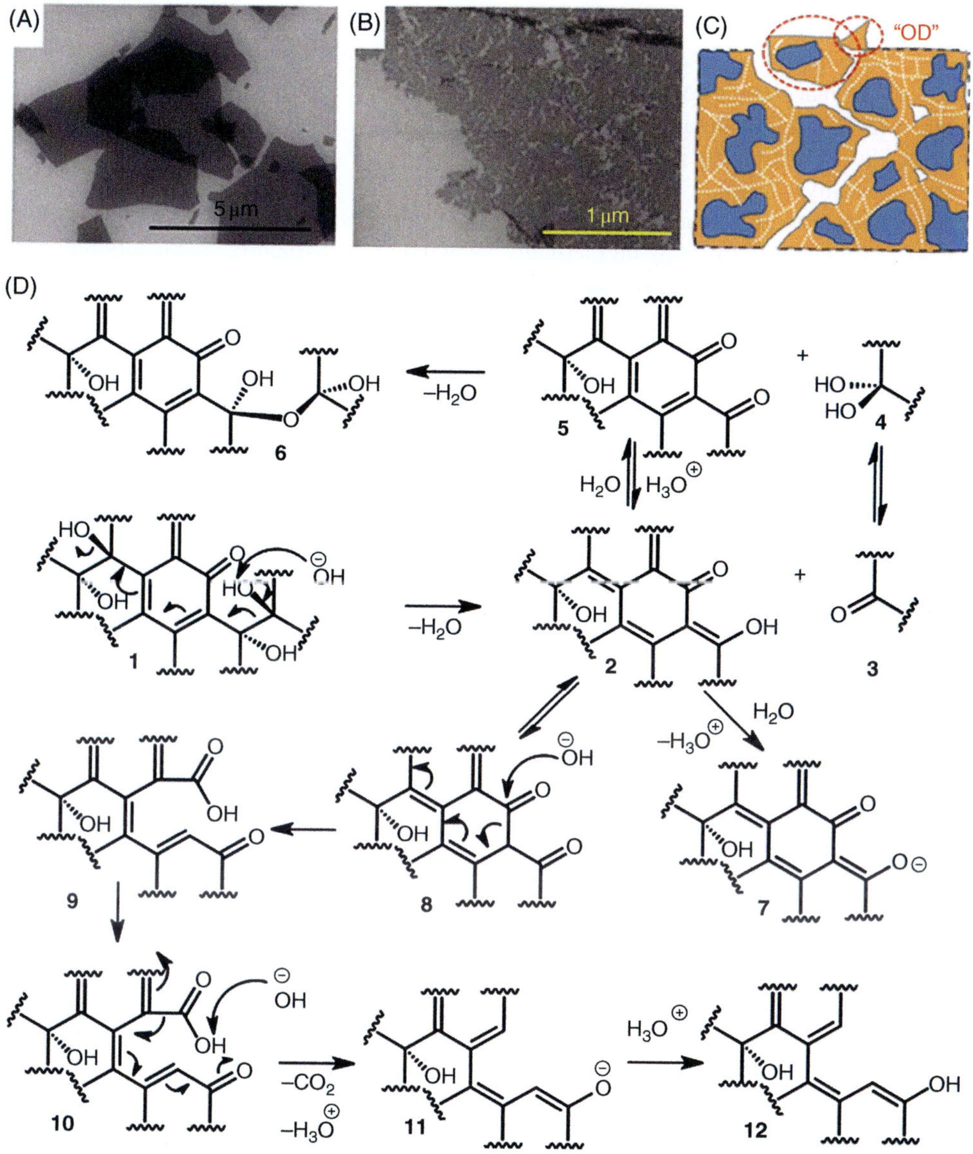

Figure 6.3 (A) Scanning electron microscope (SEM) image of flakes of GO with smooth edges. (B) SEM image of a base-treated GO flake that shows etched edges and disintegration of the flake to form oxidative debris. (C) Schematic model of the disintegration of flakes of GO with holes (white dots), graphitic domains (blue) and oxidized domains (orange). (D) Proposed reaction schemes for the degradation of GO by water and base. Plausible reaction mechanisms for CO_2 generation and the formation of acidic groups are illustrated. Base treatment of **1** can lead to extended π networks with acidic functional groups (**2–7**). Hydrates (**3, 4**) and C–O–C bonds can be formed and permanent defects (**6**). Hydroxide can react with carbonyl groups to form carboxylic acids after C–C bond breakage (**8–10**). Decarboxylation and subsequent reactions lead to further degradation [1, 28, 29]. Reproduced from [28] with permission of Elsevier

6.3.3 Stability of Oxo-Functionalized Graphene

Oxo-G_1 bears only very few defects and is therefore more stable against degradation than GO. The thermal stability of oxo-G_1 was probed in films by infrared spectroscopy and, in contrast to GO, CO_2 formation was not observed at 50 °C in iteratively heated samples. Instead, the CO_2 trapping in films started at about 90 °C, giving evidence for a higher thermal stability.

With respect to the almost intact carbon framework of oxo-G_1, it is possible to use statistical Raman spectroscopy on thermally treated and subsequently chemically reduced films of flakes. The analyses reveal that no additional defects are introduced into the σ framework of C atoms for thermal treatment up to 100 °C (Figure 6.4A). Even a slight improvement in the quality of graphene was detected after annealing oxo-G_1 at 75–100 °C. The plot of the I_D/I_G ratio versus Γ_{2D} in Figure 6.4(B) illustrates the thermal stability. Up to about 100 °C the Γ_{2D} values are between 60 and 70 cm^{-1} and remain almost constant. Treatment of oxo-G_1 at 150 °C, prior to reduction, results in the decomposition of the carbon framework and Γ_{2D} values between 90 and 300 cm^{-1}, and about 200 cm^{-1} on average, are found. The worst quality of graphene is measured for samples that had been heated to 500 °C. The carbon grid is amorphous and the 2D peak is hardly detectable with Γ_{2D}= 300 cm^{-1}. Thermal treatment of oxo-G_1 up to 1000 °C can partially restore the carbon lattice recovering the six-membered-ring structure. However, statistical Raman spectroscopy reveals Γ_{2D} between 90 and 190 cm^{-1}. Although thermal annealing at 1000 °C can partially increase the quality of graphene, compared to a sample treated at 500 °C, the original quality is not recovered (Figure 6.4B).

It can be concluded that permanent defects are introduced into the carbon lattice of oxo-G_1 by thermal treatment higher than 100 °C. These findings are beneficial for the development of the controlled chemistry of oxo-G_1, because a carbon framework that is stable up to 100 °C allows thermal activation of chemical reactants. Nevertheless, it cannot be ruled out that reagents may also disrupt the C–C bonds during thermal activation, as demonstrated in section 6.8.3.1.

Chemical reduction of oxo-G_1 that bears a density of defects of about 0.6–0.8%, prepared from large flakes of natural graphite, leads to an increase of the Raman D peak, combined with line sharpening [31]. As outlined in section 3.4.4, the I_D/I_G ratio follows a relation, and on increasing the distance between defects from 2 nm to 3 nm, an increase of the I_D/I_G ratio is expected, accompanied with line sharpening. This quality increase is high enough to pattern oxo-G_1 and to visualize the density of defects by Raman microscopy.

A flake of oxo-G_1 was chosen and probed by atomic force microscopy (AFM). Afterwards, the middle of the flake was irradiated by a focused laser pulse (532 nm at 0.06 mW). The flake was chemically reduced after irradiation and first analyzed by AFM as illustrated in Figure 6.5. The image shown in Figure 6.5(C) reveals wrinkles in the middle region of the flakes, the area where the laser pulse was applied. The same flake was scanned by scanning Raman microscopy (SRM) and the map of the I_D/I_G ratio (Figure 6.5D) reveals that the quality of the generated graphene decreases by laser irradiation. While the I_D/I_G ratio is constant for untreated samples, a decreased I_D/I_G ratio of 1.5 in the middle region and an I_D/I_G ratio of 2.2 in a loop region is found. Here the deceased I_D/I_G ratio is accompanied by line broadening, indicating the decreased quality of graphene (see section 3.4.4). Reference flakes show only little variations of the I_D/I_G ratio.

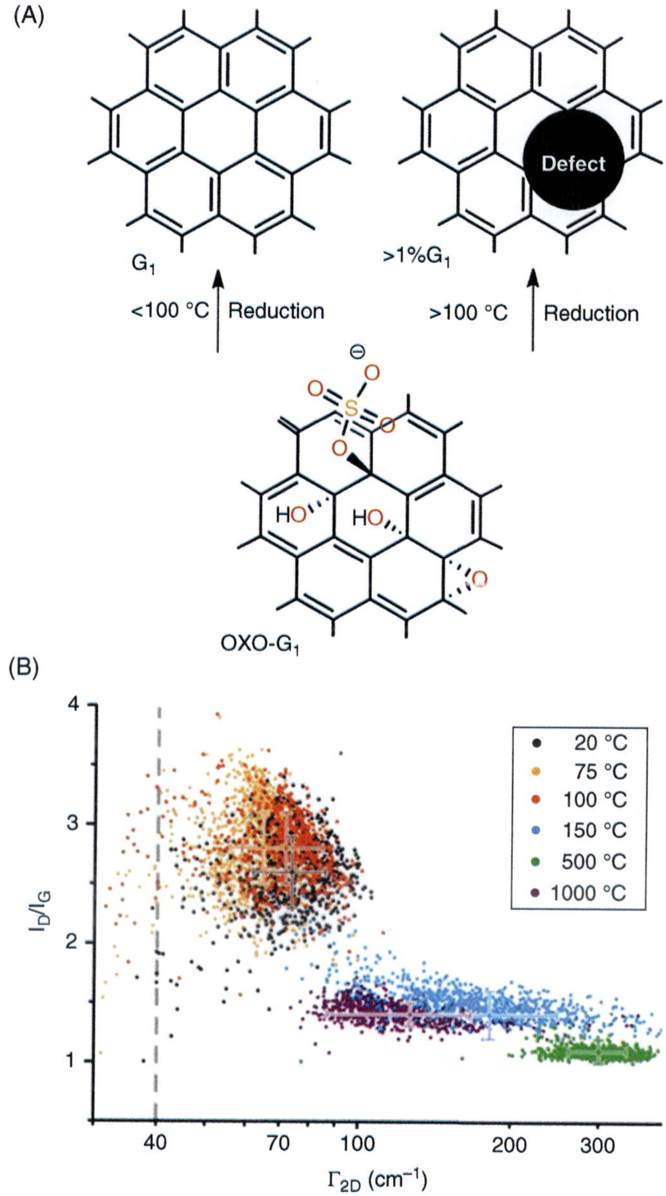

Figure 6.4 *(A) The carbon framework of oxo-G_1 is thermally stable up to 100 °C, but it is unstable at higher temperatures [30]. (B) Statistical Raman analysis of graphene derived from oxo-G_1 after thermal treatment and subsequent chemical reduction [30]. Reproduced from [30] with permission of John Wiley & Sons*

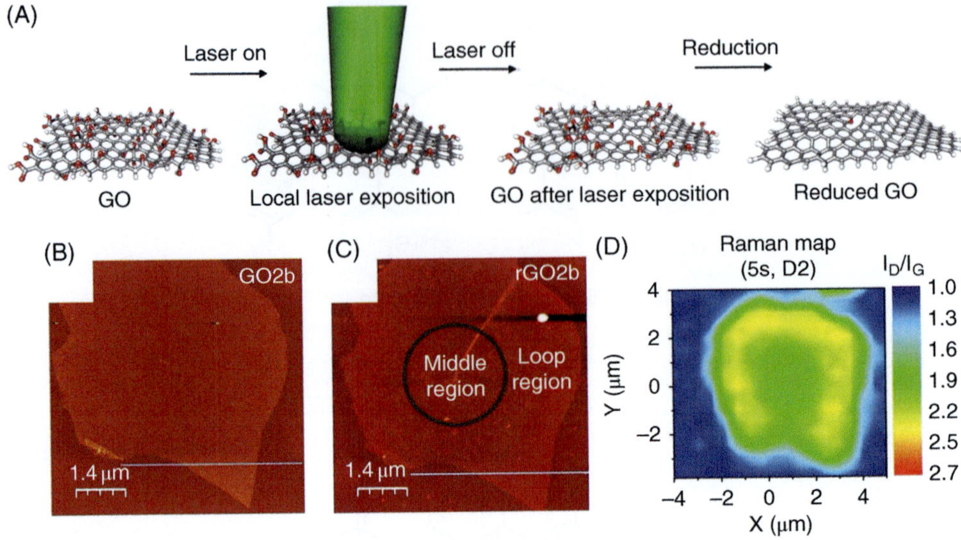

Figure 6.5 (A) Illustration of GO or oxo-G_1 that is damaged by a laser pulse and the locally thermally reduced material is further reduced chemically. AFM images of (B) GO and (C) RGO after a 5 s laser pulse at 0.06 mW applied to GO. (D) Raman map of the I_D/I_G ratio of the flake of RGO. Reproduced from [31] with permission of Elsevier

It can be concluded that the laser pulse heats GO locally and that thermal heating introduces defects in the irradiated area. The differences are detectable after chemical reduction and can be visualized by SRM. These experiments reveal that lattice defects that are already present in GO and oxo-G_1, respectively, determine the quality of graphene. Thus, the reduction method is not necessarily quality-limiting.

6.4 Non-Covalent Chemistry

Non-covalent and covalent functionalization principles were exploited to functionalize GO. Approximately 50% of the C atoms of GO are sp^2 carbon and thus a non-covalent interaction with π conjugated molecules should be possible (Figure 6.6). Also polar interactions, such as hydrogen bonding, can be exploited to molecules adsorbed on GO. Water molecules always interact by hydrogen bonding and therefore GO is dispersible in water. The literature on non-covalent functionalization by π conjugated molecules and by polymers is very rich. There is no chance to review all the published reports within one chapter of this book. That is why, in this section, we provide only a few examples to give a very general idea of this type of GO chemistry.

In one example, the polymer of methylcellulose was used to make hydrogen-bonded composites. The polymer isolates individual sheets and it was found that fluorescence properties could be adjusted over a wide range of pH [32]. In accordance with the non-covalent functionalization concept, fluorescent dye-labeled single-strand DNA (ssDNA) can interact with GO, which leads to the quenching of fluorescence. It could be observed

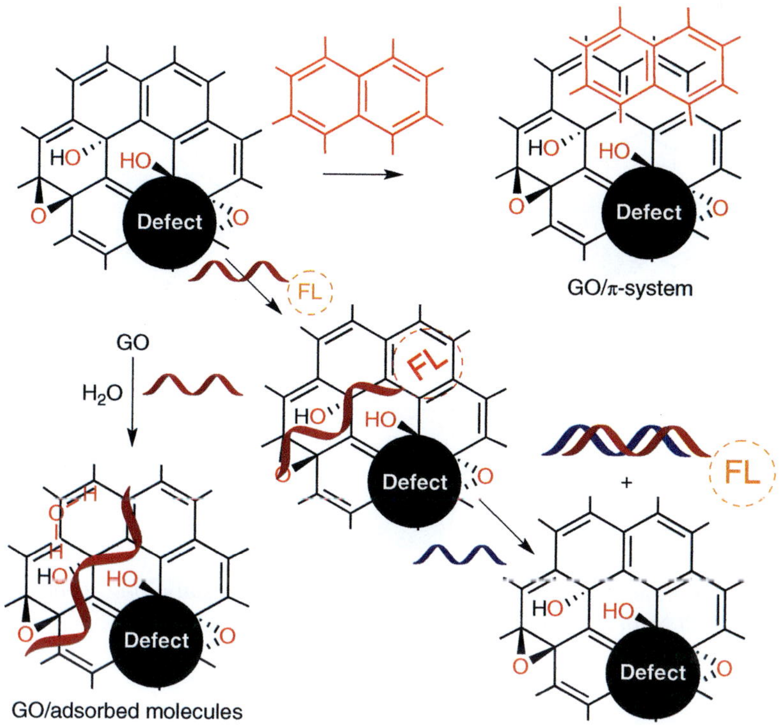

Figure 6.6 Non-covalent approaches to functionalized GO with π interactions or polar interactions. Example of ssDNA conjugated with a fluorophore (FL, quenched) adsorbed on GO exploiting polar and π interactions. A complementary target leads to desorption and restores fluorescence [33]

that the fluorescence is restored upon addition of a complementary target, because of the stronger target–ssDNA interaction compared to ssDNA–GO interaction (Figure 6.6) [33].

In another approach, a stable dispersion of RGO in water was prepared. First, GO was stabilized by non-covalent interaction with a cationically charged polymer. A dispersion of this GO/polymer is prepared by adsorption of poly(diallyldimethylammonium chloride) (PDDA) and subsequent chemical reduction of the composite with sodium borohydride led to stable dispersions of RGO/PDDA (Figure 6.7) [34]. The interaction of the polymer with GO and RGO, respectively, yields stable dispersions of single layers. Without the stabilizing polymer, the RGO aggregates forming precipitates. It can be assumed that, whereas GO and PDDA interact by polar interactions of the oxo-functional groups of GO with the cationic groups of PDDA, the interaction of RGO and PDDA is due to van der Waals interactions and in part via polar groups that are located at the edges of defects.

The non-covalent π–π interaction of pyrene derivatives with GO is, however, limited. This observation may also be due to solubility incompatibilities. However, RGO offers a huge π system and functional groups at defect sites still provide polar groups (Figure 6.8) [35]. An anchoring unit of 1-pyrenebutyric acid was functionalized with a reversible addition – fragmentation chain-transfer (RAFT) agent that was synthesized from hydroxyl

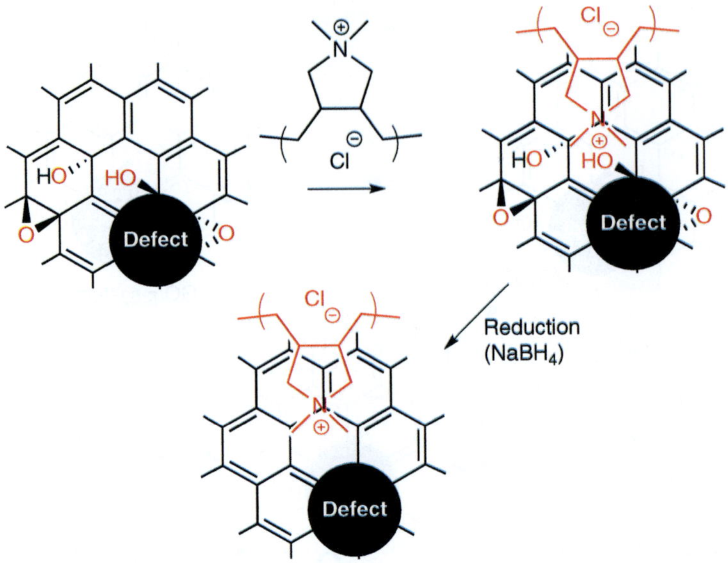

Figure 6.7 Non-covalent polar interaction of PDDA with GO as stabilizing agent, followed by deoxygenation with $NaBH_4$ to yield stabilized RGO [34]

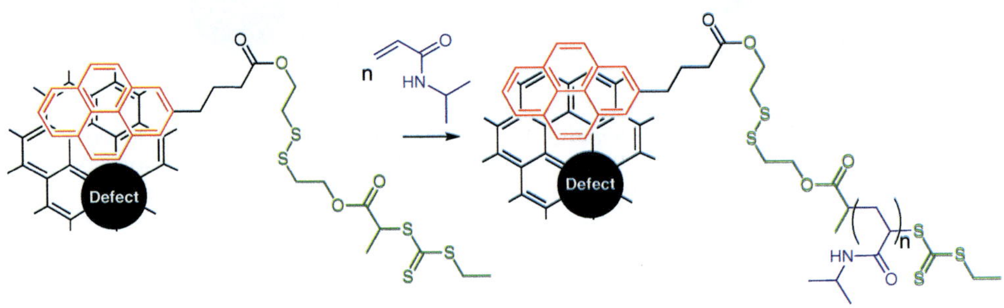

Figure 6.8 Non-covalent polar interaction of pyrene with the π system of RGO enables attachment of the RAFT agent for composite formation [35]

ethyl disulfide and 2-(((ethylthio)carbonothioyl)thio)propanoic acid. The pyrene unit binds to the π system of RGO and the RAFT agent enables the polymerization of *N*-isopropylacrylamide. This approach enables the formation of an acryl polymer with RGO sheets incorporated.

6.5 Covalent Chemistry

The covalent chemistry of GO is aimed at the reactions with functional groups that are located (i) on the basal planes of sheets of GO and (ii) at edges. These edges can either be the edges of the flakes or the edges of defect sites. Defects play an important role in GO chemistry. Thus, the chemistry of GO is often dominated by the chemistry of the edge functional groups, which is reviewed in section 6.5.3.

6.5.1 Reactions Mainly at the Basal Plane

Functional groups on the basal plane are hydroxyl groups as well as epoxy groups, and the oxidation of graphite with permanganate in sulfuric acid results roughly in a degree of functionalization of 50%. Every second C atom is therefore expected to be functionalized. Due to the homogeneous functionalization, those groups are located on both sides of the basal plane, and NMR spectroscopy reveals that epoxy C atoms and hydroxyl C atoms are roughly equally distributed. Thus, about two hydroxyl groups per one epoxy group can be expected in a typical GO. In addition, numerous defects of unspecified nature can be considered as an important part of GO. Such a structure accounts for the main features of GO as illustrated in Figure 6.9.

However, for adopting the properties of GO, it is a good idea to address hydroxyl groups. Epoxy groups are in part expected to be sterically hindered and nucleophiles may more likely react with carboxylic acids in acid–base reactions or with carbonyl groups located at edge groups.

Although the basal plane of GO is ruptured due to some over-oxidation, there are still many functional groups on both sides of the basal plane, in particular, hydroxyl groups, which are the major functional species of GO. According to Layek *et al.* [36], tertiary alcohol groups can be exploited to form ester groups, with activated carboxylic acids The example illustrated in Figure 6.10 illustrates the reaction of the carbonic acid bromide, α-bromoisobutyryl bromide, with GO [36]. The reaction occurs in the presence of pyridine to further activate α-bromoisobutyryl bromide and to act as a scavenger of the HBr formed. The formed ester groups now bear the tertiary bromide, which can be exploited for atom

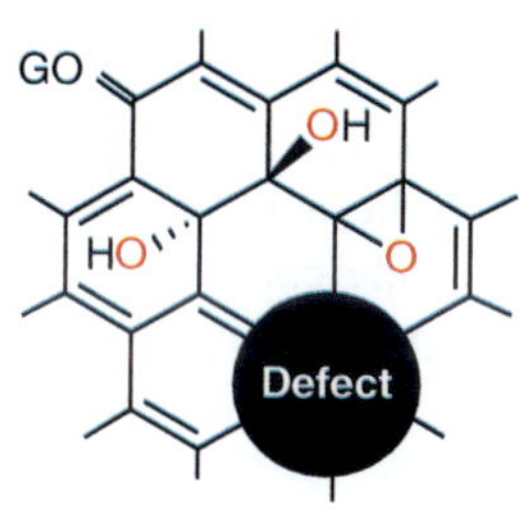

Figure 6.9 Chemical sketch of GO accounting for functional groups on both sides of the basal plane and not well-defined defects

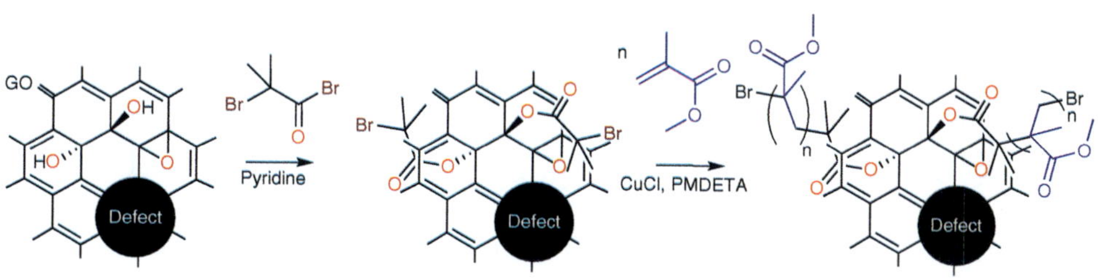

Figure 6.10 Functionalization of GO with α-bromoisobutyryl bromide to form ester groups and the bromo end-group that can be exploited for ATRP reactions. The ATRP polymerization with methyl methacrylate is subsequently initiated by CuCl and N,N,N,N,N-pentamethyl diethylene triamine (PMDETA) [36]

transfer radical polymerization (ATRP) reactions. The addition of CuCl as catalyst and the tertiary amine initiates the polymerization in the presence of methyl methacrylate. Hydrazine reduction finally yields polymer-wrapped RGO.

In fact, the chemical reactions described above might be quite different. Many reactants were involved in this process; each of them may play its own role, different from the one aimed at by the authors. First of all, it is known that, unlike primary and secondary alcohols, the tertiary alcohols are not prone to esterification with acids. It should be noted that amines are well known to chemically reduce GO, at least in part. That reduction could therefore be a side reaction of this approach. The same is true for Cu(I) that should serve as a reducing agent. Tertiary amine, used as polymerization initiator, can also be physisorbed electrostatically on GO.

In this regard, a general consideration is shown in Figure 6.11. Hydroxyl groups can be tertiary hydroxyl groups bonded either on both sides of the basal plane or in the plane bound at defect sites. Ester or ether formation of hydroxyl groups will proceed at all alcohol groups where the reaction is sterically possible. However, C–O bonds will behave differently upon chemical reduction: C–O bonds are cleaved with reduction of the carbon framework and the carbon framework becomes π conjugated. The regeneration of π conjugation is the driving force for C–O bond cleavage and in-plane functional groups can be expected to persist upon reduction. Anisole, which is the methyl ether of phenol, will not be cleaved upon treatment with hydrazine.

The aim of the approach depicted in Figure 6.12 is the binding of a functional addend that allows C–C bond formation and the introduction of further functional groups in GO derivatives. Therefore, triethyl orthoacetate is activated by an acid and then an alcohol of GO forms first a C–O bond [37]. The ether bond is formed after ethanol cleavage and formation of the C=C double bond. In the second step a [3, 3] sigmatrope rearrangement can occur if a C=C double bond of the carbon framework is in the right conformation. As illustrated in Figure 6.12, carboxylic acids form under basic conditions. However, in-plane defects may also generate some carboxylic acids. The influence of functional groups at defect sites on this reaction protocol is not clear. Nevertheless, it could be demonstrated that carboxylic acids can be formed and that the reduction product forms stable dispersions in water. Moreover, X-ray diffraction (XRD) measurements indicate an

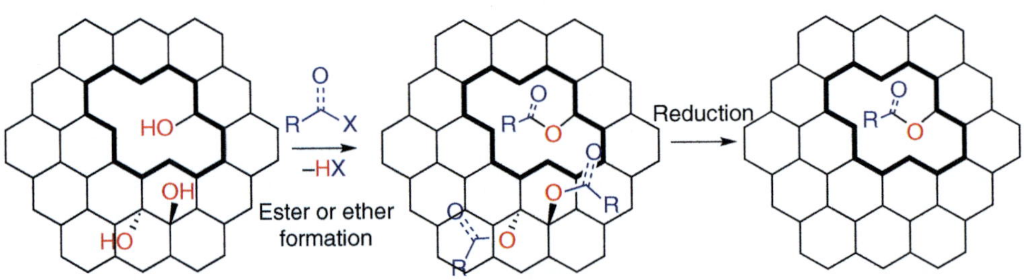

Figure 6.11 Hydroxyl groups in GO exist is arbitrary amounts on the basal plane and at defect sites. Ester or ether formation with hydroxyl groups will proceed at tertiary hydroxyl groups and the more phenol-like hydroxyl groups located at edges (if the ester and ether formation are sterically possible). Chemical reduction is expected to cleave preferably the C–O bond on the basal plane

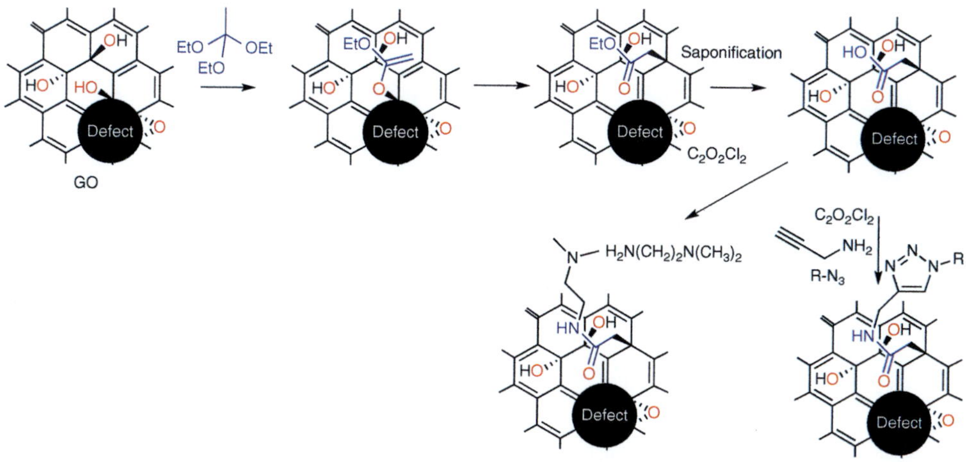

Figure 6.12 Illustration of the reaction of hydroxyl groups of GO to form C–C bonds by Johnson–Claisen rearrangement. The yielded ester can be saponified, activated and undergoes amidation with various amines, including propargyl amine, that can be further exploited for alkine azide reactions. The addition of various groups with ethylene glycol moieties or sulfonate groups was reported. Terminal amines can also be protonated to yield positively charged derivatives suitable for layer-by-layer assembly [37]

increased sheet distance in films of flakes, in contrast to as-synthesized GO. The surface properties of GO can be widely altered by this approach (Figure 6.12). The formation of a series of amide bonds was reported to be suitable to anchor carbon-rich molecules or molecules with triple bonds. The triple bonds were further used to make "click" reactions to further functionalize the surface.

However, from the perspective of organic chemistry, it seems to be illusive to prove the direct bond formation between one sp^2 carbon atom of the carbon lattice and the C atom of the addend. Analytical tools must be developed or improved to quantify the amount of formed C–C bonds and to evaluate the regiochemistry. It is desirable to be able to distinguish between edge functionalization, adsorption and functionalization on the intact carbon framework. The chemistry of GO is dominated by the intrinsic properties of GO and functionalization is mostly based on C–O bond formation. In contrast to that, the chemistry of graphene concentrates on C–C bond formation. Thus, these two worlds merge with C–C bond formation, and overcoming structural defects in GO would possibly lead to graphene derivatives that cannot be distinguished from derivatives directly prepared from graphene.

It has been reported that the hydroxyl groups of GO sheets can be cross-linked by condensation with benzene-1,4-diboronic acid, forming boronic esters. This approach yields a porous network that is attractive for gas adsorption [38, 39]. Again, the actual reaction mechanism here might be questioned because of the known resistance of the tertiary alcohols toward esterification. Furthermore, organic isocyanates were used for functionalization in the early pioneering work of Stankovich *et al.* in 2006 [40]. Porous materials are of significant importance for applications, such as charge storage or adsorption materials. Aerogels can be synthesized by various methods and those approaches are summarized in the recent literature [38, 39, 41, 42].

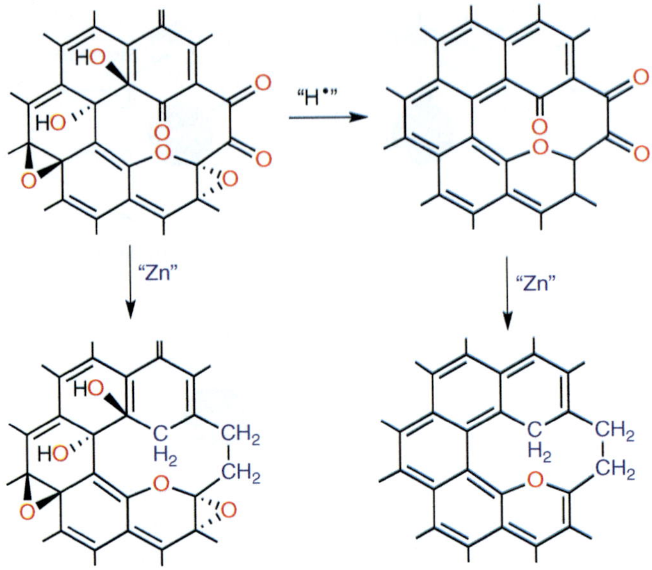

Figure 6.13 Clemmensen reduction of carbonyl groups to CH_2 groups, proceeding at the edges of flakes and defects as proposed by Sofer et al. [43]. Nascent hydrogen instead leads to defunctionalization of surface bound oxo-groups [43]

Moreover, ketone groups, which are functional groups at either the edges of flakes or the edges of defect sites, can in principle be reduced with zinc and an acid (Figure 6.13). Sofer *et al.* [43] claim that deuterium labeling leads to the conclusion that nascent hydrogen reduces functional groups on both sides of the carbon lattice. In contrast, the direct contact of ketones with zinc allegedly was reported to lead to the formation of C–H bonds. It can be assumed that selective engineering of functional groups at defect sites, e.g. by exchanging polar carbonyl with alkyl bonds, can modulate the properties of RGO. In any case, the chemical reduction of surface functional groups occurs and RGO is formed. However, it is questionable if ketones located at edges are chemically reduced to CH_2 groups. The Fourier-transform infrared (FTIR) spectra of the HCl/H_2O-RGO and the DCl/D_2O-RGO are almost identical. There is no evidence of C–D bond formation. On the contrary, there are weak but clear absorption bands at ~2800 and 2900 cm^{-1} suggesting the presence of C–H bonds. However, there is no apparent signal at roughly 1900–2050 cm^{-1}, where the C–D bond absorption bands should appear [44]. The absence of such vibrations questions the formation of any C–D bonds in Zn/HCl-reduced GO, and further clarification is needed. It remains unclear how to conduct the reduction experiment to chemically reduce ketones selectively. Otherwise, RGO is formed that may bear a few CH_2 groups instead of ketones at edges. Moreover, the concept of "nascent hydrogen" and mechanistic evolution and reactivity remains questionable [45].

There are efforts to selectively react functional groups of GO. It was found that epoxy groups that are located at the basal plane are preferably reduced using a combination of In/InCl [46]. The π system of GO can be increased by such a method and the residual functional groups make the product dispersible in solvents. However, the deoxygenation is also in part accompanied by the cleavage of hydroxyl groups, and also metal impurities may be introduced.

The direct reaction of amines with graphite oxide and GO, respectively, was also reported. This approach leads to partially reduced and functionalized materials [47, 48]. Although the products are dispersible in organic solvents, such as tetrahydrofuran, and although they can be mixed with polymers, the type of binding of amines to GO can only be assumed, and functionalized polar edges of defect sites may play a crucial role. Following this approach, stearyl amine-treated GO was incorporated in composite materials with styrene [49]. The reaction of amines with GO can, however, also lead to electrostatically bound structures of deprotonated acidic groups of GO forming alkylammonium counter-ions, as facilitated in the examples of section 6.8.3.3. Ethylene diamine and GO react with each other and form partially RGO as a hydrogel. Freeze-drying yields a porous solvent-free material that can be further processed by microwave irradiation to form a hydrogel with a density of only 3 mg cm^{-3}. The hydrogel has the ability to fully recover its original volume after 90% of compression.

Several review articles summarize approaches to functionalized and dispersible GO and RGO, respectively [17, 19, 50–52].

Also nanoparticles (NPs) can be formed and deposited onto the sheets of GO and RGO, respectively, to form metal–graphene composites [54]. In a typical approach to composites, GO is dispersed in ethylene glycol together with, for example, H_2PtCl_6 or $RuCl_3$, and after heating the mixture at 130 °C the product is isolated and tested in applications. The electrocatalytic activity of those composites is beneficial for methanol or ethanol oxidation [54]. The as-formed metal nanoparticles typically have a diameter less than 10 nm, and are tightly bound to GO/RGO flakes.

Along these lines, the binding of Fe ions with GO was facilitated to prepare magnetic Fe_3O_4 particles that are bound to GO by using NH_4OH to prepare the RGO hybrid (Figure 6.14) [53]. A magnet can attract those dispersed GO/NPs materials. However,

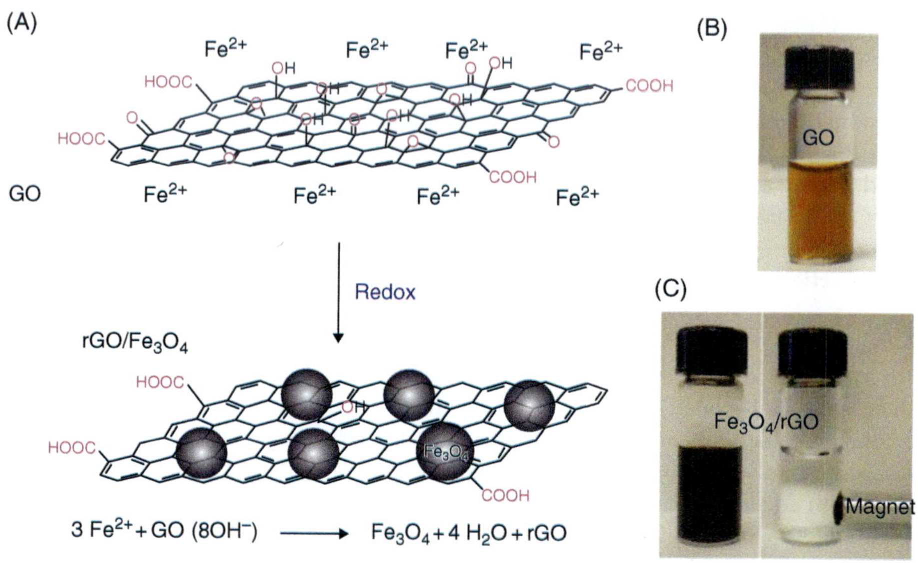

Figure 6.14 (A) GO functionalized with nanoparticles (NPs). Typically, magnetic Fe_3O_4 NPs are prepared by redox reaction. Fe^{2+} ions are adsorbed on GO and the NPs form in the presence of NH_4OH at pH = 9. (B) Dispersion of GO and (C) images of dispersion after redox reaction. Reproduced from [53] with permission of RSC

crucial questions about the binding and growth mechanism of the particles remain open. The interaction between RGO and the metal particles remains uncertain. The particles may be non-covalently bound or alternatively a metal–C bond may be formed. Residual oxygen moieties of RGO can also form metal–O–C bonds. It can be assumed that defects play a crucial role in those wet-chemical approaches to anchor nanoparticles or to limit their growth.

6.5.2 Consideration About C–C Bond Formation on the Basal Planes

It was also reported that C–C bonds could be formed to GO. This approach is facilitated because approximately 50% of C atoms in GO are sp^2-hybridized. The formation of C–C bonds on the surface of graphene is highly desired for the synthesis of molecular architectures. One approach toward C–C bond formation in GO was reported using partially RGO. The idea is to increase the π system by partial defunctionalization of functional groups from the basal plane, followed by the addition of an aryldiazonium salt that also bears a sulfonic acid to provide water solubility. Although this approach is appealing and black dispersions were yielded, crucial questions could not be addressed at that time. The quantification of defects and the determination of the degrees of functionalization remained unaddressed [55]. It is plausible that aryl ether groups can be formed by the reaction of the aryldiazonium salt and hydroxyl groups at the surface or at edges (as illustrated in Figure 6.12) [56]. Ether formation may even become the major reaction and C–C bond formation could be a side reaction. Also the edges of graphene flakes have been demonstrated to be more reactive than the graphene basal plane, an effect that has not been well investigated for GO [57]. Moreover, the adsorption of aryldiazonium salts on carbon surfaces is well known and such additional adsorption also alters the surface properties [58].

Other approaches to C–C bond formation were reported, based on, for example, carbene chemistry. Carbon–nucleophile or C–C bond formation by rearrangement reactions have also been reported [59–61]. Generally black dispersions are yielded: that observation gives evidence for defunctionalization of the surface rather than functionalization. It can be expected that the addition of any addend to the π system of GO would lead to a brighter material than a darker one. Anyhow, it was demonstrated that soluble graphene can be prepared, e.g. by a chemical approach using rearrangement reactions, based on the Claisen reaction, as illustrated in Figure 6.12 by the Johnson–Claisen reaction [37, 62].

6.5.3 Reactions at Edges

6.5.3.1 The Nature and the Fraction of Edge Atoms

The terms "flake edges", "edge atoms" and "edge functional groups" are very common in graphene- and GO-related research. However, there is some confusion in understanding these terms. Unlike graphene, with respect to GO, the term "edge" is not very straightforward, and often a writer and a reader understand it differently. For GO, the term "edge" includes both edges of the flakes along the flake perimeter, and also edges of the defect sites, or holes.

Defect sites must play an important role for the functionalization of GO, as outlined below. Let us assess the content, or fraction, of the different edge atoms in GO flakes. Figure 6.15 illustrates how the ratio of edge groups to the total number of framework atoms

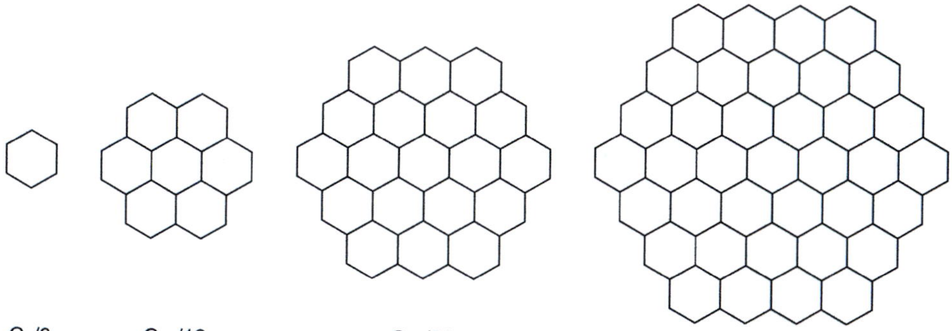

Figure 6.15 Illustration of the increase of the carbon framework of nano-graphene with zigzag edges (number of C atoms/number of edge atoms). While benzene bears only edges, coronene with its C_{24} unit has 18 edge atoms and six in-plane atoms. For the C_{96} unit there are more in-plane atoms that edge atoms

decreases with increasing flake size. A typical diameter of a flake of GO is a few micrometers and therefore the ratio of flake-edge carbon atoms to in-plane carbon atoms is low. Assuming a flake diameter of 3 μm relates to an area of 7×10^6 nm^2.

Further, functional groups at the edges of a flake may have an extent of 0.3 nm, which would relate to an area of 2.8×10^3 nm^2. Thus the ratio in this example is 0.04%. Note that, in very many studies, the size of GO flakes is even larger than 3 μm. Thus, the number of flake-edge atoms is negligible compared to the number of basal-plane atoms. Any functionalization along the flake perimeter will be undetectable by spectroscopy methods due to the low fraction. Thus it is impossible that edge functional groups could possibly dominate the overall functionalization of such a flake. In contrast, assuming a flake diameter of 20 nm, the related area is 314 nm^2 and, following the same assumptions as above, 5.7% of the area is related to the flake-edge functional groups and therefore, for example, alkylation at edges can plausibly dominate the properties of such a macromolecule. This simple consideration brings us to the conclusion that, in GO, the functionalization of defect sites (hole edges), not flake perimeter edges, plays a major role for adopting the properties of functionalized GO.

With the oxidation of the carbon framework, many carboxylic acids or ketones may be introduced. Moreover, the formation of CO_2 due to over-oxidation has been reported [29]. Ketone groups can only be located at the edges of flakes or new edges that are formed upon over-oxidation of the carbon lattice. Thus, the formation of ketones requires breaking of C–C bonds of the carbon framework. As illustrated in Figure 6.16, the removal of one C atom produces three new edge carbon atoms. Those edge atoms can be oxidized and ketones can form. However, the space is limited, and it seems unlikely that three ketones are formed. Nevertheless, less space is required if a semi-acetal was formed from the reaction of the enol form of the ketone with another ketone. Following these considerations, when more carbon atoms are missing, more options for functional groups exist. In addition, the oxidation may not stop at ketones. Breaking one additional C–C bond of the carbon framework can result in the formation of in-plane carboxylic acids.

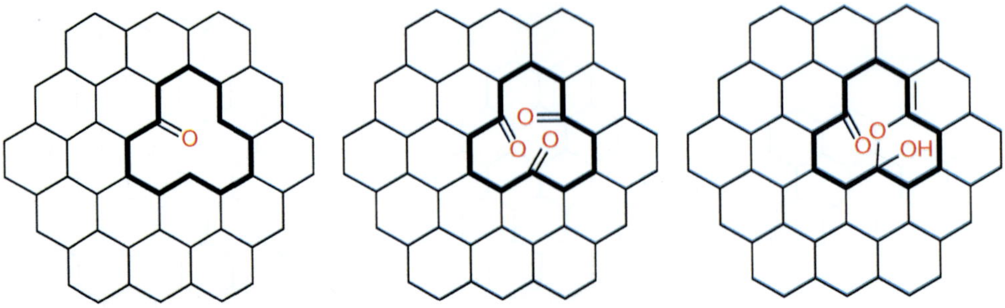

Figure 6.16 *Illustration of the hexagonal carbon framework of graphene with one missing C atom. Three new edges are produced that may be oxidized to ketone groups. The middle structure illustrates that the space required for three ketones is limited. However, the formation of an enol structure that is forming a semi-acetal may solve that problem*

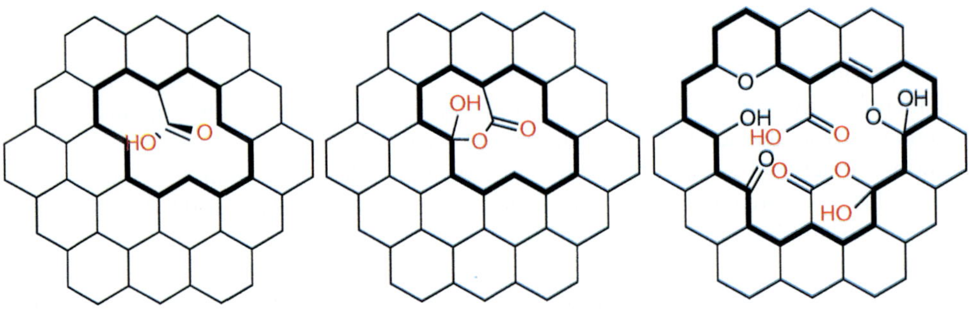

Figure 6.17 *Illustration of the hexagonal carbon framework of graphene that was over-oxidized to form carboxylic acids. A certain amount of space is required for the formation of a carboxylic acid, and therefore they are more likely to be formed at larger holes. Formation of lactol groups at edges is plausible*

As depicted in Figure 6.17, a carboxylic acid requires more space, and it does not seem possible that a carboxylic acid can be formed at a single-atom vacancy defect. The formation of lactols is also plausible. An adjacent ketone group can react with the OH group of the carboxylic acid. Thus, single and double vacancy groups may be too small to form carboxylic acids. The situation changes for a more ruptured carbon framework. Although it remains difficult to saturate all the edges of the defect site, the formation of carboxylic acids is plausible. Other functional groups at the edges of defects may include ether groups, hydroxyl groups, ketones and semi-acetals. However, it must be kept in mind that up to now the nature of functional groups at edges is not well understood. Therefore, the illustrations of Figure 6.17 are only suitable models for considering plausible possibilities. Here, it is not shown that over-oxidation may also induce the rearrangement of the carbon framework and therefore not only hexagons but also pentagons or other structure may be present. See Figure 2.14 for additional possible arrangements of edge functional groups.

6.5.3.2 Reactions Involving Carboxylic Groups

Edge groups, such as carbonyl or carboxyl groups of GO, can be used for functionalization reactions [17, 63]. Most chemical functionalization approaches use protocols suitable for reactions with carboxyl groups that are activated and subsequently converted to esters or amides [64–66].

Activation agents for carboxylic acids, such as $SOCl_2$ or carbodiimides, were reported. Especially, the reaction of $SOCl_2$ with alcohols forming eventually unstable C–Cl bonds on the surface according to the S_Ni mechanism may be a concurrent reaction that cannot be ruled out in most approaches.

After activation of carboxylic acids, alcohols or amines are added to form esters and amides (Figure 6.18). Although this approach is manifold facilitated, it remains challenging to quantify the exact fraction of established covalent bonds. Moreover, it is difficult to distinguish between carbonyl and carboxyl groups by spectroscopic techniques, and therefore quantification of ester groups and amides remains challenging, respectively. Although it is plausible that reactions can only occur at defect sites, proving the localization with precision remains challenging.

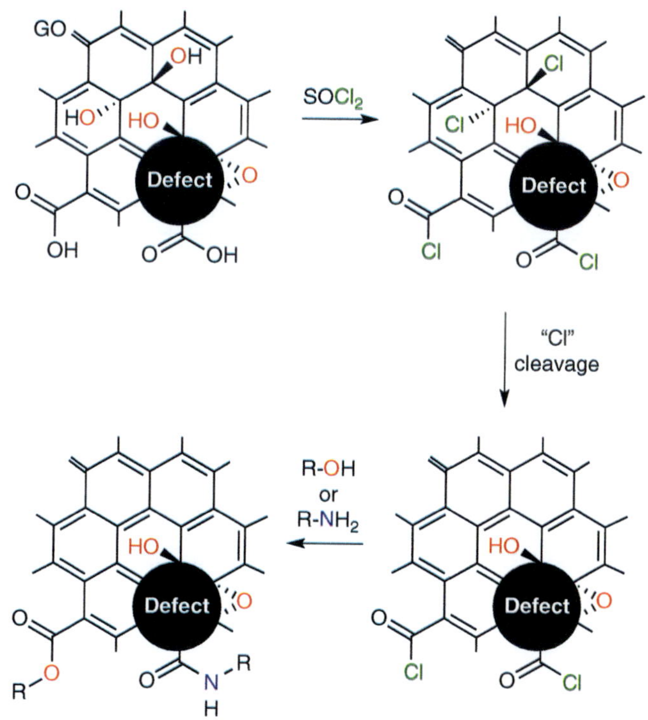

Figure 6.18 Functionalization of GO at carboxylic acids, located at the edges of flakes or defect sites. Activation of carboxylic acids by thionyl chloride can also activate surface OH groups and cause elimination or other side reactions. Acid chlorides can subsequently react with alcohols or amines to form esters and amides, respectively

It is plausible that the edges of flakes and defects of GO are chemically functionalized. Carboxyl or carbonyl groups among alcohols can be expected. It remains an unsolved problem to address those functional groups selectively without affecting surface functional groups. The type of edge groups may, however, also be heterogeneous, which prohibits the efficient functionalization of edges [67]. As introduced above, GO bears a high concentration of defect sites. Moreover, surface functional groups can be removed by reduction and functional groups remain at least in part bound to defect sites. It is well documented that carboxylic acids can be prepared at the edges of carbon nanotubes by oxidation with nitric acid. Carboxylic acids can be activated by reagents such as carbodiimides or thionyl chloride or other reagents and undergo the formation of amides by the reaction with amines.

According to this concept, preparation of the edge-functionalized RGO was described as follows. First, GO was reduced at 100 °C using hydrazine as reducing agent; boiling a brown aqueous dispersion of GO leads to aggregated RGO that can be isolated by filtration (Figure 6.19). This RGO was then oxidized at the edges of flakes and the edges of defects by treatment with HNO_3 to form carboxylic acid groups. Afterwards boiling carboxylated RGO with thionyl chloride formed the acyl chloride groups that were subsequently reacted with octadecylamine. The product is black and soluble in non-polar solvents, including lube oil.

However, from the chemical point of view it remains challenging to determine the exact degree of functionalization. It would be interesting to quantify the amount of carboxylic acid groups and formed amide groups, respectively, to calculate a reaction yield for this

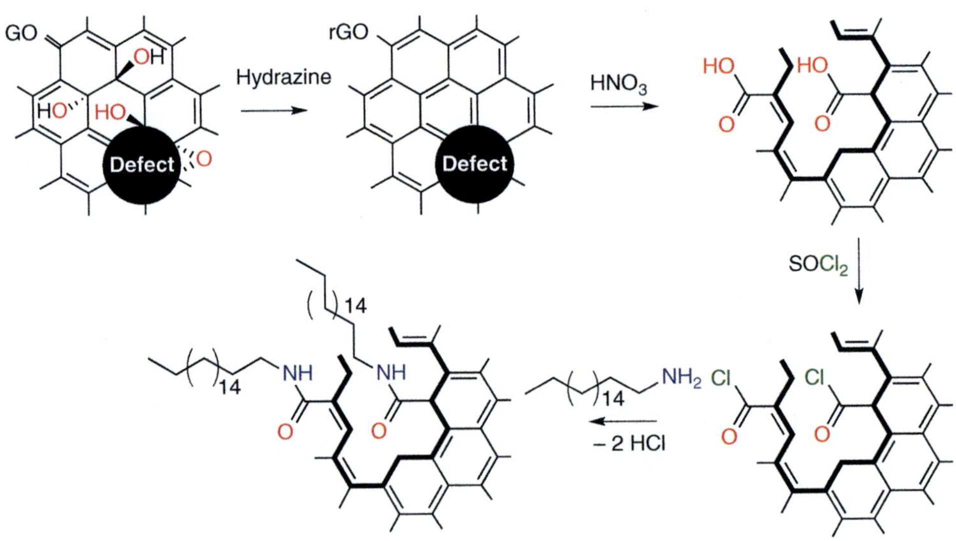

Figure 6.19 Reaction scheme for an example of edge functionalization of RGO, starting from GO with structural defects. Surface functional groups are chemically removed by reduction with hydrazine. Mild oxidation with nitric acid causes oxidation at defect sites to introduce carboxylic acids. Subsequently carboxylic acids are activated by thionyl chloride to enable the reaction with octadecylamine (the defect site is marked as thick black lines and additional functional groups are omitted for clarity) [67]

reaction sequence. Also the evolution of the size of flakes and the level of individualization of flakes in the final product are questions of interest.

Although the spectroscopic data for the intermediate and final products have been presented, the visualization and especially quantification of amides is not yet possible. This paper [67] is a typical publication claiming covalent functionalization of GO. This is why we decided to use it as a model study to show the lack of strong evidence for the claims.

The only way to claim successful reaction in this case is to present spectroscopic evidence showing the formation of the C–N bond. Two types of spectroscopic data were presented in the paper [67]: FTIR and X-ray photoelectron spectroscopy (XPS). The FTIR spectrum for octadecylated GO (ODC-GO) contains a broad band centered at 1220 cm^{-1}, that was attributed to the stretching mode of the amide C–N linkage. However, the very same peak is present in the spectrum of carboxylated RGO, before reacting with octadecyl. The other absorption bands such as the 2800–3000 cm^{-1} only prove the existence of the octadecyl on the GO, but not the formation of the amide link (Figure 6.20). No additional evidence can be obtained from FTIR in support of the amide linkage.

The strongest evidence from the XPS data is the weak nitrogen signal (content is not quantified). The position of the N 1s signal is at 399.5 eV, which was ascribed to the nitrogen atoms bonded to carbon by an amide link. However, octadecylamine itself contains the C–N bond as well, and the nitrogen atom from octadecylamine have the N 1s signal exactly at the same position at 399.5 eV. The deconvolution of the C 1s envelope, made by the authors [67] (Figure 6.21) to show allegedly existing C–N and O=C–N components, is nothing but pure speculation. Thus, the XPS data support only the presence of octadecylamine in the sample, but not the formation of the amide link.

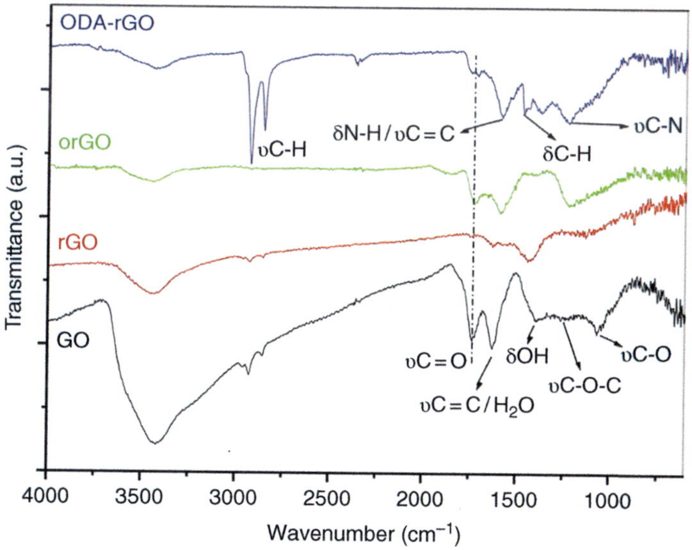

Figure 6.20 *FTIR spectra for GO (black), RGO (red), carboxylated GO (green) and ODC-GO (blue) with the bands' assignment as performed by the authors [67]. Reproduced from [67] with permission of ACS*

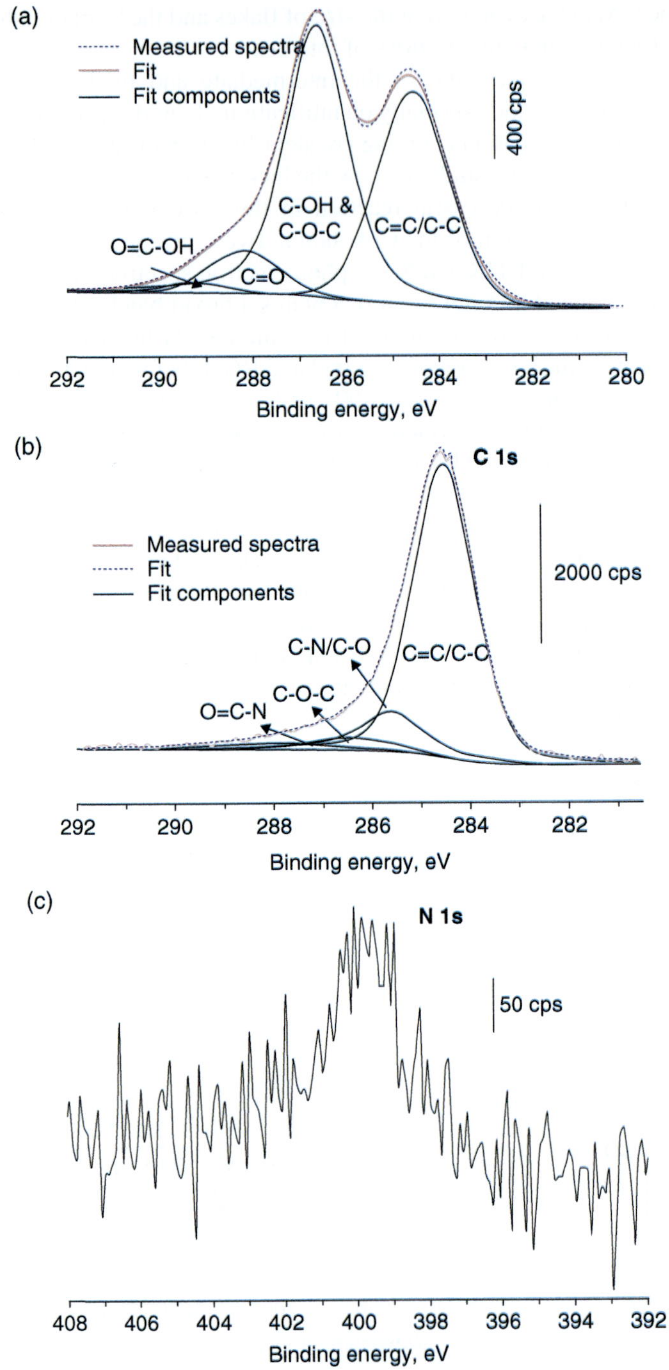

Figure 6.21 *XPS spectra for GO, RGO, carboxylated GO and ODC-GO with deconvolution of the envelope spectra as performed by the authors [67]. (a) C 1s spectrum for original GO. (b) C 1s spectrum for ODC-GO. (c) N 1s spectrum for ODC-GO. Reproduced from [67] with permission of ACS*

The provided TGA and XRD data do not prove anything either. All the provided experimental data would look absolutely the same if octadecylamine was simply physisorbed on the RGO surface. This is what most likely happens in reality. The positively charged amino group of octadecylamine strongly physisorbs onto the negatively charged GO. However, there is one piece of experimental data that evidences strongly against the covalent functionalization. The authors [67] present pictures showing the solubility of the functionalized ODC-GO in different organic solvents. ODC-GO is partially soluble in ethanol and non-soluble in hexane. This is not what one would expect if the claimed functionalization mechanism were successful. The hydrophobic hexadecyl tails would render ODC-GO more soluble in hexane, rather than in ethanol. Higher solubility in ethanol, in turn, suggests the presence of unreacted physisorbed hexadecylamine. The solubility of the latter in ethanol is three times higher than in hexane. Thus, physisorbed hexadecylamine renders ODC-GO partially soluble in ethanol.

Our discussion above does not minimize the significance of this work in terms of the properties of the final product, regardless of the ways of its production. To give credit to the authors, this is one of the best reports on the topic. Many works provide even lesser evidence. To tell the truth, proving covalent functionalization of GO is very difficult.

As shown in Figure 6.17, a functional group such as a carboxylic acid needs more space than just one carbon atom. Therefore, it is plausible that a single-atom vacancy defect would not be enough for carboxylic acid formation at a defect site. Moreover, if a carboxylic acid was formed, the neighboring edges of carbon atoms may become sterically hindered. This limits the concentration of carboxylic acids at edges. Steric effects and considerations about the quantification of functional groups must be taken into account to determine the efficiency of a reaction.

Another concept that allows the functionalization of edges, while basal-plane functional groups play a minor role, is realized by decreasing the size of GO flakes [63]. It was reported that the lateral size of GO flakes of about 50–500 nm could be further reduced to 5–50 nm by sonication. Such small GO flakes have a high contribution of functional groups at flake perimeter edges and it is plausible that carboxylic acids at these edges play a major role (Figure 6.22). Carbodiimide-mediated coupling of amino-terminated polyethylene glycol led to soluble PEGylated GO.

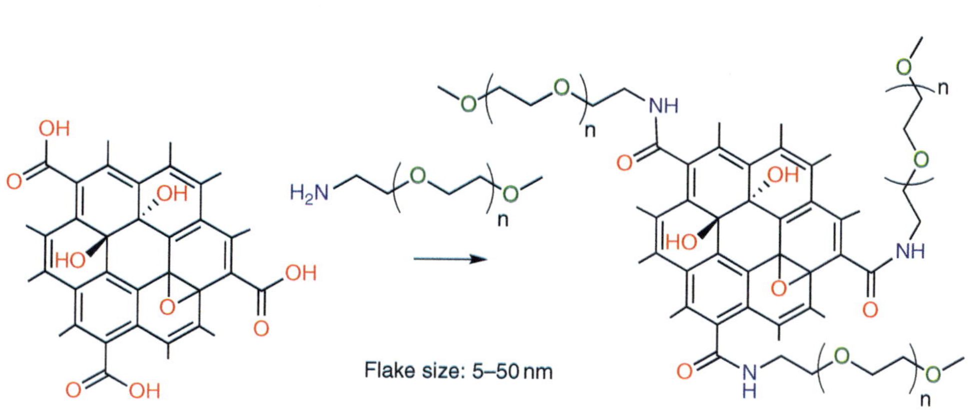

Figure 6.22 Simplified scheme of GO with a flake size of about 5–50 nm with carboxylic acids located at the edges is functionalized by carbodiimide-mediated amide bond formation to introduce polyethylene glycol groups that provide water solubility [63]

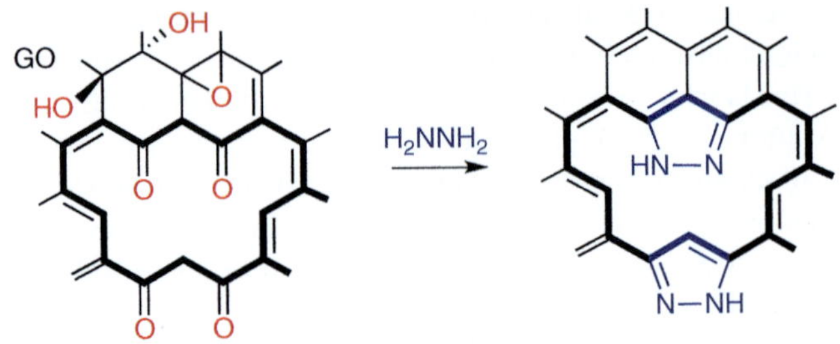

Figure 6.23 Edge functionalization after hydrazine reduction of GO. The reduction of GO with hydrazine can lead to the formation of pyrazole moieties. Epoxy and hydroxyl groups located at the basal plane are removed by this reduction method. The formation of pyrazole at the edges of flakes or defect sites gives evidence for adjacent ketones [68]

6.5.3.3 *Reactions Involving Ketones*

The functionalization of ketone groups that are located at the edges of flakes and defect sites was demonstrated by a solid-state NMR (SSNMR) study conducted on ^{15}N-labeled hydrazine-reduced ^{13}C-labeled GO [68]. The formation of pyrazole moieties was proved by this approach. As shown in Figure 6.23, pyrazole can form upon the condensation of hydrazine with two ketone groups. This experiment also gives information about the regiochemistry of ketone groups, because pyrazoles can only be formed from ketones located at edges in 1,3-position. Moreover, forming pyrazole moieties requires a certain amount of space. Single-atom vacancies would not provide enough space, which indicates that holes within the carbon framework must at least have the size of three or even many more carbon atoms.

6.6 Reduction and Disproportionation of Graphene Oxide

6.6.1 Reduction

Reduction is the best understood and the most obvious reaction in which GO can be involved. Unlike many other types of reactions reviewed in this chapter, the signs of successful reduction are obvious. The GO solution immediately changes its color from yellow-brown to black, signaling the restoration of the conjugated sp^2 network and the formation of the graphitic structure. However, with typical GO, the original graphene network is never fully restored. This is because GO itself is already highly defective: the existing defects cannot be healed during the reduction. Two different types of species remain on the RGO planes after reduction: (i) holes and (ii) some oxygen functionalities.

i. The holes are formed on GO planes already in the course of its production [29, 69] due to the removal of carbon atoms in the form of carbon dioxide. Thus, during the second step of GO formation, i.e. conversion of stage-1 GIC (graphite intercalation compound) into PGO (pristine graphite oxide), significant evolution of CO_2 gas was detected [29]. Additional defects are introduced during the third step of GO formation, i.e. washing of as-prepared PGO with water [1, 28] according to the mechanism presented in Scheme 2.3 and Figure 6.2.

ii. The oxygen functionalities remaining on GO after chemical reduction are mostly carbonyls. The epoxides and tertiary alcohols are more or less effectively reduced; however, carbonyl groups cannot be easily reduced. The formation of carbonyls requires rupture of the C–C bonds. These defects cannot be healed. The ketones in GO can react with small molecules extrinsic to GO; however, they cannot react with carbon atoms from the opposite side of the defects due to the steric hindrance of such a reaction. This is why oxygen functionalities at point defects cannot be removed, and the graphene lattice cannot be fully recovered.

The formation of such point defects is based on the divalent nature of oxygen. The monovalent elements such as fluorine can form single C–F bonds only, and the graphene plane can be almost fully restored after defluorination. Unlike monovalent fluorine, divalent oxygen introduces permanent point defects. This is why, if one aims to obtain non-defective GO, one needs to conduct the oxidation reaction in a way allowing the formation of monovalent functionalities, but avoiding the formation of divalent carbonyls. This was the strategy employed by Eigler *et al.* [70]. By conducting reactions at 0–10 °C, they obtained minimally damaged GO, which upon reduction was converted back to almost intact graphene.

In addition to the existing defects, additional holes on GO planes might form during the reduction of as-formed GO. This conclusion can be drawn from comparison of the high-resolution transmission electron microscopy (HRTEM) images of GO and RGO [71]. It is apparent from the presented TEM images that the GO samples contain fewer holes than the RGO samples (see Figure 2.9). This is why the effectiveness of the reducing agent will depend on its ability to reduce epoxides and alcohols to pure carbon with full restoration of the sp^2 system. Non-effective reductants will lead to partial reduction with the formation of carbonyls and other adducts.

Any substance with more or less pronounced reducing character will reduce GO to some extent. The first reductions of GO with copper(I) chloride and iron(II) sulfate were demonstrated back in 1919 by Kohlschütter and Haenni [72]. The graphite-like products of the reduction were isolated and characterized. The now popular reduction of GO with hydrazine was first demonstrated by Hofmann *et al.* in 1934 [73]. In the new era, reduction by several typical reducing agents has been demonstrated up to the present day. Thus, reduction was successfully conducted by: hydrazine [74–77], sodium borohydride [6, 78], hydroquinone [79], hydroxylamine [80], zinc and aluminum powder [81, 82], sodium thiosulfate [83] and hydrogen iodide [84–86]. In addition to the typical reductants, reduction with some unusual and even exotic compounds have been reported. This list includes melatonin, amino acids, vitamin C, phyto extracts, supercritical alcohols and even action by several bacteria.

Electrochemical reduction was also demonstrated [87–89]. In principle, electrochemical reduction is very attractive for its efficiency and "green" approach. However, the electrochemical reduction is limited by the fact that it might occur only on the surface of the negatively charged electrode, but not in the bulk solution. Thus it might be efficient for reduction of as-prepared thin films deposited on the surface of a current collector. This method may find application in the preparation of electrodes for batteries and supercapacitors. To reduce GO in bulk solution, one needs to find a proper system with a mediator that will transfer electrons from the electrode to GO by traveling through the solution. Until such a mediator is not found, chemical reduction is the only approach to reduce bulk quantities of GO.

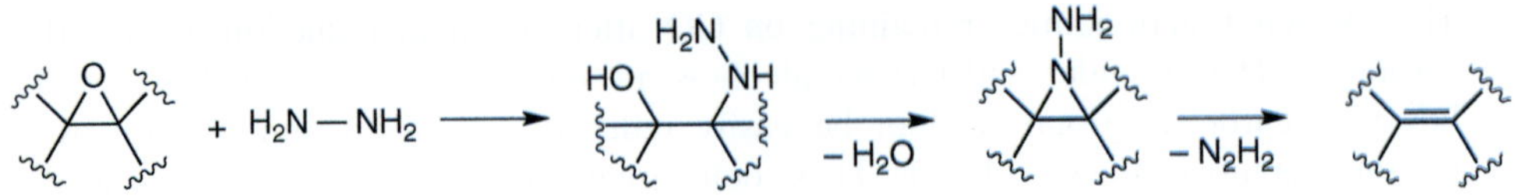

Figure 6.24 Possible reaction pathway for the reduction of epoxide groups by hydrazine

Let us consider the theoretical reaction pathways via which different reducing agents will react with epoxides and tertiary alcohols. The possible idealized reaction pathway for reducing epoxides by hydrazine is represented in Figure 6.24. However, in practice, the reaction mechanism is most likely different, because hydrazine does not afford complete GO reduction: some oxygen remains, and nitrogen is introduced.

Unlike epoxides, ketones cannot be reduced by hydrazine; reaction leads to the formation of hydrazones. This is confirmed by the presence of 3–5% nitrogen in RGO samples reduced by hydrazine.

The mechanism of reduction by other reducing agents is even less studied. It is thought, in general, that sodium borohydride ($NaBH_4$) is a more effective reducing agent than hydrazine. However, the most effective method reported to the present day is the combination of hydrogen iodide and acetic acid [85].

The effectiveness of reducing agents can be assessed based on the two main parameters: (i) C/O ratio of obtained RGO, and (ii) its electrical conductivity. The C/O ratio represents the degree of reduction, and is very straightforward. There is little room for mistake with this parameter: the higher the ratio, the deeper the reduction. However, one needs to keep in mind that elemental content data largely depend on the characterization method. Thus, elemental analysis constantly shows higher oxygen content, compared to that determined by XPS. Also, the accuracy of the analysis gets lower when the oxygen content becomes less than 3%. This is why one needs to be very cautious when reported C/O ratios become higher than 30.

The situation is not as simple with electrical conductivity. The conductivity of two-dimensional (2D) materials, including RGO, can be measured and expressed in different ways. The conductivity of bulk RGO can be measured directly if the material is in a precise geometrical form. Shaping bulk solid RGO can be achieved by cutting paper-like material into strips, or by pressing powder material into a tablet. Electrical conductivity obtained in this way will be affected by the contact resistance between RGO flakes in the bulk sample. It is a function of how densely the flakes are packed into a three-dimensional (3D) material. The contribution of these factors can be eliminated by measuring the conductivity of a single GO flake. This can be done by building a four-terminal micro-device with the assistance of photo- or e-beam lithography. The data obtained from such a device are the most accurate. Conductivity (σ) obtained in such a way is measured in siemens per meter or centimeter ($S\ m^{-1}$ or $S\ cm^{-1}$).

Since GO and RGO are truly 2D materials, their conductivity can also be expressed as sheet resistance (R_s) in Ω/□; R_s is a measure of electrical conductivity of 2D films independent of their thickness. From this perspective, it is an ideal method for measuring the conductivity of a few-layer thin film. The sheet resistance and the conductivity are related by

$$R_s = \frac{1}{\sigma d} \tag{6.1}$$

Table 6.1 *Effectiveness of reducing agents toward GO*

Reducing agent		GO	N_2H_4 [a]	$NaBH_4$ [a]	HI	G
C/O ratio		2.0–2.7	8.1–12.4	8.6	>12	>50
Conductivity ($S\ m^{-1}$)	Reduced at room temperature		1.7×10^2 to 7.2×10^3	4.5×10^2 to 9.2×10^2		
	Reduced or post-treated at 100–250 °C	1×10^{-4} to 1×10^{-2}	1.2×10^4 to 1.6×10^4		3.0×10^2 [21]	8.3×10^4 to 1.6×10^7
	Post-treated at 500–1100 °C		2.1×10^4 [12]	2.0×10^4	3.0×10^4 [22]	

[a] Average literature data are provided.

where d is the film thickness. Many researchers employ this relation to obtain conductivity (σ): they measure the sheet resistance R_s of RGO films, and divide its reciprocal by the film thickness. This is where the uncertainty comes from. Often, researchers do not actually measure the film thickness, but estimate it based on the number of layers, transparency, etc. In addition, the films are not uniform, comprising different numbers of layers in different areas. Moreover, it was shown earlier that the conductivity of bilayer and trilayer RGO does not simply double or triple with the number of layers. Thus, the conductivity of mono-, bi- and trilayer RGO nanoribbons (NRs) was respectively 35, 115 and 210 $S\ cm^{-1}$ [77].

The large difference in conductivity between mono- and bilayers is attributed to the strong interaction between the bottom layer and the SiO_2 substrate. Thus, many different factors need to be taken into account when discussing the conductivity of RGO. This is not always done in the literature. This is why, in Table 6.1, we have omitted some unreliable literature data, and provided averaged values for the methods, where multiple literature data are available.

Chemical reduction alone never yields highly conductive graphene-like material due to the highly defective nature of RGO and the presence of remaining oxygen functionalities. To improve conductivity, the as-obtained RGO needs to be annealed at around 900–1000 °C. After annealing, conductivity increases by two or three orders in magnitude and reaches values in the order of 200–300 $S\ cm^{-1}$. This occurs due to the significant reconstruction of the carbon grid at elevated temperatures.

Most importantly, the conductivity of RGO depends not only on the effectiveness of the reduction method, but also on the quality (density of defects) of the GO precursor. This is why, to compare the effectiveness of different reducing agents, one needs to subject the same GO precursor to different reducing agents. Comparing different literature data obtained by reducing different GO samples is not representative. The best data on the effectiveness of the different reducing agents can be obtained by reduction of GO with the minimal density of defects, which is how it was done by Eigler *et al.* with oxo-G_1 [90]. This will be reviewed in section 6.6.3.

6.6.2 Disproportionation

In addition to chemical reduction, deoxygenation of GO initiated by heat and light have been reported. We stress the importance of using the correct word "deoxygenation" instead of the commonly used "reduction" for this type of transformation, since it is very different from true reduction. Reduction can be conducted only by means of a reducing agent, i.e. a chemical compound that easily gives up existing electrons. The electron transfer from a reducing agent to GO should occur for a transformation to be qualified as "reduction". Electrochemical reduction is the only type of true reduction that is possible in addition to chemical reduction. The term "thermal reduction" introduced recently for the thermally initiated decomposition of GO is nonsense in a scientific sense. Heat does not have and does not supply any electrons to reduce GO. Interestingly, even people with a chemical background use the term "thermal reduction" as the established term. Unfortunately, this is not simply the wrong terminology: misuse of fundamental scientific terms inevitably leads to misunderstanding of the underlying chemical processes. The absolute majority of researchers entering the field remain under the impression that by annealing GO one can obtain nice flat graphene-like material. As an example, a colleague of one of us, with a PhD in chemistry, argued that only oxygen and hydrogen are removed from GO during thermal annealing, while the carbon grid is preserved. The argument in support of this statement was that this was written in some publications. Such a transformation, if it were ever accomplished, would be against the laws of thermodynamics: it is impossible to reduce carbon at the expense of oxygen. In an effort to overcome such a widespread misconception in the field, here we present a brief tutorial on the basics of chemistry.

In terms of redox reactions, such thermal degradation of GO during annealing can be envisioned as disproportionation. It was demonstrated by TGA-MS that carbon dioxide and carbon monoxide are formed during thermal decomposition of GO (Figure 6.2A). In this transformation, the carbon atoms from the GO network in the formal oxidation state +1 (epoxides and alcohols) are converted to elemental carbon (oxidation state 0) and carbon dioxide (oxidation state +4). The respective redox equation can be written as follows, where one carbon atom is removed from the GO network with each two oxygen atoms:

$$2R_2COC^{(+1)}R_2 \rightarrow 3R_3C^{(0)} + 2R + C^{(+4)}O_2 \tag{6.2}$$

In the respective reaction with carbon monoxide as a product, one carbon atom is removed per every oxygen atom:

$$R_2COCR_2 \rightarrow R_3C + R + CO \tag{6.3}$$

For simplicity, (6.2) and (6.3) are written for *epoxides* as the decomposing groups. Similar equations can be written for *alcohols*, for which water will be an additional decomposition product. For the decomposition of *ketones*, carbon monoxide cannot be formed, since carbon in ketones is already in the oxidation state +2, the same as in CO. Only carbon dioxide can be a decomposition product, to afford the following disproportionation reaction, which is supported by experimental data:

$$2R_2C = O \rightarrow R_2C + CO_2 \tag{6.4}$$

Considering the 30–35% oxygen content in typical GO samples, the magnitude of the carbon loss and the damage made on the carbon grid are huge. Annealing of GO introduces into GO planes numerous vacancy defects, and leads to the formation of very defective, holey, net-like GO planes, because part of the carbon grid is removed in the form of gaseous CO_2 and CO (see Figure 6.25). In addition to the numerous experimental data, this was

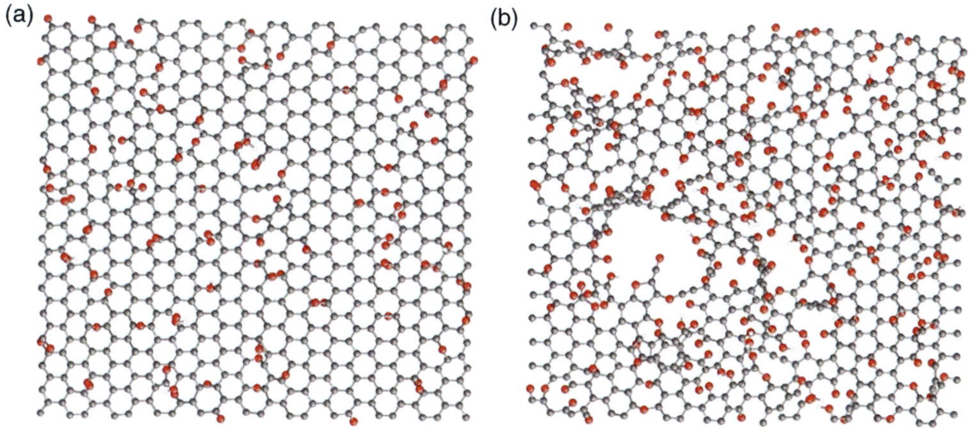

Figure 6.25 Morphology of tpGO formed during thermal annealing at 1500 K: (a) original GO contained 20% oxygen; (b) original GO contained 33% oxygen. The structures are obtained by molecular dynamics simulations. Reproduced from [91] with permission of Nature Publishing Group

also confirmed with theoretical studies [91]. Figure 6.25 demonstrates the structure of GO obtained after annealing at 1500 K, obtained by molecular dynamics simulations. The damage made on the carbon grid is proportional to the oxygen content (C/O ratio) in the original GO. While the presence of 20% oxygen causes only minor damage, 33% oxygen results in very significant degradation of the carbon grid. This is why, to differentiate chemically reduced GO (RGO) from thermally decomposed GO, we introduce the term "thermally processed GO" (tpGO) for the latter.

The main decomposition of GO occurs in the 160–300 °C temperature interval, where most of the oxygen is removed. Epoxides and alcohols are almost completely eliminated at these temperatures, while carbonyls survive. At temperatures above 500 °C, carbonyls are removed. Finally, the carbon grid is restructured at temperatures above 700 °C. Note that, after annealing above 900 °C, as-obtained tpGO does not contain as much oxygen as shown in Figure 6.25; the actual oxygen content is $<1\%$.

The same principle applies to the base of GO "reduction" by light. In the same way as heat, light triggers the disproportionation reaction, leading to the same products. In terms of thermodynamics, the reaction products, i.e. deoxygenated carbon grid and carbon dioxide, are more stable compared to GO. The role of heat, light, radiation, ball-milling, etc. in this transformation is to provide the energy to overcome the activation barrier.

The introduction of numerous holes into the tpGO planes significantly shortens the length of the ballistic transport pathway between two defects. Subsequently, the electrical properties of tpGO are inevitably inferior not only to those of pristine graphene, but also even to those of RGO. For this reason, tpGO is not a good candidate for applications where high electrical conductivity is needed. However, it is superior to RGO as an electrode material for lithium-ion batteries and supercapacitors, since holes increase the surface area and provide channels for the transport of ions.

Interestingly, the same disproportionation reaction can be triggered by the action of some chemicals, such as strong bases. GO undergoes deoxygenation when heated in aqueous solutions of strong bases such as NaOH and KOH. This phenomenon caused different speculations concerning the chemistry behind it. In addition to interpreting it as "reduction", the two-component GO structural model was introduced by Rourke [92]. In fact, NaOH and KOH do not possess any reducing nature, thus this reaction cannot be considered as reduction in principle. This is the same disproportionation reaction as the one triggered by heat. The only difference is that the presence of strong base triggers this reaction at significantly lower temperatures (70–100 °C) as opposed to the 160–300 °C for thermal decomposition.

The following transformations were proposed to explain the mechanism of deoxygenation of GO in alkaline solutions [1, 28]. Equation (6.5) represents a formal redox equation, where tertiary alcohols (formal oxidation state +1) transform to ketones (oxidation state +2) and elemental carbon (oxidation state 0):

$$2R_3C-OH \rightarrow ½(C{=}C)+R_2C{=}O+H_2O \tag{6.5}$$

The reaction mechanism for this transformation can be envisioned as in Figure 6.26. Note that, compared to the initial structure (**3**), the final structure (**4**) contains one oxygen atom less. Thus, in addition to the C–C bond cleavage, this reaction results in deoxygenation with an increase of graphitic carbon content. Interestingly, when the same functional group (vicinal diol) was used as a reactant by molecular dynamics simulations for the thermal decomposition reaction, the same functional groups (ketone and enol at cleaved C–C bond) were obtained as the products [91]. This demonstrates the possibility of some common principles in the two different routes of deoxygenation.

Conversion of alcohols to ketones (Figure 6.26) constitutes the first step of a complex reaction. Further transformation of ketones to carboxyls and carboxylates can be represented by formal equation (6.6) and Figure 6.27:

$$2R_2C=O+OH^- \rightarrow RCOOH+R^- \tag{6.6}$$

In (6.6) the minus sign on R^- represents the negative charge delocalized over the GO fragment. The reaction begins (Figure 6.27) with nucleophilic attack of a hydroxide ion on a carbonyl group of the structure (**5**), and results in C–C bond cleavage and the transformation of ketone to carboxyl group (**6**). In basic conditions, the carboxyl group of the structure (**6**) will be present in the ionized form, i.e. as a carboxylate anion.

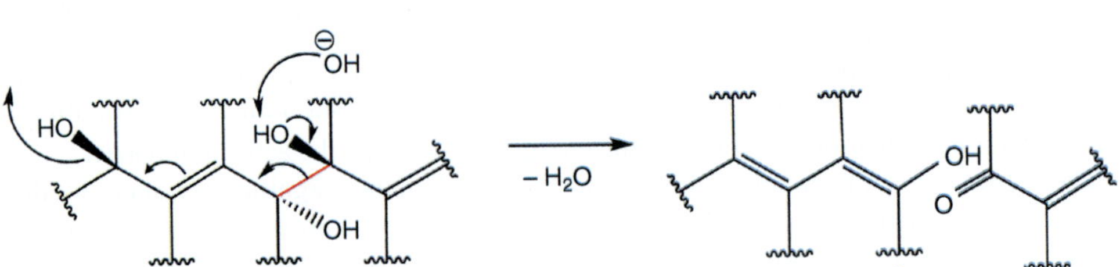

3 **4**

*Figure 6.26 The reaction scheme showing deoxygenation of GO during base treatment. Structure (**3**) is a GO fragment containing three alcohol groups. Nucleophilic attack of hydroxide ion on a hydrogen atom results in C–C bond cleavage and in the formation of a ketone. The fragment (**4**) contains one oxygen atom less compared to structure (**3**), signifying deoxygenation*

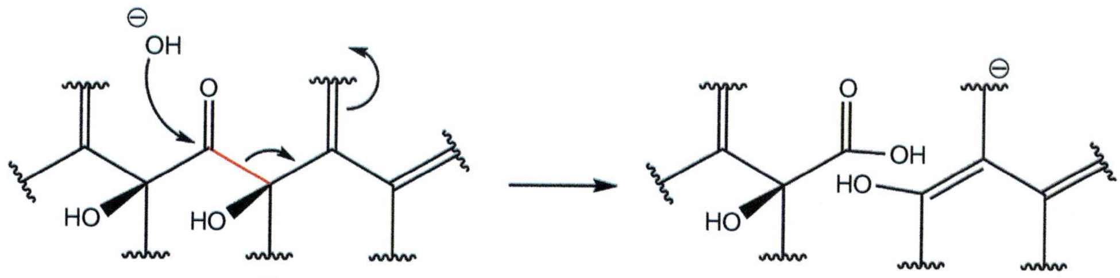

5 **6**

Figure 6.27 Conversion of ketones to carboxyls during base treatment. Structure (**5**) is a GO fragment containing one ketone and two alcohol groups. Nucleophilic attack of a hydroxide ion on a carbonyl group results in C–C bond cleavage and in the transformation of ketone to carboxyl group (**6**)

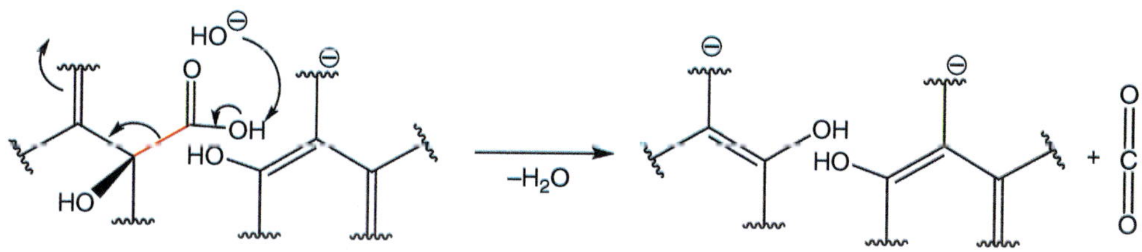

Figure 6.28 Decarboxylation of GO under strongly alkaline conditions. The reaction mechanism is typical for decarboxylation

In the extreme conditions of high base concentration and high temperature, the reaction can progress even further to lead to decarboxylation of the previously formed carboxyls, as shown in (6.7) and Figure 6.28. The nature of this transformation can still be considered as disproportionation, since here the carboxyl carbon (oxidation state +3) yields carbon dioxide (oxidation state +4) and elemental carbon (oxidation state 0):

$$RCOOH + OH^- \rightarrow R^- + CO_2 + H_2O \tag{6.7}$$

The reaction mechanism is typical for decarboxylation (Figure 6.28).

6.6.3 Reduction Strategies

The choice between the strategies to process GO (chemical reduction versus thermal annealing) depends on the application. When good electrical conductivity is critical, the two-step reduction is the only approach. On the first step, GO is chemically reduced to RGO. This step removes most of the oxygen without causing additional damage to the carbon plane. On the second step, thermal annealing at 900–1100 °C removes the remaining oxygen atoms, and restructures the carbon grid to solidify the ballistic transport channels. Because most of the oxygen is already removed during the chemical reduction step, annealing does not cause the formation of additional defects; it only removes oxygen at existing defects. Figure 6.29 exemplifies how chemical composition and electrical properties change with different steps of processing.

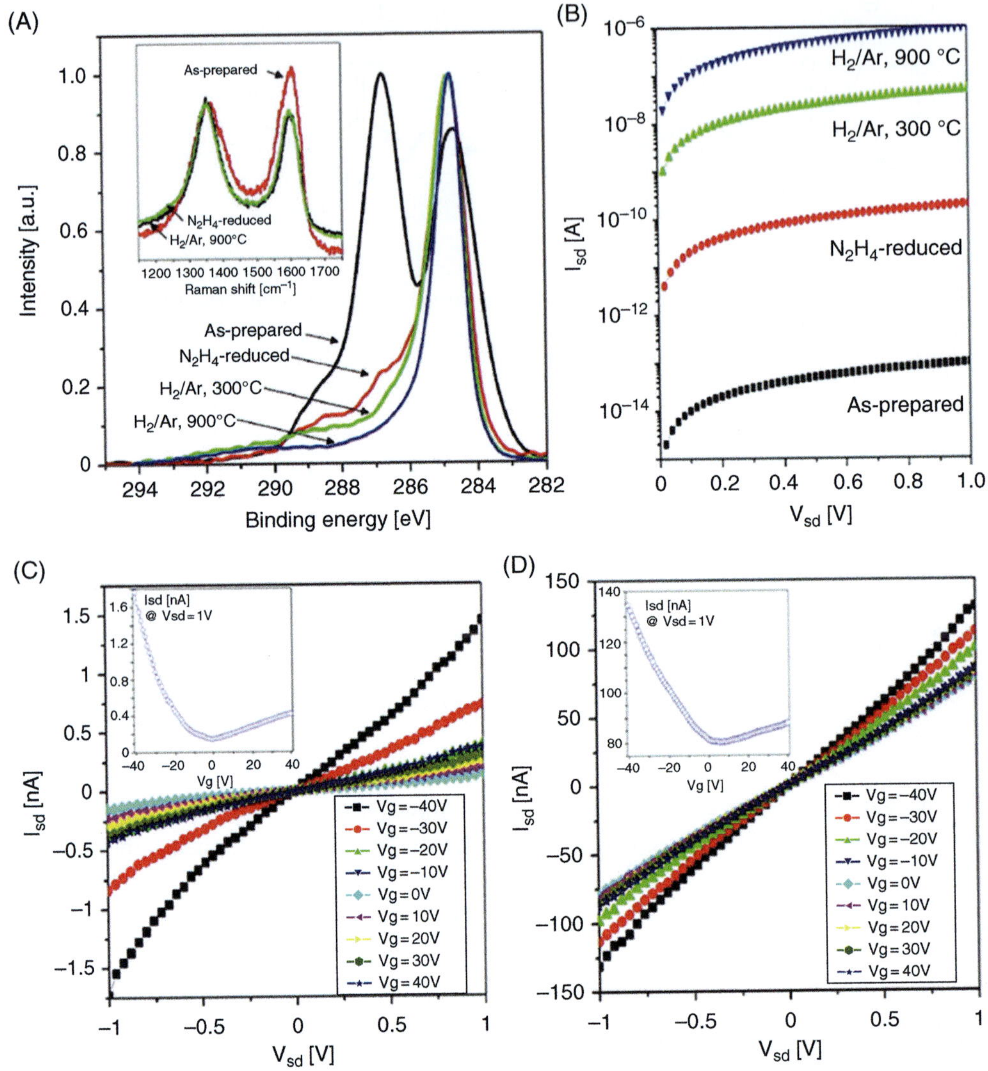

Figure 6.29 Three-step reduction of GO nanoribbons: N_2H_4 reduction, 300 °C annealing, 900 °C annealing. (A) C 1s XPS and (inset) Raman spectra. (B) Logarithmic I–V curves for the nanoribbons with different degree of reduction; the average data obtained from testing 30 different devices. (C) and (D) Source–drain current (I_{sd}), source–drain voltage (V_{sd}) and gate voltage (V_g) dependences for the same device build on an N_2H_4-reduced monolayer RGO NR before (C) and after (D) annealing in H_2/Ar at 300 °C for 30 min. Reproduced from [77] with permission of ACS

The C 1s XPS spectrum for as-prepared GO NRs (Figure 6.29A) is dominated by the component at 287 eV, which is attributed to the carbon of epoxides and alcohols. After reducing with N_2H_4 at room temperature, this component is significantly decreased. The shoulder at 289 eV is decreased but to a significantly lesser extent, suggesting that carbonyl groups mostly survived the reduction. The conductivity increased by three or four orders of magnitude compared to as-prepared GO NRs (Figure 6.29B). After the chemical reduction step, the RGO NRs were additionally subjected to annealing at 300 and 900 °C. These two steps additionally reduced the content of oxygen. After annealing at 900 °C, even the weak π–π* interaction band at 291 eV appears on the C 1s XPS spectrum, suggesting restoration of the highly conjugated sp^2 network. The conductivity increases by two orders of magnitude after annealing at 300 °C, and by one more order of magnitude for a device built on reduced graphene oxide nanoribbon (rGONR) annealed at 900 °C.

The increase in conductivity after each reduction step is further illustrated by Figure 6.29(C,D), which display the electrical properties for the same device built on an N_2H_4-reduced monolayer nanoribbon before (Figure 6.29C) and after (Figure 6.29D) annealing in H_2/Ar at 300 °C for 30 min. The electrical properties of the annealed devices were qualitatively the same, though their conductivity increased by 10- to 100-fold after annealing.

6.6.4 Reduction of Oxo-Functionalized Graphene

Oxo-G_1 exhibits an almost intact carbon framework and therefore it can potentially be chemically reduced to a high quality of graphene. Many reducing agents have been used to generate black graphite-like materials from yellowish GO [93]. However, with GO that bears too many defects, it is impossible to quantify the efficiency of the chemical reduction. An incorporation of nitrogen species is described, for example, in the literature for hydrazine RGO and proven by SSNMR on ^{13}C- and ^{15}N-labeled samples [68]. Consequently, it is difficult to rank the potential of a reducing agent. The reduction principles of GO are summarized in the sections above. Moreover, a recent review article regarding the reduction of GO summarizes various methods and chemical reduction mechanisms. The success of the reduction is thereby quantified by means of the C/O ratio according to XPS determination or by the measurement of the electrical conductivity of films of flakes [93, 94]. For RGO that bears defects of several percent, these methods are indeed suitable. However, with concentrations of defects far below 1%, Raman spectroscopy is the best technique to quantify the density of defects and consequently the quality of graphene that was obtained from oxo-G_1 that bears less than about 1% of defects. With this technique, the efficiency of some reducing agents was probed to visualize variations [90]. Moreover, a general reduction mechanism was proposed [95].

6.6.4.1 *Efficiency of Reducing Agents*

With oxo-G_1 and statistical Raman spectroscopy, the quality of generated graphene can be determined using different types of reducing agents. In principle, it is possible to rank the efficiency of the reducing agents. Plots of I_D/I_G versus Γ_{2D} values illustrate the results for four reducing methods (Figure 6.30A). A mixture of hydriodic acid (HI) and trifluoroacetic acid (TFA) was identified as the most efficient reduction method. The slightly less efficient ascorbic acid (vitamin C) can however also produce graphene of high quality. Although, hydrazine vapor is often used in the literature, it is a less effective reducing agent. The worst quality of graphene is obtained after thermal annealing of oxo-G_1 at 200 °C. The latter method produced amorphous graphene that is dominated by defect sites ($^{\mathrm{HI/TFA}}\Gamma_{2D} < {}^{\mathrm{vitamin\ C}}\Gamma_{2D} < {}^{\mathrm{hydrazine}}\Gamma_{2D} < {}^{200\ °\mathrm{C}}\Gamma_{2D}$) [90].

The reduction of oxo-G_1 with ascorbic acid yields a similar quality as for graphene obtained by reduction with HI/TFA. But, surprisingly, differences could be revealed by

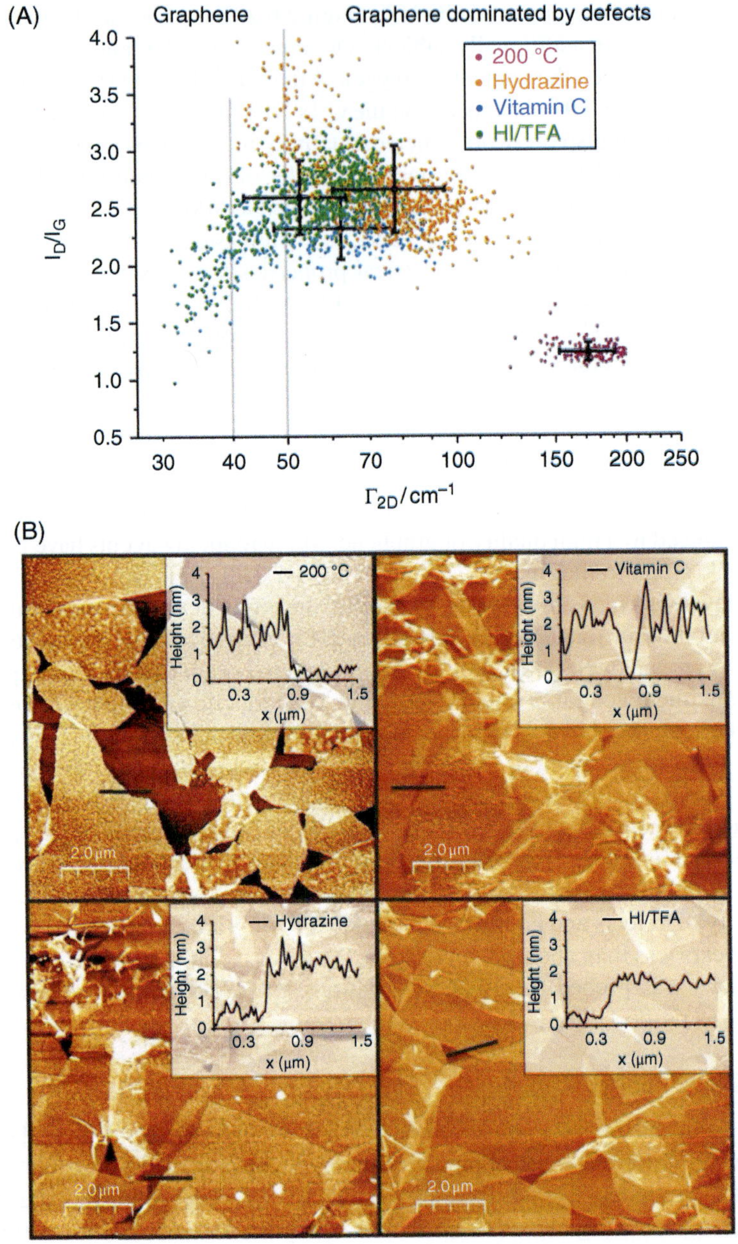

Figure 6.30 (A) Statistical Raman analysis of graphene from oxo-G_1 using different reducing agents. (B) AFM images of graphene from oxo-G_1 obtained after thermal reduction (at 200 °C) or reduction with vitamin C (ascorbic acid), hydrazine or hydriodic acid and trifluoroacetic acid (HI/TFA) [90]. Reproduced from [90] with permission from John Wiley & Sons

AFM regarding the height of samples (Figure 6.30B). A surface with a roughness of about 2 nm was found for ascorbic acid-reduced samples, and thus it can be concluded that the reaction products of ascorbic acid remain adsorbed on the surface of graphene, even after washing procedures. However, a smooth surface is required if further functionalization reactions are intended. It should be kept in mind that the stability of graphene, generated in dispersion by reduction, might also be strongly influenced by adsorbed species. Properties such as the surface tension will be very different from those of pristine graphene and adsorbed species also influence the contact resistance between flakes if the conductivity of films of flakes are measured.

6.6.4.2 *Mechanism of Reduction of Oxo-Functionalized Graphene*

With the analysis of the efficiency of reducing agents, the question arises of whether the quality of the yielded graphene is limited by the efficiency of the reducing agent or by the quality of oxo-G_1. In the study that is depicted in Figure 6.31 it is demonstrated that the quality is limited by the reducing agent. It can be concluded that an efficient reducing agent gives a better quality of graphene. However, the question arises if HI/TFA is the best reducing agent or is the quality of graphene limited by other factors, such as kinetic hindrance. Oxo-G_1 is usually deposited on a Si/300 nm SiO_2 substrate by, for example, the Langmuir–Blodgett technique. After deposition, the film of oxo-G_1 flakes is reduced by the reducing agents. However, the reducing agent can only access the upper surface of oxo-G_1, because the substrate blocks the lower side, as illustrated in Figure 6.31 [95]. It cannot be ruled out that an even better quality of graphene would be produced if the reducing agent had access to both sides of the oxo-G_1. Therefore, oxo-G_1 was placed on the aqueous subphase, as usually used for Langmuir–Blodgett coatings. In this case, the reducing agent, ascorbic acid, or alternatively hydriodic acid, was dissolved in the aqueous subphase.

The reduction starts from the lower side. The surface turns gray after several minutes and graphene starts to form and is floating on the aqueous subphase. The graphene can be transferred onto a substrate and analyzed by statistical Raman spectroscopy. The graphene, yielded after subphase reduction, can be further reduced by HI/TFA. The results compared to the direct deposition of oxo-G_1 on the substrate, followed by reduction from the top with the vapor of HI/TFA, reveals that this direct reduction procedure is the most efficient one. The subphase reduction is less efficient, and even the combination of the subphase reduction and the gas-phase reduction is slightly less efficient. This may be due to adsorption processes and hindrance of reaction pathways.

Finally, a reduction mechanism was derived for the reduction of oxo-G_1 that bears no defects. Generally, about every second carbon atom of oxo-G_1 is sp^2-hybridized, and thus a π system is still present, next to oxo-functional groups like hydroxyl, epoxy and organosulfate groups. The acid can protonate those groups to form oxonium ions. The electron affinity of oxo-G_1 increases with protonation and facilitates the electron transfer from iodide to the π system. Subsequently, water and sulfuric acid are cleaved to form graphene (Figure 6.31A). Oxo-G_1 is a 2D material and therefore it does not matter from which side electrons are transferred to the π system. Even protonation of oxo-groups between the substrate and the oxo-G_1 is possible due to the ability of proton hopping from the edges of the flakes. Thus, it can be concluded that the reduction of oxo-G_1 is highly efficient and complete using a potent reducing agent system such as HI/TFA [95].

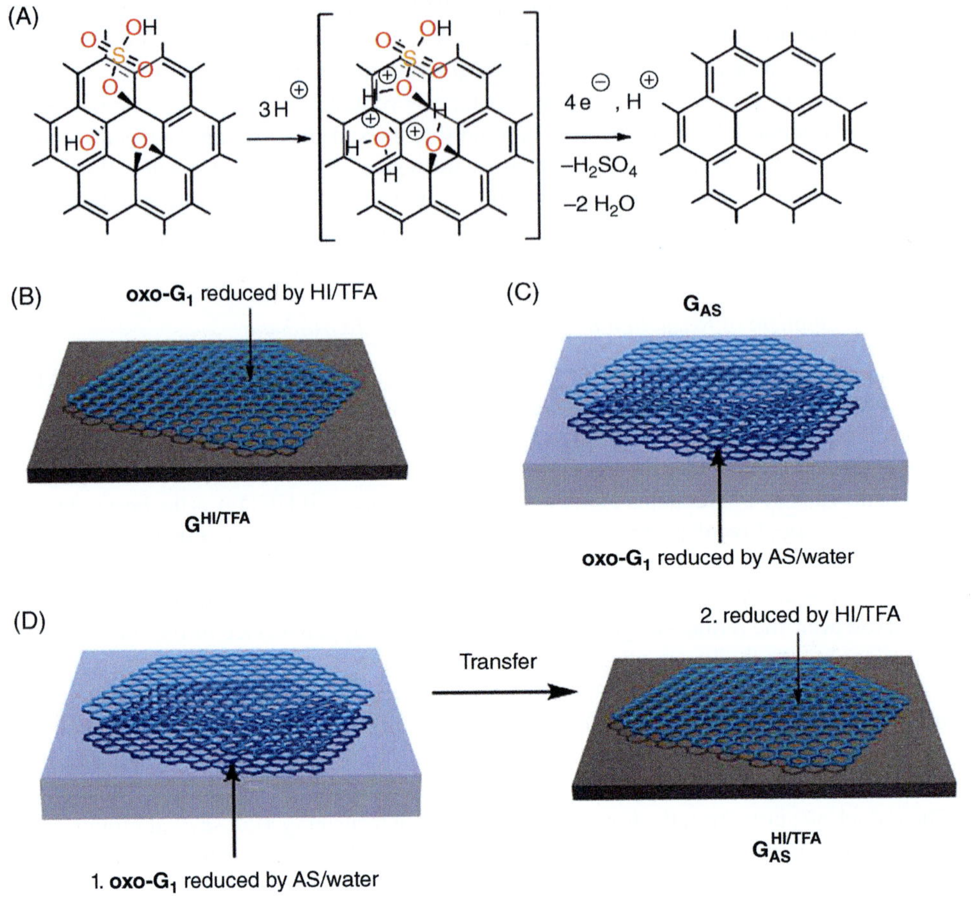

Figure 6.31 (A) Mechanism of the reduction of oxo-G_1 by an electron donor, such as iodide and a strong acid. Successive protonation and electron transfer steps remove oxygen functional groups to form graphene on a substrate. (B) Reduced on substrate by vapor of HI/TFA ($G^{HI/TFA}$). (C) Reduced by ascorbic acid (AS) from the reductive subphase (G_{AS}). (D) Reduced by the combination of both reduction methods ($G_{AS}^{HI/TFA}$) [95]. Reproduced from [95] with permission from ACS

6.7 Reactions with Reduced Form of Graphene Oxide

Here, we want to show some facile approaches toward the functionalization of RGO, although this is not the focus of this chapter. Since RGO contains very little oxygen functionalities, reactions with RGO involve the carbon grid. From this perspective, RGO chemistry is similar to the chemistry of pristine graphene. One well-demonstrated type of reaction here is diazonium chemistry. Functionalization via diazonium chemistry is normally conducted in water or in acetonitrile. The aryl diazonium salts can be either pre-synthesized or formed *in situ* from aromatic aniline derivatives and sodium nitrite. In solutions, diazonium salts generate aryl cations followed by an electron transfer from the

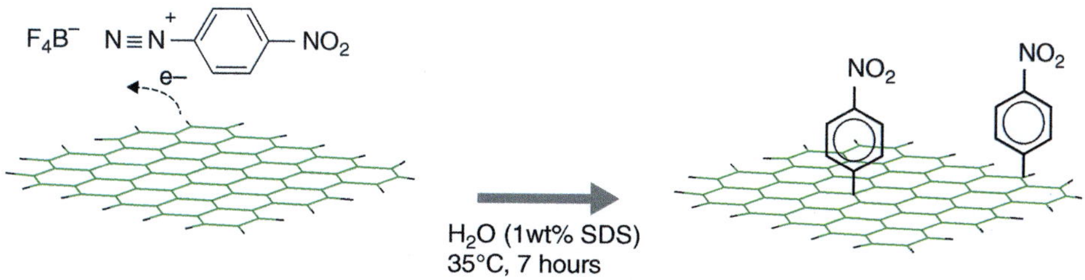

Figure 6.32 Functionalization of pristine graphene with nitro-aryl diazonium salts. Reproduced from [57] with permission of ACS

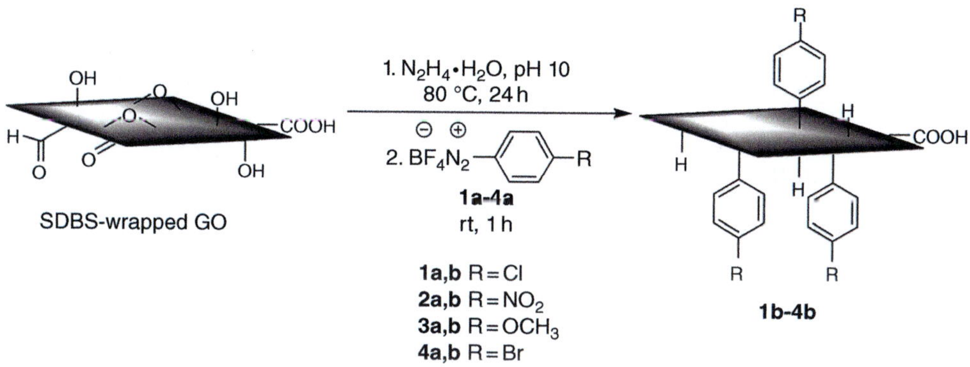

Figure 6.33 Scheme showing sodium dodecylbenzene sulfonate (SDBS)-wrapped GO, its reduction and functionalization of intermediate SDBS-wrapped RGO with diazonium salts. Reproduced from [98] with permission of ACS

graphene (or RGO) to form aryl radicals that attack the carbon grid, and establish C–C bonds between the carbon grid and the aryl radical (Figure 6.32). Thus, the carbon atom in graphene is converted into the sp^3-hybrized state.

With pristine graphene, this type of chemistry was successfully demonstrated with several diazonium salts [57, 96, 97]. It was demonstrated that the effectiveness of the reaction depends on the nature of the substrate holding graphene [97]. Reactions proceed rapidly for graphene supported on SiO_2 and Al_2O_3 (sapphire), but negligibly on alkyl-terminated and hexagonal boron nitride (hBN) surfaces. Also, a model of reactivity based on substrate-induced electron–hole puddles in graphene has been developed [97].

For RGO, this type of reaction was first demonstrated by Lomeda *et al.* [98] for RGO dispersed in 1% solution of sodium dodecylbenzene sulfonate (SDBS) in water. The SDBS-wrapped GO was first reduced by hydrazine, and next the SDBS-wrapped RGO was functionalized by four different diazonium salts as depicted on the Figure 6.33.

Unlike graphene, RGO contains some remaining oxygen functionalities. In principle, one might argue why diazonium salts indeed react with the graphenic domains of RGO, but not with those oxygen functional groups. One can only take into account that the ketones and carboxyls remaining on RGO are not prone to chemical attack by radicals generated by diazonium salts. Thus, the reaction should indeed be on the carbon grid. The challenge here is to prove successful covalent functionalization of RGO. The presence of any distinct parts of the diazonium salts on as-functionalized RGO is not yet strong evidence. As we demonstrated in section 6.5, such a presence in many cases is due to physisorption of the reactant on GO/RGO, but not due to covalent functionalization.

In the case of graphene, the fact of successful covalent functionalization is apparent from Raman spectroscopy data. The D peak arising after functionalization is indicative of introduced scattering defects, which in this case are sp^3-hybridized carbon atoms in the otherwise ideal non-defective sp^2-hybridized carbon plane. The higher the D peak, the higher the degree of functionalization. In the case of RGO, this is not an option since RGO is already highly defective with a density of scattering centers beyond the threshold when it can be quantified. Note that it is almost impossible to confirm the formation of the C–C bond spectroscopically. The only option here is monitoring the electrical conductivity with functionalization. Diazonium functionalization should in principle additionally decrease the length of ballistic transport channels; thus the conductivity of RGO should decrease after functionalization. This was the approach chosen in the work with RGO nanoribbons (NRs) [99]. In contrast to the previous work [98], RGO NRs were not only chemically reduced, but also annealed at 900 °C; thus the content of oxygen groups was very low. Functionalization was carried with RGO NRs laying flat on a substrate surface and built into a device as shown in Figure 6.34. The conductivity of RGO NRs was measured in certain functionalization time intervals. As evident from Figure 6.34, the source–drain current decreased after exposing the device to the diazonium salt solution, confirming successful functionalization. The most significant functionalization occurred in the first 5 min of grafting (Figure 6.34, right), and was almost complete within 60 min. The change in the rate of the chemical reaction, measured as the change in electrical conductivity, was typical for a first-order chemical reaction.

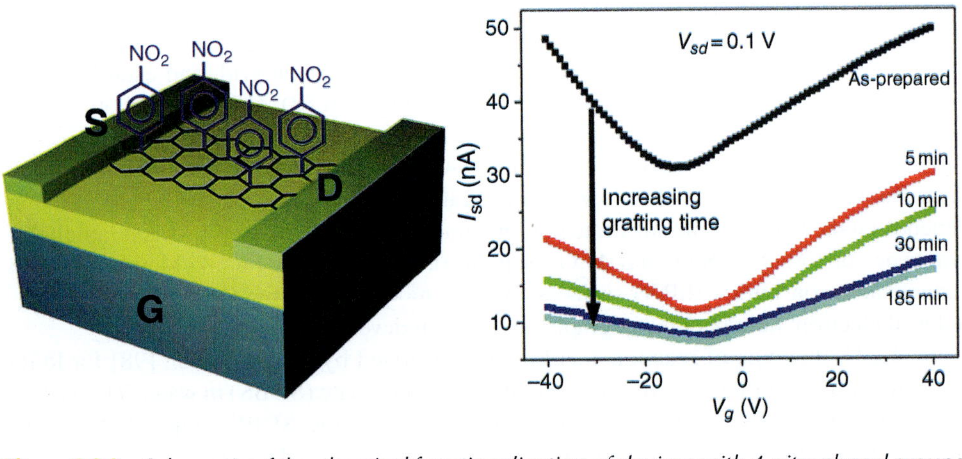

Figure 6.34 *Schematic of the chemical functionalization of devices with 4-nitrophenyl groups (left). Electronic devices consist of monolayer RGO nanoribbons contacted with Pt source (S) and drain (D) electrodes. The devices were fabricated on a 200 nm thick thermal* SiO_2 *over heavily doped p-type Si that was used as a back gate. The* I_{sd}–V_g *curves recorded at* $V_{sd}=0.1$ V *after several consecutive grafting experiments (right). The labels on the curves show the total grafting time. Reproduced from [99] with permission of ACS*

Diazonium functionalization is the best-studied type of covalent functionalization of RGO. Many reports, such as "click" chemistry on RGO, can be found in the literature [100, 101]. However, such types of reaction are conducted not on RGO itself, but on the functional groups pre-grafted on RGO by diazonium chemistry. Other functionalization concepts are plausible and have been reported in the literature. However, systematic and quantitative investigations are demanded to successfully use these chemical approaches in future applications.

6.8 Controlled Chemistry with Graphene Oxide

The structure of GO was discussed above, and extended defects in addition to various oxo-functional groups are part of the structure. Functionalities can be expected to be located not only on the basal plane but also at the edges of flakes and defects. Real GO would be best described by the structure model of Hofmann, as depicted in Figure 6.1. However, such a type of GO has not yet been prepared by wet chemistry. However, it was demonstrated that oxygen could chemically bind to graphene on SiC (G_1/SiC). A scanning tunneling microscopy investigation revealed the reversibility of that oxygen binding (Figure 6.1) [18]. Thus, functionalized graphene derivatives can in principle be synthesized while preserving the honeycomb lattice of graphene. For developing a controlled chemistry of GO and avoiding the analysis of defect structure motifs, it would be ideal to prevent the rupture of the carbon framework during oxidative functionalization. Consequently, graphene derivatives with oxo-functional groups would be yielded. Here, such systems will be called oxo-functionalized graphene, with oxo-G_1 as abbreviation. The synthetic methods that lead to oxo-G_1 are summarized in Chapter 2.

6.8.1 Nomenclature of Polydisperse and Functionalized Graphene

In contrast to monodisperse molecules such as C_{60}, graphene is a polydisperse material, especially if it was prepared from graphite. Purification procedures do not render possible separation techniques to make monodisperse graphene. Therefore, it is necessary to introduce a systematic formula that describes the functionalized graphene system. A generalized systematic formula was introduced and developed to characterize graphene and complex functionalized systems of graphene [102, 103]:

$$S/^{s,d}G_n-(R)_f/A_f \tag{6.8}$$

The notation used in (6.8) is as follows: S = substrate, no S = reactions applied in dispersion; G = graphene, s = size of graphene, d = structural defect density of graphene within the carbon framework, n = number of layers of graphene; R = addend, A = non-covalently bound molecules, and f = degree of functionalization.

The number of layers of graphene is indicated by the index n, and graphene (G_1) can be distinguished from bilayer graphene (G_2), trilayer graphene (G_3), few-layer graphene ($G_{\text{few-layer}}$ or G_{4-10}) and graphite (G_n). Graphene must be stabilized, which can be achieved, for example, by placement on a surface, such as SiO_2 (SiO_2/G_1) or boron nitride (BN/G_1). Etching Cu away from Cu/G_1 leads to graphene floating on a surface of water (H_2O/G_1). The properties of materials are well known to change with size and therefore another type of important polydispersity is the size of the flakes of graphene. The analysis of flake size can be conducted by microscopy, such as AFM, or directly from solution by analytical ultracentrifugation, a tool that was just recently developed [104]. Flakes of graphene can be on the nano-, micro- or even centimeter scales. Graphene grown on a copper foil with a size of 1 cm × 1 cm can be indicated as $Cu/^{1cm}G_1$. Delaminated graphene with a size of flakes of about 5 μm can be placed on a SiO_2 surface and this type of graphene can be described as $SiO_2/^{5\mu m}G_1$.

Defects within the carbon σ framework of graphene disturb the lattice and change the electronic properties. The density of defects can be determined by the analysis of Raman spectra. Densities of defects in the range between about 0.001% and 1–3% can be determined. This means that, for example, if one C atom of the honeycomb lattice was periodically missing and the intact areas of graphene consisted of 1000 C atoms, the density of defects would be 0.1% and would be indicated by $^{0.1\%}G_1$. The highest sensitivity of Raman spectroscopy toward defects is between 0.01% and 1%. However, not all types of defects in graphene can be identified by Raman spectroscopy, because silent defects are also known [105].

Structural defects within the honeycomb sp^2 network can be missing atoms or a rearranged carbon lattice with five-membered rings, seven-membered rings or others. Also sp^3-hybridized C atoms or clusters of sp^3-hybridized carbon can be identified as defects. However, it is not possible to distinguish between those, and therefore the synthetic procedure and origin of graphene must be taken into account to discriminate between them. Further analysis may be necessary. Thus, instead of determining the density of defects, the degree of functionalization can be derived from Raman spectra [106]. With functionalization of graphene, another type of polydispersity of graphene must be faced by means of the degree of functionalization and the regiochemistry of addends. The latter is currently difficult to determine and is therefore not indicated by the systematic formula (6.8). The product of hydroxyl functionalization, with every twentieth C atom functionalized, would be specified as G_1-$(OH)_{5\%}$. Often, *N*-methylpyrrolidone is used for the exfoliation of few-layer graphene and for a sample where elemental analysis revealed 30 mass% of adsorbed organic molecules, the composition would be indicated as $G_1/NMP_{30\%}$. With the introduction of the systematic formula (6.8), it is now possible to display the polydispersity of graphene and functionalized derivatives.

6.8.2 Organosulfate in Graphene Oxide – Thermogravimetric Analysis

The preparation of GO in sulfuric acid generally ends up in a product with a certain sulfur content that was suspected to be an impurity. However, Boehm and Scholz suggested in 1966 that some of the sulfur content may be due to sulfonic acids or sulfate esters [107]. TGA is a powerful tool to analyze chemical functionalization. The coupling of TGA with MS allows the nature of cleaved molecules to be determined. Recently, organosulfate groups (sulfate esters) were identified by thermogravimetric analysis in GO and oxo-G_1 as structure motifs [108]. Moreover, no indication was found that sulfonic acids are part of the structure.

The formation and hydrolysis of cyclic organosulfate groups in pristine graphite oxide was proposed in the literature and it seems likely that these species partially hydrolyze to form organosulfate (Figure 6.35A) [29]. Organosulfate groups are stable in water at neutral pH. Major species cleaved in the TGA experiment are water with *m/z* 18, CO (and background N_2) with *m/z* 28, CO_2 with *m/z* 44 and, surprisingly, SO_2 with *m/z* 64. As depicted in Figure 6.35(B), up to 120 °C mainly water is removed that may be adsorbed on the

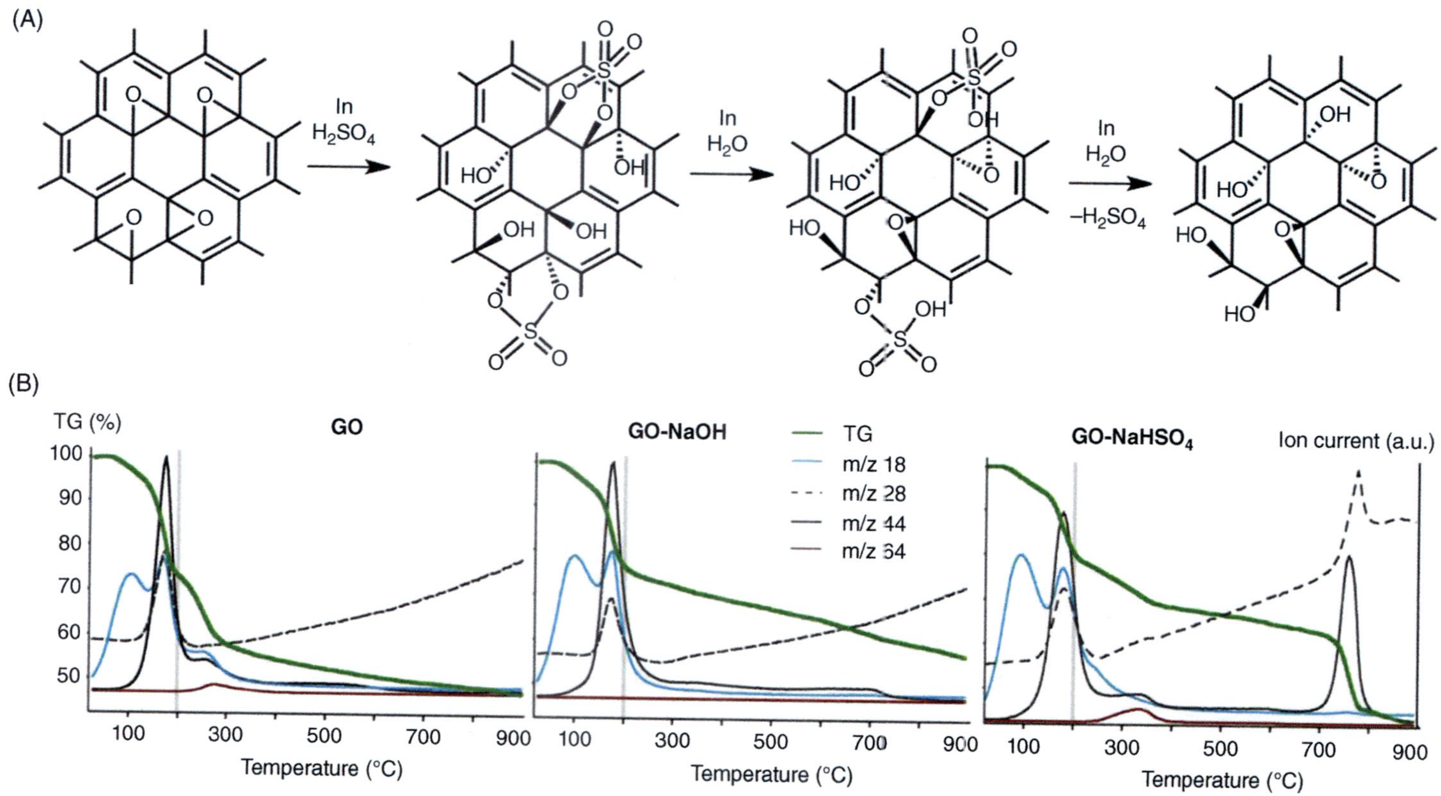

Figure 6.35 *(A) Proposed formation of cyclic organosulfate groups in GO and hydrolytic cleavage to organosulfate groups and hydroxyl groups. Acid-catalyzed epoxide formation of cis-dihydroxy groups can be assumed [17, 29]. The structure of GO therefore bears epoxy, hydroxyl and organosulfate groups as major functional groups. Defects in GO are omitted and the structure model is also valid for oxo-G$_1$ [108]. (B) Thermogravimetric analysis of (left) GO, (middle) GO treated with sodium hydroxide to cleave organosulfate groups and (right) GO treated with sodium hydroxide and mixed with sodium hydrogensulfate [108]. Reproduced from [108] with permission of John Wiley & Sons*

surface of GO, followed by the main weight loss up to 200 °C, related to the cleavage of water, CO and CO_2. Another weight loss between 200 and 300 °C can be related to organosulfate and SO_2 formed due to the decomposition of organosulfate to SO_3 that reacts with carbon to form CO_x and SO_2. The formation of SO_2 can further be identified by FTIR analysis of the evolving gases. The amount of organosulfate can be determined to one organosulfate on 20 C atoms in accordance with elemental analysis, and therefore the properties of the bulky organosulfate dominate the properties of GO. The organosulfate groups of GO can be cleaved by simple treatment with sodium hydroxide, and after purification the thermogravimetric analysis reveals that there is no additional weight-loss step between 200 and 300 °C (Figure 6.35B).

The decomposition of inorganic sulfate can be detected by thermogravimetry at 700–800 °C. Mixing of GO that was already treated by sodium hydroxide with sodium hydrogensulfate and subsequent analysis gives evidence for the decomposition temperature of inorganic sulfate in the presence of GO. A broad weight loss between 200 and 350 °C is detected also that is most likely due to *in situ* formation and decomposition of organosulfate. With respect to thermogravimetric investigations, the structural model of GO and oxo-G_1 can be extended by organosulfate moieties in addition to epoxy and hydroxyl groups, predominantly covalently bound above and below the carbon skeleton. The acidity of GO can also be explained by hydronium ions as counter-ions of organosulfate groups, as shown in Figures 6.35(A) and 6.31(A).

The identification of organosulfate in GO and oxo-G_1 is beneficial for the development of controlled chemistry because organosulfate groups can be further used to build molecular architectures, as described in section 6.8.3.3.

6.8.3 Synthetic Modifications of Oxo-Functionalized Graphene

The chemical structure of oxo-G_1 bears hydroxyl, epoxy and organosulfate groups as major functional groups. Moreover, the size of the flakes is on the micrometer scale. In addition, the density of defects is about 0.3% on average. Oxo-G_1 is stable up to 100 °C, and therefore the next step is the controlled chemical modification of the functional groups of oxo-G_1.

6.8.3.1 Reactions of Oxo-Functionalized Graphene with HCl and NaOH

For controlling the chemistry of oxo-G_1, a stable carbon framework is desirable. Chemical transformations of oxo-G_1 and subsequent analyses of the integrity of the carbon framework reveal that reactions on the basal plane are possible without rupturing the carbon framework [109].

The functional groups of oxo-G_1, such as epoxy, hydroxyl and organosulfate, are stable in aqueous HCl at the pH of about 1 at 10 °C (Figure 6.36). However, at an elevated temperature of 40 °C, organosulfate groups are cleaved, while hydroxyl and epoxy groups remain chemically bound. Infrared spectroscopy gives evidence for the formation of ether groups, due to the evolution of a strong absorption at about 1100 cm^{-1}. The carbon framework was not harmed under these reaction conditions, according to SRM investigations.

GO is not quite stable against base, as outlined in section 6.3, and therefore the chemical reactivity of oxo-G_1 against sodium hydroxide was of interest. Surprisingly, the reaction conditions can be controlled, and organosulfate and possibly also epoxy groups are cleaved to diols upon hydroxide treatment of oxo-G_1 at 10 °C. SRM proves the stability of the

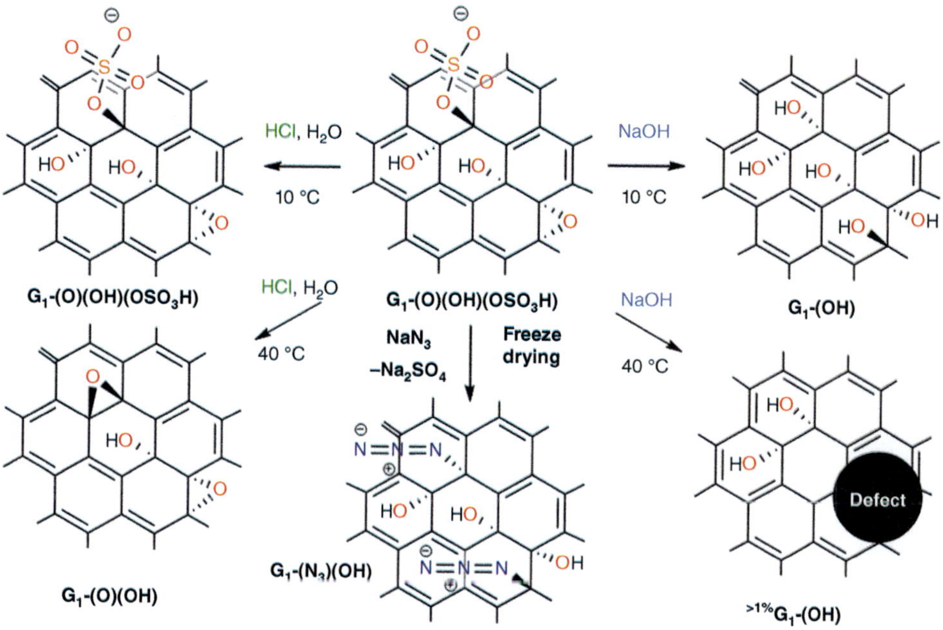

Figure 6.36 *Illustration of the reactivity of oxo-G_1. Reactivity at 10 °C (top) and 40 °C (bottom), upon treatment with HCl or NaOH is shown. Also the substitution of organosulfate by azide is possible under controlled reaction conditions [110]*

carbon framework and the cleavage of organosulfate is identified by elemental analysis and TGA. The treatment of oxo-G_1 with sodium hydroxide at 40 °C not only cleaves functional groups, but also introduces permanent defects into the carbon framework (Figure 6.36). The statistical Raman analysis reveals that Γ_{2D} increases from 72 cm^{-1} to 120 cm^{-1}. These investigations demonstrate that controlled reaction conditions must be maintained to enable the chemistry on the surface of graphene. The reaction of oxo-G_1 with sodium hydroxide is possible and can therefore give synthetic access to hydroxylated graphene (G_1-OH), a suitable precursor for further functionalization reactions.

6.8.3.2 *Controlled Synthesis of Azido-Hydroxyl Graphene*

The reactivity of organosulfate groups can be exploited to enable azide functionalization on the surface of oxo-G_1. This substitution reaction preserves thermally unstable groups due to the mild reaction with sodium azide that proceeds in the solid state upon freeze-drying (Figure 6.36). Experimental evidence indicates the substitution of organosulfate by azide anions [110]. This direct functionalization of oxo-G_1 with azide, preserving functional groups, is proved by IR spectroscopy, TGA coupled with MS, elemental analysis and ^{15}N NMR spectroscopy measured in the solid state. Azide is predominantly located on the surface of oxo-G_1 and not on the edges. The C–N stretching vibration of azide is found at 2123 cm^{-1} and shifts by 11 cm^{-1} upon binding of $^{15}N^{14}N_2$ to the carbon framework (Figure 6.37A). The substitution of organosulfate is also evident by TGA (Figure 6.37B). Also the wet-chemically cleaved sulfate can be identified after precipitation as $BaSO_4$.

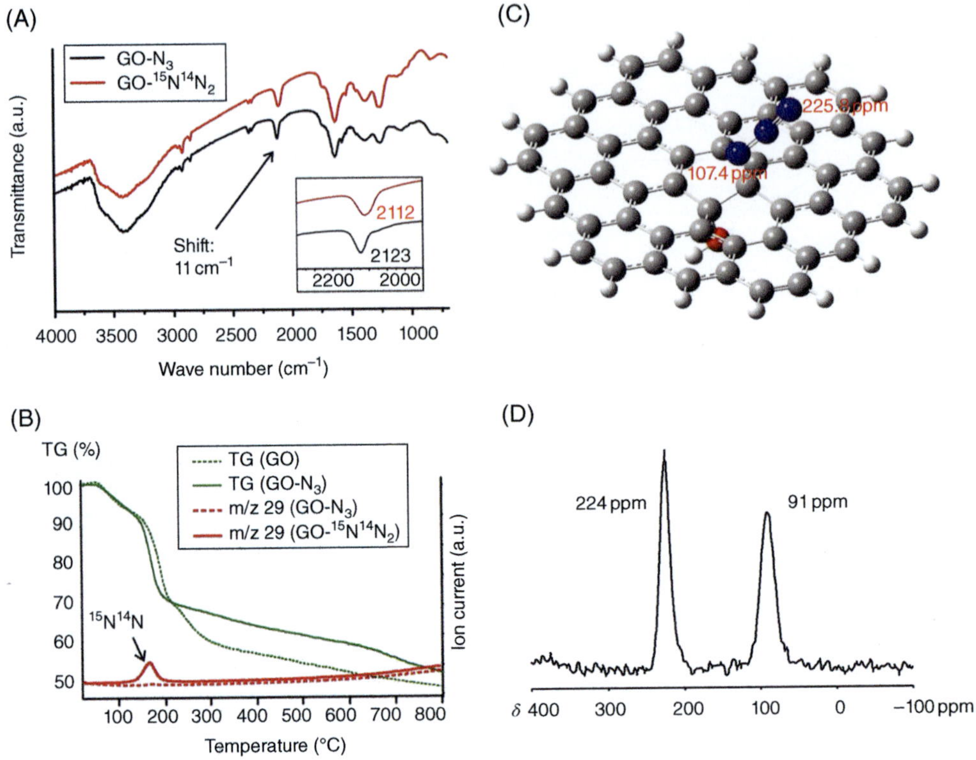

Figure 6.37 *(A) Infrared spectra of GO-N_3 and GO-$^{15}N^{14}N_2$ (here GO is oxo-G_1). (B) Thermogravimetric analysis of GO and GO-$^{15}N^{14}N_2$,* m/z 29 *of GO-N_3 and GO-$^{15}N^{14}N_2$. (C) Simplified model of GO-N_3 with an azide and a hydroxyl group connected to the carbon lattice in* trans-configuration. *^{15}N NMR shifts are calculated by* ab initio *methods (107.4 ppm and 225.8 ppm). (D) The ^{15}N SSNMR MAS (magic angle spinning) spectrum of GO-$^{15}N^{14}N_2$ with two peaks (1 : 1 ratio) [110]. Reproduced from [110] with permission from RSC*

Substitution is further proved by the absence of the weight-loss step in the temperature range that is related to organosulfate and the decomposition of azide is indicated by the signal of *m/z* 29, which is related to $^{15}N^{14}N$ formation. This investigation is only possible using labeled azide, because otherwise N_2 with *m/z* 28 is produced that cannot be distinguished from background N_2.

Also NMR spectroscopy in solids could be conducted on ^{15}N-labeled compounds (Figure 6.37D). Two signals at 224 and 91 ppm are found that relate to the two possible binding states of $^{15}N^{14}N^{14}N$ (i.e. R–$^{15}N^{14}N^{14}N$ or R–$^{14}N^{14}N^{15}N$) and they are in line with calculated predictions within the errors of the model (Figure 6.37C). Furthermore, the analyses reveal a concentration of about one azide group per 30 carbon atoms. Further on, the azide group in GO-N_3 is hydrolytically stable up to 60 °C. Therefore, GO-N_3 is a suitable precursor for other reactions, opening an avenue for the field of azide chemistry for graphene-based materials.

6.8.3.3 Organosulfate in Oxo-Functionalized Graphene as Anchor Group

The impact of the controlled chemistry of oxo-G_1 was recently demonstrated to enable the function of a floating-gate memory device working at 3 V (Figure 6.38E) [111]. Higher voltages increase the power consumption of devices. A very thin dielectric is necessary to separate the charge storage layer from the gate and the semiconductor. The thicker the dielectric layer,

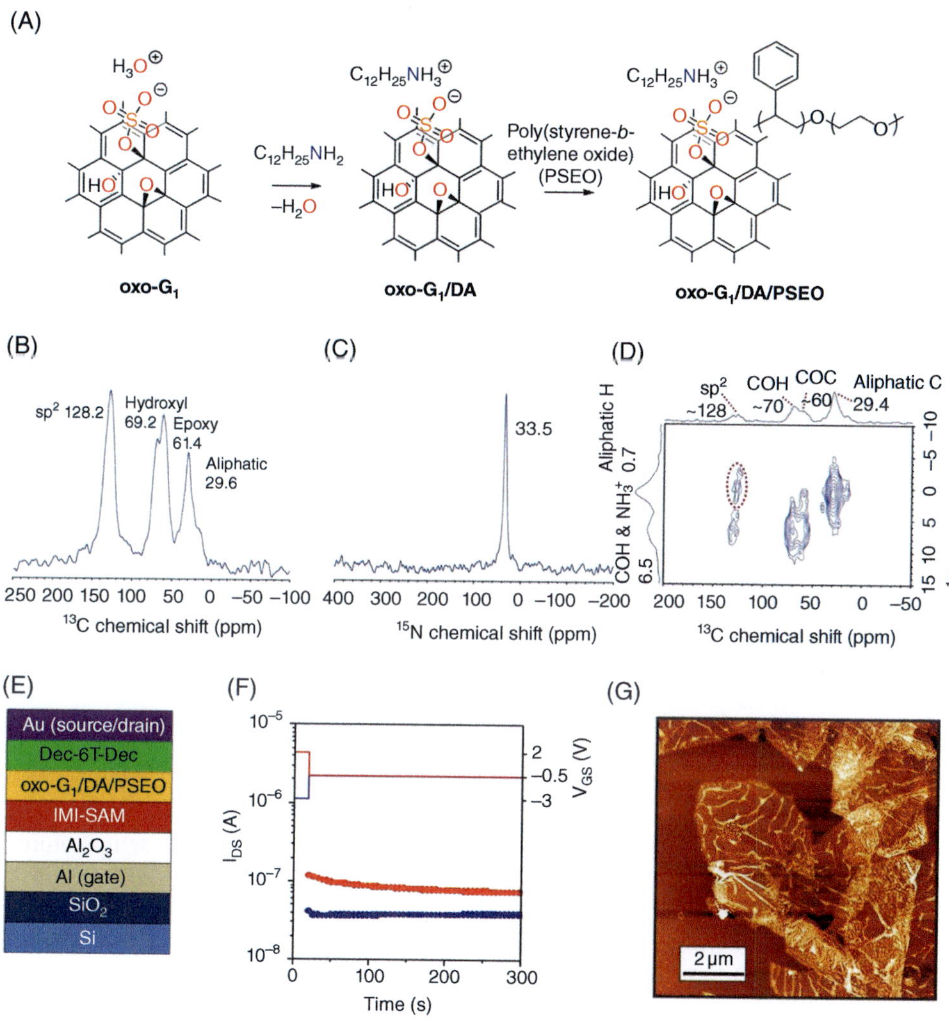

Figure 6.38 (A) Reaction scheme illustrating the reaction of oxo-G_1 with dodecylamine (oxo-G_1/DA), followed by non-covalent functionalization with a block copolymer of styrene and ethylene oxide (PSEO). Solid-state NMR spectra of oxo-G_1/DA: (B) ^{13}C NMR, (C) ^{15}N NMR and (D) ^1H–^{13}C correlation NMR. (E) Device structure of a floating-gate memory device with oxo-G_1/DA/PSEO as charge storage material. (F) Device characteristic: write signal at −3 V, erase signal at 2 V and read voltage at −0.5 V. (G) AFM image of flakes of oxo-G_1/DA/PSEO composite [111]. Reproduced from [111] with permission from RSC

the higher is the power consumption. Thus, a special molecular architecture was designed to overcome this problem. As depicted in Figure 6.38(A), the used oxo-G_1 bears epoxy and hydroxyl groups in addition to organosulfate groups that are complexed with hydronium ions. Moreover, about 50% of C atoms are sp^2-hybridized. This oxo-G_1 was titrated with dodecylamine (DA) to exchange hydronium ion by dodecylammonium that becomes electrostatically bound to organosulfate, while the other functional groups remain stable.

This oxo-G_1/DA was thoroughly analyzed by combustion elemental analysis, TGA coupled with IR spectroscopy, gas chromatography (GC) and MS. The data revealed that organosulfate and dodecylammonium are structure motifs. As shown in Figure 6.39(A), not only can the formation of CO_2 be detected by the analysis of the combustion gases, but also

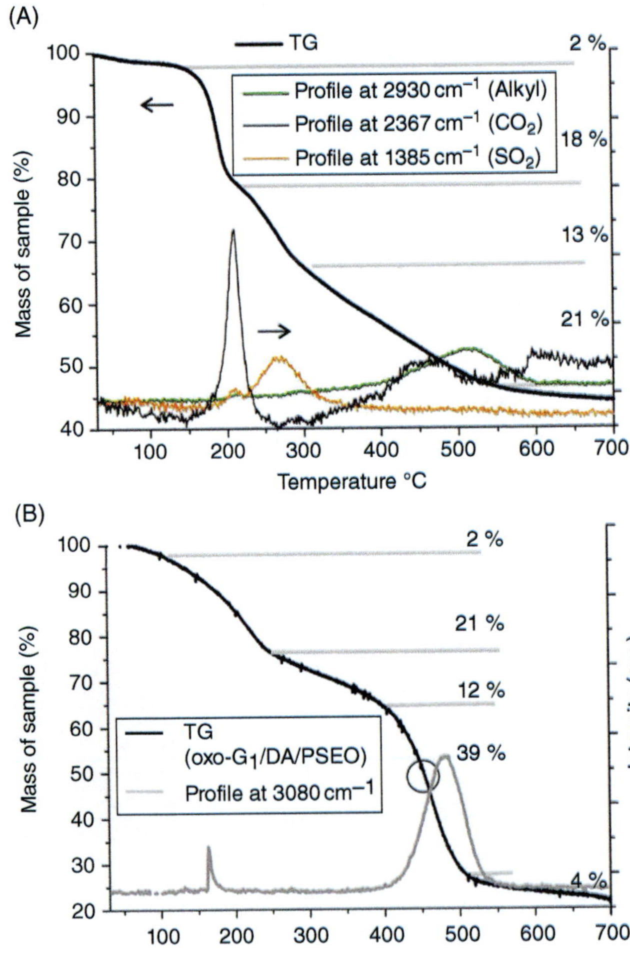

Figure 6.39 *(A) TGA of oxo-G_1/DA between 30 and 700 °C and the temperature profiles of cleaved alkyl, CO_2 and SO_2 formation, identified by FTIR. (B) TGA of oxo-G_1/DA/PSEO between 30 and 700 °C with the temperature profile of cleaved aromatic compounds (gray) identified by FTIR [111]. Reproduced from [111] with permission from RSC*

that of SO_2 originating from organosulfate and the signals of alkyl species. Moreover, GC-MS analysis revealed that the alkyl signals stem from dodecylammonium due to the identification of dodecene, which is the related elimination product. In addition, SSNMR spectroscopy was performed (Figure 6.38). In combination with *ab initio* calculations, it could be shown that only one major nitrogen-containing species is part of the structure, and the data are consistent with the proposed electrostatic binding of dodecylammonium to organosulfate (Figure 6.38C). Also information about the regiochemistry of dodecylammonium could be identified by 2D SSNMR by measuring ^{13}C–^{1}H correlation spectra (Figure 6.38D). The methylene signals of dodecylammonium were found to be close to the π system of the carbon framework. This means that the alkyl chain is masking the π system of oxo-G_1 and can therefore in part contribute to a dielectric shielding.

The complex of oxo-G_1/DA is well soluble as single layers in tetrahydrofuran and can be easily mixed with soluble polymers, such as a block copolymer of styrene and ethylene oxide (PSEO) as illustrated in the reaction scheme of Figure 6.38(A). This mixing procedure non-covalently functionalizes oxo-G_1/DA flakes to form a composite of oxo-G_1/DA/PSEO and almost no free polymer was found by AFM, as shown in Figure 6.38(G). TGA gives evidence for the decomposition of the adsorbed polymer at about 450 °C, and FTIR signals that stem from aryl groups of the decomposing polymer could be traced (Figure 6.39B). The block copolymers function as dielectric layer with a thickness of only 1–2 nm on both sides of the flakes. The oxo-G_1/DA/PSEO composite is also the charge storage material and enables the function of a floating-gate memory device. The charge storage layer is deposited as few layers on an imidazolium-terminated self-assembled monolayer as depicted in Figure 6.38(E). A sexithiophene derivative is used as semiconductor on top of oxo-G_1/DA/PSEO. A gate voltage of −3 V can introduce charge carriers and this information can be erased at +2 V. The read-out of the information can be accomplished at −0.5 V (Figure 6.38F). Reference experiments reveal that a memory device working at 3 V can only be constructed with the composite of oxo-G_1/DA/PSEO. Other materials, such as oxo-G_1, oxo-G_1/DA, defective GO, defective GO mixed with DA and PSEO, or reference systems that bear an excess of PSEO, do not enable the function of a memory device operated by a floating gate at 3 V. For example, in defective GO, about 1% of defects are introduced, organosulfate groups are cleaved and it is plausible that epoxy groups are converted in part to diols. This approach revealed that even the flat sheet structure could not be preserved and demonstrates the impact of the controlled chemistry of oxo-G_1 on the function. Other examples that relate to self-assembly processes and device fabrication with increased performance can be found in the literature [112–114].

6.9 Discussion

Organic chemistry and materials science start to mingle with conducting chemistry on GO. A literature survey rapidly reveals that GO is of interest for researchers of many disciplines. Synthetic chemists are, however, rarely involved, for one simple reason. Organic chemistry is related to molecules that can be exactly characterized. This is possible due to developed analytical methods, such as mass spectrometry, nuclear magnetic resonance spectroscopy and single-crystal X-ray analysis. When large organic molecules are synthesized and molecules become less soluble, it becomes challenging to further develop this research and come to exact conclusions on the atomic structure. The situation becomes even more complicated due to the heterogeneous nature of graphite and GO. It is clear that GO and its derivatives will never form monodisperse compounds, especially because every sp^3 carbon atom is a chiral center. Moreover, the structure of GO differs by preparation methods, and thus the reactivity and contributions of functional groups may vary. This hurdle makes it impossible to exactly compare the results of investigations based on GO. As outlined in this review chapter, it is necessary to define and simplify the chemical structure of GO to enable a controlled chemistry of GO. The first hurdles thereto are defects within the carbon framework. The degree of defects can be increased until the disintegration of the flake structure, which also results in the formation of oxidative debris. With minimizing the density of defects, it becomes possible to investigate the chemistry of the surface functional groups of graphene. Many reactions that were reported to successfully functionalize the surface of GO may only be possible due to the activation by functional groups that are located at the edges of defects. Furthermore, it must be ensured that the whole surface of GO is accessible for reagents to conduct homogeneous chemistry. A set of analytical tools has already been developed to characterize the heterogeneous material GO. Analytical ultracentrifugation was developed to determine the size of flakes directly from solution. Also thermogravimetric analysis was further developed, and coupling with infrared spectroscopy, as well as gas chromatography and mass spectrometry, give many more insights and allow conclusions on the success of functionalization of GO. Synthetic concepts that could be learned from carbon nanotube chemistry have been transferred to graphene and GO chemistry. While paramagnetic impurities in carbon nanotubes commonly preclude analysis by NMR, the analysis of GO is possible by SSNMR. Although this method developed in recent years, it is not yet fully implemented for the analysis of functionalized GO, because SSNMR is still a method for the specialist. Nevertheless, SSNMR has the potential to push the field of functionalized materials, especially when addends are labeled with NMR-active nuclei. With improving both analytical methods and synthetic protocols, the controlled synthesis of GO and its derivatives becomes possible and opens the field for the precise synthesis of novel molecular architectures with outstanding properties. With the development of the controlled chemistry of GO and oxo-G_1, respectively, it will become possible to overcome the performance limits of conventional GO.

References

[1] Dimiev, A.M.; Alemany, L.B.; Tour, J.M., Graphene oxide. Origin of acidity, its instability in water, and a new dynamic structural model. *ACS Nano* **2013**, *7* (1), 576–588.

[2] He, H.; Riedl, T.; Lerf, A.; Klinowski, J., Solid-state NMR studies of the structure of graphite oxide. *J. Phys. Chem.* **1996**, *100*, 19954–19958.

[3] Lerf, A.; He, H.; Riedl, T.; *et al.*, ^{13}C and ^{1}H MAS NMR studies of graphite oxide and its chemically modified derivatives. *Solid State Ionics* **1997**, *101–103* (2), 857–862.

[4] He, H.; Klinowski, J.; Forster, M.; Lerf, A., A new structural model for graphite oxide. *Chem. Phys. Lett.* **1998**, *287*, 53–56.

[5] Lerf, A.; He, H.; Forster, M.; Klinowski, J., Structure of graphite oxide revisited. *J. Phys. Chem. B* **1998**, *102* (23), 4477–4482.

[6] Gao, W.; Alemany, L.B.; Ci, L.; Ajayan, P.M., New insights into the structure and reduction of graphite oxide. *Nature Chem.* **2009**, *1* (5), 403–408.

[7] Cai, W.; Piner, R.D.; Stadermann, F.J.; *et al.*, Synthesis and solid-state NMR structural characterization of ^{13}C-labeled graphite oxide. *Science* **2008**, *321* (5897), 1815–1817.

[8] Casabianca, L.B.; Shaibat, M.A.; Cai, W.W.; *et al.*, NMR-based structural modeling of graphite oxide using multidimensional ^{13}C solid-state NMR and ab initio chemical shift calculations. *J. Am. Chem. Soc.* **2010**, *132* (16), 5672–5676.
[9] Kim, S.; Zhou, S.; Hu, Y.; *et al.*, Room-temperature metastability of multilayer graphene oxide films. *Nature Mater.* **2012**, *11* (6), 544–549.
[10] Hou, X.-L.; Li, J.-L.; Drew, S.C.; *et al.*, Tuning radical species in graphene oxide in aqueous solution by photoirradiation. *J. Phys. Chem. C* **2013**, *117* (13), 6788–6793.
[11] Yang, L.; Zhang, R.; Liu, B.; *et al.*, π-Conjugated carbon radicals at graphene oxide to initiate ultrastrong chemiluminescence. *Angew. Chem. Int. Ed.* **2014**, *53* (38), 10109–10113.
[12] Szabó, T.; Berkesi, O.; Forgó, P.; *et al.*, Evolution of surface functional groups in a series of progressively oxidized graphite oxides. *Chem. Mater.* **2006**, *18*, 2740–2749.
[13] Nakajima, T.; Matsuo, Y., Formation process and structure of graphite oxide. *Carbon* **1994**, *32* (3), 469–475.
[14] Clause, A.; Plass, R.; Boehm, H.P.; Hofmann, U., Untersuchungen zur Struktur des Graphitoxyds. *Z. Anorg. Allg. Chem.* **1957**, *291* (5–6), 205–220.
[15] Ruess, G., Über das Graphitoxyhydroxyd (Graphitoxyd). *Monatsh. Chem.* **1947**, *76* (3–5), 381–417.
[16] Scholz, W.; Boehm, H.P., Betrachtungen zur Struktur des Graphitoxids. *Z. Anorg. Allg. Chem.* **1969**, *369* (3–6), 327–340.
[17] Dreyer, D.R.; Todd, A.D.; Bielawski, C.W., Harnessing the chemistry of graphene oxide. *Chem. Soc. Rev.* **2014**, *43* (15), 5288–5301.
[18] Hossain, M.Z.; Johns, J.E.; Bevan, K.H.; *et al.*, Chemically homogeneous and thermally reversible oxidation of epitaxial graphene. *Nature Chem.* **2012**, *4*, 305–309.
[19] Chen, D.; Feng, H.; Li, J., Graphene oxide: preparation, functionalization, and electrochemical applications. *Chem. Rev.* **2012**, *112* (11), 6027–6053.
[20] Kozhemyakina, N.V.; Eigler, S.; Dinnebier, R.E.; *et al.*, Effect of the structure and morphology of natural, synthetic and post-processed graphites on their dispersibility and electronic properties. *Fuller. Nanotub. Carbon Nanostruct.* **2013**, *21* (9), 804–823.
[21] Wang, L.; Ambrosi, A.; Pumera, M., "Metal-free" catalytic oxygen reduction reaction on heteroatom-doped graphene is caused by trace metal impurities. *Angew. Chem. Int. Ed.* **2013**, *52* (51), 13818–13821.
[22] Feicht, P.; Kunz, D.A.; Lerf, A.; Breu, J., Facile and scalable one-step production of organically modified graphene oxide by a two-phase extraction. *Carbon* **2014**, *80*, 229–234.
[23] Boehm, H.P.; Scholz, W., Der "Verpuffungspunkt" des Graphitoxids. *Z. Anorg. Allg. Chem.* **1965**, *335* (1–2), 74–79.
[24] Kim, F.; Luo, J.Y.; Cruz-Silva, R.; *et al.*, Self-propagating domino-like reactions in oxidized graphite. *Adv. Funct. Mater.* **2010**, *20* (17), 2867–2873.
[25] Krishnan, D.; Kim, F.; Luo, J.; *et al.*, Energetic graphene oxide: challenges and opportunities. *Nano Today* **2012**, *7* (2), 137–152.
[26] Qiu, Y.; Guo, F.; Hurt, R.; Kulaots, I., Explosive thermal reduction of graphene oxide-based materials: mechanism and safety implications. *Carbon* **2014**, *72*, 215–223.
[27] Eigler, S.; Dotzer, C.; Hirsch, A.; *et al.*, Formation and decomposition of CO_2 intercalated graphene oxide. *Chem. Mater.* **2012**, *24* (7), 1276–1282.
[28] Dimiev, A.M.; Polson, T.A., Contesting the two-component structural model of graphene oxide and reexamining the chemistry of graphene oxide in basic media. *Carbon* **2015**, *93*, 544–554.
[29] Dimiev, A.; Kosynkin, D.V.; Alemany, L.B.; *et al.*, Pristine graphite oxide. *J. Am. Chem. Soc.* **2012**, *134* (5), 2815–2822.
[30] Eigler, S.; Grimm, S.; Hirsch, A., Investigation of the thermal stability of the carbon framework of graphene oxide. *Chem. Eur. J.* **2014**, *20* (4), 984–989.
[31] Eigler, S.; Dotzer, C.; Hirsch, A., Visualization of defect densities in reduced graphene oxide. *Carbon* **2012**, *50* (10), 3666–3673.
[32] Kundu, A.; Layek, R.K.; Nandi, A.K., Enhanced fluorescent intensity of graphene oxide–methyl cellulose hybrid in acidic medium: sensing of nitro-aromatics. *J. Mater. Chem.* **2012**, *22* (16), 8139–8144.

[33] Lu, C.H.; Yang, H.H.; Zhu, C.L.; *et al.*, A graphene platform for sensing biomolecules. *Angew. Chem. Int. Ed.* **2009**, *48* (26), 4785–4787.
[34] Wang, S.; Yu, D.; Dai, L.; *et al.*, Polyelectrolyte-functionalized graphene as metal-free electrocatalysts for oxygen reduction. *ACS Nano* **2011**, *5* (8), 6202–6209.
[35] Liu, J.; Yang, W.; Tao, L.; *et al.*, Thermosensitive graphene nanocomposites formed using pyrene-terminal polymers made by RAFT polymerization. *J. Polym. Sci. A: Polym. Chem.* **2010**, *48* (2), 425–433.
[36] Layek, R.K.; Samanta, S.; Chatterjee, D.P.; Nandi, A.K., Physical and mechanical properties of poly(methyl methacrylate)-functionalized graphene/poly(vinylidine fluoride) nanocomposites: piezoelectric β polymorph formation. *Polymer* **2010**, *51* (24), 5846–5856.
[37] Sydlik, S.A.; Swager, T.M., Functional graphenic materials via a Johnson–Claisen rearrangement. *Adv. Funct. Mater.* **2013**, *23* (15), 1873–1882.
[38] Burress, J.W.; Gadipelli, S.; Ford, J.; *et al.*, Graphene oxide framework materials: theoretical predictions and experimental results. *Angew. Chem. Int. Ed.* **2010**, *49* (47), 8902–8904.
[39] Burress, J.W.; Gadipelli, S.; Ford, J.; *et al.*, Graphene oxide framework materials: theoretical predictions and experimental results. *Angew. Chem.* **2010**, *122* (47), 9086–9088.
[40] Stankovich, S.; Piner, R.D.; Nguyen, S.T.; Ruoff, R.S., Synthesis and exfoliation of isocyanate-treated graphene oxide nanoplatelets. *Carbon* **2006**, *44* (15), 3342–3347.
[41] Sun, H.; Xu, Z.; Gao, C., Multifunctional, ultra-flyweight, synergistically assembled carbon aerogels. *Adv. Mater.* **2013**, *25* (18), 2554–2560.
[42] Chabot, V.; Higgins, D.; Yu, A.; *et al.*, A review of graphene and graphene oxide sponge: material synthesis and applications to energy and the environment. *Energy Environ. Sci.* **2014**, *7* (5), 1564–1596.
[43] Sofer, Z.; Jankovsky, O.; Libanska, A.; *et al.*, Definitive proof of graphene hydrogenation by Clemmensen reduction: use of deuterium labeling. *Nanoscale* **2015**, *7* (23), 10535–10543.
[44] Schäfer, R.A.; Englert, J.M.; Wehrfritz, P.; *et al.*, On the way to graphane – pronounced fluorescence of polyhydrogenated graphene. *Angew. Chem. Int. Ed.* **2013**, *52* (2), 754–757.
[45] Laborda, F.; Bolea, E.; Baranguan, M.T.; Castillo, J.R., Hydride generation in analytical chemistry and nascent hydrogen: when is it going to be over? *Spectrochim. Acta B: At. Spectrosc.* **2002**, *57* (4), 797–802.
[46] Chua, C.K.; Pumera, M., Regeneration of a conjugated sp^2 graphene system through selective defunctionalization of epoxides by using a proven synthetic chemistry mechanism. *Chem. Eur. J.* **2014**, *20* (7), 1871–1877.
[47] Bourlinos, A.B.; Gournis, D.; Petridis, D.; *et al.*, Graphite oxide: chemical reduction to graphite and surface modification with primary aliphatic amines and amino acids. *Langmuir* **2003**, *19* (15), 6050–6055.
[48] Yang, H.; Shan, C.; Li, F.; *et al.*, Covalent functionalization of polydisperse chemically-converted graphene sheets with amine-terminated ionic liquid. *Chem. Commun.* **2009**, *19* (26), 3880–3882.
[49] Beckert, F.; Rostas, A.M.; Thomann, R.; *et al.*, Self-initiated free radical grafting of styrene homo- and copolymers onto functionalized graphene. *Macromolecules* **2013**, *46* (14), 5488–5496.
[50] Kim, J.; Cote, L.J.; Huang, J., Two dimensional soft material: new faces of graphene oxide. *Acc. Chem. Res.* **2012**, *45* (8), 1356–1364.
[51] Loh, K.P.; Bao, Q.; Ang, P.K.; Yang, J., The chemistry of graphene. *J. Mater. Chem.* **2010**, *20* (12), 2277–2289.
[52] Kuila, T.; Bose, S.; Mishra, A.K.; *et al.*, Chemical functionalization of graphene and its applications. *Prog. Mater. Sci.* **2012**, *57* (7), 1061–1105.
[53] Xue, Y.; Chen, H.; Yu, D.; *et al.*, Oxidizing metal ions with graphene oxide: the in situ formation of magnetic nanoparticles on self-reduced graphene sheets for multifunctional applications. *Chem. Commun.* **2011**, *47* (42), 11689–11691.
[54] Dong, L.; Gari, R.R.S.; Li, Z.; *et al.*, Graphene-supported platinum and platinum–ruthenium nanoparticles with high electrocatalytic activity for methanol and ethanol oxidation. *Carbon* **2010**, *48* (3), 781–787.
[55] Si, Y.; Samulski, E.T., Synthesis of water soluble graphene. *Nano Lett.* **2008**, *8* (6), 1679–1682.

[56] DeTar, D.F.; Kosuge, T., Mechanisms of diazonium salt reactions. VI. The reactions of diazonium salts with alcohols under acidic conditions; evidence for hydride transfer. *J. Am. Chem. Soc.* **1958**, *80* (22), 6072–6077.

[57] Sharma, R.; Baik, J.H.; Perera, C.J.; Strano, M.S., Anomalously large reactivity of single graphene layers and edges toward electron transfer chemistries. *Nano Lett.* **2010**, *10* (2), 398–405.

[58] Pinson, J.; Podvorica, F., Attachment of organic layers to conductive or semiconductive surfaces by reduction of diazonium salts. *Chem. Soc. Rev.* **2005**, *34* (5), 429–439.

[59] Xiao, L.; Liao, L.; Liu, L., Chemical modification of graphene oxide with carbethoxycarbene under microwave irradiation. *Chem. Phys. Lett.* **2013**, *556*, 376–379.

[60] Collins, W.R.; Lewandowski, W.; Schmois, E.; *et al.*, Claisen rearrangement of graphite oxide: a route to covalently functionalized graphenes. *Angew. Chem. Int. Ed.* **2011**, *50* (38), 8848–8852.

[61] Collins, W.R.; Schmois, E.; Swager, T.M., Graphene oxide as an electrophile for carbon nucleophiles. *Chem. Commun.* **2011**, *47* (31), 8790–8792.

[62] Johnson, W.S.; Werthemann, L.; Bartlett, W.R.; *et al.*, Simple stereoselective version of the Claisen rearrangement leading to trans-trisubstituted olefinic bonds. Synthesis of squalene. *J. Am. Chem. Soc.* **1970**, *92* (3), 741–743.

[63] Liu, Z.; Robinson, J.T.; Sun, X.; Dai, H., PEGylated nanographene oxide for delivery of water-insoluble cancer drugs. *J. Am. Chem. Soc.* **2008**, *130* (33), 10876–10877.

[64] Zhang, X.; Huang, Y.; Wang, Y.; *et al.*, Synthesis and characterization of a graphene–C_{60} hybrid material. *Carbon* **2009**, *47* (1), 334–337.

[65] Xu, Y.; Liu, Z.; Zhang, X.; *et al.*, A graphene hybrid material covalently functionalized with porphyrin: synthesis and optical limiting property. *Adv. Mater.* **2009**, *21* (12), 1275–1279.

[66] Yu, D.; Yang, Y.; Durstock, M.; *et al.*, Soluble P3HT-grafted graphene for efficient bilayer-heterojunction photovoltaic devices. *ACS Nano* **2010**, *4* (10), 5633–5640.

[67] Mungse, H.P.; Khatri, O.P., Chemically functionalized reduced graphene oxide as a novel material for reduction of friction and wear. *J. Phys. Chem. C* **2014**, *118* (26), 14394–14402.

[68] Park, S.; Hu, Y.; Hwang, J.O.; *et al.*, Chemical structures of hydrazine-treated graphene oxide and generation of aromatic nitrogen doping. *Nature Commun.* **2012**, *3*, 638.

[69] Charpy, G., Sur la formation de l'oxyde graphitique et la définition du graphite. *C. R. Hebd. Séances Acad. Sci.* **1909**, *148* (5), 920–923.

[70] Eigler, S.; Enzelberger-Heim, M.; Grimm, S.; *et al.*, Wet chemical synthesis of graphene. *Adv. Mater.* **2013**, *25* (26), 3583–3587.

[71] Erickson, K.; Erni, R.; Lee, Z.; *et al.*, Determination of the local chemical structure of graphene oxide and reduced graphene oxide. *Adv. Mater.* **2010**, *22* (40), 4467–4472.

[72] Kohlschütter, V.; Haenni, P., Zur Kenntnis des graphitischen Kohlenstoffs und der Graphitsäure. *Z. Anorg. Allg. Chem.* **1919**, *105* (1), 121–144.

[73] Hofmann, U.; Frenzel, A.; Csalán, E., Die Konstitution der Graphitsäure und ihre Reaktionen. *Liebigs Ann. Chem.* **1934**, *510* (1), 1–41.

[74] Stankovich, S.; Dikin, D.A.; Piner, R. D.; *et al.*, Synthesis of graphene-based nanosheets via chemical reduction of exfoliated graphite oxide. *Carbon* **2007**, *45*, 1558–1565.

[75] Li, D.; Müller, M.B.; Gilje, S.; *et al.*, Processable aqueous dispersions of graphene nanosheets. *Nature Nanotechnol.* **2008**, *3* (2), 101–105.

[76] Park, S.; An, J.; Jung, I.; *et al.*, Colloidal suspensions of highly reduced graphene oxide in a wide variety of organic solvents. *Nano Lett.* **2009**, *9* (4), 1593–1597.

[77] Sinitskii, A.; Dimiev, A.; Kosynkin, D.V.; Tour, J.M., Graphene nanoribbon devices produced by oxidative unzipping of carbon nanotubes. *ACS Nano* **2010**, *4* (9), 5405–5413.

[78] Shin, H.-J.; Kim, K.K.; Benayad, A.; *et al.*, Efficient reduction of graphite oxide by sodium borohydride and its effect on electrical conductance. *Adv. Funct. Mater.* **2009**, *19* (12), 1987–1992.

[79] Wang, G.; Yang, J.; Park, J.; *et al.*, Facile synthesis and characterization of graphene nanosheets. *J. Phys. Chem. C* **2008**, *112* (22), 8192–8195.

[80] Zhou, X.; Zhang, J.; Wu, H.; *et al.*, Reducing graphene oxide via hydroxylamine: a simple and efficient route to graphene. *J. Phys. Chem. C* **2011**, *115* (24), 11957–11961.

[81] Mei, X.; Ouyang, J., Ultrasonication-assisted ultrafast reduction of graphene oxide by zinc powder at room temperature. *Carbon* **2011**, *49* (15), 5389–5397.

[82] Fan, Z.; Wang, K.; Wei, T.; *et al.*, An environmentally friendly and efficient route for the reduction of graphene oxide by aluminum powder. *Carbon* **2010**, *48*, 1686–1689.
[83] Chen, W.; Yan, L.; Bangal, P.R., Chemical reduction of graphene oxide to graphene by sulfur-containing compounds. *J. Phys. Chem. C* **2010**, *114* (47), 19885–19890.
[84] Cataldo, F.; Ursini, O.; Angelini, G., Graphite oxide and graphene nanoribbons reduction with hydrogen iodide. *Fuller. Nanotub. Carbon Nanostruct.* **2011**, *19* (5), 461–468.
[85] Pei, S.; Zhao, J.; Du, J.; *et al.*, Direct reduction of graphene oxide films into highly conductive and flexible graphene films by hydrohalic acids. *Carbon* **2010**, *48* (15), 4466–4474.
[86] Moon, I.K.; Lee, J.; Ruoff, R.S.; Lee, H., Reduced graphene oxide by chemical graphitization. *Nature Commun.* **2010**, *1*, 73.
[87] Zhou, M.; Wang, Y.; Zhai, Y.; *et al.*, Controlled synthesis of large-area and patterned electrochemically reduced graphene oxide films. *Chem. Eur. J.* **2009**, *15* (25), 6116–6120.
[88] Shao, Y.; Wang, J.; Engelhard, M.; *et al.*, Facile and controllable electrochemical reduction of graphene oxide and its applications. *J. Mater. Chem.* **2010**, *20* (4), 743–748.
[89] Ambrosi, A.; Pumera, M., Precise tuning of surface composition and electron-transfer properties of graphene oxide films through electroreduction. *Chemistry* **2013**, *19* (15), 4748–4753.
[90] Eigler, S.; Grimm, S.; Enzelberger-Heim, M.; *et al.*, Graphene oxide: efficiency of reducing agents. *Chem. Commun.* **2013**, *49* (67), 7391–7393.
[91] Bagri, A.; Mattevi, C.; Acik, M.; *et al.*, Structural evolution during the reduction of chemically derived graphene oxide. *Nature Chem.* **2010**, *2* (7), 581–587.
[92] Rourke, J.P.; Pandey, P.A.; Moore, J.J.; *et al.*, The real graphene oxide revealed: stripping the oxidative debris from the graphene-like sheets. *Angew. Chem. Int. Ed.* **2011**, *50* (14), 3173–3177.
[93] Chua, C.K.; Pumera, M., Chemical reduction of graphene oxide: a synthetic chemistry viewpoint. *Chem. Soc. Rev.* **2014**, *43* (1), 291–312.
[94] Pei, S.; Cheng, H.-M., The reduction of graphene oxide. *Carbon* **2012**, *50* (9), 3210–3228.
[95] Eigler, S., Mechanistic insights into the reduction of graphene oxide addressing its surfaces. *Phys. Chem. Chem. Phys.* **2014**, *16* (37), 19832–19835.
[96] Bekyarova, E.; Itkis, M.E.; Ramesh, P.; *et al.*, Chemical modification of epitaxial graphene: spontaneous grafting of aryl groups. *J. Am. Chem. Soc.* **2009**, *131* (4), 1336–1337.
[97] Wang, Q.H.; Jin, Z.; Kim, K.K.; *et al.*, Understanding and controlling the substrate effect on graphene electron-transfer chemistry via reactivity imprint lithography. *Nature Chem.* **2012**, *4* (9), 724–732.
[98] Lomeda, J.R.; Doyle, C.D.; Kosynkin, D.V.; *et al.*, Diazonium functionalization of surfactant-wrapped chemically converted graphene sheets. *J. Am. Chem. Soc.* **2008**, *130* (48), 16201–16206.
[99] Sinitskii, A.; Dimiev, A.; Corley, D.A.; *et al.*, Kinetics of diazonium functionalization of chemically converted graphene nanoribbons. *ACS Nano* **2010**, *4* (4), 1949–1954.
[100] Wang, H.X.; Zhou, K.G.; Xie, Y.L.; *et al.*, Photoactive graphene sheets prepared by "click" chemistry. *Chem. Commun.* **2011**, *47* (20), 5747–5749.
[101] Castelain, M.; Martinez, G.; Merino, P.; *et al.*, Graphene functionalisation with a conjugated poly(fluorene) by click coupling: striking electronic properties in solution. *Chem. Eur. J.* **2012**, *18* (16), 4965–4973.
[102] Koehler, F.M.; Stark, W.J., Organic synthesis on graphene. *Acc. Chem. Res.* **2013**, *46* (10), 2297–2306.
[103] Eigler, S.; Hirsch, A., Chemistry with graphene and graphene oxide – challenges for synthetic chemists. *Angew. Chem. Int. Ed.* **2014**, *53* (30), 7720–7738.
[104] Walter, J.; Nacken, T.J.; Damm, C.; *et al.*, Determination of the lateral dimension of graphene oxide nanosheets using analytical ultracentrifugation. *Small* **2015**, *11* (7), 814–825.
[105] Ferrari, A.C.; Basko, D.M., Raman spectroscopy as a versatile tool for studying the properties of graphene. *Nature Nanotechnol.* **2013**, *8* (4), 235–246.

[106] Englert, J.M.; Vecera, P.; Knirsch, K.C.; *et al.*, Scanning-Raman-microscopy for the statistical analysis of covalently functionalized graphene. *ACS Nano* **2013**, *7* (6), 5472–5482.

[107] Boehm, H.-P.; Scholz, W., Untersuchungen am Graphitoxyd, IV. Vergleich der Darstellungsverfahren für Graphitoxyd. *Liebigs Ann. Chem.* **1966**, *691* (1), 1–8.

[108] Eigler, S.; Dotzer, C.; Hof, F.; *et al.*, Sulfur species in graphene oxide. *Chem. Eur. J.* **2013**, *19* (29), 9490–9496.

[109] Eigler, S.; Grimm, S.; Hof, F.; Hirsch, A., Graphene oxide: a stable carbon framework for functionalization. *J. Mater. Chem. A* **2013**, *1* (38), 11559–11562.

[110] Eigler, S.; Hu, Y.; Ishii, Y.; Hirsch, A., Controlled functionalization of graphene oxide with sodium azide. *Nanoscale* **2013**, *5* (24), 12136–12139.

[111] Wang, Z.; Eigler, S.; Ishii, Y.; *et al.*, A facile approach to synthesize an oxo-functionalized graphene/polymer composite for low-voltage operating memory devices. *J. Mater. Chem. C* **2015**, *3* (33), 8595–8604.

[112] Wang, Z.; Eigler, S.; Halik, M., Scalable self-assembled reduced graphene oxide transistors on flexible substrate. *Appl. Phys. Lett.* **2014**, *104* (24), 243502.

[113] Kirschner, J.; Wang, Z.; Eigler, S.; *et al.*, Driving forces for the self-assembly of graphene oxide on organic monolayers. *Nanoscale* **2014**, *6* (19), 11344–11350.

[114] Wang, Z.; Mohammadzadeh, S.; Schmaltz, T.; *et al.*, Region-selective self-assembly of functionalized carbon allotropes from solution. *ACS Nano* **2013**, *7* (12), 11427–11434.

Part II

Applications

7

Field-Effect Transistors, Sensors and Transparent Conductive Films

Samuele Porro and Ignazio Roppolo

7.1 Field-Effect Transistors

Graphene was suggested as a perfect candidate for fabricating channels in next-generation field-effect (field-emission) transistor (FET) devices, due to its peculiar band structure and electronic properties, like the fact that carriers (both electrons and holes) can be continuously tuned by a gate electric field [1], and behave as massless relativistic particles, leading to mean free paths of up to 300 nm even at room temperature [2]. Moreover, due to its two-dimensional characteristics, nanometer-scale devices may be patterned to create single-electron or few-electron transistors, making graphene suitable also for ultra-high-frequency transistors [3, 4]. In particular, reduced graphene oxide (RGO) can be used for these devices, due to the simplicity of deposition and processing, starting from solution-based techniques using graphene oxide (GO) dispersions and subsequent reducing treatments [5, 6].

In principle, due to the intrinsic nature of pristine graphene, which is basically an electrical conductor, FET devices made from pristine graphene typically show low on/off current ratios [1]. The introduction of an energy bandgap in graphene can be engineered by several routes in order to achieve a semiconducting behavior of the material, which makes it more suitable for the fabrication of electronic devices such as transistors [7]. Among these routes, the most common feature is the generation of graphene strips having ultra-thin width (<50 nm), making quasi-one-dimensional structures, which are called graphene nanoribbons, by constricting the lateral dimension of the graphene sheets. Graphene nanoribbons show semiconducting properties and are assumed to represent a technological alternative to

Graphene Oxide: Fundamentals and Applications, First Edition. Edited by Ayrat M. Dimiev and Siegfried Eigler.

silicon for the fabrication of field-effect transistors less than 10 nm wide, with on/off current ratios exceeding 10^6 at room temperature [4, 8]. For instance, FET devices can be fabricated using graphene nanoribbons produced by chemical routes, sonicating exfoliated graphite in organic solvents [9], or unzipping carbon nanotubes by plasma etching [10] or chemical oxidation [11], obtaining structures with widths below 10 nm having a 0.4 eV energy bandgap and on/off current ratio of 10^7. Graphene nanoribbons are normally p-doped due to surface adsorbed water, but it is possible to achieve n-type characteristics by electrochemical reaction in a reducing environment [12]. Moreover, the synthesis of n-doped graphene-based materials by annealing GO at relatively high temperatures in a reducing environment has been demonstrated [13].

A very simple transistor device can be fabricated by depositing metallic source and drain electrodes on top of RGO sheets sprayed from a solvent dispersion on an insulating substrate, to form a continuous film of overlapping platelets [14]. In that work [14], the conductivity of the original GO was increased by four or five orders of magnitude after chemical reduction of the sheets by exposure to hydrazine vapor, on more than 100 devices. Using a Si substrate as the back gate, the field-effect responses of the devices were tested by varying the gate bias values and obtaining an increase in conductivity as the voltage lowered, as shown in Figure 7.1(a), indicating the p-type semiconducting characteristic of RGO films. Figure 7.1(b) shows that the resistance of the device increases as the temperature decreases, which is a typical semiconducting-type behavior. The I/T^2 versus inverse temperature plot fits with a Schottky contact model, confirming a semiconductor/metal interface.

GO can be used for the fabrication of field-effect transistors even without reduction. The electronic structure of GO can be simply tuned by varying the oxygen content of the layers, which can be controlled by the oxidation time during the synthesis [15], with a bandgap variation from 1.7 to 2.1 eV. In particular, GO with a bandgap of 1.7 eV shows a p-type semiconducting behavior in air and was used for the fabrication of FETs without reduction of the layers by simply spray coating a dispersion on metallic electrodes [16]. These devices show low on/off ratio due to the presence of localized defect states, with p-type behavior in air and ambipolar behavior under vacuum, due to the removal of oxygen-related functional groups and water from GO.

Another aspect of using graphene-based materials in electronic devices is their application as electrodes. Graphene-based thin films of different fabrication, including solution-processed RGO, can be employed as flexible electrodes in various optical and electronic applications, including FETs, due to their high flexibility and transparency. For example, several works report the fabrication of FETs with all-graphene electrodes using solution-processable RGO. In Chen *et al.* [17], a flexible organic FET with all-graphene electrodes obtained from chemical reduction of GO shows performances comparable to conventional silicon-based devices with metal electrodes. Wang *et al.* [18] report the use of thermally reduced GO as both source and drain electrodes (thick graphene layers) and active channel (thin graphene layer) in a FET device. In this case, a particular method for isolation of large GO flakes with lateral size above 25 μm was applied, to improve the electrical characteristics of the layers. The thin-film graphene device, obtained via solution-processed fabrication, shows intrinsic carrier mobility greater than 5000 $cm^2 V^{-1} s^{-1}$, which is in the same order of magnitude as devices based on mechanically cleaved graphene. A schematic representation of the microfabrication of this device is reported in Figure 7.2. This example is particularly interesting because a

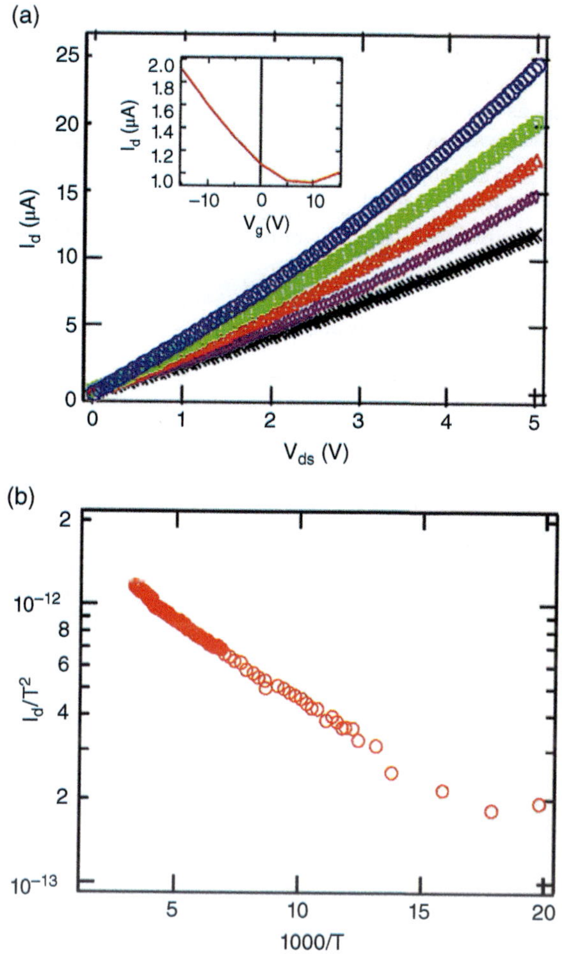

Figure 7.1 (a) The I–V characteristics of an RGO-based FET device (gate biases of 15, 0, −5, −10 and −15 V, respectively, from bottom to top). As the gate voltage changes from +15 to −15 V, the conductance of the device increases, indicating a p-type semiconductor behavior. (b) Temperature-dependent measurements, confirming the semiconducting behavior of the RGO film. Reproduced from [14] with permission of ACS

solution process is employed for the fabrication of a FET based on RGO, opening the way for the possibility to fabricate large-scale all-carbon electrodes via two-dimensional additive techniques such as inkjet printing, with the advantage of avoiding the pre-patterning of catalysts that is needed, for instance, in chemical vapor deposition (CVD). The benefits of using inkjet printing technology for direct fabrication of devices are greatly advantageous compared to conventional clean-room-based techniques [19, 20]. RGO can also be used as a conducting interlayer in organic field-effect transistors to improve the electrical characteristics at the interface between the metallic electrodes and the organic active layers [21]. The insertion of RGO layers allows one to improve the charge injection, to

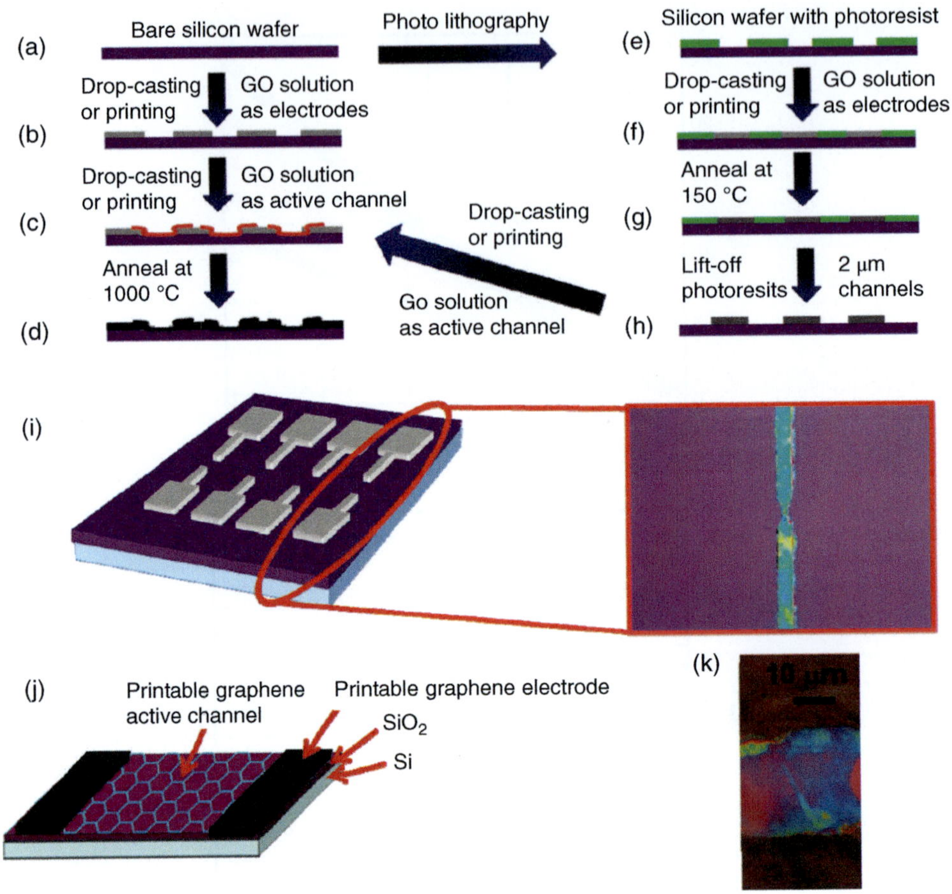

Figure 7.2 (a–d) Microfabrication of an all-carbon graphene FET device. Both GO active channel and electrodes are deposited onto an SiO_2/Si substrate by drop-casting or printing. (e–h) Device fabrication using conventional photolithography. Schemes and optical microscopy images of the RGO electrodes (i) and the whole device (j,k). Reproduced from [18] with permission of ACS

lower the losses at the metal/organic interface, and to increase the effective carrier mobility. Figure 7.3 reports a comparison of the linear mobility of a FET device made with and without an RGO interlayer between the Au electrodes and the pentacene organic active layer, showing the increased carrier mobility with the presence of RGO.

Similarly to SiO_2, which is used in silicon-based electronic devices as the gate insulator layer, GO was recently suggested as a suitable gate dielectric for graphene channels [22]. An example of an all-graphene-based thin-film transistor has been reported [23], where a flexible and transparent all-RGO device was fabricated on a polymeric substrate, making it suitable for application in flexible electronics and biomolecular sensing. In this work, relatively thick RGO film (>9 nm) electrodes were deposited by spin coating on thin RGO films (2–4 nm) patterned by microfluidic techniques and used as active channels (Figure 7.4). The average on/off ratio was measured to be about 3.8 on many so-fabricated

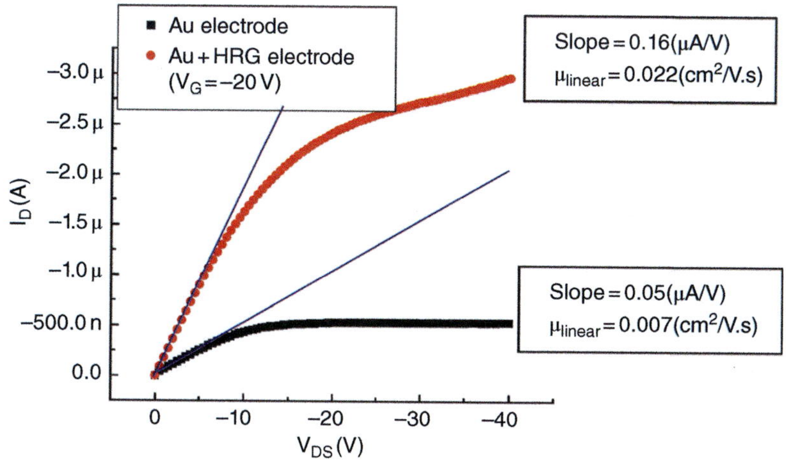

Figure 7.3 *Increase of carrier mobility in Au/pentacene organic FET devices due to the presence of a highly reduced GO (HRG) interlayer. Reproduced from [21] with permission of AIP Publishing LLC*

devices, showing field-effect characteristics similar to RGO devices with metal electrodes, with better repeatability than single-layer graphene FETs. These devices also showed excellent flexibility without modification of performances, as no variation of resistance was reported even after 5000 bending cycles. Another example was reported [24] where a 100 nm GO film is employed as a gate dielectric and graphene as a channel, source/drain and gate electrodes in a flexible device fabricated on a plastic substrate. The suitability of GO as gate material is confirmed by the very stable measured capacitance without dielectric dispersion up to 1 MHz, and a relative permittivity of about 5 at room temperature. The measured capacitance and relative permittivity confirm that GO films are suitable as gate dielectric in graphene-based FETs.

7.2 Sensors

GO and GO-based materials have an exceptional importance in current sensors science. Based on Scopus data, at the time of writing more than 1500 original research articles have been published in international journals in this field in less than 10 years, positioning GO not so far from graphene itself. One of the reasons for this explosion consists in the fact that many sensors are actually based on RGO, which for some aspects could be considered an analog of graphene, and thus used in sensors for its exceptional electrical conductivity and superior mechanical strength. Nevertheless, also unreduced GO represents an important alternative for some applications due to its high number of reactive groups, which could easily interact with the environment. Moreover, GO solution processability meets the requirements of most thin-film technologies [25].

In this section an overview of the sensors based on GO is reported, principally divided by the application rather than the mechanism, which is discussed for each case. These types of devices find their main application principally in three fields: gas, humidity and biological sensing.

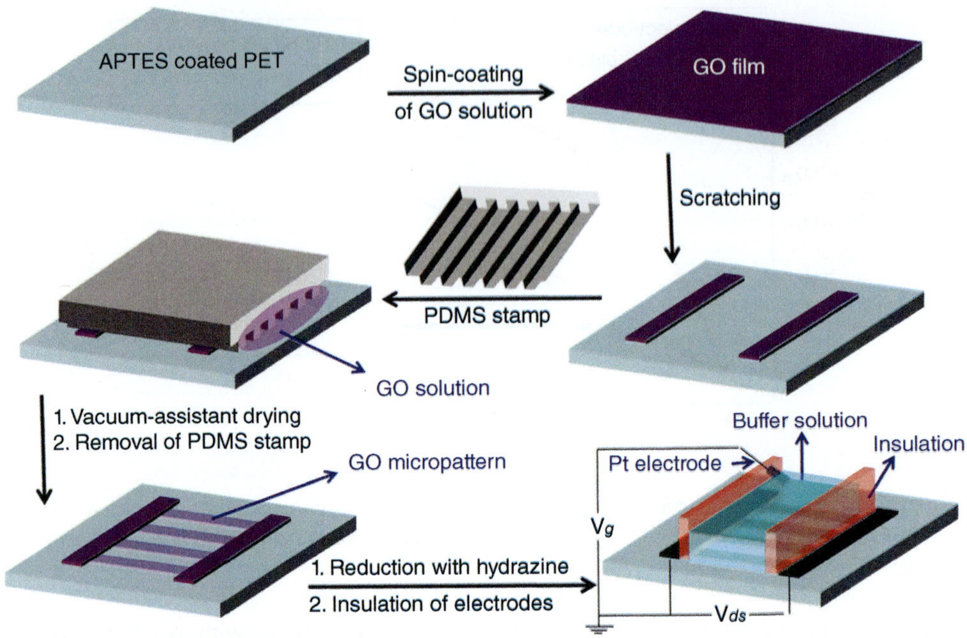

Figure 7.4 Fabrication of an all-RGO thin-film transistor. A GO solution is spin-coated onto a flexible polymeric substrate, and subsequently scratched to make two separated electrodes. Between the electrodes, another GO film is micro-patterned using the microfluidics technique. RGO films are then obtained via exposure to hydrazine vapor. A silicone rubber is finally used to insulate the RGO electrodes. Reproduced from [23] with permission of ACS

7.2.1 Gas Sensors

Graphene is known to be sensitive to both electron-acceptor (NO_2, moisture, etc.) and electron-donor (CO, alcohols, ammonia, etc.) gases [26]. Since solution-processed GO and RGO represent a relatively cheap alternative to epitaxial and CVD graphene, they were largely employed as gas sensing materials [27]. Differently from graphene, RGO still possesses oxygen defects that, interacting with gases, could enhance its sensitivity towards such molecules [28, 29].

From the early stage of graphene-based sensing, much attention was devoted to controlling the level of reduction in order to optimize the efficiency of the devices [30–32]. For instance, thermally processed GO may present p-type semiconductor characteristics, and thus the interaction with a strong oxidizer such as NO_2 leads to enrichment of the hole concentration and thus to an increase of electrical conductance [33]. In this context, sensing devices were fabricated by direct laser reduction: by tuning the laser exposure, both electrodes and sensing material can be engineered, controlling the level of reduction (Figure 7.5) [34]. The semiconductor behavior can be further improved by ozone treatment on CVD-grown graphene to tune the surface oxidation [35]. Leaving simple GO chemistry, the sensitivity toward NO_2 can also be improved by using alternative methods such as the synthesis of mesocrystals by conjugating RGO and Cu_2O [36] or RGO functionalization with different moieties such as sulfonated groups or ethylene diamine groups [37].

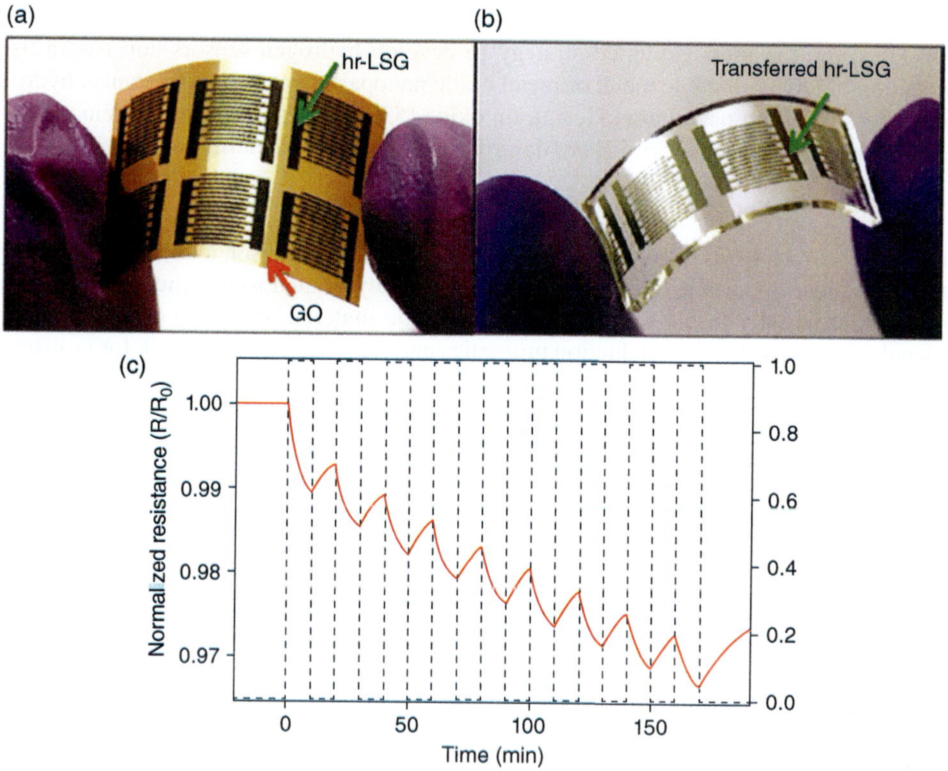

Figure 7.5 *Example of an NO_2 sensor. (a) All-organic flexible set of interdigitated electrodes generated from highly reduced laser-scribed graphene (hr-LSG). (b) The same interdigitated electrodes transferred onto polydimethylsiloxane (PDMS). (c) NO_2 detection using the all-organic flexible interdigitated electrodes. Here the sensor uses hr-LSG as the active electrodes and marginally laser-reduced graphite oxide as the detecting medium. The NO_2 concentration is 20 ppm in dry air. Reproduced from [34] with permission of ACS*

As previously mentioned, GO is used also as detecting material for electron-donor gases such as alcohols and ammonia. In particular, the presence of epoxy and hydroxyl groups on the GO surface promotes the absorption of these gases: molecules adsorbed so close to the carbon matrix enhances the charge transfer and modifies the electrical conductivity [29]. Different approaches were used in the literature for the detection of electron-donor gases: from simple chemical reduction [38–40] to nanoparticle decoration [41–43], metal organic framework (MOF) decoration [44, 45] and polymer functionalization of GO [46]. A particularly interesting alternative to resistive sensors consists in optical sensors for ammonia detection that exploit the change of reflectivity of silver-decorated RGO by absorbing ammonia molecules [47].

An important effort was also devoted to develop H_2 sensors, since hydrogen is a hazardous gas but very strategic in many industrial applications as a power source, reducing agent or carrier gas. RGO can be used as the active layer for H_2 detection in an acoustic wave sensor showing sensitivity down to 100 ppm [48]. Hydrogenated GO can also be used to realize a resistive H_2 sensor, showing different behavior depending on the reduction degree [49]. However, the most exploited strategy to develop hydrogen sensors consists in the decoration of GO with noble metal or metal oxide nanoparticles [50]. For instance, hydrogen sensors based on palladium [51, 52], tin oxide [53], titanium dioxide [54], zinc oxide [55] or composite systems [56, 57] are described in the recent scientific literature.

7.2.2 Humidity Sensors

Among the various types of gases for which GO can be used as sensor, moisture is certainly one of the most important. This property is intrinsically related to the chemistry of GO: since GO, differently from graphene, is a hydrophilic material, its interaction with water molecules could modify its conduction properties and thus be a building block for moisture sensors [29].

As previously mentioned, the ratio between oxidized reduced moieties is a key issue for the realization of moisture sensors [30]. A selective reduction was recently performed via laser-induced reduction. Using this technique, it is possible to fabricate a complete sensor from a single GO deposition: unreduced GO is used as the sensing part and RGO as the electrode material [58]. An interesting alternative simply using RGO has been proposed [59]: a thick film of GO is reduced via sunlight-induced photoreduction. By this technique it is possible to produce a gradient reduction profile in the layer inducing responsive behavior to humidity related to the different hydrophobicity/hydrophilicity along the thickness of the material.

Similarly to other gas sensors, also for humidity sensors, more sensitive materials were exploited by functionalizing RGO sheets with polymers [60, 61] or nanoparticle decoration [62]. Layer-by-layer techniques were used to improve the adhesion between the electrodes and the RGO sheets (Figure 7.6) [63, 64].

As a final note on the performance of the devices, it should be noted that this has been improved dramatically over the past few years: in one of the first literature examples of RGO sensing, the fabricated sensor showed good sensitivity only at relative humidity higher than 70%, with an electrical response time of a few seconds [65]. In more recent works the sensitivity was drastically improved, with levels of detectable relative humidity down to 30% and evaluated response time of 30 ms [66].

7.2.3 Biological Sensors

GO is a material that is particularly interesting for the development of biosensors because it shows tunable electrical characteristics (insulator, semiconductor or semimetal) depending on the degree of reduction, and many chemically active moieties at the surface can be exploited for the anchoring of suitable reactive molecules. Moreover, it presents peculiar optical properties in terms of photoluminescence, which makes it very promising for biosensor application [67, 68].

A technique that is very widespread for the development of biosensors is based on Förster (or fluorescence) resonance energy transfer phenomenon (FRET), which occurs when photoexcitation is transferred from a donor fluorophore to an acceptor molecule. FRET is nowadays applied to different fields such as *in vivo* monitoring in cellular research, structural elucidation of biological molecules and their interactions, nucleic acid analysis, signal transduction and light harvesting [69]. The efficiency of this process is six times inversely proportional to the distance of the donor/acceptor molecules and depends also on the excitation/emission spectral overlap and orientation of the dyes [70].

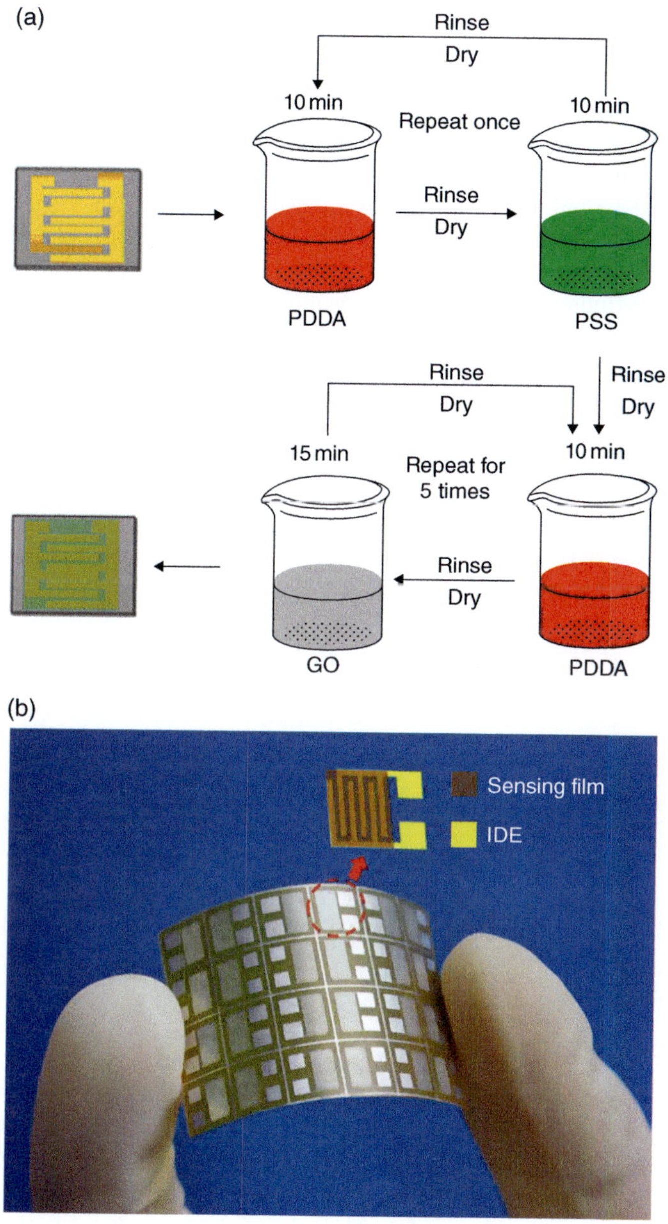

Figure 7.6 (a) Layer-by-layer fabrication process of a humidity GO nanocomposite film. (b) Optical image of 4×6 sensor array on a flexible polyimide substrate. Reproduced from [63] with permission of Elsevier

Due to its heterogeneous structure, which depends strongly on the degree of oxidation, GO can have a fluorescence [71] ranging from the near-infrared to the ultraviolet part of the spectrum [72–74]. Moreover, it can behave as donor (since it possesses photoluminescent characteristics) and as acceptor (since it can behave as a semimetal) as well for developing FRET-based sensors [67].

Regarding the interaction with biomolecules, as previously mentioned, the presence of reactive moieties (epoxy, ester, carboxylic and hydroxyl groups) on the basal planes and edges of GO allows an easy decoration with biomolecules or biomarkers. Moreover, the hydrophilic characteristic of GO sheets permits secondary interactions with polar moieties such as diols, amines and phenyls. On the other hand, the hexagonal lattice of GO allows π–π interactions with similar ring structures in nucleotides. Therefore, a large range of biomolecules can be monitored using GO as a sensing platform [75].

Going into details, GO is an excellent photoluminescence quencher and thus it has been widely used for biomolecule detection. The general approach is the following. A reactive molecule bearing a fluorescent group is adsorbed onto GO surfaces via non-covalent bonding. In the presence of the target, the molecule is desorbed from the GO surface, thus recovering the fluorescence, allowing target recognition and quantification (Figure 7.7). This approach was reported in many applications such as DNA [76, 77] or protein [78] recognition, thrombin detection [79] and imaging living cells [80].

GO photoluminescence was also used to develop label-free biological sensors for drug delivery [81] and intracellular imaging [82]. Moreover, devices for sensing other complex molecules or biological materials can be found in the literature, for instance, for polydopamine [83] or viruses [84].

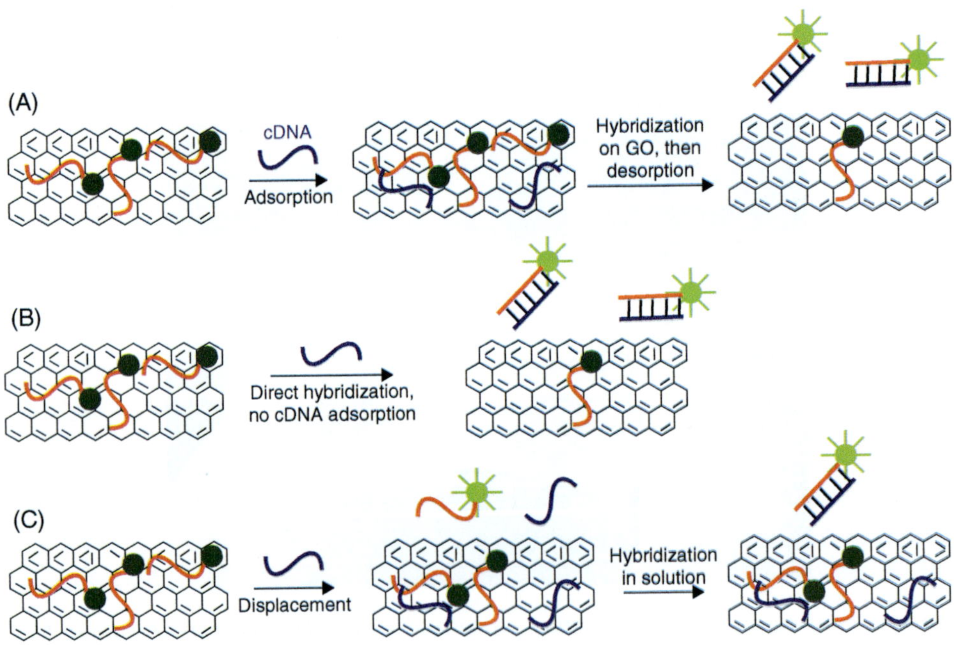

Figure 7.7 Three possible mechanisms of hybridization between a probe DNA adsorbed by GO and its cDNA (target DNA): (A) Langmuir–Hinshelwood mechanism; (B) Eley–Rideal mechanism; and (C) displacement mechanism. In all three cases, the probe DNA with a fluorophore label is pre-adsorbed and the cDNA is added afterwards. The tendency of GO to adsorb double-stranded ds-DNA is lower than that to adsorb single-stranded ss-DNA. Reproduced from [76] with permission of ACS

Finally, RGO can also be used as a transducer for bio-induced signals [85]. Examples of an electrical approach to bio-detection can be found for glucose sensing [86], H_2O_2 delivery to cancer cells [87] or uric acid [88, 89].

7.3 RGO Transparent Conductive Films

Transparent conductive films (TCFs) are of fundamental importance for producing transparent electrodes for optoelectronic devices such as organic light-emitting diodes (OLEDs), organic photovoltaic cells (OPVs), liquid crystal displays (LCDs) and touch panels. With the rapid development of the portable electronics market, TCFs are predicted to have a growing role in future technology [90]. RGO, with its electrical conductivity, high carrier mobility, optical transparency in the visible range and superior mechanical properties, is considered one of the best candidates for replacing indium tin oxide (ITO) as lead material in the realization of TCFs [91]. Moreover, methodologies based on the colloidal solution of GO are suitable for high production volumes at low cost and are compatible with flexible substrate technologies [92].

Among the various parameters necessary for producing high-quality TCFs, the control of film thickness and the homogeneity of deposition are key issues. Since an increase of RGO film thickness leads to an increase of conductivity, but, at the same time, to a decrease of transparency, the strategy consists in a trade-off between conductivity and transparency (Figure 7.8). Different techniques have been developed to produce TCFs such as spin coating [93], dip coating [94], spray coating [95], inkjet printing [96], transfer printing [97], electrophoretic deposition [98] and Langmuir–Blodgett (LB) techniques [99].

All those techniques present some advantages as well as drawbacks. Spin coating and dip coating are relatively cheap techniques and useful for mass production, but present difficulties regarding uniformity in distribution and stacking of GO sheets. Spray coating could lead to partial aggregation of sheets during heat drying. Inkjet printing does not allow close packing of sheets and involves the use of surfactants that could reduce electrical conductivity. Transfer printing, although guaranteeing high deposition quality, suffers from lack of transparency due to too thick depositions. Electrophoretic deposition is also suitable for mass production but it is limited to application on conductive substrates. Finally, LB techniques are the most sophisticated and precise, and guarantee high orientation and packing and a superior lateral resolution; moreover, by controlling properly the deposition parameters, it is possible to produce very thin TCFs. The main drawback consists in the low deposition rate, which presently makes those methodologies not desirable for large-scale production [92].

A thermal or chemical reduction process usually follows deposition in order to achieve the desired electrical conductivity. This process is clearly of fundamental importance in order to obtain sheets with low resistivity; however, it must be taken into account that in TCFs the total resistance suffers from inter-sheet carrier diffusion and thus, even if the sheets are well reduced, a very high resistance could be measured [100].

Although TCFs prepared by deposition and reduction of GO suspension possess sufficient transparency and conductivity for most applications, a large effort has been devoted to further improve these properties. The most used approaches include the realization of RGO composites with metal nanowires (NWs) or nanogrids [101] or carbon

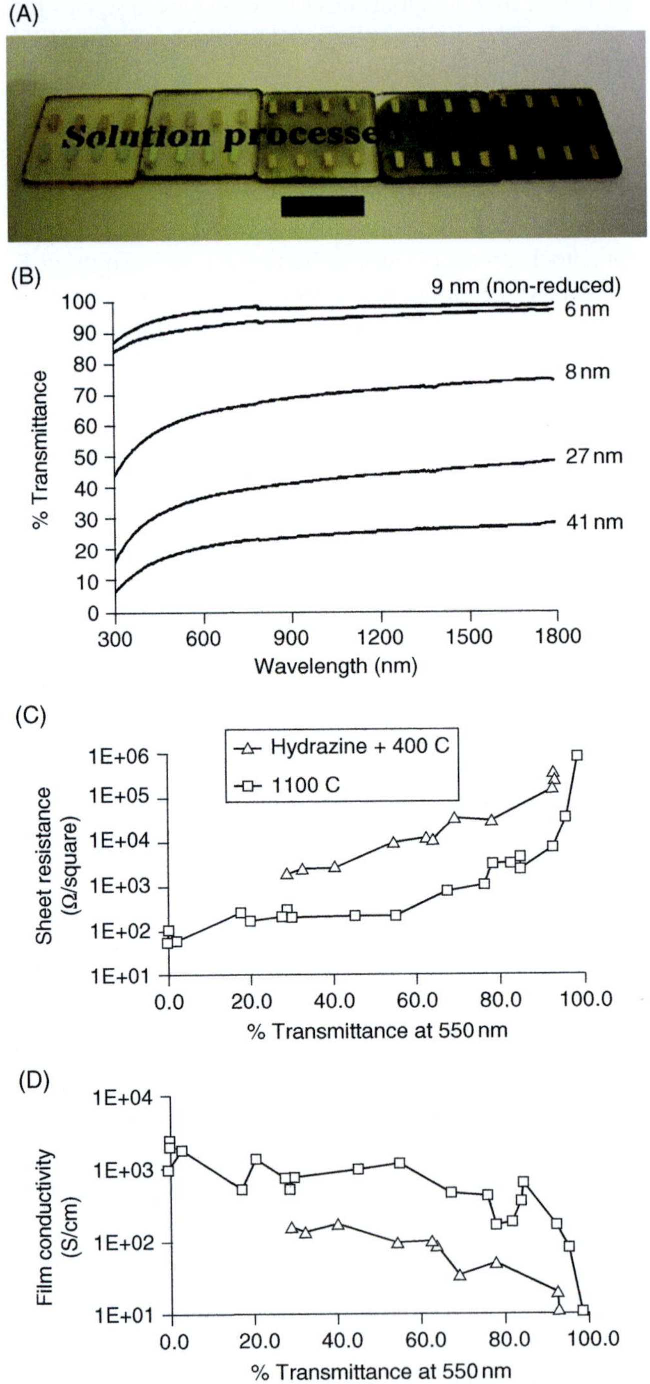

Figure 7.8 Example of dependence of different properties of RGO on film thickness. (A) Photograph of films with increasing thickness. (B) Ultraviolet–visible spectra of films with increasing thickness. (C, D) Dependence of the electrical behavior on transparency with different reduction processes. Reproduced from [93] with permission of ACS

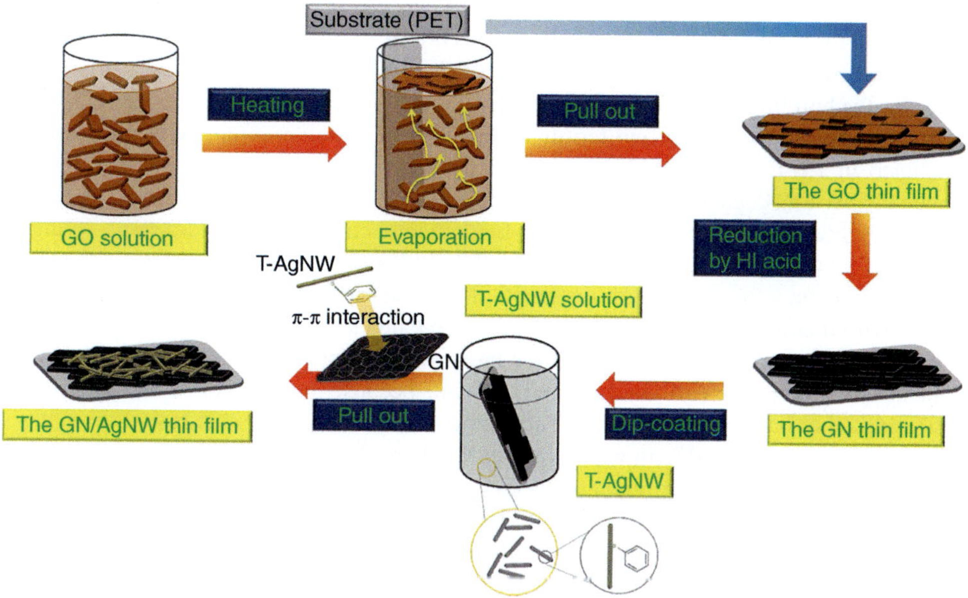

Figure 7.9 Preparation process of RGO-Ag nanowire TCFs. Reproduced from [106] with permission of RSC

nanotubes (CNTs) [102, 103], RGO electrode doping [104] and increasing the dimensions of GO sheets [105].

Going into the details, the realization of hybrid materials with metal nanowires or nanogrids allows one to decrease the contact barrier due to bridging of RGO sheets induced by the presence of the metal structures [106] (Figure 7.9). A similar effect can be achieved by creating TCFs with CNTs [107]. Different techniques for the deposition of TCFs with RGO and CNTs are reported in the literature, such as LB techniques [108], solution mixing [109], layer-by-layer self-assembly [110] and solvent–water surface self-assembly [111].

RGO sheet doping can be achieved by treatment with different acids or halogenating agents such as HNO_3 [112], $SOCl_2$ [113], $SOBr_2$ [104] or $AuCl_3$ [114].

Finally, the last approach involves the exploitation of large-area GO sheets. The RGO sheets normally used show a relatively small surface area (a few tens of square micrometers) and thus higher inter-sheet contact resistance. Different techniques can be employed to obtain large-area GO sheets (up to some square millimeters) through edge selective graphite functionalization [115], sonication in sulfuric acid [116] or other modified solution-phase processes that do not involve sonication [117].

7.4 Memristors Based on Graphene Oxide

As a last example of devices that exploit the peculiar properties of GO, a particular type of nanoscale electronic device is reported in this section: the memristor.

Memristive devices can generate intense fields by the application of relatively low voltages, relying purely on the properties at the nanoscale level of the active material layer. They are predicted to find application in the fabrication of fast, non-volatile and low-energy electrical switching. In addition, as they are able to retain the internal resistance state according to the history of the applied voltage and current, memristors can be used as resistive random access memory [118], programmable logic [119] and neural networks, and represent the building block for neuromorphic devices and artificial intelligence [120].

The existence of a fourth two-terminal fundamental passive component was already postulated in 1971 by Chua [121], to complete the picture of the three classical circuit elements (resistor, capacitor and inductor). Merely by logic deduction, he introduced a sixth differential relation to couple the four fundamental circuit variables (current, voltage, magnetic flux and charge), in addition to the already known five relations. The sixth missing relation, connecting current and magnetic flux, was sustained by the introduction of a resistor with memory, that is, the memristor. From Chua's equations, it was clear that the value of memristance depended on the time integral of the current, thus on the past history of the device. Memristors are fundamental elements because they cannot be reproduced by combining other passive components. The memristor exhibits two stable resistance states: a high-resistance state (HRS) in which the device shows low electrical conductance, and a low-resistance state (LRS) of high conductance. The device can switch from HRS to LRS (or vice versa) by applying a SET (or RESET) voltage. The performance of a memristor can be evaluated by its values of retention and endurance [122]. The retention indicates how long the internal state of a device can be held when it is not powered, with typical required values for non-volatile memory applications exceeding 10 years. The endurance indicates how many cycles can be supported by the device before reaching breakdown, with typical values being between 10^3 and 10^7 cycles depending on the application.

In order to understand the potential for an application of GO in this field, it is important to note that the most significant feature of memristors is their nanoscale dimension. Therefore, the application of relatively low voltages can generate very intense fields inducing nonlinear and reversible transport of charge carriers, which typically can be either metal ions or oxygen vacancies. The evidence for this effect increases on reducing the thickness of memristive materials. The first working memristor was fabricated in 2008 by HP Labs [123], basing the switching on the formation of a filament of oxygen vacancies in a very thin layer of TiO_2 sandwiched between two Pt electrodes. Since then, many materials and systems have been tested for the fabrication of memristors, such as various perovskite oxide materials, binary transition metal oxides (like TiO_2, Ta_2O_5, NiO), chalcogenides and organic materials [124, 125]. In the last few years, graphene-based nanostructures have also been tested in memristive switching systems [126], and in particular those based on GO [127, 128]. In fact, the oxygen functionalities randomly placed over GO basal planes (hydroxyl and epoxide) and at the sheet edges (carbonyl and carboxyl) [129] are responsible for the electrically insulating behavior of GO, and make it a suitable candidate for the fabrication of memristive devices in substitution of other thin-layered oxide materials. Moreover, being only about 1 nm thick, the scalability of GO-based devices is better than that of organic and inorganic thin-film memristors, and allows scaling beyond the current limits of semiconductor technology for high-density fabrication of memristive devices.

7.4.1 Fabrication of Devices

The most developed structure for the fabrication of memristors is the metal/insulator/metal (MIM), which can be used for an extremely wide range of insulating materials, including GO, sandwiched between two metal electrodes (Figure 7.10). GO can be easily dispersed

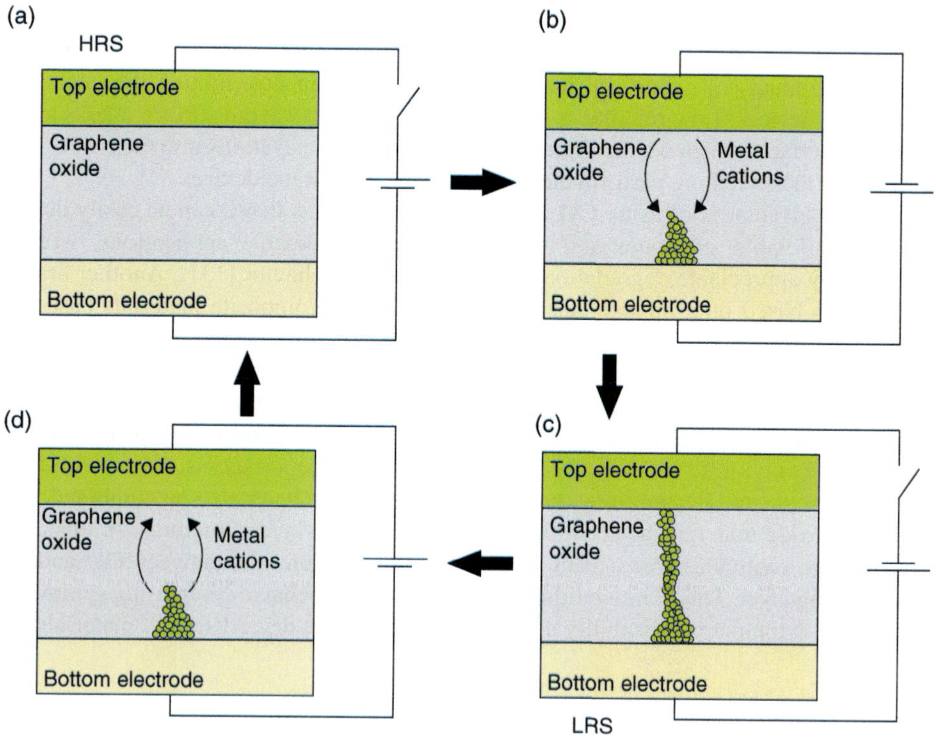

Figure 7.10 *Schematic representation of a MIM memristive structure based on a GO switching layer, with mechanism based on metal filament formation. (a) Device in pristine state. (b) A positive bias induces the oxidation of the top electrode, originating metal cations that migrate towards the bottom electrode, where the formation of metal filaments is promoted after the reduction. (c) The grown filament reach the top electrode with the generation of a local short-circuit (device switches from HRS to LRS). (d) On inverting the external bias, the opposite process is achieved. Reproduced from [127] with permission of Elsevier*

in common polar solvents, including water, and deposited on metallic electrodes with a precise control of GO layer thickness by spin coating, drop casting, vacuum filtration, inkjet printing or LB deposition. This is usually followed by thermal annealing to improve crystallinity and uniformity, at relatively low temperatures in an inert environment to avoid GO reduction. The metallic top electrode is usually deposited by thermal evaporation or sputtering through a shadow mask.

The choice of materials for the metallic electrodes is very important, because it affects the properties of the fabricated device and the switching mechanisms [130]. The metal electrodes sandwiching GO are usually asymmetric, with the work function of the active electrode lower than that of the inert electrode. Switching may also arise due to local oxidation of one of the metal electrodes at the interface with GO. MIS (metal/insulator/semiconductor) devices can also be fabricated using bottom electrodes of semiconducting nature, like p-doped silicon instead of the metal. Semiconductors are intrinsically rectifying, having the advantage of suppressing cross-talk between adjacent devices, for instance in a crossbar configuration, without the need to introduce any additional rectifying element (e.g. diodes) like in many MIM devices. In general, for the fabrication of memristive devices based on GO, particular care should be taken about the surface roughness of the metallic electrodes, because the contact resistance can have great importance in determining the switching proprieties of devices. Low roughness is preferable, in order to avoid surface cracks at the metal–GO interface, which can facilitate the penetration of metal atoms into the oxide layer and influence the switching performance or even short-circuit the device.

One of the advantages of using GO as the switching layer is that it can be easily deposited also on flexible substrates and used for electronic flexible applications, without displaying any appreciable degradation of the memristive behavior [131]. Another branch of memristors based on graphene can be developed using composite materials based on polymeric matrices filled with graphene or GO, which may improve not only the electrical but also the structural characteristics of flexible devices. Like in the case of pure GO, the composites can be used as an active layer sandwiched between conductive electrodes. The oxygen functionalities present on GO surfaces can be engineered to form covalent bonds with the polymeric matrices [132], with memristive properties associated with electron transfer between GO and the polymer [133]. Finally, GO can also be combined with other mixed oxide materials as active layers, to create multi-layer structures. In this case, the memristive switching arises from an exchange of oxygen ions between GO and the other insulating layer. This combination ensures a switching behavior even using symmetric electrodes, with temperature stability up to 85 °C and without degradation of materials.

7.4.2 Switching Mechanisms

The electrical behavior of the metal oxide-based memristors, including devices based on GO, can be explained by two different mechanisms [134]. One is related to oxygen ion diffusion driven by the electric field and Joule heating, while the other is driven by the formation of a metal nanofilament due to the diffusion of species from the top electrode under a bias voltage (Figure 7.10). Most devices show a bipolar switching behavior, where SET and RESET depend on the polarity of the applied voltage, typically in the order of a few volts, also displaying a great retention (higher than 10^4 s) and good endurance. Other devices show unipolar switching, with dependence on the amplitude of the applied voltage rather than on its polarity. An initial stage during which the formation of a percolative path inside the GO layer, named "forming", is often required to stabilize the switching operation and to determine the consequent electrical behavior (unipolar or bipolar) [135]. Generally, fabrication parameters, such as the choice of deposition technique for GO and top electrode, choice of electrode materials, etc., influence the hybridization state of GO, thus contributing to determine the differences in the behavior of devices and influencing the choice of switching mechanism.

The most frequently reported mechanism for resistive switching of GO-based devices is oxygen ion diffusion. However, different interpretations of the physical phenomenon can be given, depending on the scale at which the phenomenon is considered. The memristive behavior can be related to a transition happening in the bulk of the material, which involves a structural modification of the entire GO layer. However, this effect can happen also locally and be associated with a modification of the contact resistance with the top electrode, which greatly influences the characteristics of devices, as previously discussed. As-fabricated devices are normally in their HRS due to the presence of insulating metal oxide layers. The application of a negative bias induces a migration of oxygen ions from the oxide interface layer into the GO layer, with the formation of metal-like conductive path in the metal oxide insulating layer. The thickness reduction causes the device to switch to LRS. By applying a positive voltage, the back-diffusion of the ions from GO restores the original resistance state. A typical *I–V* characteristic curve of a GO-based MIM memristive structure is reported in Figure 7.11, together with data on endurance and retention of the device [131].

As previously explained, in a MIM structure based on GO sandwiched between two metal electrodes (e.g. aluminum [136]), the switching may not be directly related to the presence of GO, which does not participate directly in the switching phenomenon. These conversely depend on the presence of a metal oxide insulating film at the interface with one of the electrodes. This creates a contact resistance between GO and the metal, and GO acts as an oxygen ion reservoir for the switching processes. In this case, the switching occurs locally when redox reactions between GO and the metal oxide layer take place and cause physical adsorption of oxygen in the GO layers.

On the other hand, rather than being attributed to the presence of a metal oxide film, resistive switching can also be associated with local reversible rearrangement of highly resistive thin regions of the GO film near the interface with the metallic electrode. Also in this case the as-fabricated device is in HRS, with the area at the interface mainly composed of a continuous GO layer that is heavily sp^3-hybridized. The application of an external voltage causes the migration of oxygen ions, providing the formation of sp^2-hybridized cluster filaments inside the highly resistive thin regions, which allows switching to LRS.

The memristive behavior of GO-based devices can also be associated with the modification of the hybridization state of the entire GO layer, rather than to the interface between GO and electrodes, for example with the presence of oxygen vacancies and electron traps in the GO layers that act as preferential conductive paths. Differently from previously analyzed devices, in this case the as-fabricated device is in LRS, and changes to HRS after the application of an external bias. The application of a positive voltage leads to recombination between oxygen ions and vacancies with oxidation of the graphene layer. The initial conductive state can then be restored by reversing the bias polarity, achieving back-diffusion of the oxygen ions.

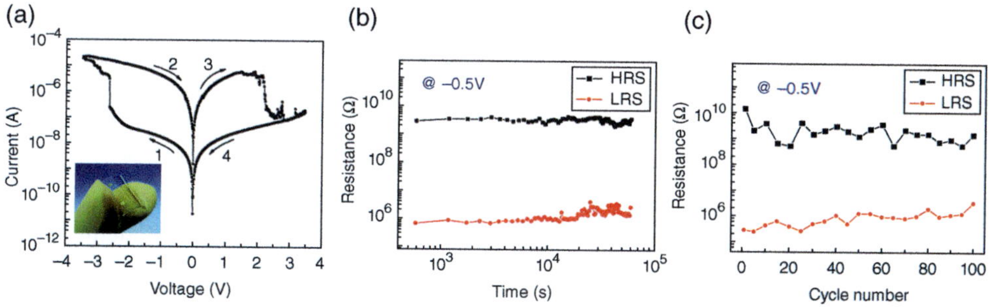

Figure 7.11 *Typical characteristics of a symmetric MIM device based on GO switching layer, showing switching based on oxygen ion diffusion and bipolar behavior: (a) I–V characteristic curve, (b) endurance and (c) retention tests. Reproduced from [131] with permission of ACS*

All the devices illustrated so far have in common the oxygen ion drift mechanism, which often shows a different speed of SET and RESET operations, with the SET speed (typical value of 100 μs) significantly slower than the RESET speed (about 100 ns) [137]. The oxygen diffusivity is strictly related to the hopping energy barrier, which depends on the injection and extraction of electrons from the GO layer, which in turn is strictly related to the electron density in the insulating layer. In particular, when a negative bias is applied to the top electrode (RESET), electrons are injected into the GO, increasing the energy barrier, and thus reducing the oxygen diffusivity and speed. In the same way, when the bias voltage is reversed (SET), the electrons are extracted from the GO, reducing the energy barrier, and thus increasing the ions movement speed [138].

The other and less common mechanism for GO-based memristors is based on the formation of a metal filament within the insulating layer [139]. In this case the HRS and LRS states show nonlinear and linear resistance variations with respect to temperature, respectively. A positive bias applied to the top electrode induces a high electric field inside the device (SET). During the oxidation of the top electrode, metal ions migrate to the bottom electrode through the GO film, and accumulate at the bottom electrode interface, inducing a reduction process through the insulating layer, with the eventual formation of a metal filament. The resistance of the device switches abruptly from HRS to LRS as soon as the growing filament reaches the top electrode. An inversion of the bias voltage causes the rupture of the filament with the back-diffusion of the metal ions, restoring the HRS.

References

[1] Novoselov, K.S.; Geim, A.K.; Morozov, S.V.; *et al.*, Electric field effect in atomically thin carbon films. *Science* **2004**, *306* (5696), 666–669.

[2] Geim, A.K.; Novoselov, K.S., The rise of graphene. *Nature Mater.* **2007**, *6* (3), 183–191.

[3] Brey, L.; Fertig, H.A., Electronic states of graphene nanoribbons studied with the Dirac equation. *Phys. Rev. B* **2006**, *73* (23), 235411.

[4] Son, Y.W.; Cohen, M.L; Louie S.G., Energy gaps in graphene nanoribbons. *Phys. Rev. Lett.* **2006**, *97* (21), 216803.

[5] Giardi, R.; Porro, S.; Chiolerio, A.; *et al.*, Inkjet printed acrylic formulations based on UV-RGO nanocomposites. *J. Mater. Sci.* **2013**, *48* (3), 1249–1255.

[6] Porro, S.; Giardi, R.; Chiolerio, A., Real-time monitoring of GO reduction in acrylic printable composite inks. *Appl. Phys. A* **2014**, *117* (3), 1289–1293.

[7] Zhu, Y.; Murali, S.; Cai, W.; *et al.*, Graphene and GO: synthesis, properties, and applications. *Adv. Mater.* **2010**, *22* (35), 3906–3924.

[8] Barone, V.; Hod, O.; Scuseria, G.E., Electronic structure and stability of semiconducting graphene nanoribbons. *Nano Lett.* **2006**, *6* (12), 2748–2754.

[9] Li, X.L.; Wang, X.R.; Zhang, L.; *et al.*, Chemically derived, ultrasmooth graphene nanoribbon semiconductors. *Science* **2008**, *319* (5867), 1229–1232.

[10] Jiao, L.Y.; Zhang, L.; Wang, X.R.; *et al.*, Narrow graphene nanoribbons from carbon nanotubes. *Nature* **2009**, *458* (7240), 877–880.

[11] Kosynkin, D.V.; Higginbotham, A.L.; Sinitskii, A.; *et al.*, Longitudinal unzipping of carbon nanotubes to form graphene nanoribbons. *Nature* **2009**, *458* (7240), 872–876.

[12] Wang, X.R.; Li, X.L.; Zhang, L.; *et al.*, N-doping of graphene through electrothermal reactions with ammonia. *Science* **2009**, *324* (5928), 768–771.

[13] Li, X.; Wang, H.; Robinson, J.T.; *et al.*, Simultaneous nitrogen doping and reduction of GO. *J. Am. Chem. Soc.* **2009**, *131* (43), 15939–15944.

[14] Gilje, S.; Han, S.; Wang, M.; *et al.*, A chemical route to graphene for device applications. *Nano Lett.* **2007**, *7* (11), 3394–3398.

[15] Jeong, H.K.; Jin, M.H.; So, K.P.; *et al.*, Tailoring the characteristics of graphite oxides by different oxidation times. *J. Phys. D: Appl. Phys.* **2009**, *42* (6), 065418.
[16] Jin, M.; Jeong, H.K.; Yu, W.J.; *et al.*, GO thin film field effect transistors without reduction. *J. Phys. D: Appl. Phys.* **2009**, *42* (13), 135109.
[17] Chen, Y.; Xu, Y.; Zhao, K.; *et al.*, Towards flexible all-carbon electronics: flexible organic field-effect transistors and inverter circuits using solution-processed all-graphene source/drain/gate electrodes. *Nano Res.* **2010**, *3* (10), 714–721.
[18] Wang, S.; Ang, P.K.; Wang, Z.; *et al.*, High mobility, printable, and solution-processed graphene electronics. *Nano Lett.* **2010**, *10* (1), 92–98.
[19] Bocchini, S.; Chiolerio, A.; Porro, S.; *et al.*, Synthesis of polyaniline-based inks, doping thereof and test device printing towards electronic applications. *J. Mater. Chem. C* **2013**, *1* (33), 5101–5109.
[20] Chiolerio, A.; Bocchini, S.; Porro, S., Inkjet printed negative supercapacitors: synthesis of polyaniline-based inks, doping agent effect, and advanced electronic devices applications. *Adv. Funct. Mater.* **2014**, *24* (22), 3375–3383.
[21] Lee, C.G.; Park, S.; Ruoff, R.S.; Dodabalapur, A., Integration of RGO into organic field-effect transistors as conducting electrodes and as a metal modification layer. *Appl. Phys. Lett.* **2009**, *95* (2), 023304.
[22] Sharma, B.K.; Ahn, J.H., Graphene based field effect transistors: efforts made towards flexible electronics. *Solid-State Electronics* **2013**, *89*, 177–188.
[23] He, Q.; Wu, S.; Gao, S.; *et al.*, Transparent, flexible, all-RGO thin film transistors. *ACS Nano* **2011**, *5* (6), 5038–5044.
[24] Lee, S.K.; Jang, H.Y.; Jang, S.; *et al.*, All graphene-based thin film transistors on flexible plastic substrates. *Nano Lett.* **2012**, *12* (7), 3472–3476.
[25] He, Q.; Wu, S.; Yin, Z.; Zhang, H., Graphene-based electronic sensors. *Chem. Sci.* **2012**, *3* (6), 1764–1772.
[26] Llobet, E., Gas sensors using carbon nanomaterials: a review. *Sens. Actuators B: Chem.* **2013**, *179*, 32–45.
[27] Kochmann, S.; Hirsch, T.; Wolfbeis, O.S., Graphenes in chemical sensors and biosensors. *Trends Anal. Chem.* **2012**, *39*, 87–113.
[28] Zhang, J.; Song, L.; Zhang, A.; *et al.*, Environmentally responsive graphene systems. *Small* **2014**, *10* (11), 2151–2164.
[29] Toda, K.; Furue, R.; Hayami, S., Recent progress in applications of GO for gas sensing: a review. *Anal. Chim. Acta* **2015**, *878*, 43–53.
[30] Robinson, J.T.; Perkins, F.K.; Snow, E.S.; *et al.*, RGO molecular sensors. *Nano Lett.* **2008**, *8* (10), 3137–3140.
[31] Fowler, J.D.; Aleen, M.J.; Tung, V.C.; *et al.*, Practical chemical sensors from chemically derived graphene. *ACS Nano* **2009**, *3* (2), 301–306.
[32] Lu, G.; Park, S.; Yu, K. *et al.*, Toward practical gas sensing with highly RGO: a new signal processing method to circumvent run-to-run and device-to-device variations. *ACS Nano* **2011**, *5* (2), 1154–1164.
[33] Lu, G.; Ocola, L.E.; Chen, J., Gas detection using low-temperature RGO sheets. *Appl. Phys. Lett.* **2009**, *94* (8), 083111.
[34] Strong, V.; Dubin, S.; El-Kady, M.F.; *et al.*, Patterning and electronic tuning of laser scribed graphene for flexible all-carbon devices. *ACS Nano* **2012**, *6* (2), 1395–1403.
[35] Chung, M.G.; Kim, D.H.; Lee, H.M.; *et al.*, Highly sensitive NO_2 gas sensor based on ozone treated graphene. *Sens. Actuators B: Chem.* **2012**, *166–167*, 172–176.
[36] Deng, S.; Tjoa, V.; Fan, H.M.; *et al.*, RGO conjugated Cu_2O nanowire mesocrystals for high-performance NO_2 gas sensor, *J. Am. Chem. Soc.* **2012**, *134* (10), 4905–4917.
[37] Yuan, W.; Liu, A.; Huang, L.; *et al.*, High-performance NO_2 sensors based on chemically modified graphene. *Adv. Mater.* **2013**, *25* (5), 766–771.
[38] Ghosh, R.; Midya, A.; Santra, S.; *et al.*, Chemically RGO for ammonia detection at room temperature. *ACS Appl. Mater. Interfaces* **2013**, *5* (15), 7599–7603.
[39] Ghosh, R.; Singh, A.; Santra, S.; *et al.*, Highly sensitive large-area multi-layered graphene-based flexible ammonia sensor. *Sens. Actuators B: Chem.* **2014**, *205*, 67–73.
[40] Kavinkumar, T.; Sastikumar, D.; Manivannan, S., Effect of functional groups on dielectric, optical gas sensing properties of GO and RGO at room temperature, *RSC Advances* **2015**, *5* (14), 10816–10825.

[41] Xia, X.; Guo, S.; Zhao, W.; *et al.*, Carboxyl functionalized gold nanoparticles in situ grown on RGO for micro-gravimetric ammonia sensing. *Sens. Actuators B: Chem.* **2014**, *202*, 846–853.
[42] Meng, H.; Yang, W.; Ding, K.; *et al.*, Cu_2O nanorods modified by RGO for NH_3 sensing at room temperature. *J. Mater. Chem. A* **2015**, *3*, 1174–1181.
[43] Lin, Q.; Li, Y.; Yang, M., Tin oxide/graphene composite fabricated via a hydrothermal method for gas sensors working at room temperature. *Sens. Actuators B: Chem.* **2012**, *173*, 139–147.
[44] Zhou, X.; Wang, X.; Wang, B.; *et al.*, Preparation, characterization and NH_3-sensing properties of RGO/copper phthalocyanine hybrid material. *Sens. Actuators B: Chem.* **2014**, *193*, 340–348.
[45] Travlou, N.A.; Singh, K.; Rodríguez-Castellón, E.; Bandosz, T.J., Cu–BTC MOF–graphene-based hybrid materials as low concentration ammonia sensors. *J. Mater. Chem. A* **2015**, *3* (21), 11417–11429.
[46] Hu, N.; Yang, Z.; Wang, Y.; *et al.*, Ultrafast and sensitive room temperature NH_3 gas sensors based on chemically RGO. *Nanotechnology* **2014**, *25* (2), 025502.
[47] Sansone, L.; Malachovska, V.; La Manna, P.; *et al.*, Nanochemical fabrication of a GO-based nanohybrid for label-free optical sensing with fiber optics, *Sens. Actuators B: Chem.* **2014**, *202*, 523–526.
[48] Arsat, R.; Breedon, M.; Shafiei, M.; *et al.*, Graphene-like nano-sheets for surface acoustic wave gas sensor applications. *Chem. Phys. Lett.* **2009**, *467* (4–6), 344–347.
[49] Zhang L.S.; Wang, W.D.; Liang, X.Q.; *et al.*, Characterization of partially RGO as room temperature sensor for H_2. *Nanoscale* **2011**, *3* (6), 2458–2460.
[50] Tan, C.; Huang, X.; Zhang, H., Synthesis and applications of graphene-based noble metal nanostructures. *Mater. Today* **2013**, *16* (1–2), 29–36.
[51] Du, Y.; Xue, Q.; Zhang, Z.; Xia, F., Great enhancement in H_2 response using graphene-based Schottky junction. *Mater. Lett.* **2014**, *135*, 151–153.
[52] Hong, J.; Lee, S.; Seo, J.; *et al.*, Highly sensitive hydrogen sensor with gas selectivity using a PMMA membrane-coated Pd nanoparticle/single-layer graphene hybrid. *ACS Appl. Mater. Interfaces* **2015**, *7* (6), 3554–3561.
[53] Russo, P.A.; Donato, N.; Leonardi, S.G.; *et al.*, Room-temperature hydrogen sensing with heteronanostructures based on RGO and tin oxide. *Angew. Chem. Int. Ed.* **2012**, *51* (44), 11053–11057.
[54] Dutta, D.; Hazr, S.K.; Das, J.; *et al.*, Studies on p-TiO_2/n-graphene heterojunction for hydrogen detection. *Sens. Actuators B: Chem.* **2015**, *212*, 84–92.
[55] Anand, K.; Singh, O.; Singh, M.P.; *et al.*, Hydrogen sensor based on graphene/ZnO nanocomposite. *Sens. Actuators B: Chem.* **2014**, *195*, 409–415.
[56] Esfandiar, A.; Ghasemi, S.; Irajizad, A.; *et al.*, The decoration of TiO_2/RGO by Pd and Pt nanoparticles for hydrogen gas sensing. *Int. J. Hydrogen Energy* **2012**, *37* (20), 15423–15432.
[57] Esfandiar, A.; Irajizad, A.; Akhavan, O.; *et al.*, Pd–WO_3/RGO hierarchical nanostructures as efficient hydrogen gas sensors. *Int. J. Hydrogen Energy* **2014**, *39* (15), 8169–8179.
[58] Guo, L.; Jiang, H.B.; Shao, R.Q.; *et al.*, Two-beam-laser interference mediated reduction, patterning and nanostructuring of GO for the production of a flexible humidity sensing device. *Carbon* **2012**, *50* (4), 1667–1673.
[59] Han, D.D.; Zhang, Y.L.; Jiang, H.B.; *et al.*, Moisture-responsive graphene paper prepared by self-controlled photoreduction. *Adv. Mater.* **2014**, *27* (2), 332–338.
[60] Yu, H.W.; Kim, H.K.; Kim, T.; *et al.*, Self-powered humidity sensor based on GO composite film intercalated by poly(sodium 4-styrenesulfonate). *ACS Appl. Mater. Interfaces* **2014**, *6* (11), 8320–8326.
[61] Hwang, S.H.; Kang, D.; Ruoff, R.S.; *et al.*, Poly(vinyl alcohol) reinforced and toughened with poly(dopamine)-treated GO, and its use for humidity sensing. *ACS Nano* **2014**, *8* (7), 6739–6747.
[62] Mao, S.; Cui, S.; Lu, G.; *et al.*, Tuning gas-sensing properties of RGO using tin oxide nanocrystals. *J. Mater. Chem.* **2012**, *22*, 11009–11013.
[63] Zhang, D.; Tong, J.; Xia, B., Humidity-sensing properties of chemically reduced grapheneoxide/polymer nanocomposite film sensor based on layer-by-layer nano self-assembly. *Sens. Actuators B: Chem.* **2014**, *197*, 66–72.
[64] Su, P.; Chiou, C., Electrical and humidity-sensing properties of RGO thin film fabricated by layer-by-layer with covalent anchoring on flexible substrate. *Sens. Actuators B: Chem.* **2014**, *200*, 9–18.

[65] Zhang, J.; Shen, G.; Wang, W.; *et al.*, Individual nanocomposite sheets of chemically RGO and poly(*N*-vinyl pyrrolidone): preparation and humidity sensing characteristics. *J. Mater. Chem.* **2010**, *20* (48), 10824–10828.
[66] Borini, S.; White, R.; Wei, D.; *et al.*, Ultrafast GO humidity sensors. *ACS Nano* **2013**, *7* (12), 11166–11173.
[67] Morales-Narváez, E.; Merkoçi, A., GO as an optical biosensing platform. *Adv. Mater.* **2012**, *24* (25), 3298–3308.
[68] Chartuprayoon, N.; Zhang, M.; Bosze, W.; *et al.*, One-dimensional nanostructures based bio-detection. *Biosens. Bioelectron.* **2015**, *63*, 432–443.
[69] Li, Z.; He, M.; Xu, D.; Liu, Z., Graphene materials-based energy acceptor systems and sensors. *J. Photochem. Photobiol. C: Photochem. Rev.* **2014**, *18*, 1–17.
[70] Selvin, P.R., The renaissance of fluorescence resonance energy transfer. *Nature Struct. Biol.* **2000**, *7* (9), 730–734.
[71] Loh, K.P.; Bao, Q.; Eda, G.; Chhowalla, M., GO as a chemically tunable platform for optical applications. *Nature Chem.* **2010**, *2* (12), 1015–1024.
[72] Luo, Z.; Vora, P.M.; Mele, E.J.; *et al.*, Photoluminescence and band gap modulation in GO. *Appl. Phys. Lett.* **2009**, *94* (11), 111909.
[73] Chen, J.L.; Yan, X.P., A dehydration and stabilizer-free approach to production of stable water dispersions of graphene nanosheets. *J. Mater. Chem.* **2010**, *20* (21), 4328–4332.
[74] Peng, J.; Gao, W.; Gupta, B.K.; *et al.*, Graphene quantum dots derived from carbon fibers. *Nano Lett.* **2012**, *12* (2), 844–849.
[75] Wang, Y.; Li, Z.; Wang, J.; *et al.*, Graphene and GO: biofunctionalization and applications in biotechnology. *Trends Biotechnol.* **2011**, *29* (5), 205–212.
[76] Liu, B.; Sun, Z.; Zhang, X.; Liu, J., Mechanisms of DNA sensing on GO. *Anal. Chem.* **2013**, *85* (16), 7987–7993.
[77] Liu, X.; Wang, F.; Aizen, R.; *et al.*, GO/nucleic-acid-stabilized silver nanoclusters: functional hybrid materials for optical aptamer sensing and multiplexed analysis of pathogenic DNAs. *J. Am. Chem. Soc.* **2013**, *135* (32), 11832–11839.
[78] Wang, X.; Wang, C.; Qu, K.; *et al.*, Ultrasensitive and selective detection of a prognostic indicator in early-stage cancer using graphene oxide and carbon nanotubes. *Adv. Funct. Mater.* **2010**, *20* (22), 3967–3971.
[79] Lu, C.H.; Yang, H.H.; Zhu, C.L.; *et al.*, A graphene platform for sensing biomolecules. *Angew. Chem. Int. Ed.* **2009**, *48* (26), 4785–4787.
[80] Shi, Y.; Pramanik, A.; Tchounwou, C.; *et al.*, Multifunctional biocompatible GO quantum dots decorated magnetic nanoplatform for efficient capture and two-photon imaging of rare tumor cells. *ACS Appl. Mater. Interfaces* **2015**, *7* (20), 10935–10943.
[81] Sun, X.; Liu, Z.; Welsher, K.; *et al.*, Nano-GO for cellular imaging and drug delivery. *Nano Res.* **2008**, *1* (3), 203–212.
[82] Peng, C.; Hu, W.; Zhou, Y.; *et al.*, Intracellular imaging with a graphene-based fluorescent probe. *Small* **2010**, *6* (15), 1686–1692.
[83] Chen, J.L.; Yan, X.P.; Meng, K.; Wang, S.F., GO based photoinduced charge transfer label-free near-infrared fluorescent biosensor for dopamine. *Anal. Chem.* **2011**, *83* (22), 8787–8793.
[84] Kumar, S.; Ahlawat, W.; Kumar, R.; Dilbaghi, N., Graphene, carbon nanotubes, zinc oxide and gold as elite nanomaterials for fabrication of biosensors for healthcare. *Biosens. Bioelectron.* **2015**, *70*, 498–503.
[85] Kotanen, C.N.; Moussy, F.G.; Carrara, S.; Guiseppi-Eli, A., Implantable enzyme amperometric biosensors. *Biosens. Bioelectron.* **2012**, *35* (1), 14–26.
[86] Luo, Z.; Yuwen, L.; Han, Y.; *et al.*, Reduced graphene oxide/PAMAM–silver nanoparticles nanocomposite modified electrode for direct electrochemistry of glucose oxidase and glucose sensing. *Biosens. Bioelectron.* **2012**, *36* (1), 179–185.
[87] Xiao, F.; Li, Y.; Zan, X.; *et al.*, Growth of metal–metal oxide nanostructures on freestanding graphene paper for flexible biosensors. *Adv. Funct. Mater.* **2012**, *22* (12), 2487–2494.
[88] Xue, Y.; Zhao, H.; Wu, Z.; *et al.*, The comparison of different gold nanoparticles/graphene nanosheets hybrid nanocomposites in electrochemical performance and the construction of a sensitive uric acid electrochemical sensor with novel hybrid nanocomposites. *Biosens. Bioelectron.* **2011**, *29* (1), 102–108.

[89] Sheng, Z.; Zheng, X.; Xu, J.; *et al.*, Electrochemical sensor based on nitrogen doped graphene: simultaneous determination of ascorbic acid, dopamine and uric acid. *Biosens. Bioelectron.* **2012**, *34* (1), 125–131.
[90] Du, J.; Pei, S.; Ma, L.; Cheng, H.M., Carbon nanotube- and graphene-based transparent conductive films for optoelectronic devices. *Adv. Mater.* **2014**, *26* (13), 1958–1991.
[91] Zhu, Y.; Murali, S.; Cai, W.; *et al.*, Graphene and graphene oxide: synthesis, properties, and applications. *Adv. Mater.* **2010**, *22* (35), 3906–3924.
[92] Zheng, Q.; Li, Z.; Yang, J.; Kim, J.K., Graphene oxide-based transparent conductive films. *Prog. Mater. Sci.* **2014**, *64*, 200–247.
[93] Becerril, H.A.; Mao, J.; Liu, Z.; *et al.*, Evaluation of solution-processed reduced graphene oxide films as transparent conductors. *ACS Nano* **2008**, *2* (3), 463–470.
[94] Dong, X.C.; Su, C.Y.; Zhang, W.J.; *et al.*, Ultra-large single-layer graphene obtained from solution chemical reduction and its electrical properties. *Phys. Chem. Chem. Phys.* **2010**, *12* (9), 2164–2169.
[95] Pham, V.H.; Cuong, T.V.; Hur, S.H.; *et al.*, Fast and simple fabrication of a large transparent chemically-converted graphene film by spray-coating. *Carbon* **2010**, *48* (7), 1945–1951.
[96] Choi, H.W.; Zhou, T.; Singh, M.; Jabbour, G.E., Recent developments and directions in printed nanomaterials. *Nanoscale* **2015**, *7* (8), 3338–3355.
[97] Yamaguchi, H.; Eda, G.; Mattevi, C.; *et al.*, Highly uniform 300 mm wafer-scale deposition of single and multilayered chemically derived graphene thin films. *ACS Nano* **2010**, *4* (1), 524–528.
[98] An, S.J.; Zhu, Y.W.; Lee, S.H.; *et al.*, Thin film fabrication and simultaneous anodic reduction of deposited graphene oxide platelets by electrophoretic deposition. *J. Phys. Chem. Lett.* **2010**, *1* (8), 1259–1263.
[99] Cote, L.J.; Kim, F.; Huang, J.X., Langmuir–Blodgett assembly of graphite oxide single layers. *J. Am. Chem. Soc.* **2009**, *131* (3), 1043–1049.
[100] Wang, S.; Ang, P.K.; Wang, Z.Q.; *et al.*, High mobility, printable, and solution-processed graphene electronics. *Nano Lett.* **2010**, *10* (1), 92–98.
[101] Kholmanov, I.N.; Magnuson, C.W.; Aliev, A.E.; *et al.*, Improved electrical conductivity of graphene films integrated with metal nanowires. *Nano Lett.* **2012**, *12* (11), 5679–5683.
[102] Hong T.K.; Lee, D.W.; Choi, H.J.; *et al.*, Transparent, flexible conducting hybrid multilayer thin films of multiwalled carbon nanotubes with graphene nanosheets. *ACS Nano* **2010**, *4* (7), 3861–3868.
[103] Huang, J.H.; Fang, J.H.; Liu, C.C.; Chu, C.W., Effective work function modulation of graphene/carbon nanotube composite films as transparent cathodes for organic optoelectronics. *ACS Nano* **2011**, *5* (8), 6262–6271.
[104] Zheng, Q.B.; Gudarzi, M.M.; Wang, S.J.; *et al.*, Improved electrical and optical characteristics of transparent graphene thin films produced by acid and doping treatments. *Carbon* **2011**, *49* (9), 2905–2916.
[105] Lin, X.Y.; Shen, X.; Zheng, Q.; *et al.*, Fabrication of highly-aligned, conductive, and strong graphene papers using ultralarge graphene oxide sheets. *ACS Nano* **2012**, *6* (12), 10708–10719.
[106] Hsiao, S.T.; Tien, H.W.; Liao, W.H.; *et al.*, A highly electrically conductive graphene–silver nanowire hybrid nanomaterial for transparent conductive films, *J. Mater. Chem. C* **2014**, *2* (35), 7284–7291.
[107] Huang, J.H.; Fang, J.H.; Liu, C.C.; Chu, C.W., Effective work function modulation of graphene/carbon nanotube composite films as transparent cathodes for organic optoelectronics. *ACS Nano* **2011**, *5* (8), 6262–6271.
[108] Zheng, Q.B.; Zhang, B.; Lin, X.; *et al.*, Highly transparent and conducting ultralarge graphene oxide/single-walled carbon nanotube hybrid films produced by Langmuir–Blodgett assembly. *J. Mater. Chem.* **2012**, *22* (48), 25072–25082.
[109] Tung, V.C.; Chen, L.M.; Allen, M.J.; *et al.*, Low-temperature solution processing of graphene–carbon nanotube hybrid materials for high-performance transparent conductors. *Nano Lett.* **2009**, *9* (5), 1949–1955.

[110] Yu, D.S.; Dai, L.M., Self-assembled graphene/carbon nanotube hybrid films for supercapacitors. *J. Phys. Chem. Lett.* **2010**, *1* (2), 467–470.

[111] Azevedo, J.; Costa-Coquelard, C.; Jegou, P.; *et al.*, Highly ordered monolayer, multilayer, and hybrid films of graphene oxide obtained by the bubble deposition method. *J. Phys. Chem. C* **2011**, *115* (30), 14678–14681.

[112] Kasry, A.; Kuroda, M.A.; Martyna, G.J.; *et al.*, Chemical doping of large-area stacked graphene films for use as transparent, conducting electrodes. *ACS Nano* **2010**, *4* (7), 3839–4344.

[113] Eda, G.; Lin, Y.Y.; Miller, S.; *et al.*, Transparent and conducting electrodes for organic electronics from reduced graphene oxide. *Appl. Phys. Lett.* **2008**, *92* (23), 233305.

[114] Gunes, F.; Shin, H.J.; Biswas, C.; *et al.*, Layer-by-layer doping of few-layer graphene film. *ACS Nano* **2010**, *4* (8), 4595–4600.

[115] Bae, S.Y.; Jeon, I.Y.; Yang, J.; *et al.*, Large-area graphene films by simple solution casting of edge selectively functionalized graphite. *ACS Nano* **2011**, *5* (6), 4974–4980.

[116] Su, C.Y.; Xu, Y.P.; Zhang, W.J.; *et al.*, Electrical and spectroscopic characterizations of ultra-large reduced graphene oxide monolayers. *Chem. Mater.* **2009**, *21* (23), 5674–5680.

[117] Zhou, X.F.; Liu, Z.P., A scalable, solution-phase processing route to graphene oxide and graphene ultralarge sheets. *Chem. Commun.* **2010**, *46* (15), 2611–2613.

[118] Chua, L.O., Resistance switching memories are memristors. *Appl. Phys. A* **2011**, *102* (4), 765–783.

[119] Borghetti, J.; Snider, G.S.; Kuekes, P.J.; *et al.*, Memristive switches enable stateful logic operations via material implication. *Nature* **2010**, *464* (7290), 873–876.

[120] Sah, M.P.; Hyongsuk, K.; Chua, L.O., Brains are made of memristors. *IEEE Circuits Syst. Mag.* **2014**, *14* (1), 12–36.

[121] Chua, L.O., Memristor – the missing circuit element. *IEEE Trans. Circuit Theory* **1971**, *18* (5), 507–519.

[122] Waser, R.; Dittmann, R.; Staikov, G.; Szot, K., Redox-based resistive switching memories – nanoionic mechanisms, prospects, and challenges. *Adv. Mater.* **2009**, *21* (25–26), 2632–2663.

[123] Strukov, D.B.; Snider, G.S.; Stewart, D.R.; Williams, R.S., The missing memristor found. *Nature Lett.* **2008**, *453* (7191), 80–83.

[124] Yang, Y.; Choi, S.H.; Lu, W., Oxide heterostructure resistive memory. *Nano Lett.* **2013**, *13* (6), 2908–2915.

[125] Chen, Y.; Liu, G.; Wang, C.; *et al.*, Polymer memristor for information storage and neuromorphic applications. *Mater. Horiz.* **2014**, *1* (5), 489–506.

[126] Chen, Y.; Zhang, B.; Liu, G.; *et al.*, Graphene and its derivatives: switching ON and OFF. *Chem. Soc. Rev.* **2012**, *41* (13), 4688–4707.

[127] Porro, S.; Accornero, E.; Pirri, C.F.; Ricciardi, C., Memristive devices based on graphene oxide. *Carbon* **2015**, *85*, 383–396.

[128] Porro, S.; Ricciardi, C., Memristive behaviour in inkjet printed graphene oxide thin layers. *RSC Advances* **2015**, *5* (84), 68565–68570.

[129] Lerf, A.; He, H.; Forster, M.; Klinowski, J., Structure of graphite oxide revisited. *J. Phys. Chem. B* **1998**, *102* (23), 4477–4482.

[130] Hong, S.K.; Kim, J.E.; Kim, S.O.; *et al.*, Flexible resistive switching memory device based on graphene oxide. *IEEE Electron Device Lett.* **2010**, *31* (9), 1005–1007.

[131] Jeong, H.Y.; Kim, J.Y.; Kim, J.W.; *et al.*, Graphene oxide thin films for flexible nonvolatile memory applications. *Nano Lett.* **2010**, *10* (11), 4381–4386.

[132] Liu, G.; Zhuang, X.; Chen, Y.; *et al.*, Bistable electrical switching and electronic memory effect in a solution-processable graphene oxide–donor polymer complex. *Appl. Phys. Lett.* **2009**, *95* (25), 253301.

[133] Zhuang, X.; Chen, Y.; Liu, G.; *et al.*, Conjugated-polymer-functionalized graphene oxide: synthesis and nonvolatile rewritable memory effect. *Adv. Mater.* **2010**, *22* (15), 1731–1735.

[134] Waser, R.; Aono, M., Nanoionics-based resistive switching memories. *Nature Mater.* **2007**, *6* (11), 833–840.

[135] Hong, S.K.; Kim, J.E.; Kim, S.O.; Cho, B.J., Non-volatile memory using graphene oxide for flexible electronics. In *Proc. 2010 10th IEEE Conf. on Nanotechnology (IEEE-NANO)*, pp. 604–606. IEEE, Piscataway, NJ, 2010.

[136] Nho, H.W.; Kim, J.Y.; Wang, J.; *et al.*, Scanning transmission X-ray microscopy probe for in situ mechanism study of graphene-oxide-based resistive random access memory. *J. Synchrotron Radiat.* **2014**, *21* (1), 170–176.

[137] Wang, L.H.; Yang, W.; Sun, Q.Q.; *et al.*, The mechanism of the asymmetric SET and RESET speed of graphene oxide based flexible resistive switching memories. *Appl. Phys. Lett.* **2012**, *100* (6), 063509.

[138] Porro, S., Graphene nanostructures for memristive devices. *Encyclopedia of Nanotechnology*, ed. B. Bhushan (Living Reference Work Entry), 10 pp. Springer Science, Dordrecht, **2015**. DOI: 10.1007/978-94-007-6178-0_101030-1; Online ISBN: 978-94-007-6178-0.

[139] Zhuge, F.; Hu, B.; He, C.; *et al.*, Mechanism of nonvolatile resistive switching in graphene oxide thin films. *Carbon* **2011**, *49* (12), 3796–3802.

8

Energy Harvesting and Storage

Cary Michael Hayner

8.1 Solar Cells

Solar cells are a very promising technology to convert light into electricity by harvesting solar energy. However, while current solar cells have become cost-competitive with conventional energy sources, there is still much room for new technologies with increased energy conversion efficiency and lower cost. Increasing the efficiency of the solar cell can provide an opportunity to lower the cost by increasing the incident spectral range available to reach the absorption layer, thereby increasing the electron collection. The conventional transparent conducting oxide films used as transparent electrodes in solar cells often struggle with transmitting light near the infrared (IR) band in the solar spectrum. Additionally, current transparent electrode materials such as indium tin oxide (ITO) are mechanically brittle, unstable at high temperatures, incompatible with strong acids and extremely expensive due to the inclusion of the rare element indium.

In order to overcome the cost and performance limitations of existing transparent conducting oxides, as well as to add new functionality, such as device flexibility, there has been a strong push to develop new transparent conducting films. Graphene is regarded as an ideal alternative to ITO due to its high electron mobility and ability to contribute to absorption from a broad spectrum. Indeed, graphene materials have already shown promise in a wide-ranging variety of solar cell types, including conventional silicon solar cells, thin-film cells and dye-sensitized solar cells (DSSCs). Within these applications, graphene-based materials can serve a variety of roles, including as transparent layers, electron acceptors, hole conductors and conductive inks, among others. Most remarkably for DSSCs, graphene materials – with their wide range of properties – have been applied to almost all aspects of

---

Graphene Oxide: Fundamentals and Applications, First Edition. Edited by Ayrat M. Dimiev and Siegfried Eigler.

the device, including as transparent conductors, in the semiconducting layer and as the sensitizer itself.

Due to the far-reaching application of graphene-based materials to solar cell technologies, a comprehensive review is beyond the scope of this chapter. Overviews of the major approaches and advancements have recently been examined in several publications. A number of concise review articles have discussed the use of graphene oxide (GO) and its reduced counterparts for solar cell applications, and the author directs the reader to those articles for more complete analysis [1–6]. Although still in a nascent stage, graphene-type materials are attractive candidates to improve the efficiency and lower the cost of solar cell technologies. Graphene-based materials show great promise for enabling practical commercialization of advanced solar cells to harvest solar energy.

8.2 Lithium-Ion Batteries

8.2.1 Introduction

Lithium-ion batteries are by far the most widespread form of electrical energy storage for portable applications such as electric vehicles, consumer devices and aerospace. The rechargeable Li-ion battery was first commercialized by Sony in 1991 and began to displace the incumbent nickel–metal hydride (NiMH) batteries that were used previously. Compared to the incumbent NiMH batteries, Li-ion batteries offered longer cycle life, no memory effect, low self-discharge rate, and higher gravimetric and volumetric energy densities. Whereas NiMH batteries could store ~60 W h kg^{-1} in the early 1990s, the lithium-ion batteries were able to store over 100 W h kg^{-1} when they first entered the market. Since they were first commercialized, the lithium-ion battery industry has grown to a market size of $10 billion dollars globally in 2014 and is expected to grow over 10% annually to over $25 billion by 2025 [7]. Every year, *four billion* lithium-ion battery cells are constructed and sold. Much of the current growth has been due to the rapid proliferation of portable electronic devices, including cameras, cell/mobile phones and laptops. In addition to the continued growth of portable electronics, the promise of clean-powered electric vehicles has the potential to completely transform the battery market, with some estimates suggesting that the electric vehicle (EV) market could grow to over $10 billion by 2030 [7].

In order to power the improvements necessary to sustain the future demands of batteries, the charge storage that can be packed into these batteries must be continually improved. Remarkably, the energy density of lithium-ion batteries has been steadily improved at a rate of 3–5% annually. Although steady improvements have been made since entering the market in the early 1990s, very little innovation has occurred within the fundamental energy storage components, termed the anode and the cathode for the negative and positive electrodes, respectively. The following section will describe the operation of a lithium-ion battery and how graphene-containing materials can enable significant advancements compared to state-of-the-art commercial cells.

8.2.2 Electrochemistry Fundamentals

In its simplest form, a lithium-ion battery consists of two electrodes – anode and cathode – and an electrolyte (Figure 8.1). These cells make use of the difference in reduction or

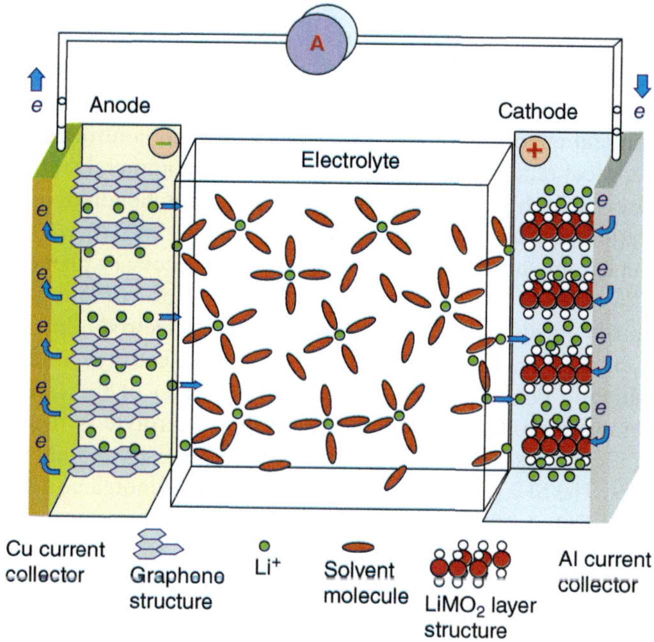

Figure 8.1 Schematic of a Li-ion battery in discharge mode (graphite anode, $LiMO_2$ cathode). Reproduced from [9] with permission of American Chemical Society

oxidation potential (the driving force to accept or give away electrons, respectively) between the two electrodes to generate electrical energy. By convention, reduction (gain of electrons) occurs at the cathode while oxidation (loss of electrons) occurs at the anode during a discharge reaction. To control the rate of electron transfer, the cathode must be physically and electrically isolated from the anode using an ionically conductive, but electrically insulating, medium, typically a liquid electrolyte. Often a polymer separator is also used for mechanical separation between the two electrodes to prevent a short-circuit [8]. Since its introduction in the early 1990s, almost all commercial lithium-ion batteries have consisted of a carbonaceous anode and a lithiated metal oxide cathode. The most commonly used anode material for commercial applications is graphite, the three-dimensional layered version of graphene. Lithium cobalt oxide, $LiCoO_2$ or LCO, is the most common choice for the cathode material. The non-aqueous electrolyte is typically composed of a lithium salt (such as $LiPF_6$) dissolved in a mixture of polar organic solvents such as ethylene carbonate (EC) and dimethyl carbonate (DMC).

Figure 8.1 shows the schematic of a Li-ion battery in standard discharge mode. For the Li-ion battery system in its charged state, Li^+ ions are initially stored in the anode material (e.g. graphite). During discharge, an external electrical connection is made between the two electrodes, allowing Li^+ ions to be removed from the anode and to travel through the electrolyte to insert into the cathode (e.g. $LiCoO_2$). Simultaneously, electrons are removed from the anode (oxidation) and transferred through an external circuit, creating current to power a load, and then enter the cathode (reduction). The discharge process will continue to occur until the potential difference between the two electrodes becomes too low, at which point the cell is fully discharged. In secondary batteries, external energy can be applied to transfer electrons (and hence Li^+ ions) from the cathode back to the anode, thereby oxidizing the "cathode" and reducing the "anode". This process is commonly known as charging the battery. Simply put, lithium-ion batteries are able to supply power by converting chemical energy into electrical energy through controlled, thermodynamically favorable chemical reactions. In order to be used for commercial applications, these reactions must also be highly reversible and repeatable for hundreds of times without noticeable degradation or changes to the material.

Additionally, during the initial discharge of a battery, a necessary phenomenon wherein a thin formation layer film (<50 nm) containing a complex combination of chemical species spontaneously forms on the surface of the anode electrode due to the reductive decomposition of electrolyte. This film is termed the solid–electrolyte interface (SEI), and is electrically insulating but ionically conductive. Formation of the layer occurs at the anode, which operates at voltage potentials where the electrolyte is thermodynamically unstable, creating a complex heterogeneous collection of phases and layers having many secondary interfaces. The SEI phase contains a combination of inorganic lithium-containing species, including Li_2CO_3, Li_2O, LiOH, LiX (X=F, Cl), as well as an organic species such as $ROCO_2Li$, ROLi, $(ROCO_2Li)_2$ and oligomeric compounds [10]. SEI formation is required for stable operation of the lithium-ion battery and consumes a portion of the limited Li^+ available for reversible storage. The amount of Li^+ ion consumption is dependent on a variety of factors, including the surface area of the electrode. To minimize the irreversible capacity loss (ICL) due to the film formation, it is critical that lithium-ion battery anodes are designed to have low accessible surface areas. Therefore, with the integration of GO-based materials into lithium-ion batteries, its high surface area potential can be seen as a negative and must be managed.

Graphite, the most common anode material, is able to store one Li^+ ion (and hence contribute one electron) for every C_6 ring, forming a lithiated material with the formula LiC_6. With this amount of Li^+ storage, the theoretical gravimetric and volumetric capacities can be calculated as 372 mA h g^{-1} and 830 mA h cm^{-3}. On the cathode side, $LiCoO_2$, a common representative cathode material, has a practical gravimetric and volumetric capacity of 140 mA h g^{-1} and ~330 mA h cm^{-3}, respectively. In general, the total capacity of a Li-ion cell (C_{total}) may be expressed in terms of anode and cathode capacities as follows:

$$C_{total} = \frac{1}{\frac{1}{C_A} + \frac{1}{C_C} + \frac{1}{Q_M}} \tag{8.1}$$

Here C_A and C_C are the theoretical capacities of the anode and cathode materials, respectively, and $1/Q_M$ is the specific mass of the other cell components (electrolyte, separator, current collectors, case, etc.) in mA h g^{-1} [11]. For anodes such as graphite, C_A is 372 mA h g^{-1}, and for $LiCoO_2$, C_C is 140 mA h g^{-1}. For a Sony cylindrical cell (2550 mA h, 46 g), Q_M may be calculated to be ~130 mA h g^{-1}. Using (8.1), it is possible to construct a plot of total cell capacity as a function of anode specific capacity (C_A) as shown in Figure 8.2. As can be seen from Figure 8.2, there are large improvements in total cell capacity with an increase in anode specific capacity up to ~1200 mA h g^{-1}.

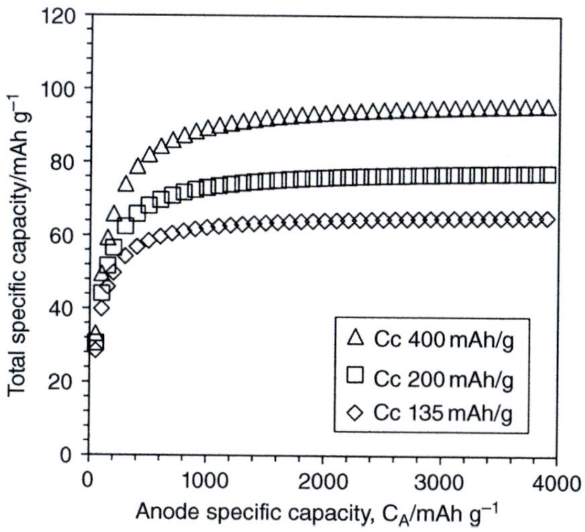

Figure 8.2 *Total capacity of a Li-ion cell as a function of anode capacity (C_A), including masses of other required internal components and case ($Q_M = 130$ mA h g^{-1}). Capacities of cathodes considered were 135, 200 and 400 mA h g^{-1}*

In order to meet the increasing demands for portable applications, new anode and cathode materials capable of storing more Li ions per unit mass and volume are highly desired. Consequently, significant attention has been paid to developing better electrode materials for advanced rechargeable Li-ion batteries. Particularly, electrode materials with significantly greater capacity, improved rate capability, better cycling stability, enhanced safety and environmental friendliness have been extensively pursued. The emergence of graphene has generated considerable attention within the Li-ion battery community, mainly due to its extremely high electrical conductivity, unique two-dimensional (2D) morphology and large surface area. Since its first reported battery application in 2008 [12], GO materials for battery applications have been extensively reported and developed. In this section, we will describe some of the important achievements and developments of GO-based materials for lithium-ion batteries and strategies that can be used to further extend their development.

8.2.3 Anode Applications

8.2.3.1 Pure Reduced Graphene Oxide Anode Materials

Graphite has been the most significant carbon form studied for commercial lithium-ion battery applications. Graphite's dominance is due to its combination of long cycle life, low cost and abundance, and decent gravimetric and volumetric capacities. However, graphite also has drawbacks that limit its use in future applications. Chiefly, the limited gravimetric and volumetric capacities afforded by graphite cannot meet the demands for next-generation lithium-ion battery applications. Therefore, in order to improve the performance of future cells, anode materials with higher capacity and discharge rate behavior must be developed.

Graphene, the highly researched monolayer of graphite, has the potential to contribute to the development of advanced anode applications due to its plethora of extraordinary properties. Most critically, graphene and its related materials are chemically stable, and exhibit high electrical and thermal conductivity, high surface area and excellent mechanical properties. Of all the possible methods to achieve "graphene-like" materials, the synthesis of GO and subsequent transformation to reduced graphene oxide (RGO) is highly regarded as one of the most promising routes due to its synthetic flexibility and scalability.

Indeed, a large number of reports on the application of GO-derived anodes exist, and the resulting performance of the materials varies greatly [13, 14]. The significant discrepancies in performance highlight the importance that the preparation methods and the resulting structural parameters play in dictating electrochemical performance. Particularly, processing parameters, such as the oxidation degree and the reduction method, as well as structural parameters, such as sheet size, functionality, surface area, electrical conductivity, dopants and porosity, are highly likely to affect the overall performance of the GO-derived material. Consequently, a critical understanding of the factors that affect electrochemical performance is required in order to understand the energy storage mechanism and advance the engineering design of graphene-based electrodes. GO reduction methods via hydrazine chemical reduction, thermal treatment and electron-beam irradiation were studied by Pan *et al.* [15]. It was determined that the thermal treatment and electron-beam irradiation processing provided higher reversible capacities for the electrode material compared to chemical reduction. The explanation for the discrepancy in performance was attributed to the degree of disorder, which arose from the reduction methods. Thermal treatment and e-beam irradiation generated structures with a higher degree of disorder, including edge and other defects that could act as additional Li-ion storage sites and provide the surplus capacity. More recently, the effect of GO oxidation degree on electrochemical performance was examined by Lee *et al.* [16] using a thermally processed GO (tpGO) electrode. To synthesize GOs with different oxidation degrees, the number of times the material was oxidized was controlled from one to three times. It was determined that, as the extent of oxidation increased, the reversible capacity of the tpGO increased from 1252 to 2311 mA h g^{-1}, for one and three oxidation cycles, respectively. The reported capacities attained were extremely large and much greater than would be expected from a pure carbon material. The large capacity values were attributed to the increased surface area as well as defect and edge sites that provide additional Li^+ storage sites.

The unique 2D sheet-like morphology of GO offers an extraordinary opportunity to form flexible and freestanding electrodes. These binder-free materials are formed via the vacuum filtration of either GO or its chemically reduced counterpart. In either case, liquid solvents of the dispersions are removed, providing a self-directed assembly of graphene-based materials along the *c*-direction. An example of the flexible, freestanding RGO paper is shown in Figure 8.3 [17]. As shown in Figure 8.3, scanning electron microscopy (SEM) shows that the freestanding RGO papers consist of densely packed, aligned sheets with a thickness on the order of 10 μm. X-ray diffraction (XRD) measurements indicated that the average interlayer spacing was 0.37 nm, which is greater than the graphite interlayer spacing of 0.334 nm. Additionally, the presence of a broad peak around 26° indicated a well-ordered but non-highly graphitized structure. When applied as an anode material, the freestanding electrode exhibited a significantly lower specific capacity of 84 mA h g^{-1} at 50 mA g^{-1} compared to 288 mA h g^{-1} for a graphene-powder control electrode. The poor performance of the freestanding paper electrode was attributed to densely packed sheets that created a kinetic barrier for the diffusion of Li ions. This behavior was later further

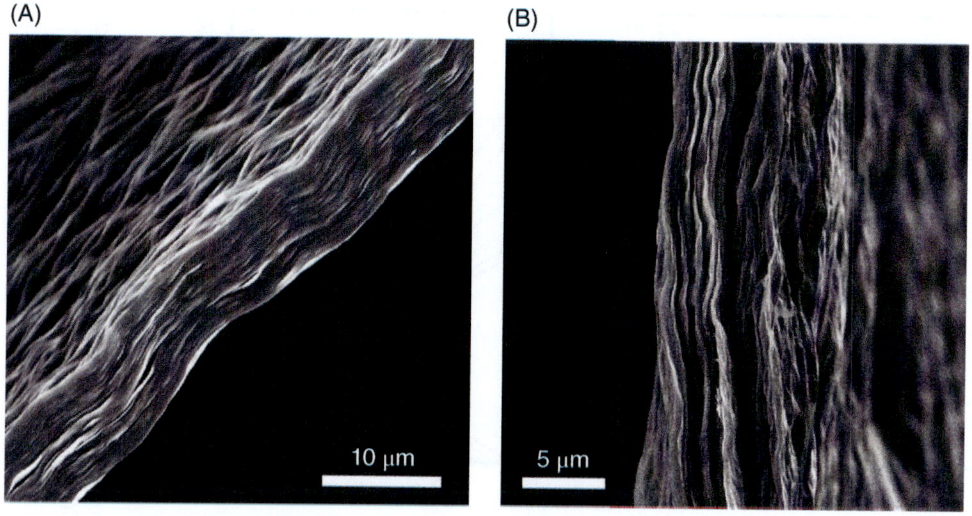

Figure 8.3 SEM images of a chemically reduced graphene oxide (RGO) paper showing the (A) rolling paper-like electrode surface and (B) non-uniformity and cracking along a fractured edge. Reproduced from [17] with permission of American Chemical Society

explored by examining the rate dependence of freestanding electrodes as a function of electrode thickness [18]. When the freestanding paper electrode thickness increased from 1.5 to 10 μm, the reversible capacity decreased from ~200 mA h g^{-1} to ~80 mA h g^{-1}, respectively. Additionally, when the cycling rate is increased from 50 mA g^{-1} to 500 mA g^{-1}, the reversible capacity of a 3 μm paper decreased from ~175 mA h g^{-1} to ~100 mA h g^{-1}, demonstrating significant rate limitations due to insufficient ion diffusion through the film. It was determined that the thinner papers always outperform the thicker films due to the dense restacking of graphene sheets and the large aspect ratio of the paper. For practical implementation into commercial applications, thick electrodes are required to provide high energy density. Therefore, it is not feasible to circumvent the issue by simply decreasing the electrode thickness.

To overcome these rate-limitation issues, researchers have developed methods to generate pores in freestanding graphene papers. With the 2D non-porous morphology, the Li^+ ion diffusion proceeds predominantly in the in-plane direction, and cross-plane diffusion is hindered due to the large aspect ratio of GO and limited defect and edge sites. By generating pores, new ion diffusion channels are generated that can substantially increase the efficiency of carbon utilization during charge/discharge. Zhao and colleagues developed a method to introduce in-plane defects into the basal planes ("holes"), providing enhanced ion diffusion pathways [19]. In-plane defects were generated by the combination of nitric acid (HNO_3) oxidation and sonication energy applied to GO suspensions (Figure 8.4).

The size and density of the pores could be tuned by the acid concentration, and could be controlled from ~5 nm to hundreds of nanometers in size. After generating the in-plane defects, the holey GO materials were vacuum filtered and thermally treated to render them as holey tpGO materials. Electrochemically, the holey tpGO electrodes exhibited higher reversible capacities of up to 403 mA h g^{-1} compared to only 336 mA h g^{-1} for the control

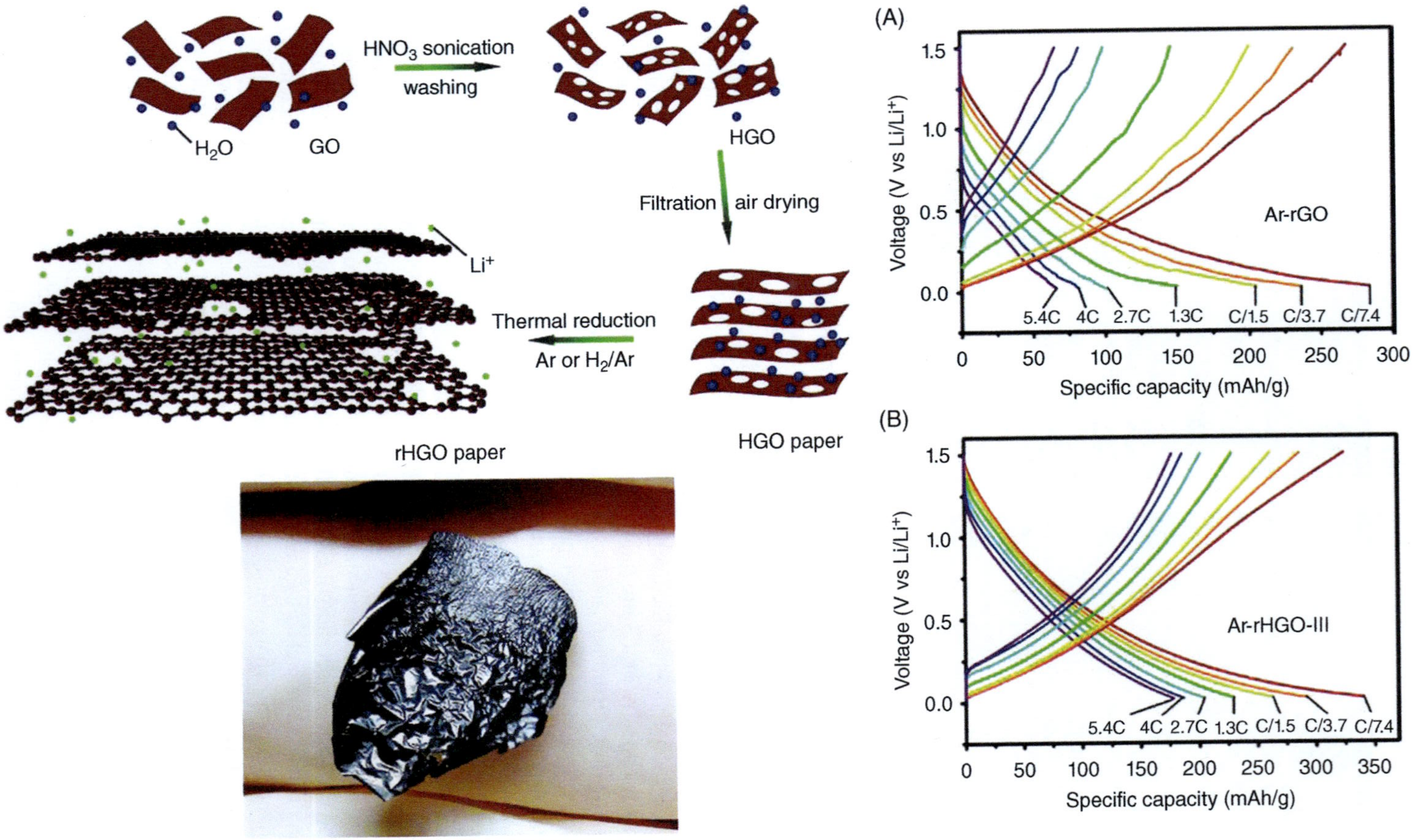

Figure 8.4 *Schematic drawing (left) of the process to introduce in-plane pores into the GO material and subsequent formation into freestanding holey tpGO paper material. The electrochemical performance (right) of (A) a traditional tpGO paper (no holes) and (B) holey tpGO paper are compared. Adapted from [19] with permission of American Chemical Society*

sample without in-plane defects. More importantly, the holey tpGO samples exhibited far superior rate performance, delivering a reversible capacity of 178 mA h g^{-1} at a 2 A g^{-1} rate compared to ~70 mA h g^{-1} for the control sample (~44% versus ~20% capacity retention). The improvement in Li-ion diffusion kinetics and charge transfer due to the introduction of the holes was confirmed via electrochemical impedance spectroscopy (EIS). Additionally, a preferred hole size of 20–70 nm was observed in order to achieve the best electrochemical performance. Further increase in hole size resulted in a comparatively lower capacity, perhaps due to decreased electrical conductivity resulting from the highly defective structure. Nonetheless, this report highlighted the novel approach to generate holes into the basal plane of graphene materials as an effective method to improve the energy and power density of lithium-ion batteries.

Another effective strategy to increase the energy and power density of graphene-based materials and to tailor the physicochemical properties is to chemically dope the materials. One simple method to accomplish this is to introduce a dopant precursor gas during a high-temperature thermal treatment step to fabricate tpGO electrodes [20]. In one instance, nitrogen and boron doping was successfully achieved using ammonia (NH_3) and boron trichloride (BCl_3) as precursor molecules. Spectroscopic analysis confirmed that the N species existed in pyridinic and pyrrolic forms, whereas the boron existed as BC_3 and BC_2O domains within the doped structure. In both cases, the dopant species predominantly substituted carbon atoms on defect or edge sites, as shown in Figure 8.5. Electrochemically, the doped materials exhibited improved performance compared to the pristine, undoped sample. The N- and B-doped electrodes achieved reversible capacities of 1043 and 1549 mA h g^{-1}, respectively, compared to 955 mA h g^{-1} for the pristine sample. Additionally, the doped samples offered significantly better rate performance and capacity retention. The improved energy and power density of the doped materials was attributed to the increased electrical conductivity of the samples, as well as due to a more disordered surface morphology that provides additional Li^+ storage sites.

At this point, it is important to recall from section 8.2.2 that, during the initial cycling formation of a practical battery, a complex thin film is formed on the anode electrode

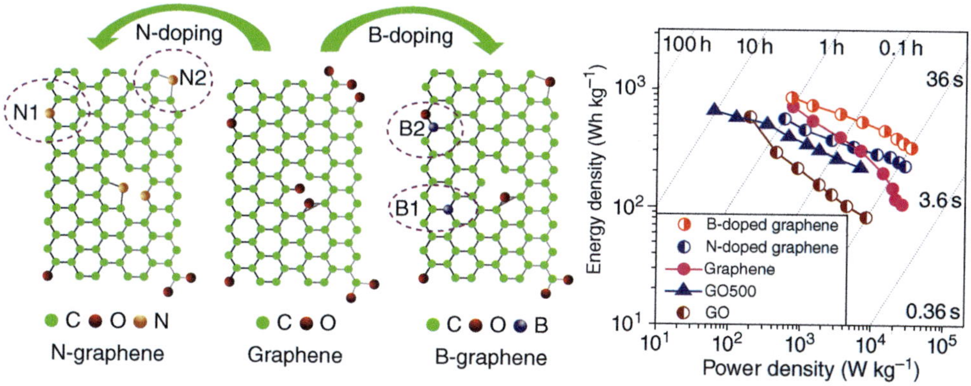

Figure 8.5 *Schematic showing the nitrogen and boron doping into the graphene plane, and resulting high energy and power performance of doped materials compared to the undoped control samples. Reproduced from [20] with permission of American Chemical Society*

surface, consuming limited Li^+ ions and leading to a decrease in overall energy density. This SEI layer contains a combination of inorganic Li^+-containing species, as well as a variety of oligomeric organic species. This film passivation layer is formed on all accessible surfaces, and, consequently, the larger the surface area of an electrode material, the greater the loss of capacity in a practical cell. Indeed, for the previously described study of oxidation extent by Lee and colleagues [16], the highest oxidized sample irreversibly consumed 55% of the Li^+ charges upon initial cycling due to its high surface area of 991 m^2 g^{-1}. In order to minimize these losses, it is important to keep the electrode surface area low, often <5 m^2 g^{-1}. Therefore, for composites that integrate graphenic materials, it is critical to determine the overall electrode surface area to ensure the values are reasonable. High-surface-area anodes will exhibit large ICLs in the initial few cycles and will translate into practical cells with low reversible energy densities. For this reason, graphene-based materials are often combined with electroactive materials that can store much greater charge in order to develop hybrid materials with synergistic properties.

8.2.3.2 Reduced Graphene Oxide Composite Anode Materials

To create optimal high-performance energy storage materials, graphene is commonly integrated as part of a composite material, wherein it serves a synergistic purpose to enhance the overall properties of an electrode. These composite structures are able to highlight and emphasize attractive GO properties while mitigating some of the shortcomings of RGO by itself. The formation of a continuous three-dimensional (3D) electrically conductive network created by the graphenic structure can substantially improve the battery's properties via enhanced electron and ion transportation within the electrode materials. Indeed, it is clear from section 8.2.3.1 that there is a critical relationship between graphenic architecture and its available chemical, conductive, mechanical and other properties. Additionally, improved electrodes containing advanced anode materials allow for greater storage of Li^+ ions and therefore increased battery capacity. Graphene's high thermal conductivity can also dissipate local hot spots and resistive heating generated within the electrode during operation, thereby stabilizing battery temperature and improving cell safety. Therefore, electrodes containing forms of graphene (included RGO, annealed tpGO and others) have the potential to last significantly longer than conventional batteries. Hybrid structures can therefore be tuned for certain applications requiring a combination of long cycle life, energy delivery and rate performance.

Silicon is considered as one of the most attractive next-generation anode materials to form composite electrodes with RGO, because it is inexpensive, abundant and possesses the highest gravimetric capacity (3579 mA h g^{-1} for $Li_{15}Si_4$) known – nearly 10 times higher capacity than the incumbent graphite anode. Unfortunately, silicon undergoes a large volume change (up to 300%) during cycling due to the accommodation of the Li^+ ions required during operation. This extreme and repeated fluctuation causes severe pulverization of silicon particles, leading to particle isolation and significant capacity fade within dozens of cycles. Therefore, the conventional cycle life of silicon-based anode materials is much too short for commercial applications. To overcome the large volume change and obtain better capacity retention for Si anodes, various approaches have been used. These approaches include the use of nanoparticles [21], thin films [22, 23], nanowires [24], dispersing into an inactive/active matrix [25, 26], preparing 3D porous structures [27, 28]

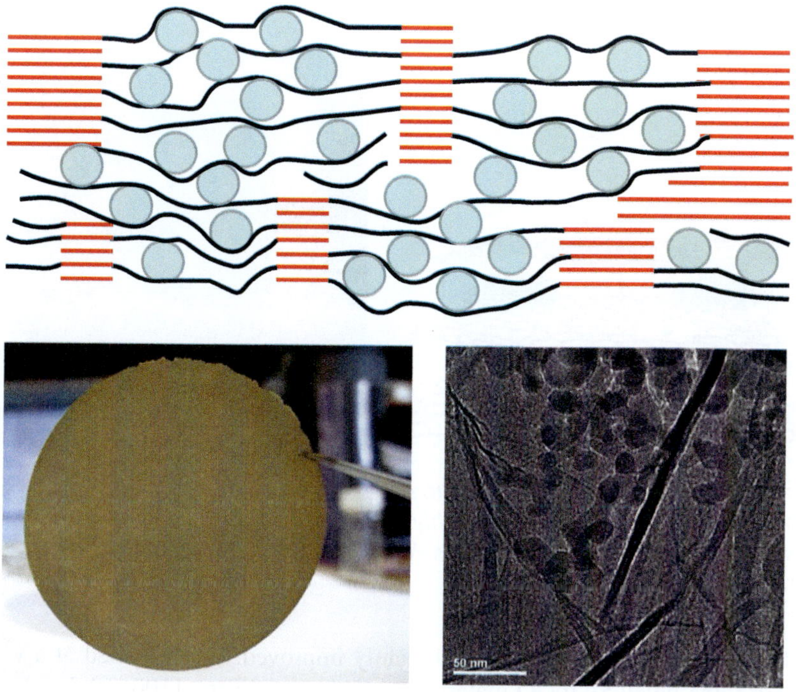

Figure 8.6 Cross-sectional schematic (not to scale) of the Si–tpGO composite structure (top). Digital image of prepared Si-GO composite paper (bottom left). Transmission electron microscopy (TEM) image of Si nanoparticles encapsulated between tpGO sheets (bottom right). Adapted from [31] with permission of The Royal Society of Chemistry

and carbon coating [29, 30]. Graphene, however, with its flexible 2D morphology, high surface area and good electrical conductivity, has many properties that lends itself to be a very promising composite material to improve the stability of the silicon system.

One of the first reported Si-RGO systems was developed by Lee and colleagues by combining silicon nanoparticles with the freestanding graphene paper structure [31]. Silicon nanoparticles were mixed with an aqueous GO suspension, vacuum filtered and thermally treated under inert gas to form flexible Si–tpGO composite paper anodes. With this format, the graphene sheets formed a 3D electrically conductive network that also provided a mechanically robust scaffold to anchor and stabilize the embedded silicon nanoparticles (Figure 8.6). The resulting Si–tpGO materials exhibited significantly enhanced cycling stability compared to bare silicon particles. The Si–tpGO materials were able to achieve reversible capacities >2500 mA h g^{-1} at 1 A g^{-1} and were able to retain nearly ~60% capacity after 300 cycles. The improved performance was attributed to the unique 3D network that buffered the volumetric expansion of Si during cycling as well as the high electrical conductivity provided by the network. Later, Zhao *et al.* introduced in-plane carbon vacancies into the GO structure prior to composite formation in order to provide additional ion diffusion pathways to increase the energy and power density of the composite material (Figure 8.7) [32]. As expected, the holey Si–tpGO composite materials offered enhanced reversible capacity compared to the non-holey counterpart. Notably, the rate performance

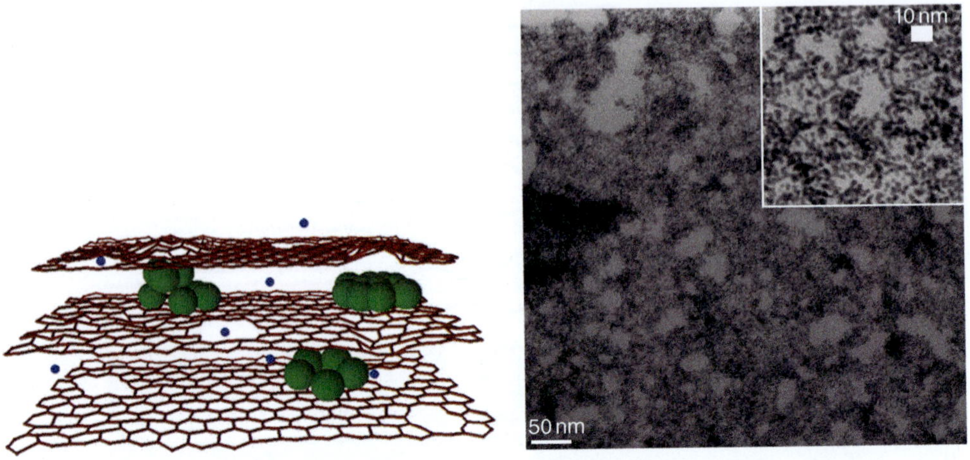

Figure 8.7 Cross-sectional schematic (not to scale) of the holey Si–tpGO composite structure (left). TEM image (right) depicting the highly defective tpGO structure (scale bar: 50 nm). Inset: Zoomed-in view of the tpGO structure. Adapted from [32] with permission of Wiley-VCH Verlag GmbH

of the holey Si–tpGO materials was significantly improved. When cycled at a very high rate of 8 A g^{-1}, the holey Si–tpGO material was able to deliver ~1100 mA h g^{-1} capacity, which was 34% of the capacity that could be delivered at 1 A g^{-1}.

Another promising route to the synthesis of composite materials is through the fabrication of RGO-encapsulated nanoparticle composites. Taking advantage of the anionic charge of GO in aqueous suspensions, co-assembly with cationically charged nanoparticles can lead to an encapsulated structure. This co-assembly process is driven by mutual electrostatic interactions and results in a flexible and atomically thin GO shell that effectively envelops the nanoparticle. This unique structure provides a number of advantages, including suppression of particle agglomeration, accommodation of volumetric change during cycling and fast electron transport via the highly conductive graphene wrapping. One approach is shown in Figure 8.8 and can be accomplished by modifying the surface of metal oxide nanoparticles with an available amine species [33]. This approach is especially useful for metal oxide anode materials, which suffer from numerous issues that limit their cyclability, including substantial volumetric change during cycling and low electrical conductivity. In one example, cobalt oxide (Co_3O_4) particles were treated with aminopropyltrimethoxysilane (APS) to generate a positively charged surface, followed by co-assembly with the negatively charged GO in aqueous suspension. The resulting GO-wrapped Co_3O_4 particles were then treated with hydrazine to chemically reduce the GO material to an electrically conductive support (Figure 8.8). The electrochemical performance of the co-assembled composites demonstrated far superior performance compared to bare Co_3O_4 particles. The Co_3O_4-RGO composite achieved a very high and stable reversible capacity of ~1100 mA h g^{-1} and were nearly completely stable for over 130 cycles. The superior performance was attributed to the synergistic effect between the flexible RGO wrapping that stabilized the particles, and EIS confirmed that the composite possessed lower contact and charge-transfer resistances than the bare Co_3O_4 particle electrode. As an additional control, a physical mixture of Co_3O_4 and RGO was constructed and showed worse capacity and stability compared to the co-assembled structure.

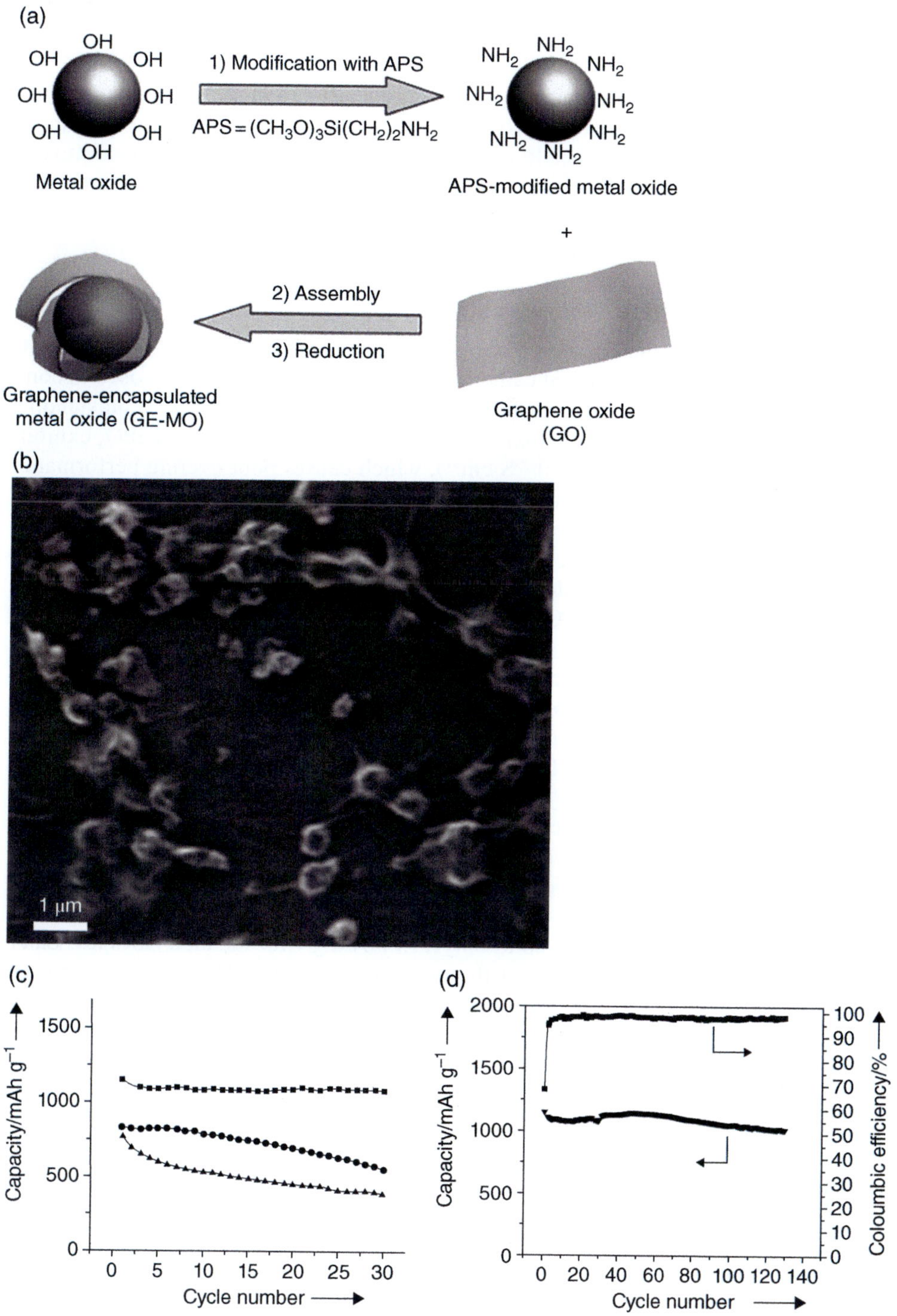

Figure 8.8 (a) Fabrication of the RGO-wrapped metal oxide particles. (b) SEM image demonstrating the Co_3O_4 particles embedded in the flexible RGO structure. (c) Electrochemical cycling of co-assembled Co_3O_4-RGO composites (squares), physically mixed Co_3O_4-RGO composite (circles) and bare Co_3O_4 electrodes (triangles). (d) Cycle performance of the co-assembled Co_3O_4-RGO electrode for 130 cycles at 74 mA g^{-1}. Adapted from [33] with permission of Wiley-VCH Verlag GmbH

8.2.4 Cathode Applications

In addition to enhancing features on the anode electrode, RGO and its panacea of attractive properties can also be applied to improve the electrochemical performance of cathode materials. For cathode applications, RGO often serves as an improved carbon additive to replace activated carbon, due to its extremely high conductivity and surface area. The combination of those two properties makes RGO a wonderful material to wire cathode materials, which typically suffer from low electrical conductivity themselves.

8.2.4.1 Transitional Metal Phosphates

One such material that has garnered significant attention is lithium iron phosphate ($LiFePO_4$). $LiFePO_4$ is an important cathode material that has been widely used in commercial lithium-ion batteries due to its good capacity (~170 mA h g^{-1}), outstanding reversibility, inexpensive cost, low toxicity and high safety. However, $LiFePO_4$ exhibits very low electrical conductivity (10^{-9} S cm^{-1}), which causes poor cycling performance if not properly designed. Traditionally, nanostructuring and carbon coating are employed in order to enhance the electrochemical performance of $LiFePO_4$. However, with traditional carbon-coating procedures, an amorphous carbon with limited electrical conductivity coats the surface of the particle. Therefore, significant room for improvement can be achieved by structuring $LiFePO_4$ with much more electrically conductive graphenic materials.

The first demonstration of a $LiFePO_4$–RGO composite was reported by Ding and colleagues in 2010 [34]. In the seminal work, $LiFePO_4$ nanoparticles were co-precipitated with RGO to form a composite structure. The material was then isolated and thermally treated at 700°C under inert gas to form the final product. It was determined that inclusion of the RGO material resulted in a decrease in the $LiFePO_4$ particle size to 20 nm, compared to 70 nm for $LiFePO_4$ particles synthesized under identical procedures except for RGO addition. Electrochemical cycling of the materials demonstrated that the reversible capacity of the $LiFePO_4$ was considerably improved with RGO addition. When cycled at a rate of 35 mA g^{-1}, the $LiFePO_4$–RGO composite achieved near-theoretical reversible capacities of 160 mA h g^{-1} compared to only 113 mA h g^{-1} for the bare particles. Later, Zhou and colleagues further improved the performance of $LiFePO_4$–RGO composites using a novel composite structure formed via spray drying and thermal annealing processes [35]. Through this process, $LiFePO_4$ primary nanoparticles were structured into micro-sized spherical secondary particles, which were wrapped by an electrically conductive thermally RGO 3D network (Figure 8.9). This unique RGO wrapping facilitated electron transport throughout the structure, while the voids between primary particles provided easy routes for Li^+ diffusion. Electrochemical analysis confirmed that the RGO-wrapped materials exhibited significantly better cycling stability and rate capability compared to an amorphous carbon-coated composite. Surprisingly, combining the amorphous carbon-coated $LiFePO_4$ material into the unique process scheme afforded an even better-performing material than just achieved with the RGO wrapping alone, demonstrating that there may have been rate limitations arising from individual $LiFePO_4$ particle connections. For the RGO-wrapped, carbon-coated $LiFePO_4$ sample, a maximum reversible capacity of ~150 mA h g^{-1} was achieved at a low rate of 0.1C (~15 mA g^{-1}), which is noticeably lower than the theoretical value (here C is the C-rate related to battery charging). However, these

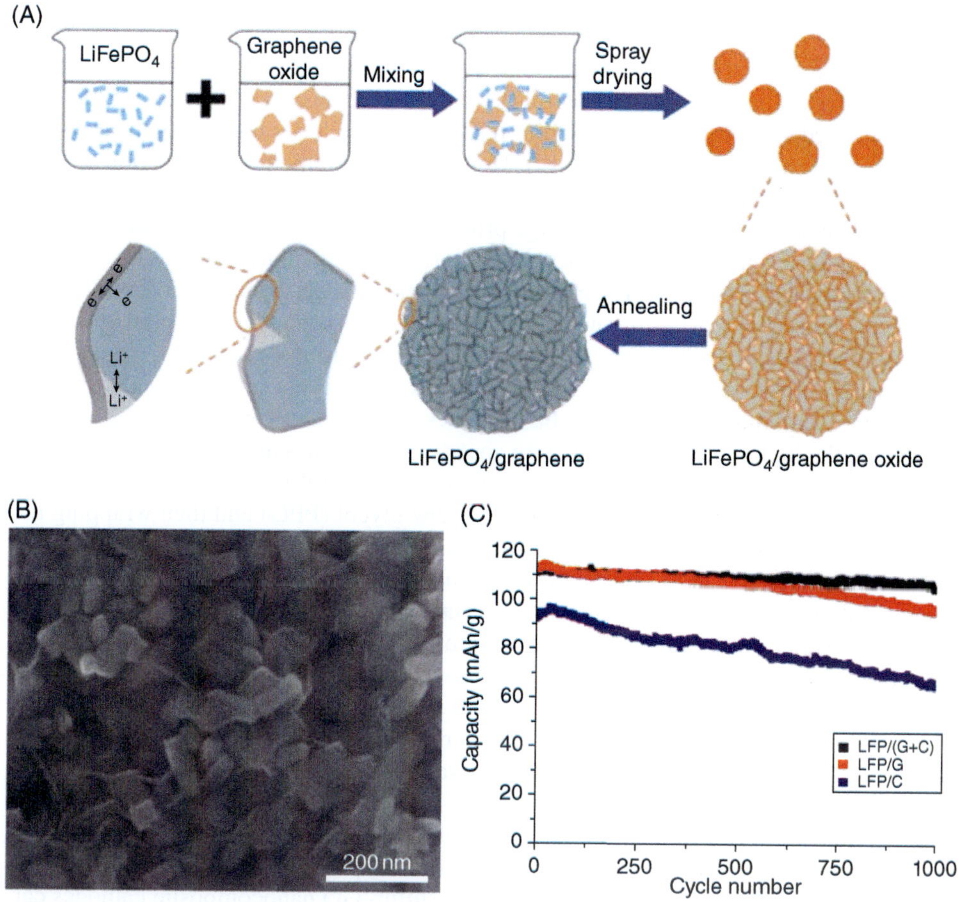

Figure 8.9 *(A) Preparation process and microscale structure of $LiFePO_4$–RGO composite. (B) SEM image showing $LiFePO_4$ primary particles wrapped by the RGO structure. (C) Electrochemical cycling of $LiFePO_4$ composite electrodes at a 10C–20C asymmetric charge–discharge process. Adapted from [35] with permission of The Royal Society of Chemistry*

composite materials exhibited an excellent rate performance, still delivering ~130 mA h g^{-1} at 10C (~1.7 A g^{-1}) and over 70 mA h g^{-1} at an extremely high discharge rate of 60C (~10 A g^{-1}). Additionally, these composites were able to retain ~95% capacity after 1000 cycles at a high rate of 10C–20C (1.7–3.4 A g^{-1}) asymmetric charging–discharging procedure.

Although these mentioned reports are limited only to $LiFePO_4$, numerous other cathode materials such as $LiMn_2O_4$, $LiCoO_2$, $LiNi_xMn_yCo_zO_2$ and $LiMnPO_4$ can also be improved using similar methodologies. In order to sufficiently improve the electrode performance in composite structures, it is critical to ensure that the graphene structure is well distributed and that electroactive particles are intimately connected to the graphene layers. Often this can be successfully achieved through *in situ* nanoparticle synthesis and thermal annealing processes.

8.2.4.2 Lithium–Sulfur Batteries

The unique properties of GO have proven to be even more useful for more advanced cathode materials such as sulfur. The lithium–sulfur (Li–S) battery is an especially attractive system that has received significant attention from researchers due to its high theoretical energy density (2600 W h kg^{-1} and 2800 W h l^{-1}). However, the commercial application of Li–S batteries has been hindered due to a few issues, including the insulating nature of sulfur, volume change during the charge–discharge process, and the severe capacity fade observed due to high polysulfide dissolution during cycling (termed the "shuttle effect").

In order to overcome these issues, the 2D structure, high conductivity, elastic and mechanically strong properties of GO and RGO have been explored as potential options to bring Li–S batteries to market. One of the most promising routes is to use the unique 2D structure as a carbon net to wrap the sulfur particles and prevent polysulfide dissolution. The oxygen-containing groups on the basal planes can further provide strong anchoring points while the high-surface-area structure absorbs polysulfides and promotes the formation of Li_2S films. In 2011, Wang and colleagues reported a sulfur–GO composite that was formed by coating sulfur particles with polyethylene glycol (PEG) and then wrapping the particles with lightly oxidized GO sheets that had been decorated with carbon particles (Figure 8.10) [36]. The PEG and GO coating layers served multiple roles, including to accommodate the volumetric expansion of sulfur, trap soluble polysulfide intermediates and provide an electrically conductive network for the sulfur particles. The resulting composites demonstrated high reversible specific capacities up to ~600 mA h g^{-1} after an initial capacity drop from 1000 mA h g^{-1}, and remained stable over 100 cycles when cycled at up to 0.5C. The performance of the composite material was superior to both sulfur–PEG and sulfur–GO electrodes alone, demonstrating the validity of the synthesized architecture to provide robust support for the sulfur cathode.

More recently, CTAB (cetyl-trimethyl ammonium bromide) has been utilized as an alternative surfactant to PEG to improve the electrochemical performance of GO/S composite cathodes (Figure 8.11) [37]. The CTAB-modified sulfur–GO nanocomposite cathodes can achieve close to theoretical capacity values of ~1400 mA h g^{-1} when cycled at 0.2C and up to ~800 mA h g^{-1} capacity at a high rate of 1C. Additionally, the composites are extremely robust and were able to cycle more than 1500 times with more than 300 mA h g^{-1} remaining capacity. It was estimated that the practical cell-level energy density of the cell could be as great as ~500 W h kg^{-1}, more than doubled compared to the existing energy density of current Li-ion cells (~200 W h kg^{-1}). The surprising rate performance of the insulating sulfur material suggests that these materials may also be suitable for high-power applications in addition to high-energy applications.

Sulfur–RGO composites have also been significantly investigated since RGO possesses similar advantages to GO while also providing the addition of electrical conductivity. Very recently, Wang and colleagues [38] reported a method to impregnate lithium polysulfide (Li_2S) into thermally annealed RGO papers (Figure 8.12). Similar to its anode counterparts, the Li_2S–tpGO papers are flexible and freestanding and can be used directly as cathode electrodes without requiring additional binders, which can lead to substantial weight savings if implemented into a commercial cell. Due to the method of impregnation, the lithium sulfide particles were formed as very small 25–50 nm nanospheres rather than the micrometer-sized particles observed without the presence of tpGO. With the freestanding

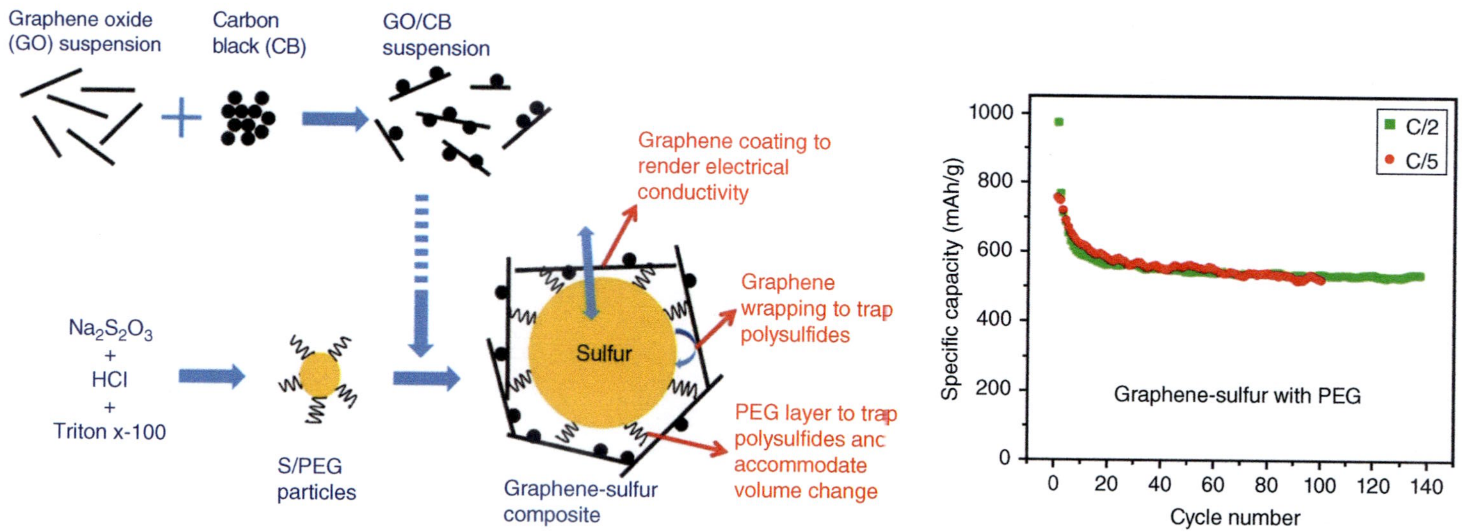

Figure 8.10 *Schematic representation of the synthesis scheme to fabricate the sulfur–PEG–GO architecture (left). Cycling performance of the composite at rates of C/5 and C/2. Reproduced from [36] with permission of American Chemical Society*

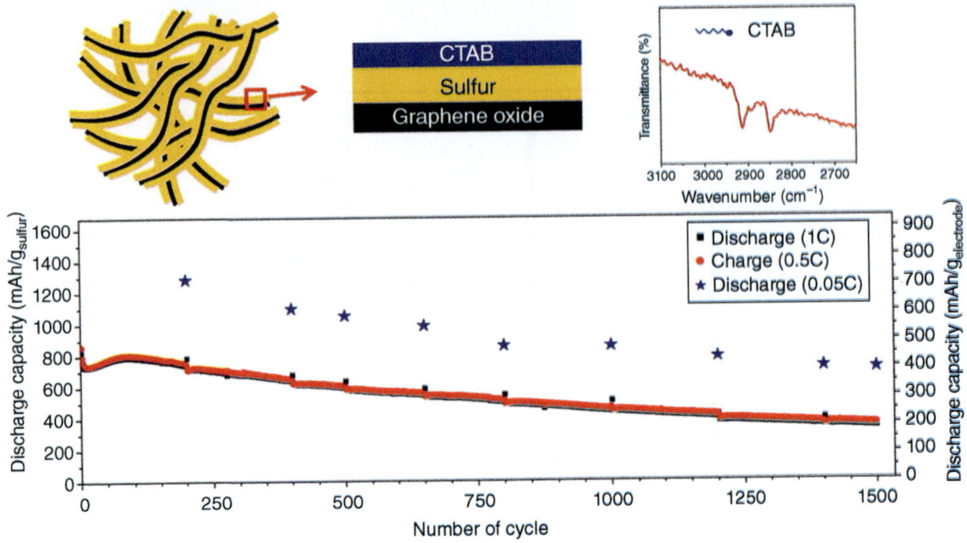

Figure 8.11 Schematic of the CTAB-modified sulfur–GO composite material (top) and accompanying composite electrochemical cycling performance (bottom). Reproduced from [37] with permission of American Chemical Society

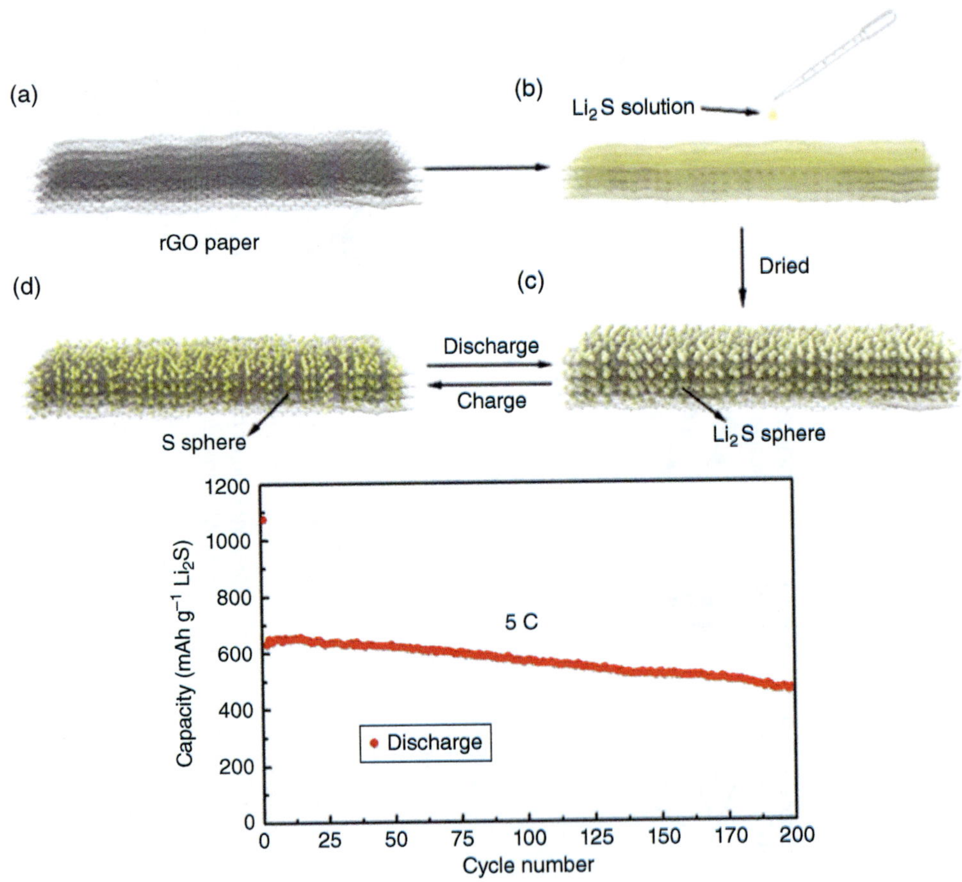

Figure 8.12 Illustrative schematic depicting the fabrication of the freestanding Li_2S–tpGO composite paper (top). Electrochemical cycling performance of the Li_2S–tpGO cathode when cycled at 5C (~6 A/g^{-1}). Reproduced from [38] with permission of American Chemical Society

structure, the high surface area of the tpGO paper provides good polysulfide retention, and the porous nature can accommodate the volumetric changes required during cycling. The prepared composites exhibited excellent cycling characteristics, initially achieving the full theoretical capacity of Li_2S (~1166 mA h g^{-1}) indicating complete participation of the electroactive material. Additionally, it was determined that the composites had excellent rate performance and could still deliver ~600 mA h g^{-1} at a very high rate of 7C (~8 A g^{-1}), more than 50% of the theoretical value. Moreover, when continuously cycled at a high rate of 5C (~6 A g^{-1}), the composite was able to cycle more than 200 times while maintaining ~70% of its initial capacity. Not only does this particular report demonstrate an impressive high-capacity and high-rate cathode material, but it also extends the tremendous versatility that can be achieved using the 3D flexible RGO paper support.

8.2.5 Emerging Applications

Besides showing great promise for conventional anode and cathode applications, GO and its derivatives have also recently been commissioned for more novel forms of lithium-ion battery applications. One such application was termed the "all-graphene battery" due to its integration of GO materials for both the anode and cathode electrodes [39]. The thermal processing temperature of GO was tuned to control the extent of oxygen moieties for each electrode. This energy storage system utilized a partially reduced tpGO as the cathode material and a more reduced tpGO as the anode material. Figure 8.13 illustrates the cell design and mechanism for charge storage on anode and cathode electrodes. The partially reduced tpGO cathode material contained a significant amount of carbonyl (C=O) functional groups that served as redox centers for Li^+ charge storage at high potentials. The electrode stability was attributed to C=O/C–O Faradaic surface reactions that resembled pseudo-capacitive storage mechanisms. This format allowed the system to utilize fast surface reactions on both electrodes, leading to high simultaneous energy and power density. The system was able to deliver an energy density of 225 W h kg^{-1}, and could also provide a high power density 6450 W kg^{-1}. When cycled at 500 mA g^{-1}, the system was able to deliver >50% of its initial capacity after 2000 cycles (Figure 8.13). The performance of this system demonstrates the use of GO to form hybrid energy storage systems that merge the conventional storage mechanisms of batteries and capacitors.

A second emerging use of GO exists as the battery separator membrane for improved safety. A battery separator is the insulating barrier that separates the anode and cathode electrodes and provides pathways for Li^+ ions to travel between electrodes. In a conventional cell, the separator material consists of a porous polymer material that allows electrolyte and Li^+ ion diffusion. As a safety feature, if a cell heats up to a predetermined elevated temperature, the polymer separator material undergoes an irreversible phase transition that closes its pores, which blocks Li^+ ion diffusion and terminates cell operation. Shen and colleagues demonstrated that an insulating GO film could also be used as a separator material [40].

Additionally, the GO films were functionalized with a temperature-sensitive polymer that can undergo a reversible conformation change based upon its temperature (Figure 8.14). The temperature-sensitive polymer, a poly(sulfobetaine) (PMABS), is grafted onto the GO sheets through surface-initiated atom transfer radical polymerization reaction. PMABS exhibits an upper critical solution temperature (UCST), and thus effectively reversibly

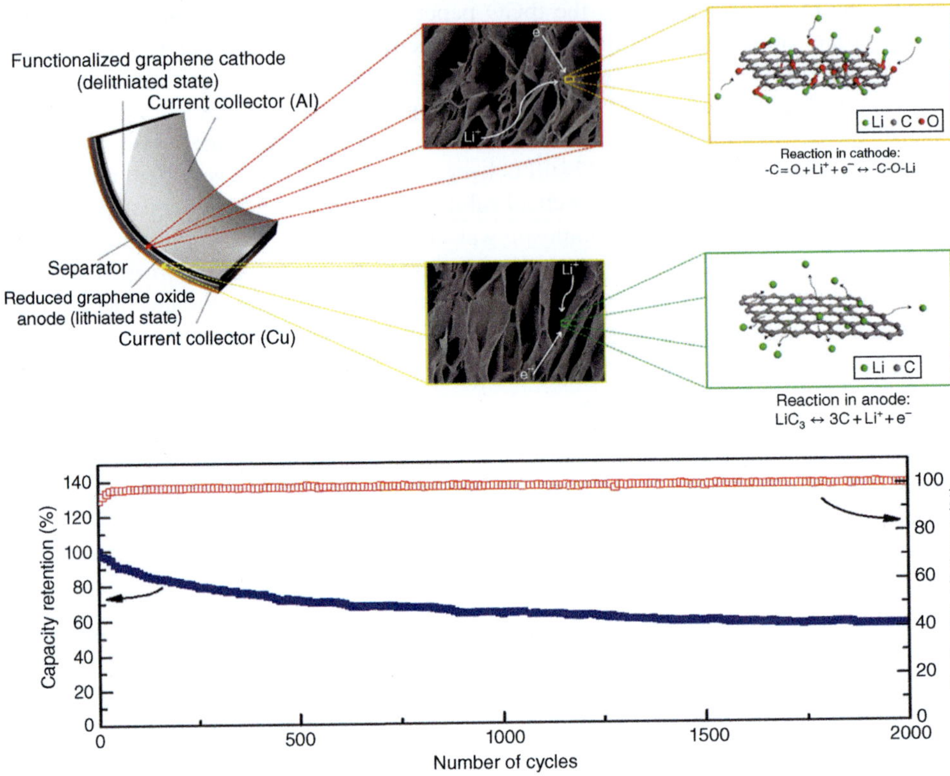

Figure 8.13 *Illustrative schematic depicting the fabrication of the "all-graphene battery" (top). Electrochemical cycling performance of the system when cycled at 500 mA g^{-1} (bottom). Reproduced from [39] with permission of Macmillan Publishers Limited*

changes its permeability and its adsorptive properties above or below its UCST point. When the temperature of the cell containing the GO-PMABS separator extends above the critical point (80°C), the permeability to Li^+ ions is considerably reduced, which leads to a significant decrease in capacity. Shen and colleagues demonstrated that this phenomenon is reversible, and lowering the cell temperature to 20°C returned the cell to its original charge capacity (Figure 8.14). Overall, it was demonstrated that the GO-PMABS membrane was capable of decreasing the storage capacity of the cell by >50% upon heating the cell from 20°C to 80°C. For energy storage applications, this novel GO membrane is useful as a means to mitigate thermal runaway in batteries by decreasing Li^+ permeability at elevated temperatures, and can also act as an internal cell balance to restrict local ion flow in regions that exhibit concentrated heating.

In summary, GO-derived electrodes offer significant promise for applications in advanced lithium-ion battery applications. An extremely eclectic mix of interesting approaches have been undertaken, ranging from flexible freestanding papers to complex hierarchical 3D porous composites. Graphene oxide's unique properties render it useful for all types of anode and cathode materials, with different structural properties being advantageous for individual applications. Additionally, its amphiphilic properties afford the unique

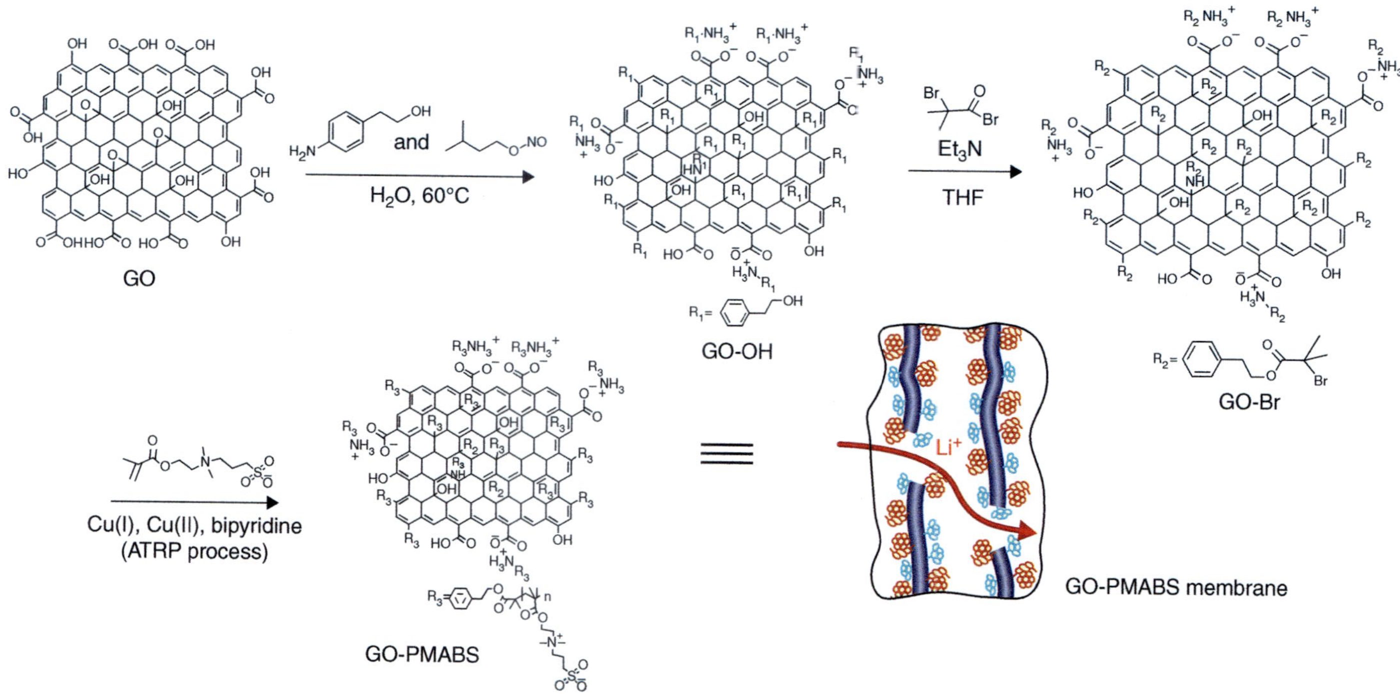

Figure 8.14 Synthetic scheme for the preparation of GO functionalized with PMABS to form a GO-PMABS membrane (top). Electrochemical cycling at 20°C and 80°C demonstrating the reversible charge storage behavior of the thermally responsive membrane (bottom) when cycled at 200 mA g^{-1}. Adapted from [40] with permission from The Royal Society of Chemistry

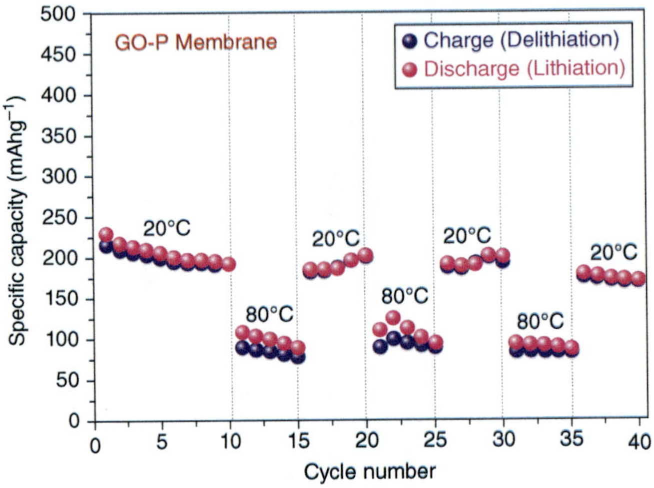

Figure 8.14 (Continued)

opportunity to conduct a variety of synthetic strategies to grow an assortment of particles. Moreover, the unique 2D structure with large sheet size can accommodate a variety of particle morphologies. Due to the extreme interest in the topic, the examples described herein represent only a fraction of the effort that has been applied to solve the world's energy storage problems. However, the overarching themes presented demonstrate the improvements that are possible when microstructural engineering approaches result in well-designed electrode materials.

8.3 Supercapacitors

8.3.1 Introduction

Supercapacitors are complementary to lithium-ion batteries in many respects and have attracted significant attention owing to their high power density and long cycle life. They are a new type of electrical energy storage system that can be applied for harvesting energy and to deliver bursts of high power within short time periods. Compared to lithium-ion batteries, which are ideal in applications required to be powered on the order of minutes and hours, supercapacitors are ideal for applications that must have very short bursts of high power delivery on the order of seconds. These properties have attracted attention for a wide variety of applications, including regenerative braking for electric vehicles, power grids and aerospace applications, as well as the growing fields of consumer devices such as wearable technology. The global market size for supercapacitors was $3.8 billion in 2014 and is expected to grow at over 15% annually into 2020 as capacitors are increasingly used in the transportation and consumer electronics markets [41].

Supercapacitors offer a combination of high power density (~10 kW kg^{-1}), excellent cycling stability and rapid response to external load sources. The rapid response of supercapacitors allows them to be implemented in conjunction with batteries where they can serve as charge conditioners assisting with load balancing and optimize the method to charge batteries. However, compared to other forms of electrical energy storage, capacitors suffer from relatively low energy density (<10 W h kg^{-1}), which is more than an order of magnitude lower than the rechargeable lithium-ion battery (200 W h kg^{-1}). Significant effort has been undertaken to increase the storage capabilities of supercapacitors. Many reports focus on investigating assorted structures of porous carbons, including activated carbons, aerogels, carbon nanotubes and mesoporous carbons. Nonetheless, practical implementation of improved supercapacitor performance has been elusive. This section will focus on the recent developments in electrodes utilizing GO materials.

8.3.2 Electrochemistry Fundamentals

Supercapacitors are a type of electrochemical capacitor that can store electrical energy via two different mechanisms – electrical double-layer capacitance (EDLC) and pseudo-capacitance. With the EDLC mechanism, electrical storage of energy is achieved via separation of charges at the surface of electrode materials in a Helmholtz double layer. Therefore, the charge is stored electrostatically and leads to performance that is extremely reversible for hundreds of thousands of charge–discharge cycles (>100 000 cycles). In order to maximize the storage by this method, the electrodes must be designed to have extremely high specific surface area and good electrical conductivity for rapid cycling. Therefore, high-surface-area porous carbons are the material of choice for EDLC capacitors.

In contrast to EDLC, with pseudo-capacitance the charge storage is achieved through Faradaic charge-transfer redox reactions between the electrolyte and electroactive species on the electrode surface. Most commonly, this phenomenon is achieved with conducting polymers such as polyaniline (PANI) and polypyrrole (PPy), or with metal oxide materials, such as Co_3O_4, MnO_2 and RuO_2. Generally, these reactions can provide significantly higher capacitance, up to 1000 F g^{-1}, which is ~10× greater than is typically achieved by the EDLC mechanism alone. Unfortunately, these reactions can often degrade the electrode material much quicker, leading to severe capacitance degradation. In addition, many pseudo-capacitive materials suffer from low power density due to the relatively low electrical conductivity compared to EDLC materials. In order to apply these materials for capacitor applications, pseudo-capacitive materials are often combined with porous carbon materials to form hybrid materials that can provide the desired combination of high capacitance, good electrical conductivity and adequate cycling stability.

It should be noted that both of the storage methods used for supercapacitors are different from the electrochemical redox reactions employed for the lithium-ion battery storage mechanism. In simple terms, lithium-ion batteries store charge in the bulk of the electrode materials, allowing for high-energy applications, whereas supercapacitors store charge predominantly on the surface only, allowing for high power applications.

Supercapacitors, similar to lithium-ion batteries, consist of two electrodes that are physically separated by a separator and are connected by an ionically conductive electrolyte. When an external voltage is applied, the electrode surfaces are polarized and the ions in the electrolyte form an electric double layer consisting of opposite-polarity ions to the electrode's polarity. As an example, a positively polarized electrode surface will attract a layer of negative ions at the electrode–electrolyte interface followed by an additional layer of positive ions (cations) to act as a charge-balancing layer (hence the "electrical double-layer" terminology). With the negatively polarized electrode surface, a layer of positive ions will be present at the electrode–electrolyte interface, followed by an additional adsorbed layer of negative ions to balance charges.

In general, the total capacitance (C_{total}) of supercapacitors can be expressed in terms of the capacitance of the two individual capacitors (C_A and C_B) as follows:

$$C_{total} = \frac{C_A \times C_B}{C_A + C_B} \tag{8.2}$$

Supercapacitors allow for either symmetric or asymmetric electrode design. For symmetric design, both of the electrodes have the same capacitance value and hence C_{total} is equal to half the value for a single electrode (i.e. $C_{total}=0.5C_A$). In the case of asymmetric design, C_{total} will be limited by the electrode with the smaller capacitance due to the mathematical equation (8.2), and is therefore roughly taken to be the capacitance value of the smaller electrode (i.e. $C_{total} \approx C_B$ if $C_A \gg C_B$).

The energy stored in a supercapacitor, E, can be calculated according to the following relationship:

$$E = \frac{1}{2}CV^2 \tag{8.3}$$

where C is the specific capacitance and V is the voltage potential window for operation. As can be seen, energy is a function of the cell capacitance and voltage. Therefore, in order to improve the energy delivery of supercapacitors, either the capacitance or the cell voltage can be modified to allow for greater performance. From (8.3), it is apparent that increasing the accessible voltage window provides a more significant opportunity to improve the energy density due to its quadratic relationship. However, the cell voltage is determined more by the electrolyte system, and the various carbon materials are found to have a lower influence. This section will not examine methods to broaden the potential window. Rather, this section will focus on improving cell capacitance by developing new electrode materials as the key to provide capacitor energy delivery. Therefore, this section will describe how various graphene-containing structures can be developed to maximize its specific capacitance (F g^{-1}), rate performance and cycle life. Particularly, this section will highlight some of the important achievements in GO-based materials for supercapacitors and strategies that can be used to further their development.

8.3.3 Carbon-only Electrodes

There are a number of crucial factors that are necessary in order to create promising electrode performance. Supercapacitor electrodes must have high specific surface area, appropriate pore size distribution, high electrical conductivity and good surface wettability. Additionally, electrodes must have high thermal stability, long-term chemical stability, high corrosion resistance, low cost and be environmentally friendly. Due to the capacitive energy storage mechanism, the entire electrode surface needs to be wetted by the electrolyte in order for it to store charge. Additionally, in order for the electrolyte to reach all the accessible surfaces, pores must be large enough to accommodate the variety of electrolyte compositions that may be employed. Therefore, pore sizes and additional ion pathways must be considered and characterized when optimizing the performance of the supercapacitor electrodes.

In general, the quantity of electrical double-layer and pseudo-capacitive charges are predominantly a function of the accessible surface area of the electrode. Therefore, supercapacitor electrodes are commonly made of porous materials with extremely high surface areas such as activated carbon. Commonly used activated carbon electrodes have specific surface areas ranging from 1000 to 3000 $m^2 g^{-1}$ and can achieve specific capacitances of 100 F g^{-1}. For reference, the electrical conductivity of activated carbon is also on the order of 1000–2000 S m^{-1}. To improve the capacitance and energy, significant efforts have been undertaken. Many studies have focused on investigating assorted structures of porous carbons, including activated carbons (ACs) [42, 43], aerogels [44–46], carbon nanotubes (CNTs) [47–51] and mesoporous carbons [52–54], among others. For practical consideration, it is important to note that, although most supercapacitor materials report specific capacitance per *mass* as the value to measure performance, the specific capacitance per *volume* is more important for commercial implementation since the space available for components is limited. Therefore, many of these materials, which offer extremely high specific surface areas, cannot be implemented commercially due to their low packing densities. This point should be taken into consideration when embarking on new electrode designs to ensure materials have both high specific surface area and high packing density.

Graphene, due to its myriad of extraordinary properties, including high surface area and broad electrochemical stability window (>4 V), has been an outstanding candidate as the electrode material for supercapacitors. Additionally, the amphiphilic surfaces provided by RGO are ideal to ensure that the entire active surface is accessible to the electrolyte. Already to date, a tremendous amount of graphene-containing materials have been reported for supercapacitors [50, 51, 55–57]. However, all of the reported specific capacitance values are significantly lower than the theoretical value of 550 F g^{-1} due to non-ideal electrode design, processing and fabrication. Strong π–π interactions between graphene sheets can cause flakes to restack to form graphite-like stacks, severely crippling the accessible surface area and reducing ion diffusion rates. This combination of properties translates to limited specific capacitances and relatively poor rate performance [58, 59]. To enhance the performance of the capacitance of GO-derived materials, researchers have utilized different morphologies, such as wrinkled and 3D networks, composites with different carbon materials such as CNTs or polymers, and forming in-plane pores within the graphene basal planes.

In order to maximize performance, electrode architectures must be well designed to provide the right combination of high specific surface area, appropriate pore size distribution, good electrical conductivity, high rate performance, high packing density and stable long-term cycling, among others. In this section, rather than providing a comprehensive enumeration of all the approaches to RGO-based supercapacitor materials, we will highlight and summarize a few key examples of methods employed to design high-performance electrodes that solve many of these key issues.

One of the first studies of RGO as a supercapacitor electrode was reported in 2008 by Stoller *et al.* [60]. In that seminal study, GO was reduced using hydrazine hydrate to form agglomerated particles ranging from 15 to 25 μm in diameter with a specific surface area (SSA) of 705 $m^2 g^{-1}$. However, due to the partial agglomeration and limited surface area, the specific capacitance of the material was relatively low, only reaching 135 and 99 F g^{-1} in aqueous and organic electrolytes, respectively. One of the reasons that this structure did not exhibit higher capacitance was due to the inability of the electrolyte to penetrate all available active graphene sheet surfaces due to the agglomeration. Additionally, the lack of engineered pore structure further hindered ion diffusion throughout the structure and prevented electrolyte access to the interior surfaces. Nonetheless, the RGO material exhibited good electrical conductivity, large surface area and chemical stability, positioning RGO as an exciting candidate for supercapacitor applications.

In order to improve the accessible surface areas, Ruoff and his team later developed a chemical activation transformation of exfoliated graphite oxide to synthesize a porous carbon material referred to as "activated graphene" [61]. Chemical activation is a common approach that is used to obtain porous carbon-based materials, and generally involves mixing the carbon precursor with chemical activating agents such as KOH [62], $ZnCl_2$ [63] and H_3PO_4 [64] followed by carbonization at high temperatures (400–900°C). Although this activation process had been applied to many carbon materials previously, Ruoff and his coworkers were one of the first to apply this treatment to GO materials. Remarkably, the unique structure consisted of a continuous 3D network of highly curved, atomically thin walls that formed meso- and micropores. These materials offered an extremely high SSA of up to 3100 $m^2\ g^{-1}$, greater than the theoretical value for graphene. The higher-than-theoretical surface area was ascribed to the generation of the extremely small (<5 nm) pores due to the KOH activation process. When tested as supercapacitor electrodes, one of the activated RGO materials with ~2400 $m^2\ g^{-1}$ SSA was able to achieve 166 F g^{-1} capacitance at a high current density of 5.7 A g^{-1} in an ionic liquid with 3.5 V working voltage, far superior to earlier reports. Additionally, the volumetric capacitance was calculated to be 60 F cm^{-3}. This material was also extremely stable and retained 97% capacitance after 10000 cycles. Surprisingly, increasing the SSA of 3100 $m^2\ g^{-1}$ resulted in a lower capacitance of 150 F g^{-1}, perhaps limited by the electrical conductivity of the highly porous structure (reported as 500 S m^{-1}). It was estimated that these materials could provide a practical packaged energy density of 20 W h kg^{-1} and a power density of 75 kW kg^{-1}, a significant improvement over the existing values for AC-based supercapacitors (5 W h kg^{-1}, ~7.5 kW kg^{-1}). Thus, Ruoff and coworkers demonstrated that the simple and well-known chemical activation process already employed for activated carbons can be applied to GO-based materials with the potential to realize advanced electrochemical energy storage devices in a short period of time.

Later, Ruoff and his team further refined their processing to fabricate the modified GO materials into hollow spheres in order to generate macropores prior to the chemical activation step [65]. Figure 8.15 outlines the improved route to the formation of highly porous

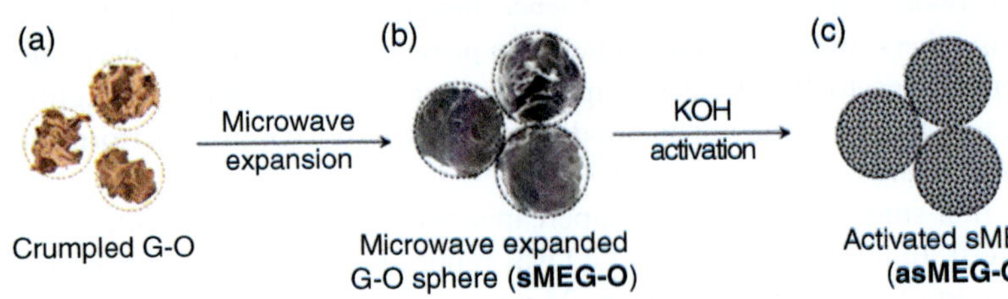

Figure 8.15 Fabrication schematic for the formation of highly porous graphene-derived carbons with hierarchical pore structures. Reproduced from [65] with permission of American Chemical Society

GO-derived carbons with hierarchical pore structure. With the improved method, the SSA of the carbon material could be further increased upon previous reported results to 3290 m^2 g^{-1}. These carbons exhibited an improved gravimetric and volumetric specific capacitance of 174 F g^{-1} and 100 F cm^{-3}, respectively, and retained 94% capacitance after 1000 cycles. With a high packing density of 0.59 g cm^{-3} and an estimated volumetric energy and power density of 44 W h l^{-1} and 210 kW l^{-1}, these materials represent a promising route for advanced supercapacitor applications. From the analysis, it is clear that control over the structure and morphology of carbon electrode materials is an effective strategy to develop high-surface-area materials with efficient paths for ion diffusion into the interior surfaces of bulk electrode particles.

Another method to increase the surface area of the electrode is through the development of novel 3D graphene architectures. Huang and his team demonstrated that transforming 2D GO sheets into a 3D crumpled ball structure was an attractive method to prevent sheet aggregation and maintaining high electron and ion transport [66]. As shown in Figure 8.16, crumpled materials are generated through an aerosol spray pyrolysis process wherein a GO suspension is atomized via ultrasonication and then carried into a heated electric furnace for thermal processing [67]. During the atomization process, crumpled GO balls are formed by isotropic capillary compression due to the rapid evaporation of the carrier solvent. The morphology of the resulting crumpled ball can be controlled by varying the GO concentration and atomization parameters. Crumpled tpGO materials exhibited relatively limited SSA of 567 m^2 g^{-1}. However, it was noted that, unlike other GO sheet morphologies, which tend to undergo substantial decrease in accessible surface areas due to the compressive forces required during electrode processing (e.g. from 407 to 66 m^2 g^{-1} before and after compression, respectively), the crumpled morphology still maintained an accessible surface area of 255 m^2 g^{-1} after compression. Additionally, after compression, the crumpled tpGO materials had relatively high density of 0.5 g cm^{-3}, demonstrating that the crumpled morphology can consist of well-wrinkled GO sheets with uniformly distributed free space inside and between particles. Crumpled tpGO materials offered good specific capacitance values of up to 150 F g^{-1} and 60 F cm^{-3} when cycled at 0.1 A g^{-1}. For reference, similar materials formed with a 2D flat structures and 2D wrinkled structures exhibited lower values of 122 and 142 F g^{-1}, respectively. Compared to flat or wrinkled morphologies for tpGO electrodes, the crumpled morphology provides higher energy and power densities. Additionally, unlike with the flat and wrinkled morphologies, the crumpled materials can provide high performance at high electrode mass and areal loading without substantial decrease in energy or power output. EIS further showed that the 3D crumpled porous structure exhibited improved charge-transfer kinetics compared to the flat and wrinkled morphologies. Compared to other graphene structures, the crumpled tpGO architecture demonstrates an encouraging route to 3D structures with high packing density and therefore high volumetric energy and power density.

Another route to the formation of high-surface-area RGO-containing electrodes for supercapacitor applications was accomplished using a hydrothermal treatment combined with a thermal activation step [68]. With this method, dispersed GO was first hydrothermally treated with biomass and/or polymer materials to form a partially reduced 3D hybrid precursor composite. This material was then mixed with potassium hydroxide (KOH) and thermally activated at 800°C to form a highly porous 3D structure. Figure 8.17 depicts the processing to form the material, as well as the electron microcopy images characterizing

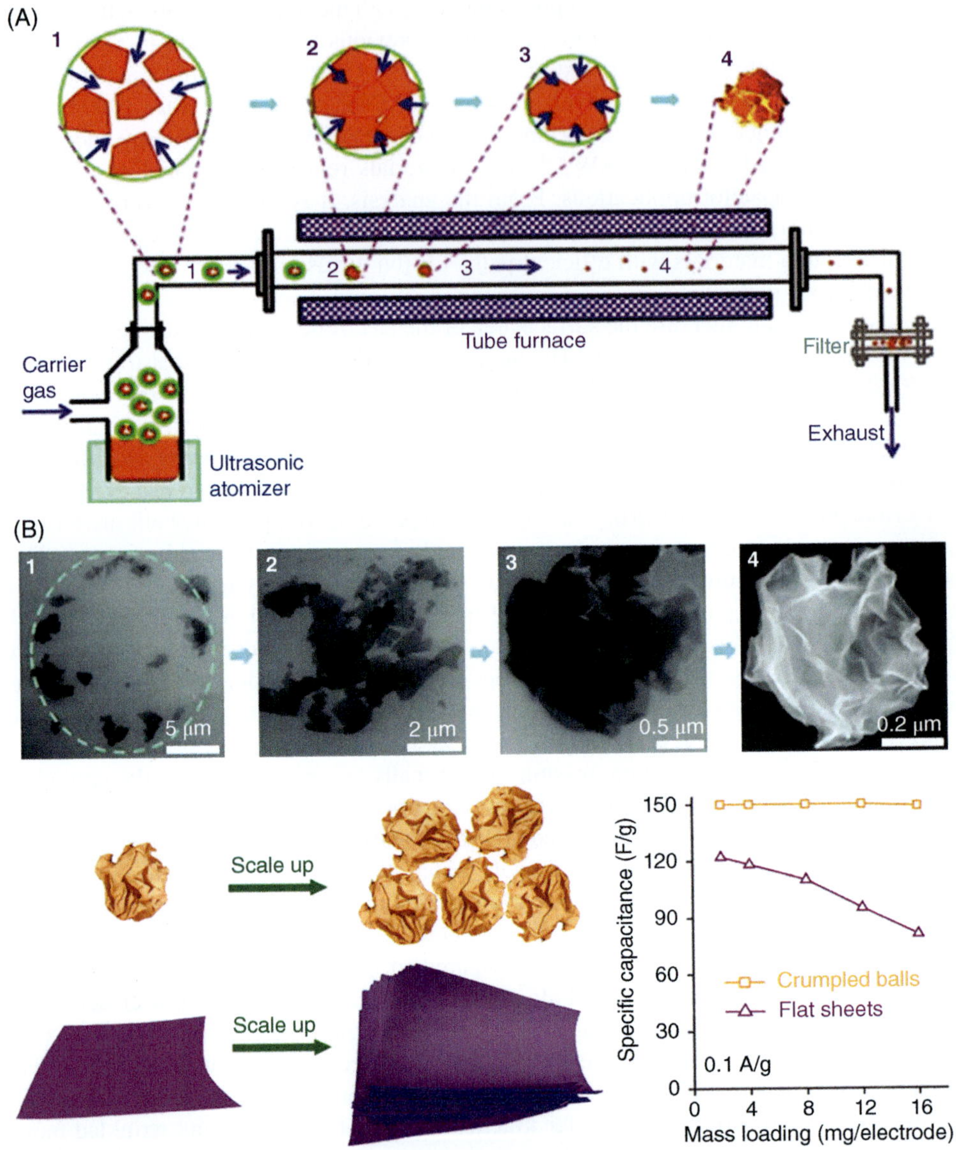

Figure 8.16 (A) Schematic illustrating the formation of (B) the crumpled sheet morphology. Illustration (bottom) of the flat sheet (purple) and crumpled ball (golden) morphologies and their respective capacitive performance. Reproduced from [66, 67] with permission of American Chemical Society

the morphology. The resulting material exhibited an unusually high surface area of 3523 m^2 g^{-1}, which is higher than the theoretical surface area for pristine graphene. Electron microscopy demonstrated that the structure is highly distorted, with carbonized material providing separation for individual graphene layers. The 3D interconnected layers provided a high

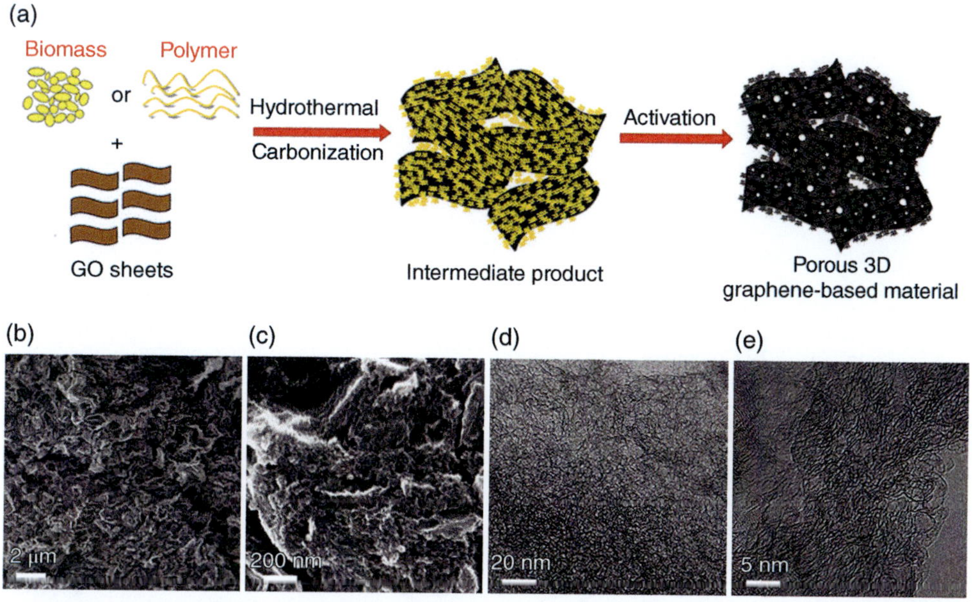

Figure 8.17 *(a) Schematic detailing the synthesis method for the combined hydrothermal and chemical activation processes. (b, c) SEM images demonstrate that the material has a sponge-like and porous structure. (d, e) TEM images show the material is composed of a dense 3D porous structure with a highly curved or wrinkled surface. Reproduced from [68] with permission of Macmillan Publishers Limited*

electrical conductivity up to 303 S m^{-1}. When examined as a supercapacitor electrode, these materials exhibited up to 231 F g^{-1} specific capacitance and exhibited almost no fade after 5000 cycles at 1 A g^{-1}. Compared to the porous 3D structure, a reference treated RGO material without the addition of the extra carbon (either biomass or polymer) exhibited a surface area of 1810 m^2 g^{-1} and correspondingly achieved a specific capacitance of only 132 F g^{-1}. This transformation demonstrates the benefit of high surface area and electrode design for achieving high-performance supercapacitors.

Very recently, another approach to the application of "holey graphene" has been developed for supercapacitor applications. Three-dimensional porous tpGO materials with in-plane pores that generated a "holey" structure were developed by Xu and colleagues [69]. A schematic of the material is shown in Figure 8.18. The structure was prepared using hydrogen peroxide (H_2O_2) to etch nanopores into the GO material while a simultaneous hydrothermal treatment was employed to self-assemble the materials into a 3D high-surface-area, interconnected hydrogel structure. The nanopores were controlled by the amount of hydrogen peroxide and exhibited pore sizes on the order of a few nanometers that were distributed throughout the basal plane of GO. In order to highlight the impact of the nanopores, a control material processed without the use of the hydrogen peroxide etchant was also fabricated. The tpGO macropores generated via the hydrothermal reaction ranged from submicrometers to several micrometers and the walls consisted of one- or few-layer tpGO sheets with defective carbon sites in the basal plane. In order to improve the packing

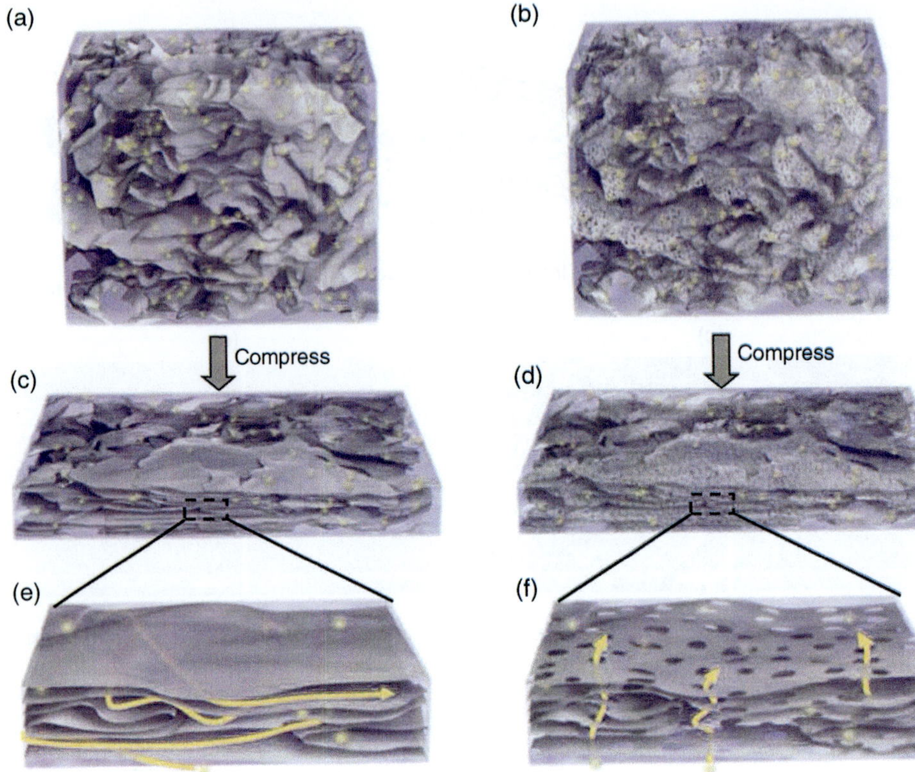

Figure 8.18 Illustration of the "holey graphene" material (b, d, f) and reference non-holey material (a, c, e) as idealized electrodes for supercapacitor applications. Reproduced from [69] with permission of Macmillan Publishers Limited

density to a level that would be required for commercial applications and to demonstrate the adaptability of the holey scaffold, the highly porous hydrogel material was compressed using a hydraulic press to form a compact "holey" material with significantly improved packing density. After compression, the surface area of the holey materials was measured to be 810 $m^2\ g^{-1}$, compared to 830 $m^2\ g^{-1}$ before compression, demonstrating that almost no graphene restacking occurred during the compression step. The lack of graphene restacking was attributed to the robust interconnection of graphene sheets that permitted mechanical compression without requiring sheet agglomeration. After compression, the packing density of the material was determined to be 0.71 g cm^{-3}. Comparing the holey material to its control counterpart, it was determined that the holey structure provided much greater surface area (810 versus 260 $m^2\ g^{-1}$), but with a slightly lower electrical conductivity (1000 versus 1400 S m^{-1}) due to the presence of carbon defects. Although the electrical conductivity of the non-porous material was greater than the porous material, the introduction of pores decreased the voltage drop observed in the electrode during cycling due to improved ion diffusion. Additionally, EIS indicated that the addition of the nanopores led to more efficient electrolyte diffusion within the electrode, as demonstrated by a lower charge-transfer resistance. When examined as a supercapacitor electrode in aqueous electrolyte, the porous holey tpGO material achieved extremely high gravimetric capacitance of up to 310 F g^{-1} at 1 A g^{-1}, in contrast to the control case, which demonstrated 208 F g^{-1} capacitance under equivalent test conditions. Therefore, the introduction of pores into the basal plane provided a 50% boost to capacitance through the improved diffusion pathways. Upon increasing the charge–discharge current density by a factor of 100 (100 A g^{-1}), the holey material was able to retain 76% of its initial capacitance, while the control sample only retained 63%, demonstrating improved rate retention upon the introduction of the pores into the basal planes. Additionally, these materials exhibited extreme cycling stability, retaining ~90% capacitance after 20 000 charge–discharge cycles at 25 A g^{-1}. Even more impressive, these materials were able to achieve an unprecedented volumetric capacitance of 212 F cm^{-3}, one of the highest reported values to date. It was calculated that these highly densified materials, when assembled into a fully packaged capacitor cell, would be able to provide 35 W h kg^{-1} energy density and a 49 W h l^{-1} volumetric density, approaching the value provided by the lead–acid battery. This report demonstrates the significant improvement available by increasing pathways for ions to migrate and access electrode surface areas as well as the extremely high energy densities that are feasible for supercapacitors when high packing densities are taken into consideration.

Overall, RGO electrodes have been shown to be stable but exhibit relatively limited capacitance (200–300 F g^{-1}) since the main charge storage method is through the EDLC mechanism. Additionally, since activated carbon and related materials can achieve similar results due to high surface area but at a much lower cost, pure RGO electrodes have not yet replaced activated carbon as the material of choice for commercial applications. In order to increase the capacitance of RGO-containing supercapacitor electrodes, hybrid materials containing pseudo-capacitive materials are often investigated as alternatives.

8.3.4 Pseudo-Capacitive GO–Composite Electrodes

As mentioned in section 8.3.2, pseudo-capacitive materials offer significantly higher specific capacitance values due to the addition of Faradaic reactions for charge storage. However, these materials also suffer from significantly worse cycling stability. Combining RGO with these promising materials offers a route to combine the advantages of both types to form hybrid structures with combined high capacitance and good cycling stability. Performance improvements that arise from this synergy include significantly increased capacitance values, faster charge–discharge rates, improved stabilization at high-voltage operation and extended cycle life. Most commonly this can be achieved using conductive polymers such as polyaniline (PANI) [70, 71] and polypyrrole (PPy) [72, 73], as well as metal oxide materials such as Co_3O_4 [74, 75], MnO_2 [75, 76], $Ni(OH)_2$ [77], RuO_2 [78] and others. With these hybrid composite materials, many of the design schemes discussed in section 8.3.3 can be applied to these systems, except for the addition of the pseudo-capacitive component material. Therefore, this section will highlight specific examples of how different types of pseudo-capacitive materials can be integrated into carbon architectures in order to develop high-performance composite supercapacitors.

PANI is an electrically conductive polymer that is commonly employed for supercapacitors due to its pseudo-capacitive behavior. PANI is considered the most promising conductive polymer for supercapacitors due to its combination of good environmental stability, ease of synthesis, doping chemistry and unique method of electron conduction. PANI, like other pseudo-capacitive materials, also suffers from poor stability during the charge–discharge process. To overcome this problem, PANI is generally combined with carbon-based materials such as activated carbons and CNTs to prepare composite electrode materials. GO provides a good scaffold for PANI synthesis due to the presence of a large amount of oxygen functional groups that can help act as nucleation sites for *in situ* polymerization and doping mechanisms [71]. Although reports exist for the use of GO in its insulating state with PANI [79], the oxidative state of GO is not an appropriate choice to improve the electrochemical performance of PANI. Rather, the reduced, electrically conductive RGO material offers much more favorable interactions with PANI composites.

Recently, Meng and colleagues reported the use of calcium carbonate ($CaCO_3$) as a sacrificial template to synthesize a 3D PANI–RGO composite material, as shown in Figure 8.19 [80]. With their method, $CaCO_3$ particles were first formed *in situ* in the GO dispersion and then dried in order to template GO into a 3D porous structure. After chemical reduction via hydrazine vapor treatment, the calcium carbonate template was removed, providing a porous RGO skeleton structure. The resulting freestanding film had high flexibility and interconnected pores that facilitated electrolyte ion transport within the structure and provided high rate performance. After forming the porous RGO structure, PANI nanowires were then grown on the surface of the structure, creating a hierarchical PANI–RGO composite material. When tested as a supercapacitor material, the presence of the pseudo-capacitive PANI material afforded a high specific capacitance of up to 385 F g^{-1} and excellent rate performance. Additionally, due to the stability provided by the RGO structure, the composite was able to retain 90% of its initial capacitance after 5000 cycles, demonstrating cycling stability similar to the bare RGO porous skeleton. The performance of the structure is attributed to the interconnected porous structure of the film, highlighting the importance of 3D structure when designing a supercapacitor electrode material.

In addition to PANI, numerous other polymer–RGO composites have recently been explored for supercapacitor applications. PPy is another attractive conductive polymer that has been used as a pseudo-capacitive material due to its similar properties of high charge capacitance, good electrical conductivity and inexpensive synthesis. Due to the many

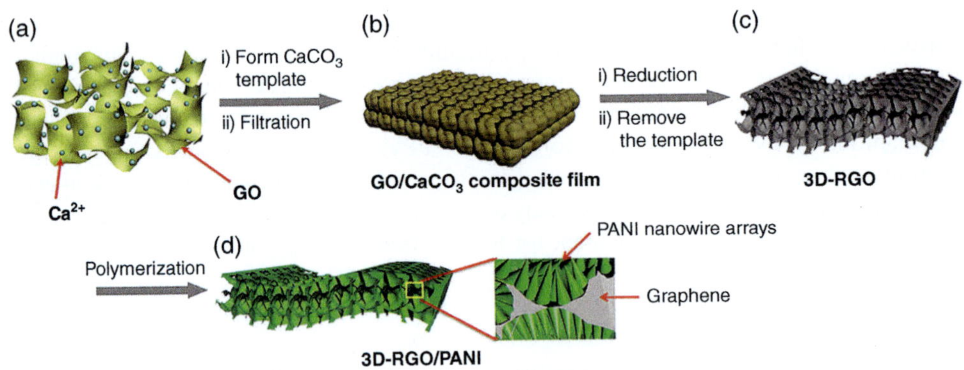

Figure 8.19 Synthetic formation of hierarchical PANI–RGO composites with 3D interconnected pores. Reproduced from [80] with permission of Wiley-VCH Verlag GmbH

similarities between PANI and PPy, it is expected that many of the synthetic schemes developed for the former polymer can be applied to the latter system. Indeed, multi-layer graphene–PPy composites have already been reported, with promising results [72].

Transition metal oxide materials are another promising class of pseudo-capacitive materials for supercapacitor applications. The mechanism of energy storage in metal oxides is mainly attributed to a combination of fast and reversible electron transfer combined with an electroadsorption of protons on the surface of electroactive materials. Among the metal oxide particles, nickel hydroxide has received a significant amount of attention for use in supercapacitors due to its high theoretical capacitance value (~3750 F g^{-1}), ease of synthesis and low cost [81]. Unfortunately, nickel hydroxide suffers from a few challenges that limit its performance, including low rate capability and therefore low capacity due to the inability to provide adequate electron transport. Due to these shortcomings, combining this material with a highly conductive support such as GO is extremely attractive. Indeed, numerous studies have been performed focusing on the integration of nickel hydroxide with graphene-based electrodes [82–86]. In many of these reports, the synthesis method is varied in order to achieve the best dispersion and size control of particles within the structure. However, in some of these structures, the graphenic material suffers from sheet aggregation, leading to poor particle dispersion and low accessible surface area. This in turn can hinder ion and electron diffusion through the structure and thereby limit the improvement observed under electrochemical testing.

To date, the highest specific capacitance reported was accomplished using a nickel hydroxide-coated 3D porous tpGO, hollow-sphere composite material [87]. In order to create this material, as shown in Figure 8.20, Zhang and colleagues layered GO sheets around a sacrificial silica (SiO_2) spherical template that had been modified with 3-aminopropyltrimethoxysilane (APS) to improve GO adhesion, followed by thermal reduction and template removal. Afterwards, nickel hydroxide nanoparticles were electrochemically deposited onto the porous tpGO scaffold using a nickel nitrate ($Ni(NO_3)_2$) precursor solution. The resulting thermally processed GO formed a porous 3D scaffold that provided rapid electron and ion transport to the pseudo-capacitive material. The engineered nickel hydroxide–tpGO composite achieved a very high specific capacitance of 2815 F g^{-1} at a scan rate of 5 mV s^{-1}, and high stability for 1000 cycles with less than 8% deterioration. Although the surface area of the composite material, which was 1159 m^2 g^{-1}, was not as high as reported for other 3D RGO structures, the excellent performance was attributed to the high dispersion of small, 4 nm $Ni(OH)_2$ nanoparticles that were anchored to the graphene structure. Additionally, to demonstrate that the high specific capacitance is due to the metal oxide material, control 3D porous tpGO scaffolds prior to nickel hydroxide coating were examined as a supercapacitor material. With this control, the maximum achievable specific capacitance was limited to ~210 F g^{-1}, more than an order of magnitude lower than the capacitance reached with the addition of the nickel hydroxide material.

In addition to $Ni(OH)_2$, MnO_2 is another outstanding metal oxide candidate due to its significant environmental abundance, low cost and safety. However, without proper nanostructuring, MnO_2 suffers from low specific capacitance due to poor electrical conductivity and low surface area. Even with nanostructured MnO_2 composites, the microstructure can be easily damaged during cycling, leading to rapid capacity fade. In a hybrid system with RGO, the graphene scaffold can impart high electrical conductivity and a robust 3D network to protect the microstructure and prevent capacity fade. Therefore, properly designed

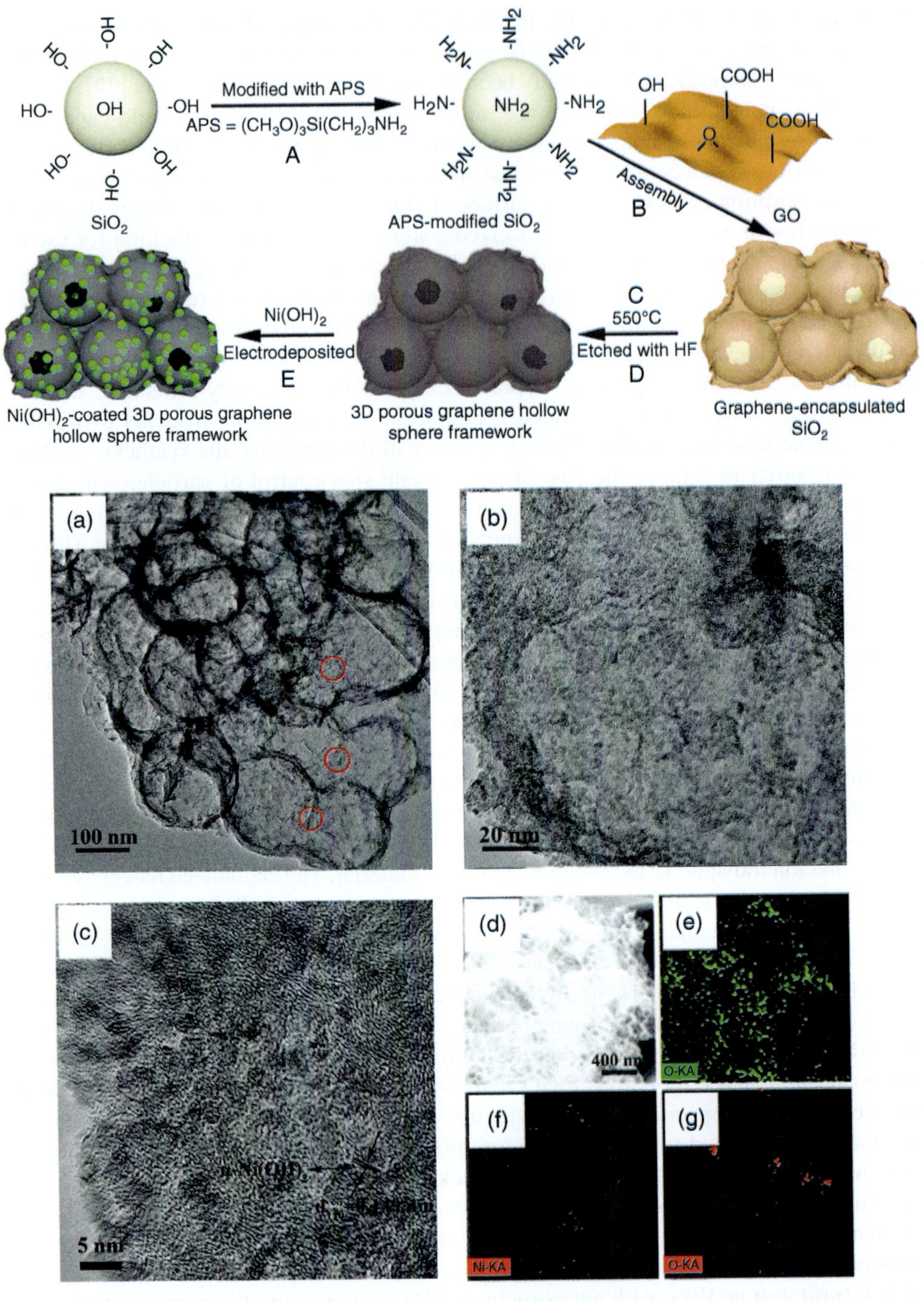

Figure 8.20 *Synthetic scheme (top) for the template procedure to form* $Ni(OH)_2$*–tpGO composite framework; and (bottom) TEM images at different magnifications (a–c) and STEM mapping of the elemental components (d–g). Reproduced from [87] with permission from the PCCP Owner Societies, Copyright 2014*

MnO_2–RGO composites are considered a promising route for improved pseudo-capacitance and cycling stability. Indeed, many reports on MnO_2–RGO hybrid composites exist and have demonstrated methods to successfully overcome the low accessible capacitance and cycling stability of MnO_2 materials. One of the strengths of the MnO_2 material is the extensive list of known chemical routes to generate nanomaterials with a wide variety of morphologies, including nanowires, nanosheets, needle-like and hollow-sphere morphologies. In one example, MnO_2–RGO composites with different morphologies were obtained by a self-assembly method [88]. From electrochemical testing, it was determined that a flower-like nanosphere shape exhibited better capacitive performance (405 F g^{-1}) than that of the nanowire shape (318 F g^{-1}) when examined at a current density of 1.0 A g^{-1}. At a lower current density of 0.2 A g^{-1}, the specific capacitance of the flower-like MnO_2 composite further improved to ~510 F g^{-1}. Additionally, it was determined that the flower-like shape provided improved rate performance and cycling stability compared to the nanowire counterpart. After 1000 cycles at 1 A g^{-1}, the flower-like composite retained ~350 F g^{-1}, corresponding to a retention of ~86%. Therefore, in addition to the importance of 3D structuring of the RGO scaffold, this example demonstrates the importance of morphological control and manipulation of the pseudo-capacitive material in order to extend the cycle life and capacitance of the hybrid composite material.

In summary, these examples highlight the need for microscopic control of electrode design in order to ensure high surface area, uniform particle dispersion and porous 3D structuring that can lead to much greater supercapacitor performance. Pseudo-capacitive materials such as conductive polymers and metal oxides offer attractive routes to high-performance advanced supercapacitor materials. Conductive polymers offer advantages such as flexibility, high conductivity and high specific capacitance, but still suffer from limited cycle life compared to their EDLC counterparts. Metal oxide materials suffer from poor rate performance and limited cycle life as well. Although the presence of graphene significantly improves the cycling stability of pseudo-capacitive materials, the performance is not satisfactory for commercial integration. As future work, further improvements in electrochemical stability of pseudo-capacitive composite materials are required for enabling high-energy and high-power supercapacitors.

8.4 Outlook and Future Development Opportunities

Graphene oxide possesses a wide variety of useful properties derived from its unique atomically thin structure, which separates itself from other multi-layer graphene platelet and graphite systems. For electrochemical systems such as lithium-ion batteries and supercapacitors, the myriad of properties includes high surface area, mechanical strength, electrical conductivity, variable sheet size, flexible nature and tailorable morphology. Particularly, the oxygen group present on GO and its reduced versions provides a unique surface for a rich variety of chemical reactions for further functionalization or to synthesize unparalleled hybrid materials. The combination of these properties paired with advanced materials offers a promising route to accelerate the development of next-generation energy storage technologies. Since its emergence in the literature less than a decade ago, a number of dramatic innovations have already been demonstrated and limitless exciting discoveries are still on the horizon.

References

[1] Bonaccorso, F.; Sun, Z.; Hasan, T.; Ferrari, A.C., Graphene photonics and optoelectronics. *Nature Photon.* **2010**, *4* (9), 611–622.

[2] Roy-Mayhew, J.D.; Aksay, I.A., Graphene materials and their use in dye-sensitized solar cells. *Chem. Rev.* **2014**, *114* (12), 6323–6348.

[3] De, S.; Coleman, J.N., Are there fundamental limitations on the sheet resistance and transmittance of thin graphene films? *ACS Nano* **2010**, *4* (5), 2713–2720.

[4] Wassei, J.K.; Kaner, R.B., Graphene, a promising transparent conductor. *Mater. Today* **2010**, *13* (3), 52–59.

[5] Brownson, D.A.C.; Kampouris, D.K.; Banks, C.E., An overview of graphene in energy production and storage applications. *J. Power Sources* **2011**, *196* (11), 4873–4885.

[6] Vilatela, J.J.; Eder, D., Nanocarbon composites and hybrids in sustainability: a review. *ChemSusChem* **2012**, *5* (3), 456–478.

[7] Pillot, P., The rechargable battery market and main trends, 2012–2025. Presentation at *Batteries 2012*. Avicenne Energy, **2012**.

[8] Winter, M.; Brodd, R.J., What are batteries, fuel cells, and supercapacitors? *Chem. Rev.* **2004**, *104* (10), 4245–4269.

[9] Xu, K., Nonaqueous liquid electrolytes for lithium-based rechargeable batteries. *Chem. Rev.* **2004**, *104* (10), 4303–4417.

[10] Nie, M.Y.; Chalasani, D.; Abraham, D.P.; *et al.*, Lithium ion battery graphite solid electrolyte interphase revealed by microscopy and spectroscopy. *J. Phys. Chem. C* **2013**, *117* (3), 1257–1267.

[11] Kasavajjula, U.; Wang, C.S.; Appleby, A.J., Nano- and bulk-silicon-based insertion anodes for lithium-ion secondary cells. *J. Power Sources* **2007**, *163* (2), 1003–1039.

[12] Yoo, E.; Kim, J.; Hosono, E.; *et al.*, Large reversible Li storage of graphene nanosheet families for use in rechargeable lithium ion batteries. *Nano Lett.* **2008**, *8* (8), 2277–2282.

[13] Zhu, J.X.; Yang, D.; Yin, Z.Y.; *et al.*, Graphene and graphene-based materials for energy storage applications. *Small* **2014**, *10* (17), 3480–3498.

[14] Mahmood, N.; Zhang, C.Z.; Yin, H.; Hou, Y.L., Graphene-based nanocomposites for energy storage and conversion in lithium batteries, supercapacitors and fuel cells. *J. Mater. Chem. A* **2014**, *2* (1), 15–32.

[15] Pan, D.Y.; Wang, S.; Zhao, B.; *et al.*, Li storage properties of disordered graphene nanosheets. *Chem. Mater.* **2009**, *21* (14), 3136–3142.

[16] Lee, W.; Suzuki, S.; Miyayama, M., Lithium storage properties of graphene sheets derived from graphite oxides with different oxidation degree. *Ceram. Int.* **2013**, *39* (Suppl. 1), S753–S756.

[17] Abouimrane, A.; Compton, O.C.; Amine, K.; Nguyen, S.T., Non-annealed graphene paper as a binder-free anode for lithium-ion batteries. *J. Phys. Chem. C* **2010**, *114* (29), 12800–12804.

[18] Hu, Y.H.; Li, X.F.; Geng, D.S.; *et al.*, Influence of paper thickness on the electrochemical performances of graphene papers as an anode for lithium ion batteries. *Electrochim. Acta* **2013**, *91*, 227–233.

[19] Zhao, X.; Hayner, C.M.; Kung, M.C.; Kung, H.H., Flexible holey graphene paper electrodes with enhanced rate capability for energy storage applications. *ACS Nano* **2011**, *5* (11), 8739–8749.

[20] Wu, Z.S.; Ren, W.C.; Xu, L.; *et al.*, Doped graphene sheets as anode materials with superhigh rate and large capacity for lithium ion batteries. *ACS Nano* **2011**, *5* (7), 5463–5471.

[21] Sandu, I.; Moreau, P.; Guyomard, D.; *et al.*, Synthesis of nanosized Si particles via a mechanochemical solid–liquid reaction and application in Li-ion batteries. *Solid State Ionics* **2007**, *178* (21–22), 1297–1303.

[22] Maranchi, J.P.; Hepp, A.F.; Evans, A.G.; *et al.*, Interfacial properties of the a-Si/Cu:active–inactive thin-film anode system for lithium-ion batteries. *J. Electrochem. Soc.* **2006**, *153* (6), A1246–A1253.

[23] Zhang, T.; Zhang, H.P.; Yang, L. C.; *et al.*, The structural evolution and lithiation behavior of vacuum-deposited Si film with high reversible capacity. *Electrochim. Acta* **2008**, *53* (18), 5660–5664.

[24] Chan, C.K.; Peng, H.L.; Liu, G.; *et al.*, High-performance lithium battery anodes using silicon nanowires. *Nature Nanotechnol.* **2008**, *3* (1), 31–35.
[25] Ahn, H.J.; Kim, Y.S.; Kim, W.B.; *et al.*, Formation and characterization of Cu–Si nanocomposite electrodes for rechargeable Li batteries. *J. Power Sources* **2006**, *163* (1), 211–214.
[26] Liu, W.R.; Wu, N.L.; Shieh, D.T.; *et al.*, Synthesis and characterization of nanoporous NiSi–Si composite anode for lithium-ion batteries. *J. Electrochem. Soc.* **2007**, *154* (2), A97–A102.
[27] Kim, H.; Han, B.; Choo, J.; Cho, J., Three-dimensional porous silicon particles for use in high-performance lithium secondary batteries. *Angew. Chem. Int. Ed.* **2008**, *47* (52), 10151–10154.
[28] Magasinski, A.; Dixon, P.; Hertzberg, B.; *et al.*, High-performance lithium-ion anodes using a hierarchical bottom-up approach. *Nature Mater.* **2010**, *9* (4), 353–358.
[29] Beattie, S.D.; Larcher, D.; Morcrette, M.; *et al.*, Si electrodes for Li-ion batteries – a new way to look at an old problem. *J. Electrochem. Soc.* **2008**, *155* (2), A158–A163.
[30] Xu, W.L.; Flake, J.C., Composite silicon nanowire anodes for secondary lithium-ion cells. *J. Electrochem. Soc.* **2010**, *157* (1), A41–A45.
[31] Lee, J.K.; Smith, K.B.; Hayner, C.M.; Kung, H.H., Silicon nanoparticles–graphene paper composites for Li ion battery anodes. *Chem. Commun.* **2010**, *46* (12), 2025–2027.
[32] Zhao, X.; Hayner, C.M.; Kung, M.C.; Kung, H.H., In-plane vacancy-enabled high-power Si–graphene composite electrode for lithium-ion batteries. *Adv. Energy Mater.* **2011**, *1* (6), 1079–1084.
[33] Yang, S.B.; Feng, X.L.; Ivanovici, S.; Mullen, K., Fabrication of graphene-encapsulated oxide nanoparticles: towards high-performance anode materials for lithium storage. *Angew. Chem. Int. Ed.* **2010**, *49* (45), 8408–8411.
[34] Ding, Y.; Jiang, Y.; Xu, F.; *et al.*, Preparation of nano-structured $LiFePO_4$/graphene composites by co-precipitation method. *Electrochem. Commun.* **2010**, *12* (1), 10–13.
[35] Zhou, X.F.; Wang, F.; Zhu, Y.M.; Liu, Z.P., Graphene modified $LiFePO_4$ cathode materials for high power lithium ion batteries. *J. Mater. Chem.* **2011**, *21* (10), 3353–3358.
[36] Wang, H.L.; Yang, Y.; Liang, Y.Y.; *et al.*, Graphene-wrapped sulfur particles as a rechargeable lithium–sulfur battery cathode material with high capacity and cycling stability. *Nano Lett.* **2011**, *11* (7), 2644–2647.
[37] Song, M.K.; Zhang, Y.G.; Cairns, E.J., A long-life, high-rate lithium/sulfur cell: a multifaceted approach to enhancing cell performance. *Nano Lett.* **2013**, *13* (12), 5891–5899.
[38] Wang, C.; Wang, X.; Yang, Y.; *et al.*, Slurryless Li_2S/reduced graphene oxide cathode paper for high-performance lithium sulfur battery. *Nano Lett.* **2015**, *15* (3), 1796–1802.
[39] Kim, H.; Park, K.Y.; Hong, J.; Kang, K., All-graphene-battery: bridging the gap between supercapacitors and lithium ion batteries. *Sci. Rep.* **2014**, *4*, 5728.
[40] Shen, J.M.; Han, K.; Martin, E.J.; *et al.*, Upper-critical solution temperature (UCST) polymer functionalized graphene oxide as thermally responsive ion permeable membrane for energy storage devices. *J. Mater. Chem. A* **2014**, *2* (43), 18204–18207.
[41] China Market Research Reports, *Global supercapacitor market to see 21% CAGR: China Focus 2014–2020 research report.* PR Newswire. See http://www.prnewswire.com/news-releases/global-supercapacitor-market-to-see-21-cagr-china-focus-2014-2020-research-report-290009141.html (accessed 9 May **2016**).
[42] Raymundo-Pinero, E.; Leroux, F.; Beguin, F., A high-performance carbon for supercapacitors obtained by carbonization of a seaweed biopolymer. *Adv. Mater.* **2006**, *18* (14), 1877–1882.
[43] Qu, D.Y., Studies of the activated carbons used in double-layer supercapacitors. *J. Power Sources* **2002**, *109* (2), 403–411.
[44] Fischer, U.; Saliger, R.; Bock, V.; *et al.*, Carbon aerogels as electrode material in supercapacitors. *J. Porous Mater.* **1997**, *4* (4), 281–285.
[45] Probstle, H.; Wiener, M.; Fricke, J., Carbon aerogels for electrochemical double layer capacitors. *J. Porous Mater.* **2003**, *10* (4), 213–222.
[46] Liu, D.; Shen, J.; Liu, N.P.; *et al.*, Preparation of activated carbon aerogels with hierarchically porous structures for electrical double layer capacitors. *Electrochim. Acta* **2013**, *89*, 571–576.
[47] Xu, G.H.; Zheng, C.; Zhang, Q.; *et al.*, Binder-free activated carbon/carbon nanotube paper electrodes for use in supercapacitors. *Nano Res.* **2011**, *4* (9), 870–881.

[48] Frackowiak, E.; Beguin, F., Electrochemical storage of energy in carbon nanotubes and nanostructured carbons. *Carbon* **2002**, *40* (10), 1775–1787.
[49] Frackowiak, E.; Jurewicz, K.; Delpeux, S.; Beguin, F., Nanotubular materials for supercapacitors. *J. Power Sources* **2001**, *97* (8), 822–825.
[50] Yan, J.; Wei, T.; Fan, Z.J.; *et al.*, Preparation of graphene nanosheet/carbon nanotube/polyaniline composite as electrode material for supercapacitors. *J. Power Sources* **2010**, *195* (9), 3041–3045.
[51] Yan, J.; Wei, T.; Shao, B.; *et al.*, Preparation of a graphene nanosheet/polyaniline composite with high specific capacitance. *Carbon* **2010**, *48* (2), 487–493.
[52] Zhao, J.C.; Lai, C.Y.; Dai, Y.; Xie, J.Y., Pore structure control of mesoporous carbon as supercapacitor material. *Mater. Lett.* **2007**, *61* (23–24), 4639–4642.
[53] Lei, Z.B.; Christov, N.; Zhang, L.L.; Zhao, X.S., Mesoporous carbon nanospheres with an excellent electrocapacitive performance. *J. Mater. Chem.* **2011**, *21* (7), 2274–2281.
[54] Wang, Q.; Yan, J.; Wei, T.; *et al.*, Two-dimensional mesoporous carbon sheet-like framework material for high-rate supercapacitors. *Carbon* **2013**, *60*, 481–487.
[55] Du, X.A.; Guo, P.; Song, H.H.; Chen, X.H., Graphene nanosheets as electrode material for electric double-layer capacitors. *Electrochim. Acta* **2010**, *55* (16), 4812–4819.
[56] Du, Q.L.; Zheng, M.B.; Zhang, L.F.; *et al.*, Preparation of functionalized graphene sheets by a low-temperature thermal exfoliation approach and their electrochemical supercapacitive behaviors. *Electrochim. Acta* **2010**, *55* (12), 3897–3903.
[57] Chen, S.; Zhu, J.W.; Wang, X., One-step synthesis of graphene–cobalt hydroxide nanocomposites and their electrochemical properties. *J. Phys. Chem. C* **2010**, *114* (27), 11829–11834.
[58] Zhu, Y.W.; Murali, S.; Cai, W.W.; *et al.*, Graphene and graphene oxide: synthesis, properties, and applications. *Adv. Mater.* **2010**, *22* (35), 3906–3924.
[59] Sun, Y.Q.; Wu, Q.O.; Shi, G.Q., Graphene based new energy materials. *Energy Environ. Sci.* **2011**, *4* (4), 1113–1132.
[60] Stoller, M.D.; Park, S.J.; Zhu, Y.W.; *et al.*, Graphene-based ultracapacitors. *Nano Lett.* **2008**, *8* (10), 3498–3502.
[61] Zhu, Y.W.; Murali, S.; Stoller, M.D.; *et al.*, Carbon-based supercapacitors produced by activation of graphene. *Science* **2011**, *332* (6037), 1537–1541.
[62] Kierzek, K.; Frackowiak, E.; Lota, G.; *et al.*, Electrochemical capacitors based on highly porous carbons prepared by KOH activation. *Electrochim. Acta* **2004**, *49* (4), 515–523.
[63] Yue, Z.R.; Mangun, C.L.; Economy, J., Preparation of fibrous porous materials by chemical activation 1. $ZnCl_2$ activation of polymer-coated fibers. *Carbon* **2002**, *40* (8), 1181–1191.
[64] Toles, C.; Rimmer, S.; Hower, J.C., Production of activated carbons from a Washington lignite using phosphoric acid activation. *Carbon* **1996**, *34* (11), 1419–1426.
[65] Kim, T.; Jung, G.; Yoo, S.; *et al.*, Activated graphene-based carbons as supercapacitor electrodes with macro- and mesopores. *ACS Nano* **2013**, *7* (8), 6899–6905.
[66] Luo, J.Y.; Jang, H.D.; Huang, J.X., Effect of sheet morphology on the scalability of graphene-based ultracapacitors. *ACS Nano* **2013**, *7* (2), 1464–1471.
[67] Luo, J.Y.; Jang, H.D.; Sun, T.; *et al.*, Compression and aggregation-resistant particles of crumpled soft sheets. *ACS Nano* **2011**, *5* (11), 8943–8949.
[68] Zhang, L.; Zhang, F.; Yang, X.; *et al.*, Porous 3D graphene-based bulk materials with exceptional high surface area and excellent conductivity for supercapacitors. *Sci. Rep.* **2013**, *3*, 1408.
[69] Xu, Y.X.; Lin, Z.Y.; Zhong, X.; *et al.*, Holey graphene frameworks for highly efficient capacitive energy storage. *Nature Commun.* **2014**, *5*, 4554.
[70] Wang, Y.G.; Li, H.Q.; Xia, Y.Y., Ordered whiskerlike polyaniline grown on the surface of mesoporous carbon and its electrochemical capacitance performance. *Adv. Mater.* **2006**, *18* (19), 2619–2623.
[71] Zhang, K.; Zhang, L.L.; Zhao, X.S.; Wu, J.S., Graphene/polyaniline nanofiber composites as supercapacitor electrodes. *Chem. Mater.* **2010**, *22* (4), 1392–1401.
[72] Biswas, S.; Drzal, L.T., Multilayered nanoarchitecture of graphene nanosheets and polypyrrole nanowires for high performance supercapacitor electrodes. *Chem. Mater.* **2010**, *22* (20), 5667–5671.

[73] Fan, L.Z.; Maier, J., High-performance polypyrrole electrode materials for redox supercapacitors. *Electrochem. Commun.* **2006**, *8* (6), 937–940.

[74] Yan, J.; Wei, T.; Qiao, W.M.; *et al.*, Rapid microwave-assisted synthesis of graphene nanosheet/Co_3O_4 composite for supercapacitors. *Electrochim. Acta* **2010**, *55* (23), 6973–6978.

[75] Liu, J.P.; Jiang, J.; Cheng, C.W.; *et al.*, Co_3O_4 nanowire@MnO_2 ultrathin nanosheet core/shell arrays: a new class of high-performance pseudocapacitive materials. *Adv. Mater.* **2011**, *23* (18), 2076–2081.

[76] Yan, J.; Fan, Z.J.; Wei, T.; *et al.*, Fast and reversible surface redox reaction of graphene–MnO_2 composites as supercapacitor electrodes. *Carbon* **2010**, *48* (13), 3825–3833.

[77] Zhang, X.J.; Shi, W.H.; Zhu, J.X.; *et al.*, Synthesis of porous NiO nanocrystals with controllable surface area and their application as supercapacitor electrodes. *Nano Res.* **2010**, *3* (9), 643–652.

[78] Wu, Z.S.; Wang, D.W.; Ren, W.; *et al.*, Anchoring hydrous RuO_2 on graphene sheets for high-performance electrochemical capacitors. *Adv. Funct. Mater.* **2010**, *20* (20), 3595–3602.

[79] Wang, H.L.; Hao, Q.L.; Yang, X.J.; *et al.*, Graphene oxide doped polyaniline for supercapacitors. *Electrochem. Commun.* **2009**, *11* (6), 1158–1161.

[80] Meng, Y.N.; Wang, K.; Zhang, Y.J.; Wei, Z.X., Hierarchical porous graphene/polyaniline composite film with superior rate performance for flexible supercapacitors. *Adv. Mater.* **2013**, *25* (48), 6985–6990.

[81] Jayashree, R.S.; Kamath, P.V., Suppression of the $\alpha \rightarrow \beta$-nickel hydroxide transformation in concentrated alkali: role of dissolved cations. *J. Appl. Electrochem.* **2001**, *31* (12), 1315–1320.

[82] Wang, H.L.; Casalongue, H.S.; Liang, Y.Y.; Dai, H.J., $Ni(OH)_2$ nanoplates grown on graphene as advanced electrochemical pseudocapacitor materials. *J. Am. Chem. Soc.* **2010**, *132* (21), 7472–7477.

[83] Lee, J.W.; Ahn, T.; Soundararajan, D.; *et al.*, Non-aqueous approach to the preparation of reduced graphene oxide/α-$Ni(OH)_2$ hybrid composites and their high capacitance behavior. *Chem. Commun.* **2011**, *47* (22), 6305–6307.

[84] Yan, J.; Fan, Z.J.; Sun, W.; *et al.*, Advanced asymmetric supercapacitors based on $Ni(OH)_2$/graphene and porous graphene electrodes with high energy density. *Adv. Funct. Mater.* **2012**, *22* (12), 2632–2641.

[85] Yan, J.; Sun, W.; Wei, T.; *et al.*, Fabrication and electrochemical performances of hierarchical porous $Ni(OH)_2$ nanoflakes anchored on graphene sheets. *J. Mater. Chem.* **2012**, *22* (23), 11494–11502.

[86] Yang, S.B.; Wu, X.L.; Chen, C.L.; *et al.*, Spherical α-$Ni(OH)_2$ nanoarchitecture grown on graphene as advanced electrochemical pseudocapacitor materials. *Chem. Commun.* **2012**, *48* (22), 2773–2775.

[87] Zhang, F.Q.; Zhu, D.; Chen, X.; *et al.*, A nickel hydroxide-coated 3D porous graphene hollow sphere framework as a high performance electrode material for supercapacitors. *Phys. Chem. Chem. Phys.* **2014**, *16* (9), 4186–4192.

[88] Feng, X.M.; Chen, N.N.; Zhang, Y.; *et al.*, The self-assembly of shape controlled functionalized graphene–MnO_2 composites for application as supercapacitors. *J. Mater. Chem. A* **2014**, *2* (24), 9178–9184.

9

Graphene Oxide Membrane for Molecular Separation

Ho Bum Park, Hee Wook Yoon and Young Hoon Cho

9.1 Rise of Graphene-Based Membranes: Two Approaches

Recently, membrane science and technology have played a significant role in many separation industries, such as water purification and gas separation processes [1], and new membrane materials and processes have been extensively developed due to urgent needs for much more economic separation processes. Definitely, to separate specific molecules and ions from mixtures with high productivity, two intrinsic properties of any membrane materials should be taken into account in corresponding membrane separation performance: one is membrane permeability and the other is membrane selectivity. Typically, polymeric membranes dominate the current global membrane market because of their low cost, mass production by roll-to-roll process, and ease of membrane formation and modulation. However, theoretical and semi-empirical upper bounds exist in terms of permeability and selectivity [2, 3]. Namely, highly permeable polymeric membranes show low selectivity, and vice versa. To overcome such strong trade-off relations between permeability and selectivity in polymeric membranes, many new membrane materials such as carbons, ceramics and inorganics have been developed for better membrane separation performance and new membrane processes. However, scale-up into membrane modules is still problematic, compared with polymeric membranes. For example, carbon molecular sieve (CMS) membranes, which can be prepared from inert pyrolysis of organic polymers, commonly show much better membrane-based gas separation performance than organic polymer membranes. However, thermal conversion is an energy-intensive process, and also natural

Graphene Oxide: Fundamentals and Applications, First Edition. Edited by Ayrat M. Dimiev and Siegfried Eigler.

cracks in and the brittleness of the resultant CMS membranes prevent the membrane industries from adopting these membranes in the practical field.

Basically, graphene is one type of carbon materials, and graphene and its derivatives are kinds of new membrane materials to be actively considered in the present day. So why are graphene and graphene-based materials now popular in the membrane science community? This is mainly due to the fact that graphene has many merits as a membrane material, such as one-atom thickness (note that membrane flux is inversely proportional to membrane thickness), scalable two-dimensional (2D) nature (related to membrane area for high flux) and excellent mechanical properties (related to membrane formation). However, perfect graphene (here "perfect graphene" means a single-crystal graphene sheet without any structural defects, such as point and line defects) cannot allow any gases or ions to penetrate through the graphene plane due to the hexagonal aromatic rings with high electron density [4]. That is to say, to make use of graphene as a selective membrane material, it is certainly necessary to create nanoscale or subnanoscale pores in a graphene membrane (Figure 9.1a). Such porous graphene would be the most ideal porous membrane that has ever existed in the world if precise pore engineering techniques were developed. In fact, porous graphene membranes with a few nanometric pores have been reported, largely by ion-beam irradiation or oxidation etching [5–7], showing ultrapermeable and more selective properties. However, the pore size is still too large to separate gases or ions selectively, and engineering subnanoscale pore size on a graphene sheet, direct synthesis of a porous graphene membrane [8] and scale-up will be a great challenge for its practical use in the future.

As an alternative to such porous graphene, permselective membranes can be prepared by using a concept to use interlayer space (slit-like pores) between graphene sheets (Figure 9.1b). However, the interlayer distance between graphitic layers is as narrow as 0.334 nm, so any gases or ions of interest in separation science cannot diffuse thorough the interlayer space. In this regard, graphene oxide (GO), a highly oxidized graphene sheet, can be considered as a graphene-based, new membrane material that employs the 2D interlayer spacing for a selective diffusion or sorption channel, because the interlayer distance between GO sheets is in the range from 0.6 to 1.2 nm, which can allow specific molecules

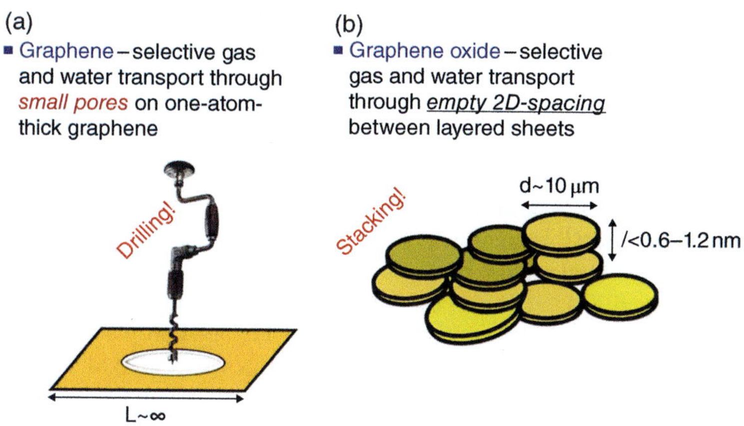

Figure 9.1 *Two distinct approaches to use graphene-based materials as selective membrane: (a) porous graphene and (b) layered GO membrane*

or ions to travel through the free space. In addition to this, GO has many features that enhance its potential utilization as a membrane material: (i) easy membrane formation as either freestanding or thin-film composite, (ii) solution-processable casting (GO is well dispersed in water), (iii) size controllable, (iv) possible pore engineering, and (v) further surface or edge modification. In general, GO has also been studied as a reinforcing nanomaterial in the form of polymer composites to give excellent gas-barrier properties because of its high aspect ratio [9]. However, rather than graphene itself, GO will also be a promising candidate for a membrane material if the pore width, diffusional channel and porosity can be easily engineered from the practical point of view.

9.2 GO Membrane: Structural Point of View

Basically, GO is a monolayer of graphite oxide. Graphite oxide was already synthesized a century ago via strong oxidation processes (Chapter 2) [10, 11]. GO sheets can be easily obtained by exfoliating graphite oxide in polar solvents such as water. Since monolayer graphene was experimentally discovered in 2004, the study of GO has become more popular because GO is an important raw material to mass produce graphene by eliminating the oxygen-containing functional groups via chemical or thermal reduction. In general, graphite oxide can be synthesized by various methods. Three major methods, in particular, namely (i) Brodie's method [12], (ii) Staudenmaier's method [13] and (iii) the Hummers–Offeman or Hummers method [14], are well known, and these methods lead to different oxidation levels. These days, the Hummers method has been widely used, with many advantages, but impurities such as permanganate ions should be completely removed by hydrogen peroxide treatment and then further dialysis to obtain high-purity GO sheets. Still, there are no clear comparative studies of GOs obtained from different methods for desirable applications, including membranes, and different oxidation levels, amounts of impurities and the presence of defects might arise from the various methods, and also lead to different properties.

Studies on the membrane applications of GO seem to be triggered by the fact that thick GO papers (or membranes) can be easily prepared from well-dispersed, aqueous GO solution via vacuum filtration [15]. Ruoff's group reported the preparation of strong, robust, thick GO membrane from well-dispersed GO solution by focusing on the mechanical properties of the GO membranes, not any transport studies. However, in the early 1990s, Japanese scientists reported the presence of graphite oxide membranes (they called them polymerized membranes using graphite oxide suspension) [16] and carbon nanofilms (CNFs) of nanometer thickness, starting from the oxidation of graphite [17]. Moreover, in 1960 Boehm, Clauss and Hofmann published the membrane properties of graphite oxide and found that the membranes are permeable to water [18].

The various physical and chemical properties of GO are mainly due to its complex and variable structure [10]. Many sketch models for GO have been proposed, but they do not define the exact chemical model because of the inherent heterogeneity due to the complex oxidation mechanism. In general, a structural model with epoxides and hydroxyls arranged in an orderly way in chain form on the graphene basal plane, while carboxylic groups lie at the GO edges, is regarded to be thermodynamically stable for GO. Basically, it is hard to assure or predict the intrinsic permeation properties of any gases or molecules through

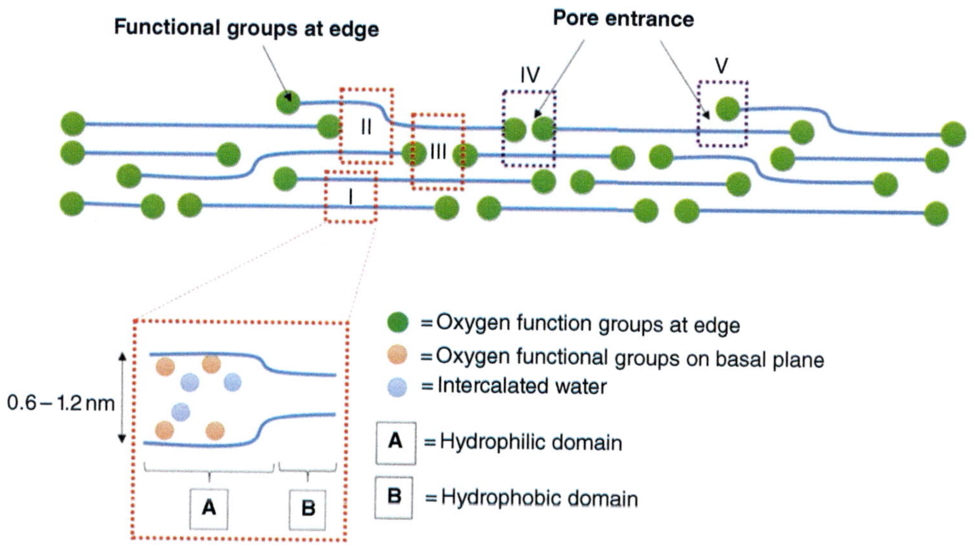

Figure 9.2 Simplified GO membrane structure to be considered for molecular separation

GO membranes because the chemistry of GO is not as simple as in some proposed structural models. For example, the coverage rate of hydroxyl and/or epoxide on the GO basal plane and the amount of carboxylic groups at the GO edges may affect the diffusion or sorption of molecules or ions, competing with water molecules having strong affinity for such oxygen-containing groups. Definitely, there should be still more systematic studies on structure–property relations for membrane applications, but in this chapter we simplify the layered GO membrane structure as shown in Figure 9.2. That is, we assume that a GO membrane has slit-like pores formed by randomly stacking GO sheets, and that a GO sheet consists of sp^2 (hydrophobic graphitic regions) to sp^3-hybridized carbon atoms (hydrophilic oxidized regions) by degree of oxidation. More generally speaking, highly oxidized GO shows three major regions, which are defects (or holes) (ca. 2%), graphitic regions (ca. 16%) and highly oxidized regions (ca. 82%). The defects in GO are due to the evolution of CO and CO_2 during the aggressive oxidation and exfoliation. The graphitic regions are due to incomplete oxidation of the graphene basal plane, while the amorphous regions of the basal plane result from abundant oxygen-containing groups (e.g. hydroxyls, epoxides and carbonyls) aggregated in these domains. Such different structures and the chemistry of GO may result in many variations for the diffusion, sorption and permeation properties of the resultant GO membranes.

9.3 GO Membrane for Gas Separation

Common sense suggests that 2D layered materials such as GO are usually considered for barrier applications [9] because of their high aspect ratio and tortuosity for molecular diffusion, even if room for molecular diffusion exists in the interlayer spacing. A simple molecular diffusion model for regularly stacked 2D platelets was proposed by Nielsen [19]

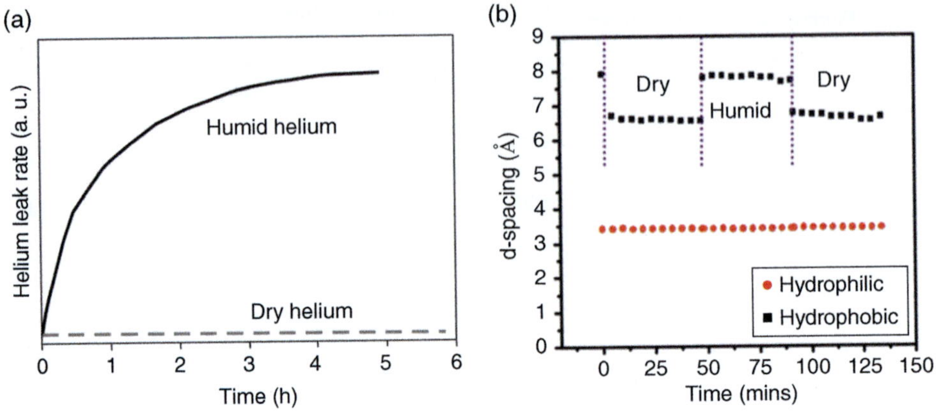

Figure 9.3 (a) Illustration (generalized) of the helium leak rate through GO film under dry and humidified feed conditions [20]. (b) Change of interlayer distance of hydrophilic and hydrophobic domains in GO under wet–dry switching. (b) Reproduced from [21] with permission of ACS

for the case where gas molecules diffuse through a layered structure in the perpendicular direction. Perpendicularly oriented layered platelets increase the gas diffusion path length, giving rise to significant reduction of gas permeability through layered composite membranes. As such, numerous studies to improve the gas-barrier properties (e.g. oxygen barrier) of polymeric films have been extensively made by incorporating GO (or reduced GO) into the polymer matrix, and, in many cases, the gas-barrier properties are significantly improved by adding a small amount of GO [9]. In 2012, Geim's group reported that a dry (or anhydrous) GO thick membrane (thickness ~ 500 nm) is completely impermeable to some liquids, vapors and gases, including helium [20], while very fast permeation of water through the same GO membranes was observed (Figure 9.3a) because GO is hydrophilic due to oxygen-containing groups and the water molecules can be readily sorbed into GO layers in the humidified feed conditions. As the relative humidity increases, the interlayer spacing of GO layers also increases due to more intercalation of water molecules, as a result, allowing the diffusion of small gas molecules through the vacant space by a water-driven diffusion process. It was assumed that the GO laminates consist of crystallites stacked on top of each other, and hydroxyl and epoxy groups attached to the graphene sheets increase the spacing between the GO sheets. Thus, the pristine graphene capillaries are wide open to accommodate the water molecules and the capillaries becomes narrow at a low relative humidity, which does not provide enough van der Waals distance to accommodate water molecules. As a result, water molecules can permeate through GO layers much faster relative to the dry state. More specifically, according to a study on the state of water in GO membranes, revealed by X-ray diffraction (XRD) [21], non-oxidized and hydrophobic graphitic and oxidized and hydrophilic regions coexist in GO sheets. In addition, the interlayer distance of oxidized and hydrophilic regions in GO are very susceptible to relative humidity during the wet–dry switching, but non-oxidized and hydrophobic domains are unchanged, as shown in Figure 9.3(b). Generally, the interlayer distance of graphite is about 0.334 nm while that of GO is 0.6–1.2 nm due to the epoxy, hydroxyl and carbonyl groups attached to the graphene surface. Therefore, it is hard to overcome the van der Waals force between hydrophobic graphene sheets, so water molecules cannot diffuse through the capillary at a low relative humidity. However, when the hydrophilic regions in GO are widely opened at high relative humidity, mono- and bilayers of water networks would be formed in GO layers.

In sharp contrast, in 2013, Kim *et al.* [22] and Li *et al.* [23] simultaneously reported that ultrathin GO membranes, prepared on different porous media, exhibit the desired gas separation characteristics, and selective gas diffusion can be achieved by controlling the gas flow channels and pores via different stacking arrangements. At this point, many questions arise: Where does the difference come from? Is dry GO membrane a barrier or a selective membrane? The answer will be different, depending on GO size, GO thickness, amount of intercalated water, GO stacking manner, degree of oxidation and measuring conditions, which strongly affect the transport properties of the resultant GO membranes. Average GO platelet size can be tuned by original graphite size, oxidation level [24] and sonication [25]. Commonly, sonicated graphite oxide in water can be much exfoliated, but causes substantial damage to the GO platelets, which rather helps to make more permeable GO membranes by reducing the diffusional path in GO membranes. As such, to improve gas permeability through GO membranes, small GO platelet sizes are much preferred for minimizing the pathway for permeating gases. Kim *et al.* [22] reported the change of gas permeability (in barrer unit, 1 barrer $=1\times10^{-10}$ cm^3(STP) cm cm^{-2} s^{-1} $cmHg^{-1}$) of thick GO membranes (~5 μm) in the dry state as a function of average GO size (300, 500 and 1000 nm). They measured the gas permeability by using a high-vacuum time-lag method, which is usually used for measuring the intrinsic gas permeability of dense membranes. As shown in Figure 9.4(a), the permeability of gases of interest significantly increases as the GO size decreases, mainly due to the much shortened diffusional pathway for gases. In most cases, the order of gas permeability in these GO membranes follows the kinetic diameter of the gases ($He>H_2>CO_2>O_2>N_2>CH_4$), indicating that the separation mechanism can be elucidated by molecular sieving as in typical carbon membranes. Only the GO membrane prepared with the GO size of 300 nm shows that the H_2 permeability is slightly higher than the He permeability, meaning that the stacking of smaller GO platelet size leads to high porous structures in GO membrane. This behavior is often observed for CMS membranes. Based on a solution–diffusion mechanism, the diffusion coefficients of He and H_2 are very similar, but the solubility coefficient of H_2 in porous media is slightly higher than that of He because He is an inert gas. Roughly, by reducing the size by half, gas permeability becomes one order of magnitude higher. Moreover, the gas permeability through GO membranes with slit-like pores or channels should be carefully measured by considering the potential energy barrier for gas diffusion. When the pressure difference (or pressure ratio) between the feed and the permeate side is smaller than the potential resistance for gas diffusion in the pore width, the driving force for gas entrance is insufficient to overcome the resistance for gas diffusion. Figure 9.4(b) shows H_2 permeance (in GPU unit, 1 GPU $=1\times10^{-6}$ cm^3(STP) cm^{-2} s^{-1} $cmHg^{-1}$) through thin GO-coated microporous membranes as a function of applied feed pressure with different platelet GO sizes. In regions with low pressure difference, even small gases are hard to diffuse into the GO membrane. In addition, steric hindrance at the entrances to small pores significantly affects the surface barrier resistance to mass transfer, depending on the barrier height on the slit width.

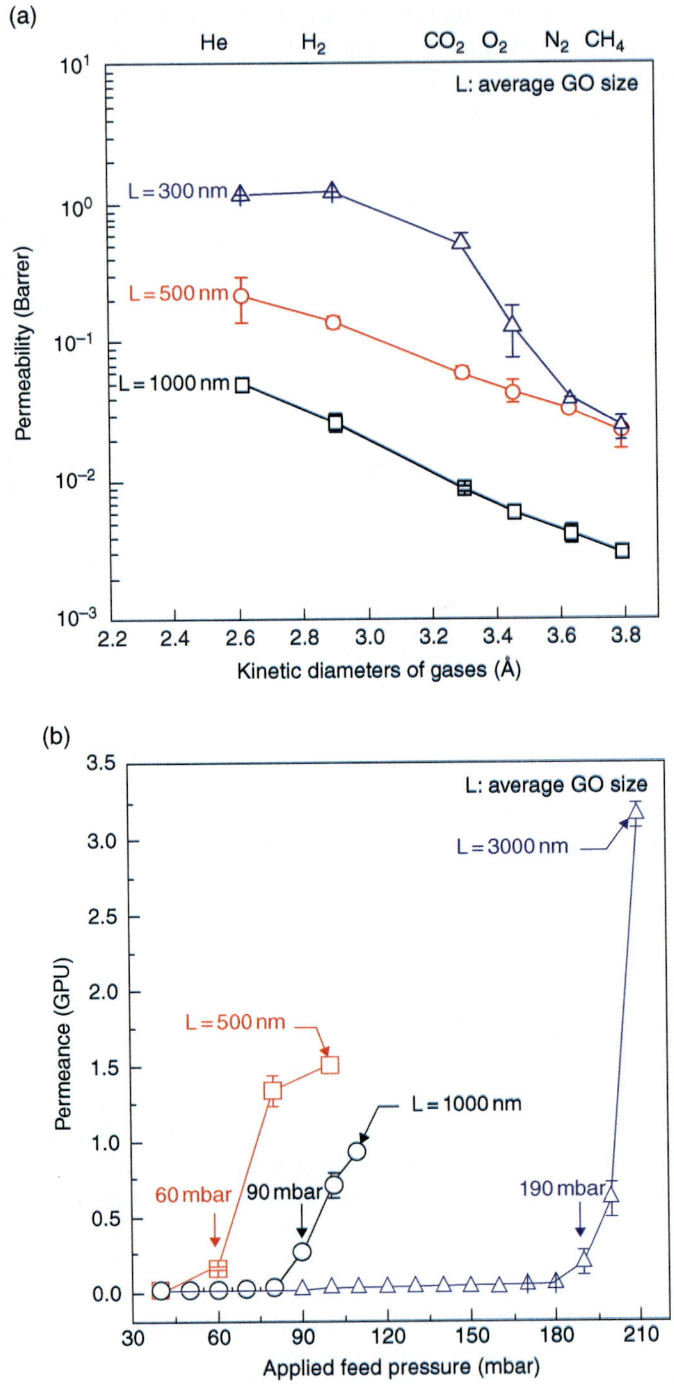

Figure 9.4 *(a) Gas permeability of thick GO membranes with different GO platelet sizes [22]. (b) H_2 permeance of thin GO-coated microporous membranes as a function of applied feed pressure [22]. Reproduced from [22] with permission of AAA Science*

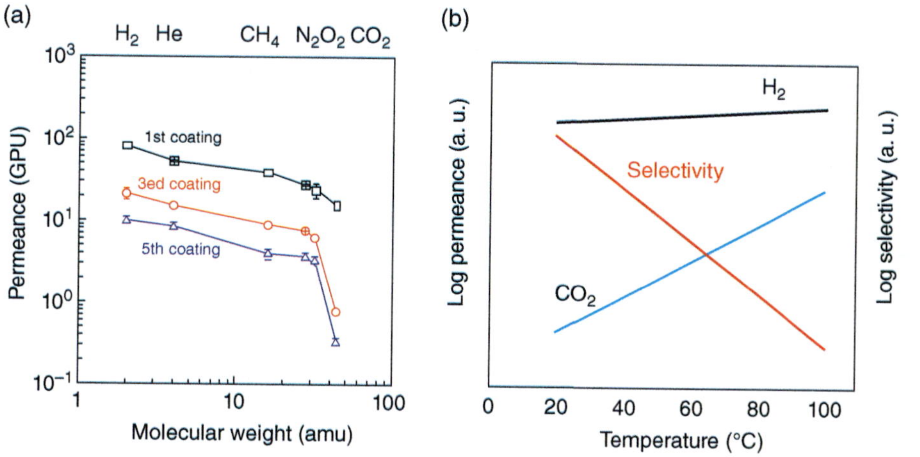

Figure 9.5 (a) Change of gas permeances of GO membrane prepared by method 1 as a function of coating times [22]. (b) Illustration of the H_2 and CO_2 permeances and selectivity of GO as a function of temperature [23]. (a) Reproduced from [22] with permission of AAA Science

The GO membranes with several GO layers exhibit different selective gas transport behaviors, depending on the stacking arrangement of the platelets and the presence of intercalated water molecules in GO layers [22]. To obtain ultrathin GO layers (3–10 nm), a microporous polymeric membrane was used as a mechanical support layer. Also, to control stacking manner and the amount of intercalated water, two different coating methods were used. Several layered GO thin-film composite membranes were prepared by contacting the support membrane surface to the air–liquid interface of a GO solution, followed by spin casting (method 1). The gas transport mechanism through the GO membrane shows typical Knudsen diffusion for which the permeance order is H_2>He>CH_4>N_2>O_2>CO_2, and selectivity is proportional to the square root of the molecular weight ratio, except for CO_2 (Figure 9.5a). Interestingly, the permeance of CO_2 is drastically reduced and reaches a constant value during the measurement. The GO membrane prepared by method 1 includes nanopores surrounded by carboxylic group-functionalized GO edges, due to a less interlocked layer structure caused by preferential face-to-face stacking of different GO sizes [26, 27], meaning that most of the gas molecules diffuse through such nanopores rather than through the interlayer space between the GO layers. Free carboxylic acid groups or carboxylic anions at edges play a significant role in the strong binding with CO_2 molecules in the presence of water, thus leading to low CO_2 permeance at a low feed pressure (~1 bar). Although the theoretical Knudsen selectivity of H_2/CO_2 is 4.67, in the case of GO, the permselectivity increases up to 30. Similar results with GO membranes made via method 1 were also demonstrated by other groups. Li *et al.* prepared an 18 nm thick GO membrane on an anodic aluminum oxide (AAO) support layer with 20 nm pore size by a filtration method, and they reported that H_2/CO_2 selectivity increases up to 3400 due to low CO_2 permeance, as shown Figure 9.5(b) [23]. Such low CO_2 permeance is seldom observed in microporous or nanoporous membranes without any strong adsorption sites [28]. Polar groups such as carboxyl or hydroxyl at GO edges provide a relatively strong interaction with CO_2, giving rise to the retardation of CO_2 transport.

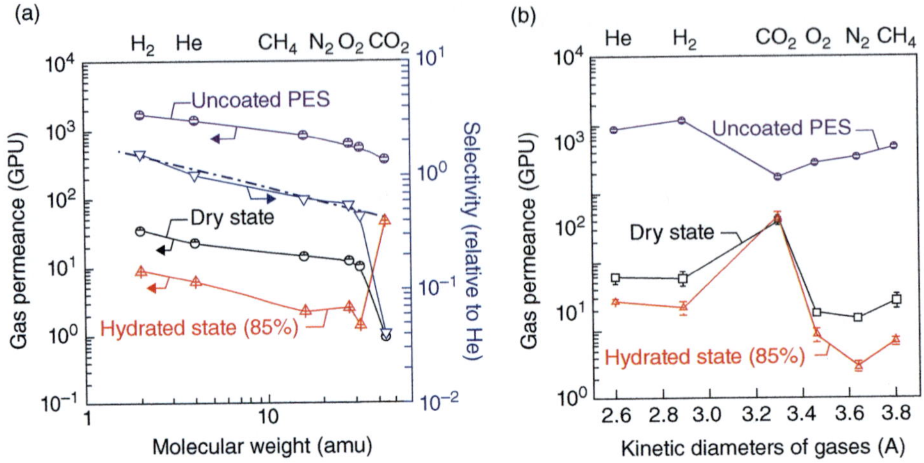

Figure 9.6 *Gas permeance of GO membranes prepared by (a) method 1 and (b) method 2 under dry and humidified feed conditions (based on single gas permeation measurement) [22]. Reproduced from [22] with permission of AAA Science*

On the other hand, other GO thin-film composite membranes were prepared by dropping a constant volume of GO solution directly onto spinning polymeric support (called method 2). The GO membrane formed by this method possesses a highly interlocked structure, like a brick model, including a relatively higher amount of intercalated water molecules as compared to GO membranes prepared via method 1. In Figure 9.6(b), the gas transport mechanism shows the highest CO_2 permeance, and the gas permeance order is $CO_2 > He > H_2 > CH_4 > O_2 > N_2$. The CO_2 permeance through the highly interlocked layered GO membrane is the most permeable as compared to other gases due to the water-enhanced separation process and CO_2/N_2 selectivity reaches about 20. Interestingly, by increasing the relative humidity (RH) in the feed gas (0 to 85% RH), water molecules can interact preferentially with polar groups via hydrogen bonds, and then nanopores and interlayers between GO sheets can be filled with water molecules. As a result, the CO_2 permeance drastically increases in the hydrated state because CO_2 is more soluble in water than any other gases, based on solution–diffusion mechanism (Figure 9.6a). In addition, CO_2/N_2 selectivity increases up to 60 with increasing CO_2 permeance at 85% RH, as shown in Figure 9.6(b). That behavior is not often observed in conventional gas separation membrane materials. Since many industrial gas streams (e.g. post-combustion, natural gas purification and syngas adjustment) contain water vapor, the effect of water vapor on membrane separation performance still remains a significant issue to be considered in the evaluation of the mass transport properties of polymeric membranes. In general, water vapor strongly deteriorates conventional membrane performance [29]; with reduction of both permeability and selectivity by water condensation on membrane surfaces or in membrane pores, an energy costly water vapor removal unit before the membrane unit is necessary. In this regard, water-enhanced CO_2 separation in GO membranes will be a great advantage in post-combustion CO_2 capture design because dehydration by condensation in the permeate side is easier than dehydration in the high-pressure feed side [30, 31].

9.4 GO Membrane for Water Purification and Desalination

Owing to the hydrophilic nature of GO, GO membranes have also been extensively investigated for water purification and desalination applications. Joshi *et al.* reported the permeation of different ionic charges through micrometer-thick, freestanding GO membranes [32]. Using a simple two-chamber diffusion cell, they initially measured permeation rates with different liquids such as water, glycerol, toluene, ethanol, benzene and dimethyl sulfoxide. No permeation was measured over a long time by monitoring liquid levels and using chemical analysis. However, water molecules can permeate rapidly through the GO membrane by osmotic pressure when the permeate side is filled with water and the feed side is filled with a 1 M sucrose solution. In addition, the permeability of various ions and organic molecules was measured by using a conductivity meter and total organic carbon analyzer. Several salt solutions with different ion sizes were employed to examine the salt permeability through GO membranes as a function of ionic size. As shown in Figure 9.7, relatively small sized ions such as potassium or magnesium ions can permeate fast through GO membranes and their permeabilities are almost the same, whereas ions with larger sizes and organic molecules showed no detectable permeation rate through the same GO membranes. They proposed that ions or molecules dissolved in water can permeate rapidly through the graphene nanocapillaries until the physical size of the ions or molecules exceeds the critical one, which is approximately 4.5 Å. To explain this observation, they used the same model previously suggested to elucidate the unimpeded evaporation of water through GO membranes [20]; in a GO crystallite, highly oxidized domains act as spacers that keep adjacent crystallites apart and also prevent them from being dissolved. These spacers let water molecules insert easily in the interlayer of GO, but hydrophobic graphitic domains provide a network of capillaries that allow nearly frictionless flow of correlated water. Similar results can be found in the water transport through carbon nanotubes [33].

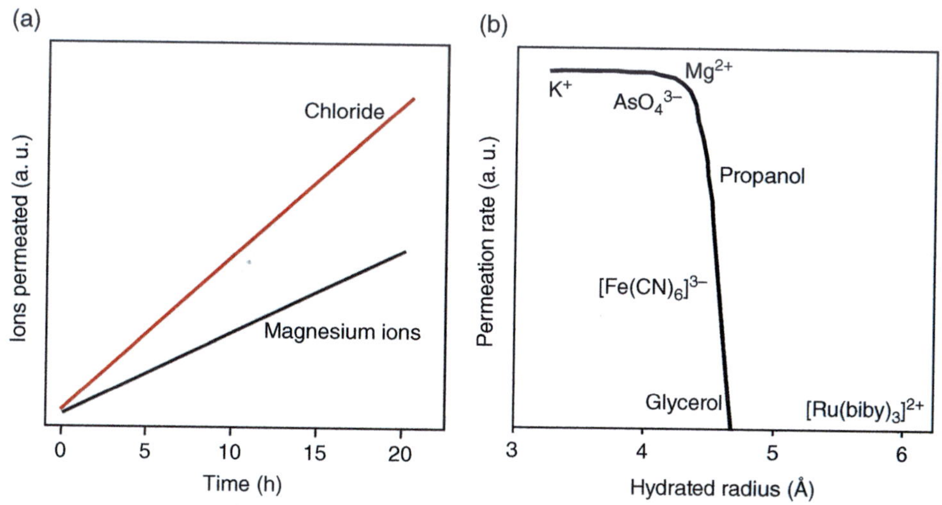

Figure 9.7 *(a) Illustration of the permeation through a 5 mm thick GO membrane from the feed compartment with a 0.2 M solution of* $MgCl_2$*. (b) Illustration of the permeation rate through the GO membrane as a function of hydrated ion radius [32]*

Generally, the model seems to be in a good agreement with their experimental data. However, this model is still too simplified to reflect the complicated chemistry and structure of the real GO membrane structure related to water and ion permeability.

Hu *et al.* prepared a GO thin-film composite (TFC) membrane for nanofiltration (NF) applications. Thin GO was coated on a microporous polysulfone support membrane that was treated with polydopamine to become hydrophilic and then cross-linked by trimesoyl chloride (TMC) [34]. Because the polysulfone membrane is hydrophobic, hydrophilic surface modification is necessary to coat the hydrophilic GO layer on the polysulfone membrane. Polydopamine is a bio-inspired material that is widely used for hydrophilic surface modification, resulting in much improved adhesion properties with the hydrophilic GO layer.

In addition, for strong integrity of the GO thin layer and increasing interlayer space, TMC was used as a cross-linking agent. The water flux though the cross-linked GO TFC membrane was in the range from 8 to 27.6 lm^{-2} h^{-1} bar^{-1}, as shown in Figure 9.8(a).

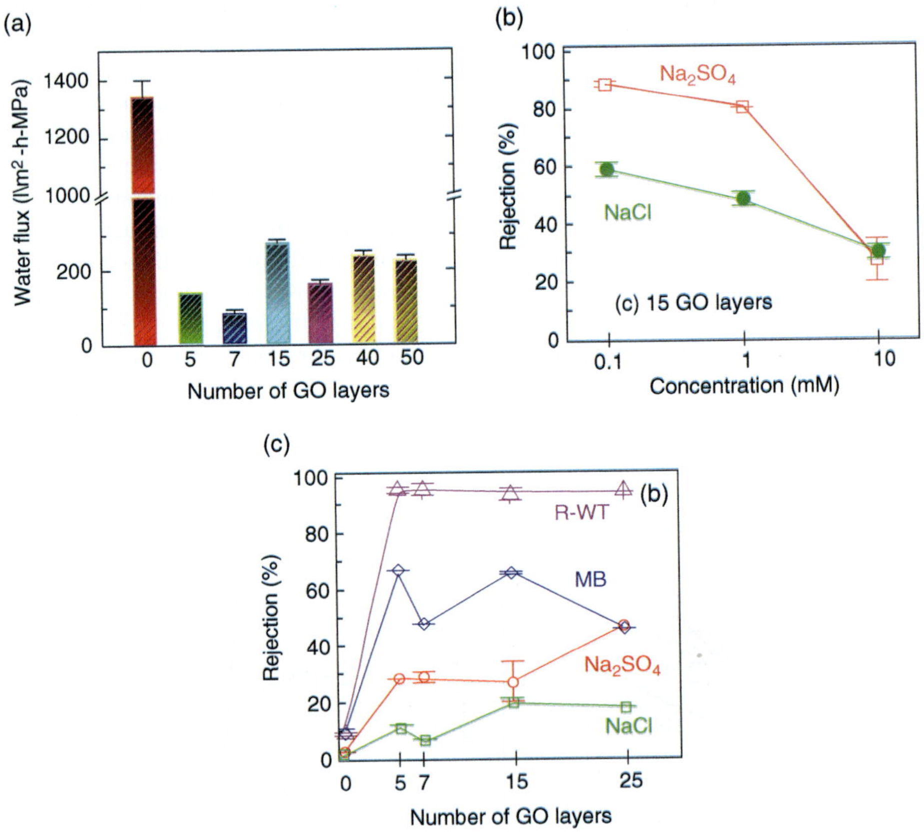

Figure 9.8 *(a) Water flux through a GO thin membrane as a function of number of GO layers. (b) Rejection of monovalent and divalent ions in a GO thin membrane as a function of salt concentration. (c) Rejection of dye (MB, methylene blue; R-WT, rhodamine-WT) and salts in a GO thin membrane as a function of number of GO layers. Reproduced from [34] with permission of ACS*

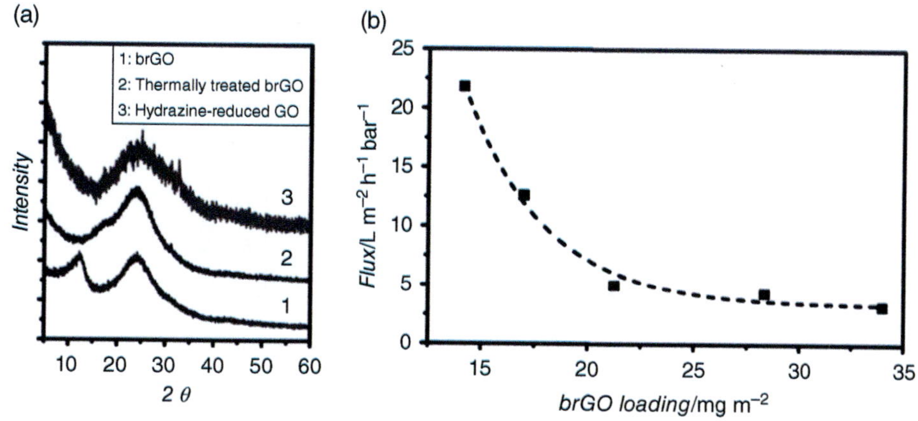

Figure 9.9 *(a) XRD patterns of base-refluxing reduced GO, hydrazine-reduced GO and thermally treated bRGO at 220 °C in vacuum. (b) Variation of pure water flux as a function of bRGO loading coated on the membranes. Reproduced from [35] with permission of John Wiley & Sons*

Those values are roughly 4–10 times higher than that of most conventional NF membranes. However, as depicted in Figure 9.8(b) and (c), the rejection of monovalent and divalent salts is relatively low (6–46%), while exhibiting moderate rejection of methylene blue, and a high rejection of rhodamine-WT.

Thin base-refluxing reduced GO (bRGO) membranes were prepared by filtering bRGO flakes on porous support substrates [35]. Differing from RGO synthesized using a strong reduction agent such as hydrazine, the bRGO was partially reduced by adding base solution such as NaOH or KOH. Based on XRD data, as shown in Figure 9.9(a), there are two distinguishable peaks of GO and RGO, meaning that hydrophilic and hydrophobic domains coexist in bRGO. Although hydrophobic domains are increased by partial reduction, thin bRGO membranes still exhibit high water flux (21.81 m^{-2} h^{-1} bar^{-1}, as shown in Figure 9.9b), with high dye rejection. Supposedly, the high water flux through partially reduced GO membranes might be due to slip flow between hydrophobic graphene sheets, as happens inside carbon nanotubes. To determine the effect of charge, the fluxes of solvents such as isopropanol, ethanol, hexane, cyclohexane and toluene were measured. Among them, hexane, the most hydrophobic liquid, shows the lowest flux, mainly due to the slip flow theory that a more hydrophobic liquid would lead to lower flux under the same pressure owing to greater interaction between the hydrophobic liquid and the hydrophobic graphene planes. However, the molecular size of the liquid, as compared with that of water, should be considered along with the interlayer spacing (or pore width), and also hydrophilic–hydrophobic repulsion should be considered for that phenomenon. In addition, oxygen moieties in hydrophilic domains would impede the transport of water due to the strong interaction between them. On the other hand, the graphene domains without any oxygen functional groups would be responsible for the rapid water transport.

The nanostrand-channeled GO (NSC-GO) membranes for ultrafiltration (UF) prepared by Huang *et al.* [36] were formed by filtering a mixed dispersion with electronegative GO and positively charged copper hydroxide nanostrands (CHNs). The 3–5 nm of nanostrand

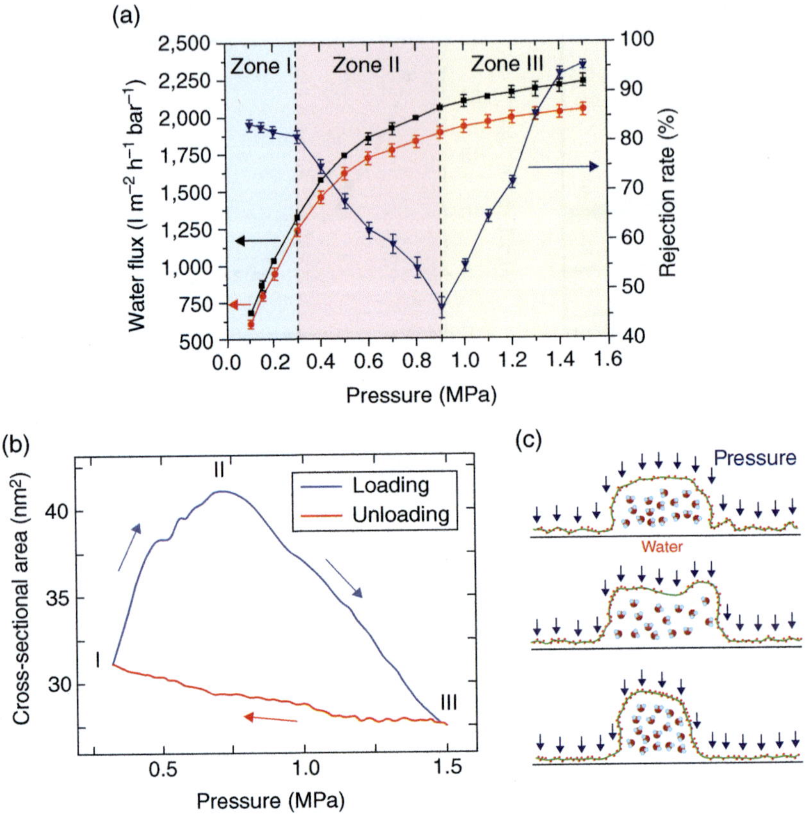

Figure 9.10 *(a) Pressure-dependent flux and rejection of EB molecules of NSC-GO membrane under different pressure. The curves marked by black solid squares and red solid circles represent the flux variation during the first and third pressure-loading processes, respectively. The curve marked by blue solid triangles denotes the rejection rate of EB during the first pressure-loading process. (b) Simulated changes in the cross-sectional area of a nanochannel by varying the applied pressure. (c) The response of a half cylindrical GO nanochannel modelled in molecular dynamics simulation. Reproduced from [36] with permission of Nature Publishing Group*

channels are created in GO layers by removing the CHNs. The membrane performance showed extremely high water flux (695 l m^{-2} h^{-1} bar^{-1}) with high rejection properties (Figure 9.10a). Interestingly, water flux and rejection in GO membranes exhibited pressure-dependent character. Water flux does not increase linearly with increasing feed pressure, and also the dye rejection of Evans blue (EB) gradually decreases with increasing feed pressure, and the rejection increases again at a certain pressure. Since water molecules are trapped between GO layers, water pockets are formed by increasing pressure. The shape of the water pockets changes from round into a flattened rectangle as the pressure is increased as high as 7.5 bar. The flattened rectangle is continuously compressed into round ripples as the pressure further increases to as high as 15 bar, as shown in Figure 9.10(b) and (c). When the channel is flattened, the cross-sectional area increases, and thus the permeation of EB molecules is significantly enhanced. During further compression, consecutive shrinkage of the nanochannels will reduce the cross-sectional area and increase the rejection rate again. Also, the compressed small nanochannels will almost recover to the original state after the pressure is released, which is reasonable because the strain energy stored in the water-containing curved GO sheets is high enough to overcome the van der Waals forces at the contact region of the newly formed smaller nanochannels.

GO composite polyamide (PA) TFC membranes were prepared by Bano *et al.* [37] for NF applications. In membrane preparation, GO was dispersed in aqueous amine monomer (e.g. 1,3-phenylene diamine (MPD)) solution. After soaking a porous support layer in MPD/GO solution, the excessive solution on the support layer was removed. Then, the interfacial polymerization was performed by contacting the membrane surface with TMC solution. As a result, GO composite PA TFC membranes with various GO compositions were obtained. The water flux of GO composite PA TFC membrane increases with increasing GO composition, while the salt rejection was maintained at a high level due to the hydrophilicity of GO. In addition, hydrophilic GO composite PA TFC membranes showed improved anti-fouling properties compared to pristine PA TFC membranes.

A few simulation studies on mass transport through GO membranes as well as porous graphene membranes have been reported. However, it is hard to perform simulations on GO membranes since the structure and chemistry of GO are much more complicated than those of graphene or porous graphene. As shown in Figure 9.11(a), a simulation on water and ion transport through graphene oxide framework (GOF) membranes were performed by Nicolaï *et al.* [38]. GOF is a recently synthesized nanoporous material prepared with GO layers cross-linked by linear linker molecules (boronic acid). In Figure 9.11(b), the effect of linear linker concentration and membrane thickness on the transport properties of GOF membranes were investigated by using molecular dynamics simulations. The simulated ultrathin GOF membranes show better water desalination properties with excellent water flux and salt rejection properties compared to conventional desalination membranes. However, the preparation of such thin GOF membranes might be a great challenge; current polymeric desalination membranes are still keeping their top position in existing desalination technologies.

9.5 Other Membrane Applications

9.5.1 Fuel Cell Membrane

In general, every fuel cell consists of an anode, cathode and electrolyte. Currently, polymer electrolyte fuel cells (so-called PEMFC) are one of the most commercially available fuel cell technologies owing to their simplicity, rapid start-up and low operation temperature. A polymer electrolyte membrane (PEM) is a key component in a PEMFC, and most PEMs are sulfonated ionomers, such as Nafion, because they have high proton conductivity and act as a separator between anode and cathode. Recently, GO has been considered as a proton-exchange membrane for fuel cells by using pure GO membranes or by incorporating GO into conventional ionomers. Interestingly, Tateishi *et al.* [39] reported the fuel cell performance by employing a GO thick membrane, showing proton conductivities of

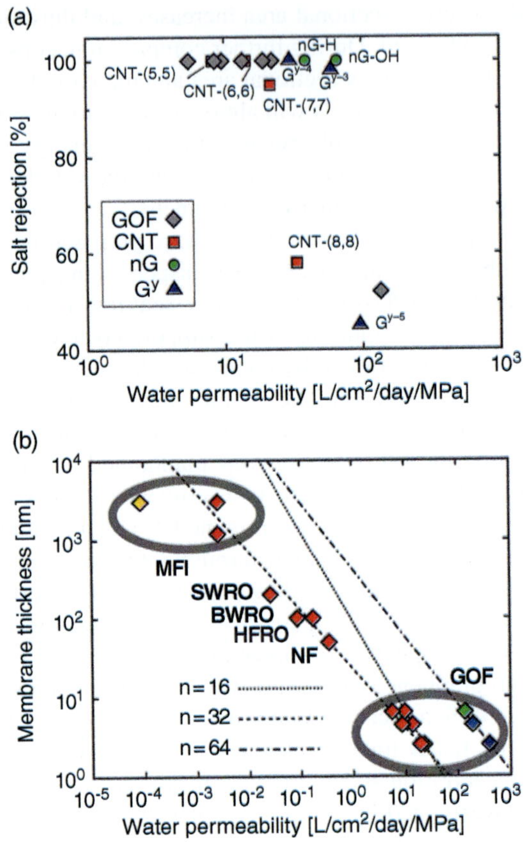

Figure 9.11 (a) Water desalination performance chart for GOF membranes compared to existing theoretical work on carbon-based materials. The labels are: CNT, carbon nanotube; nG, nanoporous graphene; and Gy, graphyne. (b) Water desalination performance chart for GOF membranes compared to existing organic and inorganic membrane technologies. Reproduced from [38] with permission of RSC

10^{-2} to 10^{-1} S cm^{-1} at 60–100% RH at room temperature, which is comparable to that of current Nafion membranes. Of course, proton conduction mechanisms through dry or hydrated GO membranes are not clear yet and also partial reduction during electrochemical reaction or at elevated temperature could be problematic. In addition, chemical, mechanical and long-term stability in severe fuel cell operation conditions should be further checked for practical applications, but the early result is very promising to explore a new type of fuel cell based on GO.

9.5.2 Ion-Selective Membrane for Next-Generation Batteries

Electrochemical energy storage systems such as rechargeable batteries have been extensively used to store electrical energy. Every electrochemical energy storage device generally consists of electrodes (i.e. cathode and anode), electrolytes and separators (i.e. ion-selective membrane). Since the great success of Li-ion batteries in the 1990s, many portable electronic devices and electrical vehicles have been developed, and in the future the development of rechargeable energy storage systems with high energy density will be urgently needed for next-generation transportation and energy harvesting/storage from renewable energy resources. Recently, lithium–air and lithium–sulfur batteries have been considered for next-generation high-energy-density batteries. However, these batteries suffer from rapid capacity degradation and poor cycling stability. To solve this problem, Huang *et al.* [40] used permselective GO membranes in a lithium–sulfur battery configuration for high stability. They reported that high permselectivity of lithium ions over polysulfide anions of GO membranes helps to improve anti-self-discharging properties, mainly due to both the physical barrier effect and the chemical barrier effect with oxygenated functional groups of GO membranes.

9.5.3 Dehydration

Pervaporation (PV) is an effective membrane separation process for diluting solutions containing trace or minor amounts of the component to be removed. Based on the solution–diffusion mechanism, the PV process can separate mixtures of liquids by partial vaporization through non-porous membranes. Usually, hydrophilic polymeric membranes are used for dehydration of alcohols containing small amounts of water, while hydrophobic polymeric membranes are employed for removal of trace amounts of volatile organic compounds from aqueous solutions. Differing from other experiments, Huang and coworkers [41] prepared GO-coated ceramic hollow fiber membranes by a vacuum suction method and they tried to separate water from dimethyl carbonate (DMC)/water mixtures by using a PV process. As a result, the GO-coated hollow fiber membranes showed efficiently selective water permeation performance for low-toxicity and degradable DMC. To explain the high water/DMC selectivity, they applied a typical solution–diffusion mechanism, that is, GO can sorb water more preferentially than DMC because of the presence of polar groups for hydrogen bonding with water, and also the water diffusivity is much faster than for other molecules; thus, selective and faster water permeation is the result of contributions of both enhanced diffusivity and solubility. As such, a solution–diffusion mechanism is effective to elucidate current results. That is, the permeants dissolve in the membrane material and then diffuse through the membrane down a concentration gradient. An effective separation can be achieved between different permeants because of differences in the amount of material that dissolves in the membrane and the rate at which the material diffuses through the membrane. Here, diffusion through the membrane is the rate-limiting step in the mass transport across a membrane.

9.6 Conclusions and Future Prospects

Membrane-based gas and liquid separation processes have several advantages over conventional separation technologies in industry. However, new membrane platforms still need further development due to exacting criteria for competing with other separation technologies. High permeability and selectivity of separation are the keys to membranes suitable for applications. At this point, graphene and its derivatives are promising membrane materials among various nanomaterials due to their 2D platform, ease of thin-film formation and scalability. In particular, GO has emerged as a rising star material that can separate molecules and ions with high precision. Owing to its excellent membrane properties, we believe that GO will be a promising carbon material for a wide variety of membrane applications, as several research groups have reported outstanding separation properties related to GO or modified GO membranes to support the versatile future of the GO membrane. Compared with other membrane materials, the GO membrane is definitely quite a new and promising membrane material due to easy membrane formation, scalability and versatile applications. However, several important issues need to be solved before membrane applications become reality. Precise diffusion channel engineering with high porosity and coating on large-area porous mechanical support layers are key challenges. Both theoretical and experimental mass transport or separation properties of GO membranes have to be studied more intensely, considering thicker and thinner membranes. Although GO membranes are more feasible for practical membrane applications, long diffusion paths due to high aspect ratio lead to low permeance. Moreover, the membrane performance of GO can be significantly affected by platelet size, thickness, the presence of intercalated water and surface chemistry. To improve the separation performance of current GO membranes, smaller flake size, reduced thickness, controlled interlayer spacing and precise pore engineering with high porosity on GO sheets should be further considered for real applications. Furthermore, the chemical and mechanical stability of GO membranes should be further investigated under some harsh membrane operation conditions.

References

[1] Baker, R.W., *Membrane Technology and Applications*, 3rd edn. John Wiley & Sons, Ltd, Chichester, 2012.

[2] Robeson, L.M., The upper bound revisited. *J. Membrane Sci.* **2008**, *320* (1–2), 390–400.

[3] Geise, G.M.; Park, H.B.; Sagle, A.C.; *et al.*, Water permeability and water/salt selectivity tradeoff in polymers for desalination. *J. Membrane Sci.* **2011**, *369* (1), 130–138.

[4] Bunch, J. S.; Verbridge, S. S.; Alden, J. S.; *et al.*, Impermeable atomic membranes from graphene sheets. *Nano Lett.* **2008**, *8* (8), 2458–2462.

[5] Fischbein, M.D.; Drndić, M., Electron beam nanosculpting of suspended graphene sheets. *Appl. Phys. Lett.* **2008**, *93* (11), 113107.

[6] Bell, D.C.; Lemme, M.C.; Stern, L.A.; *et al.*, Precision cutting and patterning of graphene with helium ions. *Nanotechnology* **2009**, *20* (45), 455301.

[7] Celebi, K.; Buchheim, J.; Wyss, R.M.; *et al.*, Ultimate permeate across atomically thin porous graphene. *Science* **2014**, *344* (6181), 289–292.

[8] Bieri, M.; Treier, M.; Cai, J.; *et al.*, Porous graphenes: two-dimensional polymer synthesis with atomic precision. *Chem. Commun.* **2009**, *2009* (45), 6919–6921.

[9] Yoo, B.M.; Shin, H.J.; Yoon, H.W.; Park, H.B., Graphene and graphene oxide and their uses in barrier polymers. *J. Appl. Polym. Sci.* **2014**, *131* (1), 39628.

[10] Dreyer, D.R.; Park, S.; Bielawski, C.W.; Ruoff, R.S., The chemistry of graphene oxide. *Chem. Soc. Rev.* **2010**, *39* (1), 228–240.

[11] Chua, C.K.; Pumera, M., Chemical reduction of graphene oxide: a synthetic chemistry viewpoint. *Chem. Soc. Rev.* **2014**, *43* (1), 291–312.

[12] Brodie, B.C., On the atomic weight of graphite. *Phil. Trans. R. Soc. London* **1859**, *14*, 249–259.

[13] Staudenmaier, L., Verfahren zur Darstellung der Graphitsäure. *Ber. Dtsch. Chem. Ges.* **1898**, *31* (2), 1481–1487.

[14] Hummers, W.S.; Offeman, R. E., Preparation of graphitic oxide. *J. Am. Chem. Soc.* **1958**, *80* (6), 1339.

[15] Dikin, D.A.; Stankovich, S.; Zimney, E.J.; *et al.*, Preparation and characterization of graphene oxide paper. *Science* **2007**, *448* (7152), 457–460.
[16] Hwa, T.; Kokufuta, E.; Tanaka, T., Conformation of graphite oxide membranes in solution. *Phys. Rev. A* **1991**, *44*, R2235–R2238.
[17] Horiuchi, S.; Gotou, T.; Fujiwara, M.; *et al.*, Carbon nanofilm with a new structure and property. *Jpn. J. Appl. Phys.* **2003**, *42*, L1073–L1076.
[18] Boehm, H.P.; Clauss, A.; Hofmann, U., Graphite oxide and its membrane properties. *J. Chim. Phys.* **1960**, *58* (12), 110–117.
[19] Nielsen, L.E., Models for the permeability of filled polymer systems. *J. Macromol. Sci. Chem.* **1967**, *1* (5), 929–942.
[20] Nair, R.; Wu, H.; Jayaram, P.; *et al.*, Unimpeded permeation of water through helium-leak-tight graphene-based membranes. *Science* **2012**, *335* (6067), 442–444.
[21] Zhu, J.; Andres, C.M.; Xu, J.D.; *et al.*, Pseudonegative thermal expansion and the state of water in graphene oxide layered assemblies. *ACS Nano* **2012**, *6* (9), 8357–8365.
[22] Kim, H.W.; Yoon, H.W.; Yoon, S.-M.; *et al.*, Selective gas transport through few-layered graphene and graphene oxide membranes. *Science* **2013**, *342* (6154), 91–95.
[23] Li, H.; Song, Z.; Zhang, X.; *et al.*, Ultrathin, molecular-sieving graphene oxide membranes for selective hydrogen separation. *Science* **2013**, *342* (6154), 95–98.
[24] Eda, G.; Chhowalla, M., Graphene-based composite thin films for electronics. *Nano Lett.* **2009**, *9* (2), 814–881.
[25] Zhang, L.; Liang, J.; Huang, Y.; *et al.*, Size-controlled synthesis of graphene oxide sheets on a large scale using chemical exfoliation. *Carbon* **2009**, *47* (14), 3365–3368.
[26] Cote, L.J.; Kim, F.; Huang, J., Langmuir–Blodgett assembly of graphite oxide single layers. *J. Am. Chem. Soc.* **2008**, *131* (3), 1043–1049.
[27] Kim, J.; Cote, L.J.; Kim, F.; *et al.*, Graphene oxide sheets at interfaces. *J. Am. Chem. Soc.* **2010**, *132* (23), 8180–8186.
[28] Boffa, V.; ten Elshof, J.E.; Petukhov, A.V.; Blank, D.H., Microporous niobia–silica membrane with very low CO_2 permeability. *ChemSusChem* **2008**, *1* (5), 437–443.
[29] Scholes, C.A.; Kentish, S.E.; Stevens, G.W., Effects of minor components in carbon dioxide capture using polymeric gas separation membranes. *Sep. Purif. Rev.* **2009**, *38* (1), 1–44.
[30] Park, H.B., Graphene-based membranes – a new opportunity for CO_2 separation. *Carbon Managem.* **2014**, *5* (3), 251–253.
[31] Kim, H.W.; Yoon, H.W.; Yoo, B.M.; *et al.*, High-performance CO_2-philic graphene oxide membranes under wet-conditions. *Chem. Commun.* **2014**, *50* (88), 13563–13566.
[32] Joshi, R.; Carbone, P.; Wang, F.; *et al.*, Precise and ultrafast molecular sieving through graphene oxide membranes. *Science* **2014**, *343* (6172), 752–754.
[33] Holt, J.K.; Park, H.G.; Wang, Y.; *et al.*, Fast mass transport through sub-2-nanometer carbon nanotubes. *Science* **2006**, *312* (5776), 1034–1037.
[34] Hu, M.; Mi, B., Enabling graphene oxide nanosheets as water separation membranes. *Environ. Sci. Technol.* **2013**, *47* (8), 3715–3723.
[35] Han, Y.; Xu, Z.; Gao, C., Ultrathin graphene nanofiltration membrane for water purification. *Adv. Funct. Mater.* **2013**, *23* (29), 3693–3700.
[36] Huang, H.; Song, Z.; Wei, N.; *et al.*, Ultrafast viscous water flow through nanostrand-channelled graphene oxide membranes. *Nature Commun.* **2013**, *4*, 2979.
[37] Bano, S.; Mahmood, A.; Kim, S.-J.; Lee, K.-H., Graphene oxide modified polyamide nanofiltration membrane with improved flux and antifouling properties. *J. Mater. Chem. A* **2015**, *3* (5), 2065–2071.
[38] Nicolaï, A.; Sumpter, B.G.; Meunier, V., Tunable water desalination across graphene oxide framework membranes. *Phys. Chem. Chem. Phys.* **2014**, *16* (18), 8646–8654.
[39] Tateishi, H.; Hatakeyama, K.; Ogata, C.; *et al.*, Graphene oxide fuel cell. *J. Electrochem. Soc.* **2013**, *160* (11), F1175–F1178.
[40] Huang, J.-Q.; Zhuang, T.-Z.; Zhang, Q.; *et al.*, Permselective graphene oxide membrane for highly stable and anti-self-discharge lithium–sulfur batteries. *ACS Nano* **2015**, *9* (3), 3002–3011.
[41] Huang, K.; Liu, G.; Lou, Y.; *et al.*, A graphene oxide membrane with highly selective molecular separation of aqueous organic solution. *Angew. Chem. Int. Ed.* **2014**, *53* (27), 6929–6932.

10

Graphene Oxide-Based Composite Materials

Mohsen Moazzami Gudarzi, Seyed Hamed Aboutalebi and Farhad Sharif

10.1 Introduction

The discovery of the extraordinary properties of two-dimensional (2D) materials, especially those of graphene, in 2004 shook the physics community, and soon after the world of materials science [1, 2]. The recent graphene-related research in chemistry was inspired by the earlier research on carbon nanotubes (CNTs) and graphite [2, 3]. Digging through those studies led the Nobel Prize winners to the first concrete observation of "real atomically thin" graphene. Similar material had been reported well before 2004 by chemists who were working on graphite oxide. However, at that time, the significance of their report was not recognized [3]. Graphene research in chemistry has roughly 150 years of historical background but it has only been in the past 10 years that research in this area has boomed. We begin this chapter with a review of the earlier research on graphite (oxide). We then briefly discuss research on graphene oxide (GO). Next, the research on GO is compared with that of CNTs. This brief review at the beginning of the chapter aims to provide an insight into the opportunities and challenges we are dealing with in GO-based materials and composites.

Graphite is a natural form of carbon, which was discovered more than five centuries ago. The unique physical and mechanical properties of graphite create many opportunities for this form of carbon to be used in a wide range of applications, such as refractories, batteries, lubricants, etc. [4, 5]. The cornerstone of the unique properties of graphite lies in its

Graphene Oxide: Fundamentals and Applications, First Edition. Edited by Ayrat M. Dimiev and Siegfried Eigler.

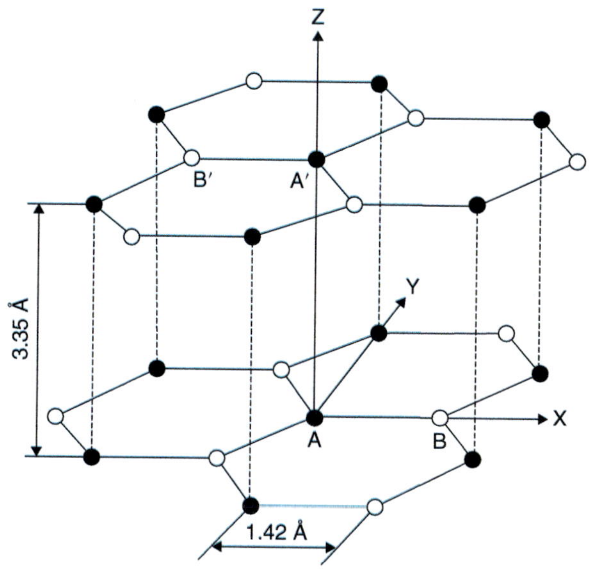

Figure 10.1 *The crystal structure of graphite. Reproduced from [6] with permission of Springer*

atomic structure, with layers of honeycomb carbon atoms. Carbon atoms bond together with sp^2 hybridization, on a layer plane (graphene), generating π electrons with high mobility. As a result, the carbon atoms in the basal plane bond covalently whereas the carbon layers in the *c*-direction (perpendicular to the basal plane) are held together by van der Waals interactions, which are orders of magnitude weaker than covalent bonds [6]. Figure 10.1 illustrates the atomic structure of graphite. More information on the structure and physical properties of graphite can be found elsewhere [4–6].

The aim of this section is to provide an oversight on the evolution of graphite-based composite materials, due to their potential impact on the development of the next generation of this class of materials, i.e. graphene-based composites. To this end, a brief history is presented on the attempts made to harness the unique properties of graphite to improve a wide range of polymeric matrices. In fact, many recent studies on graphenic composites are similar to the previous attempts [7–9]. Developing new routes to synthesize and functionalize graphene sheets is just a new tool for the same design principles and purposes.

Introducing fullerenes and CNTs also had a tremendous impact on the field of polymer nanocomposites [10–13]. Over the past three decades, efforts on designing and developing composite materials based on these two forms of carbon have provided generic principles that can be employed for formulating their graphene-based counterparts [12, 13]. This is even more valid for the functionalization and assembly of carbonic particles [12–15]. There are many strategies for treating graphene that are simply duplicated from the work on CNTs [9, 15]. Reviewing and analyzing these similarities and differences give more insights, in general, into material design, and, more specifically, into carbon-based composites.

10.1.1 How Graphite Met Polymers?

Graphite's physical properties, such as very high thermal and electrical conductivity, were known even before the advent of synthetic polymers [4, 5]. With the advent of polymers, polymeric resins and graphite were mixed together with the goal of producing novel composite materials. Much focus in this regard was made on the fabrication of electrically and/or thermally conductive composites or improving the tribomechanical properties of polymeric resins [4, 5, 16, 17]. These efforts led to the development of commercially available graphite-based composites.

Graphite is the most abundant natural form of carbon (the annual production of graphite exceeds 10^6 tonnes), and as a result it is an inexpensive mineral (a few US dollars per kilogram) [18]. The in-plane conductivity of graphite is in the order of 10^4 S cm^{-1}, which is more than 10 orders of magnitude higher than that for general-purpose polymers [6]. Other common choices that are used to improve a polymer's conductivity include metal particles, carbon black or carbon fibers. However, graphite exhibits some advantages over those options. It has lower density, higher chemical resistivity and lower cost compared to metal particles. It is also less expensive compared to other carbon forms [6, 16, 17]. This makes graphite an attractive candidate for the fabrication of electrically conductive resins and adhesives. Graphite-based conductive coatings and inks are the most commercially available form of this class of composites [19].

The first effort to make polymer–graphite composites dates back to 1915 by Aylsworth [20, 21]. As a general rule, in particulate composites, the use of finer particles, which are less agglomerated and well exfoliated, results in better properties. This was true even for the earliest attempt in which they used expanded graphite instead of graphite flakes [21]. The main objective of the later attempts was to improve the state of dispersion of graphite flakes in polymer matrices, and the interfacial interaction between the two phases [21, 22]. Difficulties in tuning the interaction of graphite and polymers originate from the relatively inert nature, and also high surface energy of the graphite. The latter became a topic of general interest because the principles of engineering the graphite–polymer interface could be applied to other classes of carbon-based composites [23].

From the very beginning of the research on graphite, the flaky appearance of graphite was attributed to its layered structure [24]. However, it was not till the 1930s that the X-ray diffraction (XRD) technique was introduced and confirmed the layered structure of graphite, with an interlayer distance of ~3.4 Å [25]. Similar to many layered materials, graphite can be intercalated with small molecules or atoms. These intercalated structures are called graphite intercalated compounds (GICs). Progress in the synthesis and understanding of the chemical structure of GICs made a big contribution to the current understanding of chemically derived graphene [3, 26, 27]. Wide ranges of chemicals can intercalate into the graphite interlayer, such as inorganic acids, alkali metals and halogens [26]. Observation of peculiar electronic and physical properties, such as superconductivity in alkali-metal-based GICs, attracted a great deal of attention from the physics community [26, 28]. These studies brought inspiration and the basis for the graphene era after 2004 [3].

The synthesis of acid-based GICs, which dates back to 1841, is done by treating graphite flakes with strong acids or mixtures of strong oxidizers [29]. It was then discovered that the sudden vaporization of the intercalate molecules results in the expansion of graphite [30]. The expansion ratio (the ratio of the initial volume to the final volume) can easily exceed

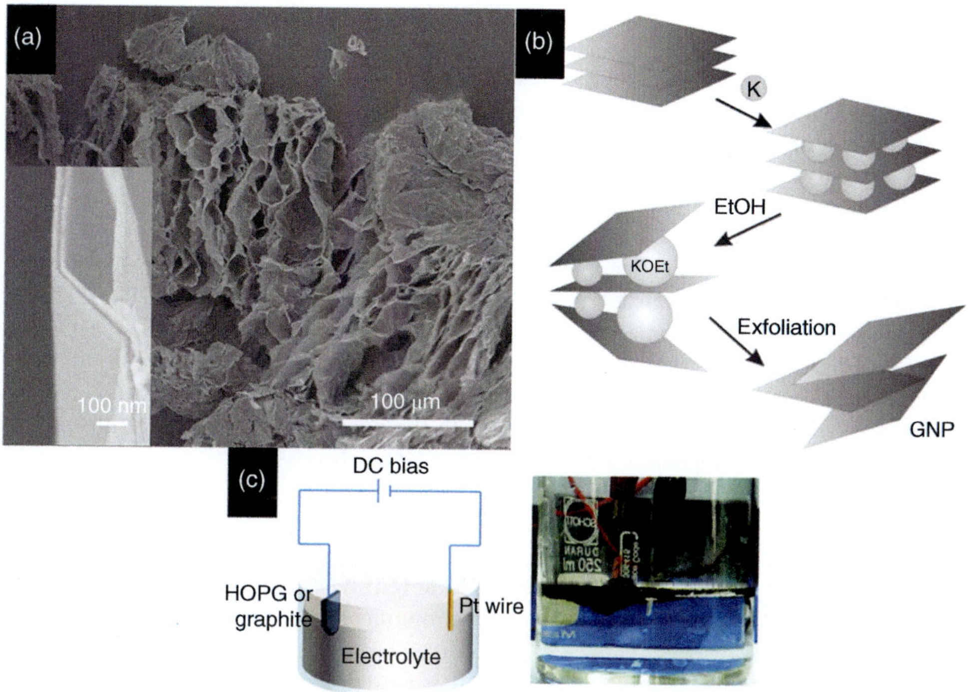

Figure 10.2 *Graphite exfoliation. (a) A scanning electron microscope image of thermally expanded graphite, resulting in "accordion-like" morphology consisting of graphite nanoplatelets with a thickness of a few nanometers (inset). (b) Intercalation of graphite with metal atoms and subsequent chemical reaction result in exfoliation and delamination of graphite into thin nanoplatelets. Reproduced with permission from Ref. [34]. Copyright 2005 Royal Society of Chemistry. (c) A typical setup for electrochemical exfoliation of a graphite electrode. Reproduced from [35] with permission of ACS*

300. In fact, due to the confinement of the intercalate within the graphene layers, a thermal shock provides significant pressure, which eventually can overcome the van der Waals interaction and exfoliate the graphite into thinner pieces [31]. It should be noted that GICs can be exfoliated by other chemicals or by electrochemical methods, but thermal expansion is the most common route for the exfoliation of graphite [32] (Figure 10.2). More details about GICs and the exfoliation of graphite can be found elsewhere [30–34].

Although the fabrication of polymer–graphite composites began in the mid-nineteenth century [20, 21], it was the renewed interest and research of the early 1990s inspired by Toyota's breakthroughs in nanocomposites that introduced a new paradigm to polymer composites [36, 37]. With the goal of perfect exfoliation and dispersion of graphite monolayers, i.e. graphene, in polymeric matrices, many studies were conducted mostly by chemical intercalation and exfoliation of graphite [22, 33, 38]. At the same time, the discovery of CNTs triggered the second hit for the development of science and technology needed for polymer nanocomposites [11]. Therefore, in the early 2000s, the reasons for studying composites based on graphite nanoplatelets (GNPs) were motivated by the following fact: GNPs are layered and inexpensive, like clay, and at the same time they are thermo-electrically conductive like CNTs [22]. However, despite some promising results, the performance of GNP-based composites did not rule out the use of its two counterparts because of the lack of systematic routes to disintegrate graphite into graphene and incorporate it into polymers [22, 39, 40]. In fact, many of those recent works on GNPs borrowed the dispersion and functionalization strategies from the two other counterparts. More details on these classes of composites are now available in several books and review reports [22, 39–42].

During the 1960s, a German team led by Ulrich Hofmann performed systematic studies on an old form of graphite (even at that time, it was *old*!), i.e. graphite oxide [42–44]. Those studies barely caught the attention for the fabrication of composite materials and if it were not for the "discovery" of graphene, they would still have been forgotten [3, 44, 45]. In fact, graphene is an example in the history of science where its first observation was too early to be digested by the scientific communities [3]. In the next subsection, a brief history of graphite oxide-based composites and realization of GO is discussed.

10.1.2 Graphite Oxide-Based Composites

Since the report on the synthesis of graphite oxide (GrO) in the mid-nineteenth century [24], it has been found that GrO is much more hydrophilic than graphite. It was well known that GrO can be easily dispersed in water especially in basic media [43, 44]. Further studies also showed that the layered structure of graphite is retained in GrO but the interlayer distance is much higher than that of graphite [45, 46]. Due to the heavily oxidized nature of GrO, it does have acidic properties and GrO is deoxygenated by chemical reducing agents or by "thermal reduction" [43, 44]. The exact chemical structure of GrO was not clear (which it is still not – recent studies propose that GrO is chemically amorphous and metastable!), but the nature of the main oxygen groups and the level of oxidation were known by the 1990s [43–49]. Despite of very old historical background and having a general picture of the chemical structure and properties of GrO, the amount of work on GrO-based composites was not noticeable till recently. However, these early works and ideas on GrO-based materials were revived in the graphene era [44, 45]. There were two main reasons for the lack of success of GrO-based composites: cost and performance.

Although GrO was synthesized more than a century and a half ago, to the best of our knowledge, there was no large-scale (even pilot-scale) production of this form of graphite derivative. A back-of the-envelope calculation shows the price of GrO can easily be one order of magnitude higher than that of graphite. This probably explains why the use of GrO did not attract a great deal of attention. The recent interest in GrO-based composites is mainly due to the improvement in the performance of the so-called high-tech products, which makes using GrO economically feasible [45].

As mentioned before, great interest in layered silicate-based nanocomposites moved researchers' attention to employing other layered materials, including GrO [37, 45]. Similar to the layered silicates, the main objective of GrO–polymer hybrid synthesis was gaining the maximum level of exfoliation and dispersion. Another similarity is the hydrophilic nature of both GrO and clays, which make them incompatible with organic polymers. As a result, the dispersion of oxidized layered graphite created a bottleneck in the fabrication process.

To make GrO less hydrophilic, surface functionalization was used. Due to the acidic nature, basic functional molecules, such as amines, are proper candidates for tailoring the surface properties. Several studies showed that GrO can be intercalated and functionalized by aliphatic amines (again, a common surface modifier for clays) [37, 50, 51]. Increasing the interlayer distance facilitates the diffusion of polymer molecules where the presence of organic surface modifier increases the compatibility. By reviewing the reports on this class of composites, one can realize that the generic feature of the research is the same as it is for the clay-based composites. Therefore, it is not surprising that, if the performance of GrO for improving the mechano-physical properties of the matrix is lower or the same as clays, the choice will be clay, as it is less expensive and does not blacken the matrix. Therefore, using GrO for making composite materials was not considered a breakthrough.

Another merit of using a carbon-based filler is improving the electrical and transport properties of insulating matrices. Traditionally, graphite and conductive forms of carbon black were the candidates. Studies by Hofmann and his colleagues showed that GrO assumes a conductive form upon chemical or thermal reduction [43, 44]. However, using GrO to improve the electrical conductivity of polymer matrices was rare till 2006 [52]. As a result, GrO rarely surfaced as a material for making electrically conductive composites.

Probably the main contribution of the pre-graphene era research on GrO-based materials was the discovery of "ultra-thin" layers of oxidized graphene in basic aqueous solutions [43–45, 53]. Such dispersions were used to fabricate layered nanocomposites through layer-by-layer assembly [53]. The presence of oxygen functional groups, especially carboxyl groups, makes GrO particles negatively charged and they can be adsorbed on positively charged substrates via electrostatic interaction [53, 54]. These facts were then widely employed for the assembly of GrO though the electrostatic interaction.

Suggested by Boehm, research on GrO (and its composites) was originally more a laboratory curiosity [44], which later became the basis for recent studies. The realization of the superior and unusual properties of 2D materials expanded the interest in many forms of layered materials [3, 55, 56]. Due to the ease of processing and the rich functionality of GrO, it became the centre of attention for the materials science community. Revisiting the synthesis approach of GrO composites and creating novel strategies for the assembly of monolayers of carbon into other matrices revealed the high potential of these old graphite-based products [45].

10.1.3 CNTs Versus Graphene (Oxide)

From the very beginning, graphene was introduced as a rival to CNTs, with many shared potential applications [1]. By the time of the arrival of graphene in 2004, research on CNTs was gaining a hold in different fields, from nanoelectronics to photonics, biomedical applications and advanced materials [57]. Despite the fact that the pioneers of graphene research, Geim and Novoselov, were newcomers to the carbon nanomaterials field, the expert groups on carbon nanomaterials followed the lead soon after [58]. Many of the ideas and methods developed for the synthesis, analysis and application were borrowed from, or at least inspired by, research on CNTs [45, 57].

One of the general challenges in the field of particulate composites is proper functionalization and dispersion of the particles in matrices [22, 37, 57]. In particular, it was the discovery of a systematic route of "dissolution" of carbon sheets, i.e. graphene, which heated up the field of graphene-based composites [52]. According to the science and technology of CNT dispersion, in order to disperse graphene, it is necessary to decorate it with proper functional groups, depending on the target matrix, to keep them separated and compatible with the host matrix [59]. Soon after, due to the similar surface chemistry of CNTs and graphene (and also GO), many of the chemical protocols used for functionalization of nanotubes were tried for their 2D counterpart [1, 60, 61] (Figure 10.3). This includes, but is not limited to, aryl diazonium functionalization [62], acylation of carboxyl groups and subsequent esterification [63], radical addition [64, 65], polymer grafting [66], amidation [67], non-covalent π–π bonding [68, 69], halogenation [70] and subsequent modification, and oxidation [71, 72]. One can tailor the surface polarity and functionality of carbon nanomaterials to almost any degree.

The dispersion of graphene in liquid media was inspired by the successful work on CNTs, with a few exceptions [75–78]. The types of surfactant and organic solvents that are suitable for CNTs are also used for graphene [75–79]. The liquid-phase exfoliation of expanded graphite into graphene happens in media also suitable for CNTs. Due to the same surface chemistry and assuming that CNTs (especially single-walled CNTs) and graphene "dissolve" in the solvent, the "good solvent" for both classes should have the same surface energy [75, 76]. This turns out to be true, as most of the good solvents for CNTs can dissolve graphene as well. Polar aprotic solvents such as *N*-methylpyrrolidone (NMP), dimethyl sulfoxide (DMS) or *N*-cyclohexyl-2-pyrrolidone (CHP) are found to be proper media for dispersing both graphene and CNTs [75, 76, 79, 80]. With the same reasoning, surfactants that can interact and stabilize the surface of CNTs were found to be effective for exfoliation and stabilization of graphene as well [77, 78]. Generally, surfactants having benzene rings are more effective due to the capability of forming π–π interactions with the surface [77, 78, 81, 82]. However, it appears that the surfactants with planar molecular structure, like pyrene-based surfactants, are more effective for exfoliation of graphite [81]. The story is also valid for the case of graphene structures derived from GO. Chemical reduction and thermal annealing of GO usually leaves GO derivatives, which cannot be dispersed easily [31, 52, 83, 84]. Similar classes of solvents and surfactants work well in further processing.

The term "graphene" can be seen in many post-2004 publications where the nanotubes are described as a rolled form of *graphene*, and the well-known example is the scheme for the description of single-walled nanotube (SWNT) chirality (Figure 10.4). Many of the physical properties of CNTs are similar to those of graphite (and/or graphene), namely their thermal and electrical conductivity, mechanical properties, thermal stability and chemical properties [6, 9]. As a result, many applications that were forecast for CNTs were potential application fields for graphene-based materials as well [3, 6, 9, 57]. These applications include energy storage, flexible electronics, composite materials, optics, biomedical applications, etc. [85].

Realization of chemical methods for the production of bulk quantities of GO and its conductive forms, i.e. reduced GO (RGO) and thermally processed GO (tpGO), during 2006–2008 turned graphene into a practical material rather than a laboratory material available for physicists [3, 45, 52, 58]. Many of the initial efforts on the acknowledgment of the above-mentioned applications was based on the employment of GO [87–90]. For instance, both CNTs and GO can be used in the fabrication of transparent conductors through the wet transfer technique [87, 91]. In this technique, a dispersion of CNTs or GO

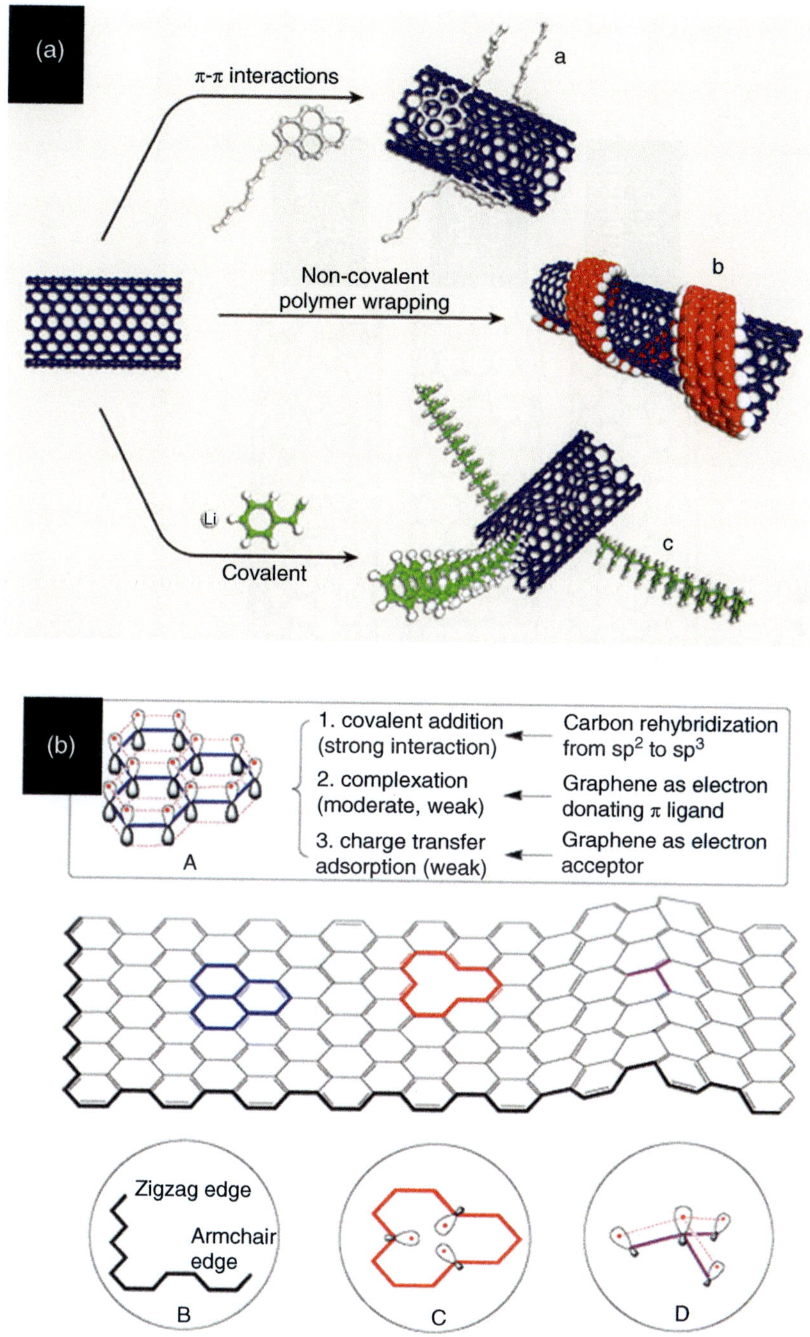

Figure 10.3 Chemical pathways for functionalization of CNTs and graphene. (a) CNTs can be modified through different interactions. (b) Origin of chemical reactivity of graphene and possible strategies for functionalization of graphene. (a) Reproduced from [73] with permission of Nature Publishing Group. (b) Reproduced from [74] with permission of RSC

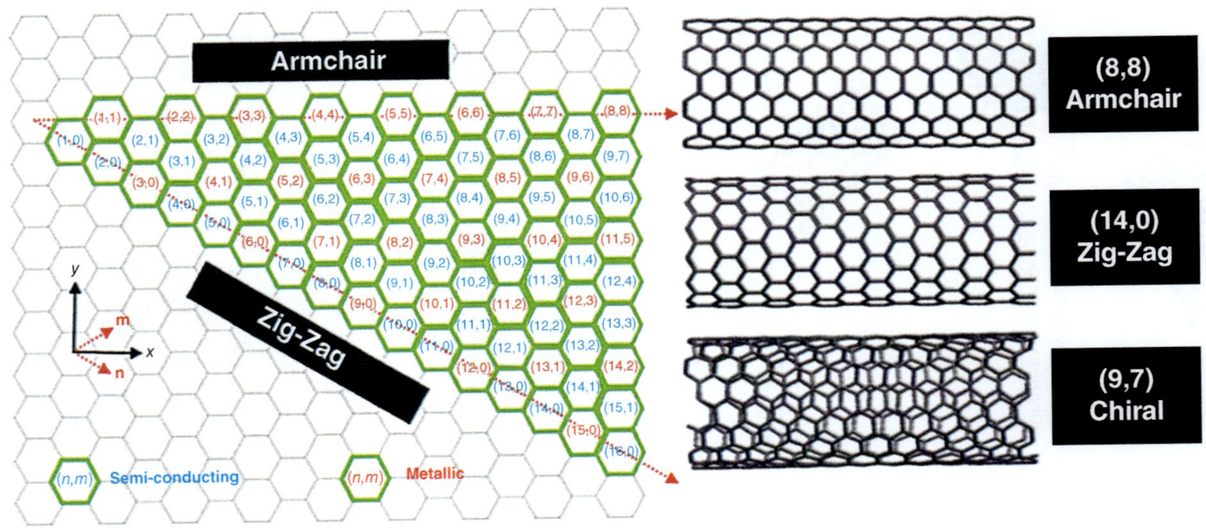

Figure 10.4 *SWNT helicity map on graphene sheet and examples of* (n,m) *chiral vectors that give rise to armchair, zigzag and chiral nanotube structures. Reproduced from [86] with permission of RSC*

(or reduced forms) is filtered through a membrane and the thin film on the membrane is then transferred onto the target substrate [87, 90, 91] (Figure 10.5a,b). Another example is "graphene oxide paper" inspired by "buckypapers" [91, 92] (Figure 10.5c,d). For making both, dispersions of GO or CNTs pass through a porous membrane filter, leaving a paper-like material on top. Many other fabrication and processing methods are also related to the synthesis of graphene-based materials, such as doping and post-treatment of graphene thin films [93–95], sorting and fractionation of graphene using density gradient centrifugation [96, 97], spinning into fibers [98–101], and mixing and dispersion protocols [102, 103].

Focusing more on composite materials, by the time of the arrival of graphene, research on CNT-based composites had matured and many challenges in design and processing had been solved [59, 102]. One of the main challenges for employing CNTs as filler was the high cost of their production, which made them less attractive for low-tech applications. The processing of CNTs is not straightforward and normally requires timely and costly purification and post-treatments [102]. Making graphene from inexpensive graphite gave hope that graphene could be a cost-effective replacement [84]. However, when the performance of graphene was evaluated and compared with that of CNTs, it was found that the cost/performance of graphene-based composites is not superior in all the cases. Although graphene brings some new functionality due to the planar geometry, depending on the problem and application that is to be dealt with, graphene still might be not the best candidate [87–103, 106].

Apart from the cost/performance ratio, another concern with CNTs and graphene (objectively, any nanomaterials) is their safety and toxicity. It is now widely accepted that the level of toxicity of graphene is lower than that of nanotubes [107–109]. Moreover, graphite is an inert "natural" material, which lowers the concern in this regard if one assumes that graphene and its derivatives are naturally derived. However, as discussed in more detail in Chapter 11, the toxicity of different forms of graphene, namely monolayer graphene (G_1), few-layer graphene (G_{few}), multi-layer graphene (G_{multi}), RGO, tpGO, GO and GrO, with different lateral sizes, has not been thoroughly researched, and, before the toxicity level and danger are established, it is not wise to introduce these materials into commercial production and real applications, especially in large quantities like composite fabrication [18, 110].

10.2 Why Mix Graphene Oxide and Polymers?

In the realm of materials science, the discovery and development of synthetic polymers made the breakthrough in the design of novel functional materials possible. Polymers are in general less expensive than ceramics and metals, they are easy to manufacture, chemically resistant and lightweight, and their mechanical and physical properties can be altered widely [22]. The molecular design of the polymeric backbone gives chemists the opportunity to systematically control the final properties of polymers. For instance, polymers with a "soft and flexible" backbone (like hydrocarbon) show rubber-like behaviour (if the polymer molecules do not crystallize), whereas polymers with a "rigid" backbone (like aromatic rings) are tough, such as polyimides [111]. Polymers with conjugated backbones can be electrically conducting, whereas typical polymers are insulating. Great advances in polymer chemistry in the twentieth century made molecular design principles available for polymer design [112]. However, these sorts of structural manipulations of polymeric

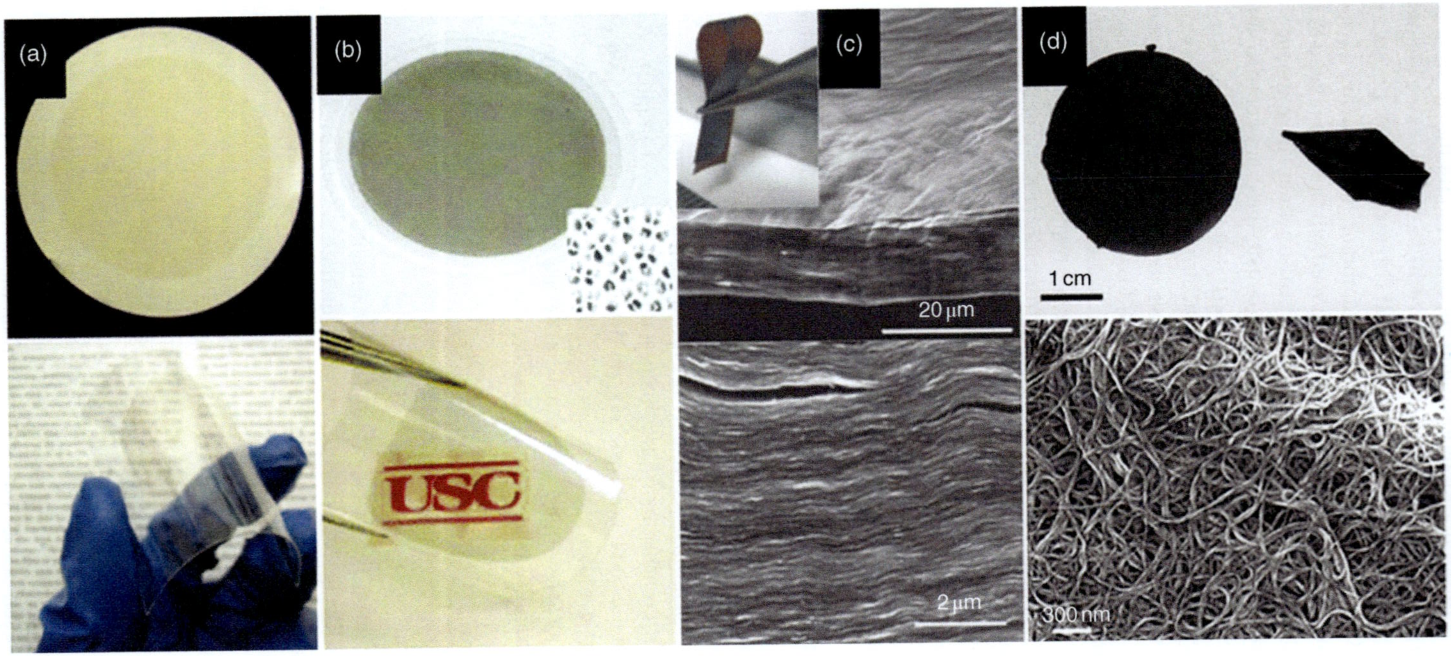

Figure 10.5 *Analogous concepts for the fabrication of GO- and CNT-based materials. (a) Fabrication of GO-based transparent conductive films by filtering a GO dispersion through a filter (top) and then wet transfer to a plastic substrate. (b) Thin film of SWNTs using filtration and wet transfer on different substrates with controlled thickness. (c) Fabrication of GO paper by filtration and drying of a GO dispersion through a filter, which results in strong flexible paper-like material (inset, top). The GO paper possesses a layered structure due to the stacking of GO sheets over each other. (d) Photograph of "buckypaper" (top); the paper (left) is tough and flexible enough to fold into an origami plane (right). Scanning electron microscope images of SWNT-based buckypaper fabricated by filtration of an aqueous dispersion of SWNTs over a filter. (a,b) Reproduced from [92] with permission of Nature Publishing Group. (c,d) Reproduced from [105] with permission of Nature Publishing Group*

materials are costly, not efficient and more importantly cannot assure multi-functionalities. As an example, making polymers conductive through the chemical pathway (conjugated polymers) results in fragile compounds that are difficult to process [113]. Therefore, incorporating new functionalities into polymer matrices without significant changes of chemistry becomes interesting.

As a rule of thumb, in an ideal case, mixing two (or more) compounds should produce a "composite" compound that possesses the properties of both (or all) ingredients at least partly. In the simplest case, if the rule of mixing is valid, the final property of the composite is linearly related to the content of the reinforcing phase, although, in reality, this is not often the case. With the goal of approaching this ideal case, compounding polymers with other materials becomes a centre of attention for materials scientists to develop functional composites with improved properties while saving the advantages of the matrices [114]. For instance, in line with the above examples, another possible route to make polymers electrically conducting is to mix them with a conductive material such as metals or carbon (nano)materials [52, 57, 103, 115].

The superior properties of graphene and its derivatives make them a tempting option for making composite materials [52]. Being the strongest, toughest and thinnest material ever known, being impermeable (even to helium) and having an extremely high thermo-electrical conductivity are some of the most prominent characteristics of graphene [3, 41]. However, putting these outstanding properties into polymeric matrices is not straightforward and requires the development of new routes of production and, more importantly, careful assembly of atomically thin sheets in matrices. The ideal chemistry and arrangement of graphenic sheets depend very much on functionalities. For instance, usually perfect dispersions and perfect alignment of nanosheets is desired when reinforcing matrices [116]. For improving electrical conductivity, it may be more efficient to have segregated networks of graphene and not necessarily homogeneous dispersions [117, 118]. The following discusses the function of graphenic sheets as a component for polymer-based composites on the basis of these functionalities.

GO is nominally considered as the best candidate for synthesizing chemically derived single-layered graphene on large scales [18, 61, 84]. As a result, most of the attempts to incorporate graphene into polymeric matrices revolve around graphite oxide and GO [52, 106]. The method used to disperse and functionalize graphite oxide depends very much on the chemistry of the matrix, the mixing procedure and the desired morphology of the final composite [103, 119]. In the following section, we discuss in detail the functionalities and properties that can be incorporated into polymers upon the addition of GO.

10.2.1 Making Stronger Polymers: Mechanical Properties

Polymers have wide ranges of mechanical properties, from soft and elastic (e.g. rubbers), through tough and strong (e.g. ultrahigh-molecular-weight polyethylene), to hard and brittle (e.g. polystyrene) [111]. These properties give rise to different classes of polymers depending on requirements. For instance, where damping of continuous loads is desired, rubbers are employed; whereas, for structural applications, tough and strong resins such as epoxy are used. Regardless of the final application, if the mechanical properties are the key parameters in functioning, improving those properties becomes a necessity. This becomes more significant if this improvement pushes the limit of the current state of the art. For

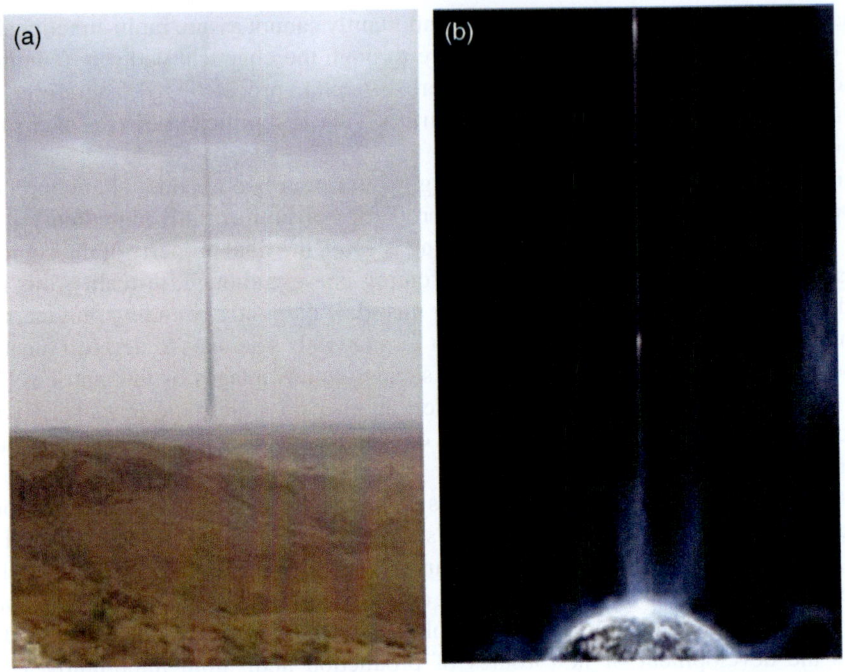

Figure 10.6 The extraordinary mechanical properties of CNTs raised hopes to achieve a "space elevator", a conceptual infrastructure that turns out to be unrealistic, at least at the current level of technology. Reproduced from [124] with permission of Elsevier

example, increasing the toughness of epoxy resins, which already have high toughness, is fascinating as it introduces new paradigms for materials design [120, 121]. For instance, this may lead to the fabrication of carbon-fiber-reinforced polymers with higher mechanical performance [122]. Reaching this goal is challenging in the sense that carbon-fiber-reinforced composites already occupy the top spot in the map of mechanical performance of all materials, and higher performance means pushing this boundary further [123].

Along the same line of aspiration, the design of some sort of "dream materials" is the final destination of this kind of effort. The prominent example in the realm of nanoscience is the "space elevator", which is made of a super-strong fiber, probably carbon-based, which can enable the transfer of cargo directly into space [124] (Figure 10.6). Apart from these science fiction-like examples, filling the gap between particulate composites and fiber-based composites is another paradigm of design with novel super-materials like GO [92, 125]. In fact, fiber-based composites are difficult to manufacture and expensive. From the mechanical properties point of view, having a non-fibrous composite with the same mechanical properties might bring new opportunities to fabricate cost-effective substitutes. For instance, the Young's modulus of pristine graphene is about 1 TPa [126]. By assuming the mixture law and perfect bonding and unidirectional dispersion, at 10 vol.% loading of graphene, a composite with modulus in the order of 100 GPa is expected. This is far beyond the modulus of conventional polymers (>5 GPa) and close to those of general-purpose fiber-based composites [111, 122]. Another goal (closer to achieving) of strengthening polymers is reducing the cost of manufacturing and/or improving the performance of polymeric products bearing loads and mechanical stress. In fact, upon increasing the critical stress that a polymeric matrix can withstand, one can reduce the size, weight and cost of the final product [114]. However, it is a prerequisite that hybridizing the polymer with a nanomaterial should be economical. This requirement was indeed fulfilled from the very beginning of the research on GO-reinforced composites, as GO seems to be a cost-effective replacement for expensive CNTs [84]. This becomes more crucial if one can significantly enhance the mechanical properties of inexpensive polymers such as polyolefins, which are inherently mechanically weak. In this case, improving the stiffness without degrading the toughness is very favorable. Another challenging task would be increasing the mechanical performance of polymers that are already considered as high-performing, such as polyimides and epoxies [120, 121, 127].

The core principles behind reinforcing the matrix with GO (or any other reinforcing phase) are high intrinsic mechanical properties of the nanosheets themselves and good load transfer from the matrix to the reinforcing phase [128–130]. In an ideal case, the "mixture law" should predict the final properties of the composites. However, in reality, there is deviation, which can be modeled by an "efficiency coefficient" (ξ) as follows [116, 131, 132]:

$$E_{\text{composite}} = E_{\text{matrix}} v_{\text{matrix}} + \xi E_{\text{GO}} v_{\text{GO}} \tag{10.1}$$

where E and v are the investigated mechanical property (e.g. modulus) and volume fraction, respectively. Normally, ξ varies between zero and one, and the higher the ξ value, the higher the reinforcing efficiency. However, ξ can be negative, which means that the addition of the filler degrades the matrix properties. On the other hand, in some rare cases, ξ can be larger than one, which indicates a synergistic effect of the addition of the filler to the polymer matrix [131, 133]. The most common explanation for such an observation is the formation of some "inter-phase" at the boundary of the filler and the matrix (usually due to strong interaction or orientation of the polymer chains at the interface), which provides higher mechanical properties compared to the base matrix [133]. Such a synergistic effect is very favorable to achieve, but requires molecular design of the interface. However, for the most frequent cases ($0<\xi<1$), answering the question about what causes the deviation from perfect additivity ($\xi=1$) would help us to understand the principles of the design of GO-based composites.

Analyzing composite materials attracted attention in the middle of the previous century with the advent of high-performance fibers such as glass and carbon fibers. Continuum mechanics can be employed to model the behavior of fiber-based (both long and short fibers) composites to a high degree [114, 116, 133]. However, going to the nanoscale and eventually, with the advent of graphene, to the atomic scale may break down the validity of continuum models [130]. Using a more sophisticated computational model such as molecular dynamics is very expensive and does not guarantee accuracy at the macroscopic level. Many studies have shown that continuum models work well for predictions of mechanical properties of graphene-based composites (or at least can capture the general picture of the behavior) [130–133]. In this regard, one simply needs to view graphene (or GO) as a rigid disk with a certain thickness that is dispersed in a continuum phase. Then some geometrical and material parameters are ascribed to each phase (filler and matrix). There are a number of models used for the modeling of particulate composites (as is the case for GO) [114, 119, 130–133]. However, the Halpin–Tsai model is the most common, which works quite well [133–139].

Now let us assume a case where nanosheets (with a length of L and thickness of t) are unidirectionally dispersed (with a volume fraction of v_s) in a matrix with a Young's modulus of E_m. Then the Young's modulus of the composite is given as [136, 137]

$$\frac{E_c}{E_m} = \frac{1+\alpha\beta v_s}{1-\beta v_s} \tag{10.2}$$

$$\beta = \frac{\frac{E_p}{E_m} - 1}{\frac{E_p}{E_m} + \alpha} \tag{10.3}$$

where α is $2L/t$ and is a measure of the aspect ratio. It is apparent from these equations that increasing α improves the reinforcing efficiency. This critical aspect ratio depends very much on the ratio of the modulus of the particles E_p to that of the polymer matrix [133, 135]. It should be noted that the same principles can be applied for one-dimensional (1D) fillers such as CNTs. In the case of GO, E_{GO} is about 250 GPa [128]. Now we consider a typical Young's modulus of 1 GPa for the polymer matrix. Then, if the aspect ratio of the GO sheets exceeds 1000 (equivalent to a lateral size around 1 μm), the modulus of the composites is just 8% lower than that calculated from the mixture law; whereas for an aspect ratio of 100, it is 40% lower. This implies the same concept of the "critical length" for fiber composites where the center of fibers can reach the highest strength [133, 140]. This observation suggests that, for efficient load transfer from GO (or in general any elongated particles), the size of the sheets should be in the order of 10 μm (this is true for a wide ranges of polymers) and monolayer GO and not GrO is preferred [130, 133, 134, 136]. Despite the fact that such gigantic monolayer GO sheets can now be synthesized, their subsequent processing and accommodation into polymer matrices are difficult tasks. It should be noted that the mixing process is associated with some obstacles, such as breakage of the sheets, gelation and the rise in viscosity of the composite [100, 116, 141–143]. However, large size GO (lateral size of a few micrometers at least) is recommended in order to stiffen the polymeric matrix in general.

Before discussing the sources of the deviation from the above prediction, it is interesting to discuss another subject on graphenic sheets for reinforcements, i.e. the flake thickness. The above analysis implies that a higher aspect ratio is always preferred to improve modulus (and some other properties). Therefore, one may consider single-layer sheets as the ultimate candidate in this regard. However, the ultra-thin nature of graphene (oxide) induces huge confinement to polymer chains when the filler content increases [136, 137]. Despite the fact that the exact behavior of confined polymer chains is not known, confined polymers have different mechanical properties and usually make the composites fragile [134, 135, 144]. In addition, there is a limit for confining the polymer between the graphenic sheets. When there is no strong interaction between the polymer chains and graphenic sheets, the polymer chains prefer to phase-separate from graphene [134, 144]. On the other hand, increasing the number of layers of graphene sheets will decrease the effective modulus of the nanosheets themselves [134]. As a result, these two effects are opposing each other, as few-layered graphene induces less confinement at the same volume fraction. Such a simple analysis suggests that the monolayer graphene (oxide) is not necessarily the best candidate for reinforcing.

In fact, depending on the confinement limit and the efficiency of the stress transfer between the layers, the optimum number of layers may vary from bilayer up to five-layer sheets [134, 135]. However, it is difficult to prove this statement experimentally, since making mono-sized graphene (oxide) (both lateral and thickness) is an impossible task. Layer-by-layer assembly may work to design layered polymer–graphene arrangements, if one deposits a favorable number of graphene sheets on top of each other, and then deposits a polymer layer on that [145, 146]. Beside the experimental proof, looking carefully at the analysis done by Gong *et al.* [135] demonstrates that this issue is important at very high concentration of graphene and the difference between monolayer and few-layer graphene is not significant (Figure 10.7). Subsequently, one need not worry about "full exfoliation" into single-layer level, but instead focus more on preserving the lateral size, even at the expense of partial exfoliation. The assumption of Gong *et al.* [135] made in the above analysis (i.e. decreasing effective modulus by increasing the number of layers) is actually opposing the experimental observation of Suk *et al.* [129] where they found that the effective modulus of mono-, bi- and trilayer GO remains constant. However, this observation actually supports the above statement. We are not aware of any experimental evidence whether few-layer GO is a more effective reinforcing agent than its monolayer counterpart, but still this analysis suggests that preserving the lateral size is more important than exfoliation down to a single layer. This fact is not usually considered in the design principles for nanocomposites.

All the analyses above are for the ideal case, where graphene sheets are aligned and the load transfer from the matrix is perfect. However, in reality, the composite behavior deviates from the ideal case. The first and most clear reason for this is the misalignment issue. The Halpin–Tsai analysis shows that the modulus of a composite becomes around three times lower when the platelets are dispersed randomly in the matrix [136–140]. Therefore, there have been numerous efforts to align graphene (oxide) in the polymeric matrix [138, 141, 142, 146–148]. This is important especially where the load to the composite is maximum in one direction. In general, the alignment of graphene (or 2D fillers) just requires putting them in parallel with each other, whereas for CNTs the ideal alignment requires alignment in two dimensions [132, 133]. In reality, perfect alignment of graphenic sheets is difficult to achieve and some deviation from the ideal case is inevitable. The situation is worse when polymer melt is used, which is indeed the most common system for general-purpose thermoplastic polymers. In these cases, mixing the polymer melt leads to random dispersion of sheets and, due to the very high viscosity of the polymer melt, alignment is not possible except by inducing shear [149, 150]. (Shear-induced alignment is very common for making polymeric fibers or sheets, but has not been explored widely in the case of graphene.)

Another very important source of deviation of the mechanical properties of graphene composites from the ideal situation is the "agglomeration" or bad dispersion of GO in the host matrix. This must be differentiated from the incomplete exfoliation to monolayers. As stated above, not having perfect exfoliation may even be useful for stiffening the matrix. However, agglomeration means inhomogeneous distribution of the filler throughout the matrix. In fact, one could have a perfect "dispersion" but not perfect exfoliation

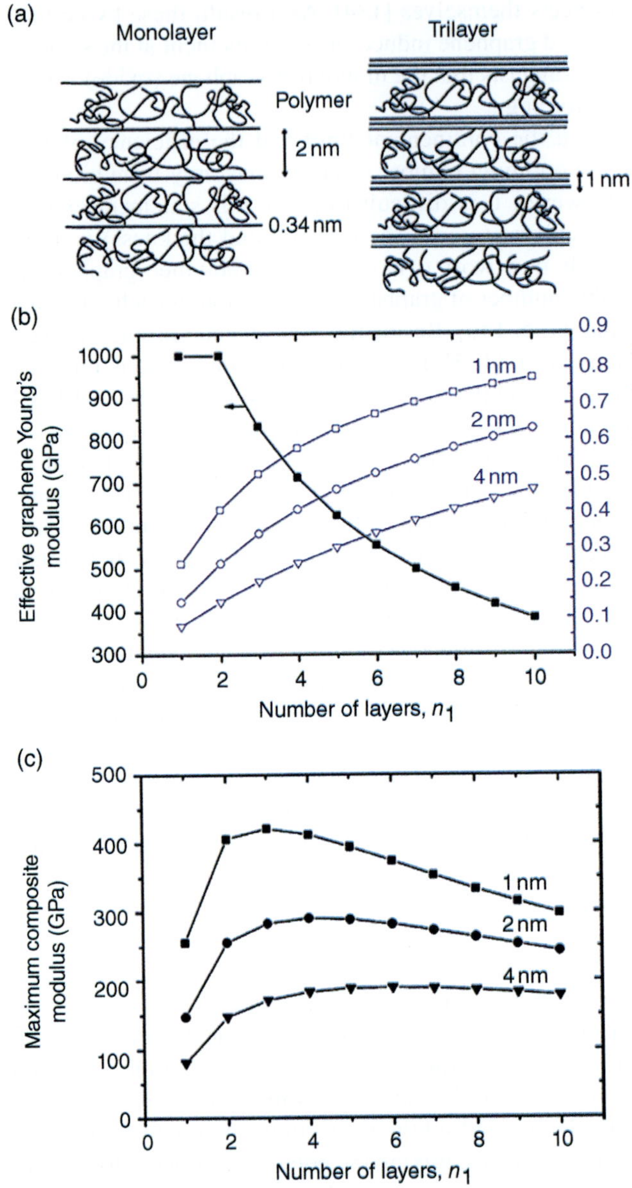

Figure 10.7 *(a) Graphenic sheets with lower thickness induce higher confinement to polymer chains. By keeping the inter-sheet distance constant, the effective volume fraction of graphene increases on increasing the number of layers. (b) Effective graphene Young's modulus and maximum graphene volume fraction for different indicated polymer layer thicknesses, as a function of the number of layers in the graphene flakes. (c) Maximum nanocomposite modulus predicted for different indicated polymer layer thicknesses as a function of the number of layers in the graphene flakes. Reproduced from [135] with permission of ACS*

(e.g. homogeneous dispersion of few-layered graphene). On the other hand, monolayer graphene (oxide) can form agglomerates. The task of homogeneous dispersion of nanosheets in the matrix is one of the main objectives of the production process [52, 106, 121]. Due to the very high surface area of GO (~2600 m^2 g^{-1}) [151] and high surface energy, GO flakes tend to stack together in non-polar media. Apart from the techniques that are used for proper dispersion, for matrix reinforcement, agglomeration of GO usually results in the catastrophic degradation of the toughness of the matrix, especially if the matrix itself possesses high toughness [121, 134, 137]. In fact, the matrix starts to fail upon inducing the load due to the failure of the stress transfer. This sort of failure is very common when the loading of GO increases [121]. As a result, researchers are often looking for an "optimum loading" of graphene where the stiffness and toughness are maximized [39–41, 119–121]. This optimum loading can be as low as a few tenths of one percent of GO [120, 121]. It is not clear if the degradation of the mechanical properties (mainly toughness and stress at break) is due to the poor dispersion, or if other parameters are involved. For instance, one can assure perfect dispersion (or "molecular-level" dispersion) of GO in a polymer matrix through "wet-chemical" methods even at very high loadings [152]. However, degradation of the toughness and brittleness of composites is still observed.

Interfacial interaction between the graphene and the polymer matrix is the dominating parameter that controls the level of reinforcement [119, 130]. Imagine a single sheet of GO in a polymer medium. The work that is required to pull it out is related to the interfacial strength [130, 135]. The action of pulling out requires an amount of energy that depends on the type of interaction at the interface. Weak interfacial energy leads to the failure of the composite due to fracture at the interface and its subsequent propagation into the matrix [137]. Subsequently, engineering of the interface is an important part of composite design. Usually there is a common tendency among researchers to make the interfacial strength as high as possible. The typical way of doing that is grafting chemical moieties with the same or similar chemical structure and polarity on the surface GO sheets (or its other derivatives) [52, 65, 66, 137, 153, 154]. In the case of thermoplastic polymers, if the so-called moieties are polymer chains, entanglement between the grafted chains and polymer chains in the matrix provide such a strong bond that de-bonding at the GO–polymer interface is unlikely [65, 66, 153]. In the case of thermoset polymers, attaching "reactive" moieties is a more popular strategy. These moieties are "reactive" in the sense that they can participate in the reaction of prepolymers (e.g. epoxy resins and hardeners), and as a result they become part of the polymer network [137]. This also leads to covalent bonding of the filler to the matrix. GO itself possesses some reactive groups such as epoxide, carboxyl and hydroxyl and even conjugated C–C bonds; this makes engineering of the interface less complicated [60]. This is especially apparent once we deal with polar matrices. We do not intend to expand on GO chemistry here, and refer the reader to Chapter 6 and also to additional reviews on this subject [22, 41, 59–61, 102, 119].

GO and its derivatives exhibit high potential as reinforcing agents in high-performance composites. The ease of processing and better dispersion are advantages compared with other functional nanomaterials. In addition, tunable surface chemistry and size can overcome classical challenges in particulate composites, i.e. weak interfacial interaction and load transfer. However, enjoying these profits is usually limited to low loading of GO and, moreover, the highly polar nature of GO makes it challenging to hybridize it with low-polarity polymers [52, 102, 103, 106, 119–121]. Thanks to the rich history of work on tailoring the interface, compatibilization of GO and graphene with almost any polymer is possible, unless the processing costs increase [41, 45, 57, 102]. While discussing the mechanical performance of GO-based composites, one may wonder what the optimal interface would look like? The typical strategy that is followed by many researchers is to increase the bonding between graphene and the matrix, preferably by covalent bonding [66, 153]. This indeed leads to enhancement of many mechanical properties, especially toughness. However, when it comes to some practical applications, enhancing stiffness or even toughness is not always the priority. For instance, when load damping is favorable, the stiffness of the polymer (composite) matters less [155–157]. In such cases, incorporation of new modes of energy dissipation is more welcome. As a result, having a strong interface (and even perfect dispersion) might not be preferred, as energy damping at the polymer–filler interface would be more helpful [157]. Another example could be the fabrication of tough composites out of stiff constituents. In such cases, creating sacrificial bonds (or interactions) is more appreciated [158, 159]. However, as the molecular origin of the mechanical properties is still not clear, usually fabricating even more complex composites demands creativity in the molecular design of composites. The advantage of GO over many nanoparticles is the ease of processing and the rich chemistry toward functionalization whether with chemical species or even with nanoparticles. By giving a challenging example on complex composite design, we finish the discussion on mechanical properties of GO-based composites. However, further details on this issue can be found in some recent reviews [7, 22, 39, 41, 119, 132, 134].

Inspired by the microstructure of the naturally occurring composite, mother of pearl (nacre), many efforts have been focused to turn the mixture of a soft binder and brittle particles into stiff and tough composites [160–164]. The first step would be to mimic the microstructure of nacre, which looks like "brick and mortar" [164] (Figure 10.8a,b). GO has played the role of "brick" in such structures with different polymers and other molecules have been used as the binder [165–167]. The first challenge in this regard is the assembly of GO into such a layered structure. To perform this task, methods such as layer-by-layer assembly [146, 163, 164], vacuum filtration [92, 148, 166], evaporation induced-orientation [165, 167] and liquid-crystal-mediated assembly [138, 141, 142, 168] have been used. Apart from the level of GO orientation (Figure 10.8c), in almost all cases the final composites suffered from brittleness, unless the loading level of GO is low (<5 vol.%) [136, 138, 152]. Although the mechanical properties of those nacre-mimicking GO-based composites are impressively higher than that for conventional polymers, they are however still far behind the main goal, i.e. high toughness [169]. The key element of success in designing such a structure is to design inter-brick interaction, making the constituent parts capable of sliding over each other, at the expense of energy dissipation [160, 161]. Following nature's way, inducing some asperities on the surface of GO and using the proper binder with an internal means of energy dissipation (such as auxetic molecules) might help to toughen the GO-based layered composites [160, 170, 171] (Figure 10.8d). This will not only lead to the fabrication of ultra-strong artificial composite materials, but also help develop new principles for strain engineering of composites materials [169–173]. Filling the gap between the principles at the nanoscale and the classical macroscale mechanics will enable us to enjoy the outstanding mechanical properties of GO and graphene, not only under the tiny probe of an atomic force microscope (AFM) [128, 129] but also under macroscopic loads in real life!

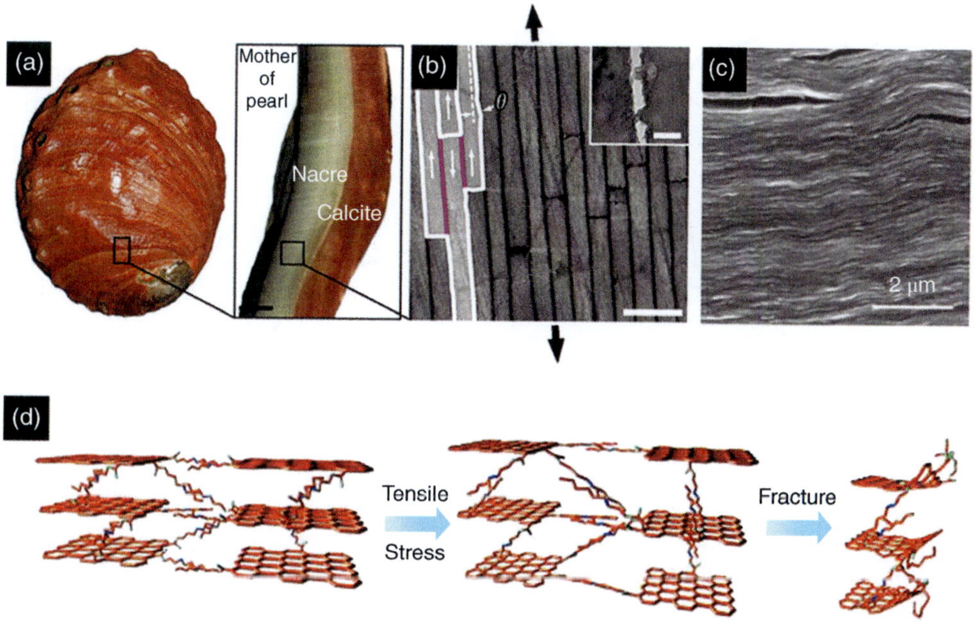

Figure 10.8 *Nacre-inspired design of GO-based composites. (a) Red abalone shell. Inset: A cross-section cut from the shell. (b) Scanning electron micrograph showing the micro/nanostructure of natural nacre (the so-called "brick and mortar" structure). Scale bar, 1 μm. (c) Scanning electron micrograph of a GO paper resembling the layered structure of nacre. (d) Improved fracture mechanism of GO paper by inserting flexible polymer molecules at interlayer. More complex strain engineering of such composites is presumed to result in tough and strong GO-based composites. (a,b) Reproduced from [160] with permission of Nature Publishing Group. (c) Reproduced from [92] with permission of Nature Publishing Group. (d) Reproduced from [166] with permission of Wiley-VCH Verlag GmbH & Co*

10.2.2 Electrical Properties

As pointed out in the introduction, increasing the electrical conductivity of polymeric matrices has been a subject of interest for many studies on GO-based composites [41, 52, 84]. However, in the pre-graphene era, due to the intrinsically low conductivity of GO, this aspect was not pursued [45]. Effective reduction of GO into a conductive form is necessary to produce conductive composites. The reduction methods are not the main focus of this section, as reduction is discussed in more detail in Chapter 6. The main concern here is the factors controlling the final conductivity of the resulting composites [154, 174]. Additional properties such as electron mobility [175], dielectric constant [176] and electrochemical capacitance [177] might also be interesting to manipulate depending on the final use of the composite. Applications such as organic electronics [175], energy storage (batteries and supercapacitors) [177], sensors [178], electro(chemical)-responsive materials [179] or actuators [180] are some examples where GO-based composites may have a chance to be employed as functional materials. However, readers are referred to other parts of this book (Chapters 7 and 8) and other references [175–181] to follow the principles of design in those areas.

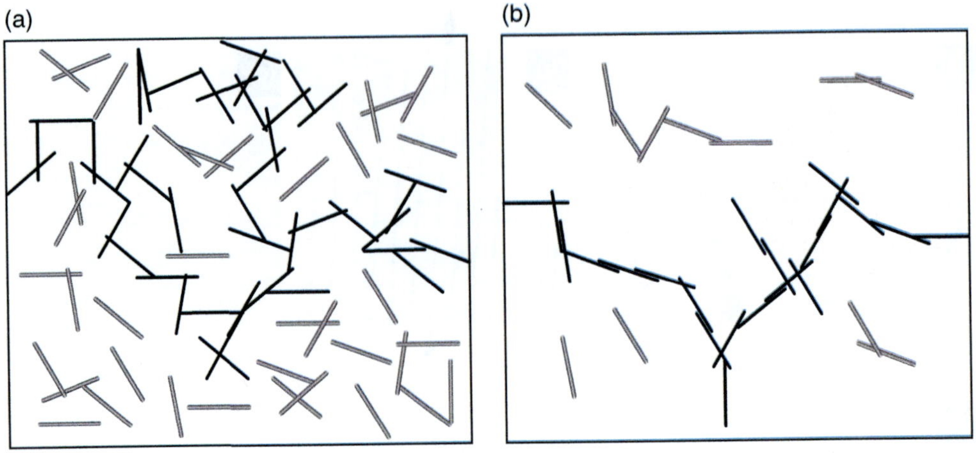

Figure 10.9 *Schematic representation of percolated networks of anisotropic filler. (a) A random network of filler results in a lengthy path for electrons and so a higher resistance is expected. (b) Segregation of filler into wire-like structure lowers the percolation threshold and enhances the conductivity at lower filler content. Reproduced from [182] with permission of RSC*

Adding conductive fillers to the insulating matrix is the most common approach that can be efficiently used to design and produce a new set of functional composite materials. In this regard, graphene offers great opportunities, owing to its remarkable conductivity. The core idea of fabrication of conductive graphene-based composites is to let electrons pass through the insulating matrix with much less resistance [52]. Therefore, the graphenic sheets act as conductive wires, and the main task is the proper "wiring" of them through the matrix. The proper way of wiring (or, in other words, the assembly of graphenic sheets in the matrix) very much depends on the type of final application and the processing method. Although homogeneous dispersion of conductive sheets is required for the formation of interconnected networks of the filler, it is not the most efficient state in most cases [117, 118] (Figure 10.9). However, homogeneous dispersion of graphene in the matrix is still critical for improving other properties such as thermo-mechanical ones [106, 119–121]. In addition, from the processing point of view, the design of three-dimensional (3D) networks of graphenic sheets is costly and may not be compatible with many matrices.

The conductivity of a composite increases suddenly once the graphene sheets percolate throughout the matrix and form a so-called interconnected network, which is called the "percolation threshold" [52, 116, 118, 137, 183]. At this point, the conductivity obeys the scaling law

$$\sigma_c = \sigma_f \left(\varphi - \varphi_c\right)^t \tag{10.4}$$

where φ is the filler volume fraction, φ_c is the percolation threshold, σ_f is the nominal filler conductivity, σ_c is the composite conductivity and t is a scaling exponent. There is a rich theoretical background on the percolation theory and on composite materials [183, 184]. Generally, we would like to have lower φ_c and higher σ_f in order to achieve higher conductivity at the lowest possible graphene content. Simple analytical modeling suggests that, in the case of perfect dispersion of the 2D filler in the matrix, φ_c depends very strongly on the aspect ratio of the filler [184]. For instance, Celzard *et al.* [185] predicted an experimental

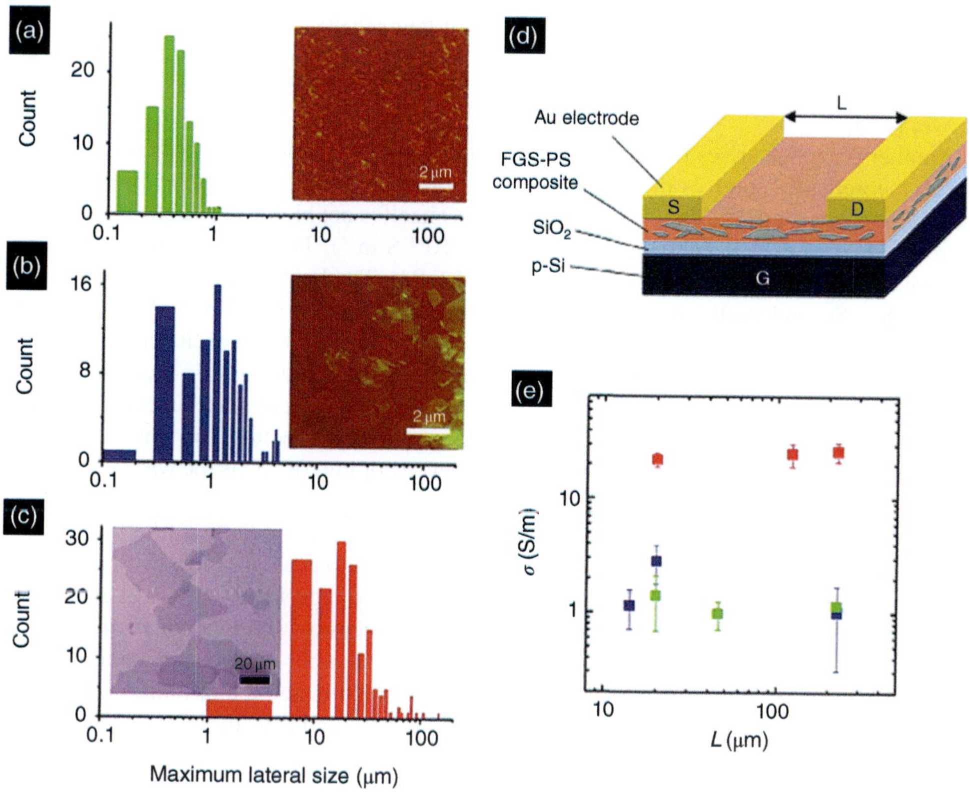

Figure 10.10 *Impact of GO aspect ratio on the electrical conductivity of PS–RGO composites. (a–c) Size distribution of GO sheets used for composite fabrication. (d) Schematic of PS–RGO composite thin-film field-effect devices employed for conductivity measurements. (e) Device conductivities measured in ambient conditions as a function of channel length for different aspect ratios, clearly demonstrating the high potential for large GO sheets to improve the electrical conductivity of the composites. Reproduced from [175] with permission of ACS*

dependence of φ_c on the aspect ratio of the filler. The higher the aspect ratio, the lower the φ_c attained. As discussed previously, the aspect ratio is a crucial parameter for the mechanical properties as well [134–138]. An aspect ratio in the order of 10^4 is favorable. Later in this section, we will discuss that this value of the aspect ratio is also a threshold to gain high barrier resistivity from graphenic sheets in composites. At this value of the aspect ratio, both theoretical and experimental observations report φ_c values of less than 0.1 vol.% [52, 106, 116, 175, 183–185]. Therefore, the task of preserving the size of graphene (or GO) during processing becomes vital from other perspectives.

The effect of GO size on the electrical properties of composites (and even GO itself) has also been examined experimentally. For example, Eda and Chhowalla [175] showed that both the electrical conductivity and mobility of polystyrene (PS)–RGO composites increase by around an order of magnitude when the aspect ratio increases by a factor of 50 (Figure 10.10). Although the composites were studied at RGO concentration much higher than φ_c, this still indicates that breakage of nanosheets creates a huge obstacle for electron conduction. Also, though they did not discuss the optimum (or saturation) aspect ratio, the best result was achieved at an aspect ratio of 2×10^4, the golden regime (see above) [175].

Another issue that arises from (10.4) is the influence of the conductivity of the filler (σ_f), which theoretically controls the absolute value of conductivity [183]. In the case of GO-based composites, this seems to be very true, as GO itself is not highly conductive (although more conductive than insulating polymers, $\sigma_{GO} \approx 10^{-3}$ S m^{-1}). Therefore, to fabricate conductive composites, reduction of GO is critical [52, 84]. The level of reduction determines the σ_f value. Several studies proved that better reduction of GO at the same volume fraction of GO leads to higher electrical conductivity [142, 186–188]. However, it seems that the maximum conductivity (the saturation value that is obtained after passing the percolation threshold) of composites does not depend solely on σ_f, and it is a complex function of additional parameters, such as the type of matrix, filler–matrix interaction, the state of dispersion, etc. [183]. However, efficient reduction (and even doping) of GO is the way to enhance the final conductivity.

As we talked about the "proper wiring" of the matrix with conductive filler, it is very insightful to elaborate upon this general idea. Consider two pieces of blocks with the same size, one being conductive and the other not. Applying a voltage when the blocks are wired in parallel to the current results in easy passage of the electrons. But when they are placed in series, the insulator stops the current. This simple example explains graphene-based composites. On top of that, graphite itself follows this picture, and the in-plane conductivity is more than 10^4 higher than the conductivity perpendicular to the plane [189]. This anisotropy can be incorporated into the composites of graphene when the nanosheets assemble in parallel [116, 147]. Parallel assembly of graphene sheets not only provides anisotropy in conductivity, but also enhances the efficiency of the filler and improves the final conductivity. The level of anisotropy depends on the level of alignment and indirectly on the aspect ratio [116]. Electrical anisotropy is desirable for some applications [149, 190] and, due to the very high aspect ratio of GO, it is an excellent candidate for this purpose.

When there is a current of electrons traveling through the composite, it mainly passes through the conductive part (filler). When two conductive particles are close enough (typically 10 nm), electrons can hop from one particle to another, and continue to pass (namely as they are in contact) [184]. If the particles are distributed homogeneously, the density of these contact points decreases. In fact, these contact points form resistances for the current and we do not want them to form "series". The first option is to use longer fillers (which act as wires). However, for macroscale samples, the presence of these so-called resistances is inevitable. Another solution is to decrease the overall resistance by "parallel circuit" design! In practice, to do so, one can add additional anisotropic conductive filler, such as CNTs or metal nanowires, to the graphene composites [191–194]. This could help improve the linkage throughout the entire conductive network of fillers. This strategy has been shown to enhance the electrical conductivity and reduce the percolation threshold.

In all the above cases, we assumed that the composites possess a homogeneous distribution of the filler. However, if the increase of the electrical conductivity is the main goal, is it a good idea to have such a morphology? To give an illustrative example, we do not make a whole building conductive to protect it against lightning, we just need to use a lightning rod. The same logic can be applied for composites. If the graphene sheets (or any other conductive filler) segregate into a continuous interconnected network, they still increase the conductivity [117, 187, 188]. In this situation, the rich phase (segregated network) contains much more filler and subsequently has higher conductivity. If this phase is continuous through the matrix, the overall conductivity is determined by this phase. Therefore, a much lower amount of filler is required to enhance the conductivity [117, 149, 190]. Fabricating the segregated network of graphene in a matrix usually demands innovative design of the composite-making process. Here, we briefly introduce three methods: latex blending, polymer blending and using foam-like graphene.

i. If GO or RGO are dispersed in water and mixed with polymer latex particles, graphene sheets cannot enter into the polymer beads, and form segregated networks upon drying (if polymer particles are glassy at processing temperature) [187, 188, 195, 196]. It is noteworthy that polymer particles with a size comparable to the size of GO (in the micrometer range) are preferred. This method has been widely used for many fillers, including RGO, to fabricate conductive composites [195, 197] (Figure 10.11). Normally this method gives one of the lowest percolation thresholds and very high conductivity after passing the threshold [188].
ii. The second method is polymer blending. As discussed above, the idea is to force the conductive filler to jam into the 3D interconnected network in the polymer matrix [188]. From the pioneering works of Pickering [198] and Ramsden [199], we know that solid particles have a high tendency to accumulate at an interface, which can also be a polymer–polymer interface [200, 201]. From the polymer physics we know that a blend of two

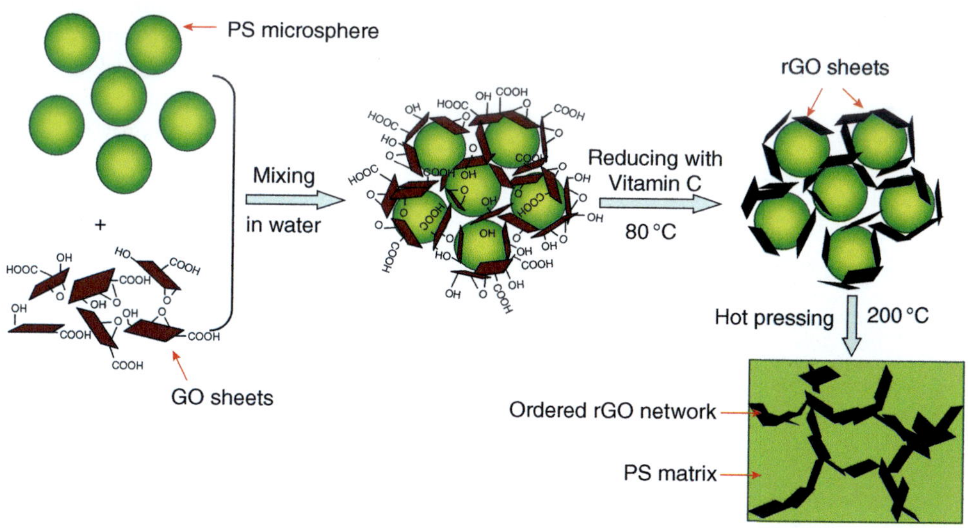

Figure 10.11 Schematic illustration for the preparation of PS–RGO composites with an ordered three-dimensional segregated network, by mixing GO sheets and PS microspheres, followed by GO reduction with vitamin C and hot pressing. Reproduced from [197] with permission of RSC

immiscible polymers can form a co-continuous phase in which, our goal will be achieved if the graphenic sheets (or any other conductive filler) go to the interface [111]. This has indeed been proven experimentally [200]. Adding an immiscible polymer to a composite of conductive filler–polymer lowers the percolation threshold [202, 203]. Getting a co-continuous morphology may not even be necessary if the filler has higher affinity to the matrix polymer in the blend. From a technical point of view, it is better not to look at the situation as "adding another polymer" to a polymer–graphene composite, but rather to keep the traditional view of "adding graphene" to a polymer blend. Polymer blends are promising materials in the polymer industries [204]. Overcoming the challenges associated with their processing is industrially very attractive. Graphene and especially GO derivatives can be considered as multi-functional additives (probably inducing conductive filler), as they can help to overcome such challenges as weak polymer–polymer interfacial interaction, proper dispersion of polymers in each other, reducing the cost of processing and improving the final properties. But graphene's capacity as a multi-functional additive has not been well explored in this area.

iii. Cheng *et al.* [205] introduced another method to make polymers conductive, by producing foam-like graphene. They synthesized a 3D interconnected graphene network first, and then incorporated an elastomer to gain an impressive conductivity of 1000 S m^{-1} at a graphene loading of ~0.5 wt.% [205] (Figure 10.12). In this approach, the continuity of the conductive network is not a concern as it is macroscopically

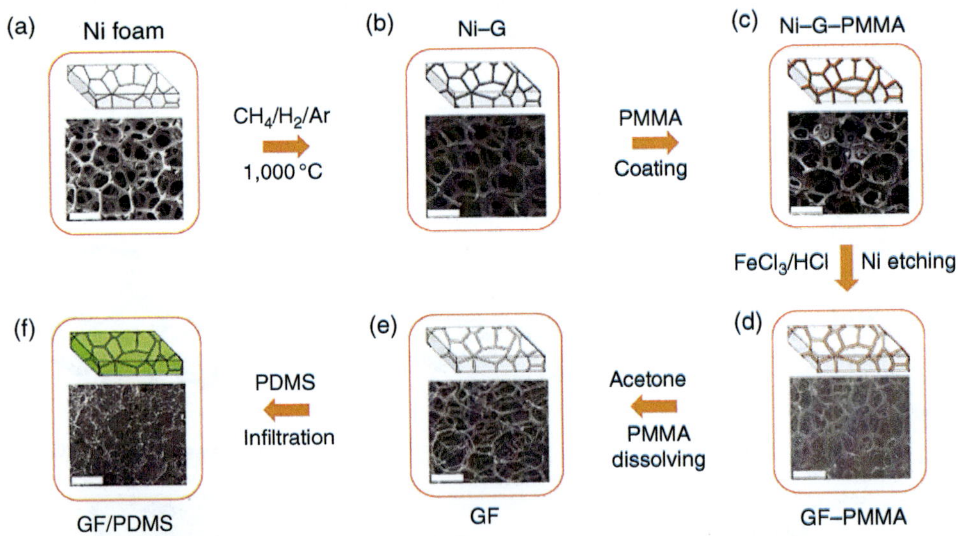

Figure 10.12 *Incorporating graphene foam (GF) into a polymer matrix with high conductivity at very low loading. (a,b) CVD growth of graphene films (Ni–G) using a nickel foam as a 3D scaffold template. (c) An as-grown graphene film after coating a thin polymethyl methacrylate (PMMA) supporting layer (Ni–G–PMMA). (d) A graphene foam coated with PMMA (GF–PMMA) after etching the nickel foam with hot HCl (or $FeCl_3$/HCl) solution. (e) A freestanding GF after dissolving the PMMA layer with acetone. (f) A GF–PDMS composite after infiltration of polydimethylsiloxane (PDMS) into a GF. All the scale bars are 500 μm. Reproduced from [205] with permission of Nature Publishing Group*

robust from the beginning. Despite the fact that this method uses graphene grown by chemical vapor deposition (CVD), the concept can be used for GO-based composites as well. GO-based aerogel can be a candidate for this purpose [206]. However, one should bear in mind that there are processing limits for this method (at least at the current state of technology!).

The above-mentioned short guidelines may assist us in formulating electrically conductive composites; however, the final design depends on the application of the composites. For instance, the requirement for antistatic coatings is to achieve a certain surface resistance (10^5–10^9 Ω/ϒ) [183]. In other words, the task is to decrease the resistance to that level. However, if the goal is to make a strain sensor, the network of graphene sheets in the matrix should be sensitive to mechanical stress and probably break and re-form (or deform) reversibly [207]. In this situation, the level of conductivity should be high enough that detectors can sense the current. However, even more important is that the response, which is usually the change of resistivity, should be high enough upon inducing strain [207]. This tells us that the design principles of composites should be innovatively implemented in the development of the final composites.

10.2.3 Thermal Conductivity

Graphene possesses the record thermal conductivity (κ) of ~5000 W m^{-1} K^{-1} [208]. Graphene is stable in air atmosphere up to 450 °C and has a low anisotropic thermal expansion coefficient [41]. These properties motivate us to mix graphene and its derivatives with polymers for improving thermal properties such as thermal conductivity, thermo-mechanical properties and dimensional stability [119]. Graphene inherits these properties from its native form, graphite [209]. As mentioned in the introduction, the thermal properties of graphite are at the core of many its commercial applications [209–211]. Can these thermal properties be translated into GO-based composites?

First we discuss κ of composites based on GO. Usually, polymer materials have low κ (in the range of 0.1–0.5 W m^{-1} K^{-1}), which is two orders of magnitude lower than that for metals. There is a demand for composite materials with moderately high κ (~10 W m^{-1} K^{-1}), high processability and chemical resistance (and in some cases with low electrical conductivity) [9, 209–212]. These criteria eliminate the metals and many ceramic materials from the list and make polymeric materials potential candidates. In addition, coolant fluids need to possess high κ, and the performance of cooling systems can be significantly improved by increasing the thermal properties of the fluids used [213, 214].

A simple rule of mixing tells us that κ of polymeric composites containing materials such as CNTs or graphene can easily attain the above-mentioned limit (~10 W m^{-1} K^{-1}) even at loadings as low as 1 vol.%. However, in reality, the final results show that κ is much lower than the value predicted from the rule of mixing [212]:

$$\kappa_c = \kappa_f \varphi_f + \kappa_m \varphi_m \tag{10.5}$$

On the other hand, there is a lower boundary for hybrid properties of composites, which is expressed as the series model, as follows [212]:

$$\kappa_c = 1/\left(\varphi_f/\kappa_f + \varphi_m/\kappa_m\right) \tag{10.6}$$

This tells us that, even at loadings as high as 50 vol.% (close to the packing volume fraction!), the value of κ_c cannot be higher than twice the value of κ_m, which is much lower than the goal of 10 W m^{-1} K^{-1}. Fortunately, the experimental results fall between these two predictions, and closer to what the series model suggests [9, 209, 212, 215]. As a result, to attain high conductivity values, very high loadings of fillers are inevitable. This is why more than half of the content of commercially available thermal greases are conductive fillers [216, 217].

Now, let us focus on the question of why κ_c is far from the value suggested by the ideal rule of mixing. The answer lies in thermal conduction at the interface. In fact, during the transport of phonons through the composite, there is a strong resistance as they meet the interface [212–214, 216]. Although one expects to observe a percolating behavior and a sudden jump of conductivity upon the formation of a 3D network in the matrix, the experimental data do not support this, at least at lower filler contents [213–216]. Mainly κ_c is a linear function of φ_f rather than exponential [216]. There are still no certain rules and mechanisms that determine the slope of this linear function. But to understand it, or at least use it as a starting point, one needs to understand which factors dominate the interfacial (Kapitza) resistance (R_K) [9, 209, 212–214]. One can imagine the interfacial resistance as follows. At a fixed heat flux (Q) applied to an interface, the temperature drop at the interface (ΔT) is proportional to the interfacial resistance ($R_K \sim Q\,\Delta T$) [218]. The Kapitza resistance R_K depends on the composition of the interface and the lattice parameters of the materials at the interface [219]. For instance, the resistance between highly conductive metals (e.g. Cu/Al) is low because both possess similar heat carriers (electrons) [219]. However, the story about graphenic compounds becomes more complicated. The sp^2 structure places them at the top of the κ table, but this unique atomic structure results in huge mismatches with other materials, especially insulating ones [209, 214, 219]. Therefore, despite the extremely high κ for graphenic materials, the κ_c of their composites is often dominated by their interfacial resistance with the matrix. Engineering these interfaces remains a challenge [212–214, 216, 217, 219].

Now, let us go back to the main topic of the chapter, GO-based composites. Severe oxidation of the graphenic lattice makes GO not only electrically but also thermally insulating. The κ_{GO} value is less than a few W m^{-1} K^{-1}, yet κ_{GO} can be recovered upon efficient reduction. For instance, Balandin and coworkers [220] reported an enhancement of κ_{GO} from 3 to 61 W m^{-1} K^{-1} upon annealing at 1000 °C, and predicted κ as high as 500 W m^{-1} K^{-1} with more efficient reduction. In another effort, Wallace and coworkers [100] reported an impressively high κ_{RGO} of 1435 W m^{-1} K^{-1} at room temperature for RGO fibers (along the axis) spun from a liquid crystalline phase of large-sized GO (Figure 10.13). This is one of the highest values reported for any macroscopic carbon material [221]. These reports give us hope that GO can also be treated as a filler for enhancing κ. However, additional work is needed on the factors controlling κ_{GO} by making functionalized graphenic structures.

Looking at the thermally conductive composites, most of the studies on carbon materials have been focused on CNTs and graphite nanoplatelets (GNPs) [9, 209, 212, 217]. GNPs were found to be the most efficient filler to improve κ. For instance, Shahil and Balandin [216] report a 23-fold increase in κ for an epoxy matrix merely by adding 10 vol.% of a mixture of graphene and GNPs. This level of enhancement is only achieved at loadings of around 50 vol.% of a conventional filler (such as metal (oxide) particles) [216, 217]. There is huge interest in the hybridization of graphene (and RGO) with other conductive fillers such as

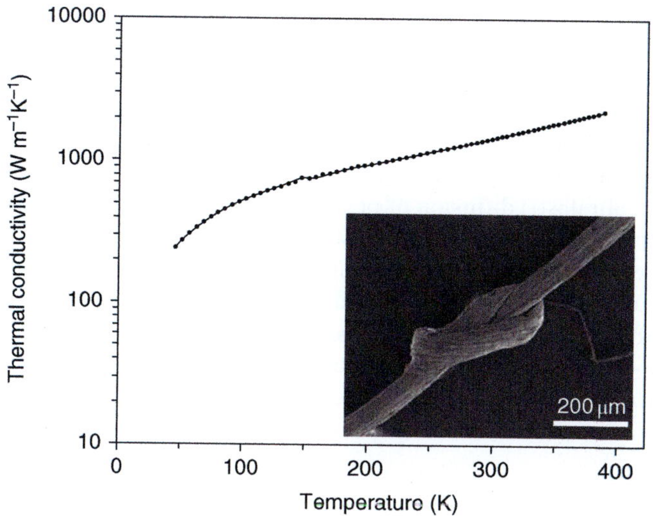

Figure 10.13 Temperature-dependent thermal conductivity of RGO fiber fabricated from wet spinning of liquid crystalline dispersion of ultra-large GO. Inset: Scanning electron micrograph of RGO fiber showing its flexibility. Reproduced from [100] with permission of Wiley-VCH Verlag GmbH & Co

CNTs or metal (oxide) particles [212, 215, 216, 222, 223]. This can lead to synergistic increase of κ or in some cases to reduction of electrical conductivity and retaining high κ [222].

It is also proven experimentally that fillers (CNTs, GNPs and probably RGO) with higher aspect ratio have a higher efficiency for increasing κ [214, 222]. As the filler–filler interface (especially in the case of graphenic fillers) has very high interfacial resistance, it is advantageous to use high-aspect-ratio fillers. It is proposed that engineering the interface by chemical bonding of polymer molecules to fillers decreases the interfacial resistance [209, 219], but at the same time the intrinsic κ of the filler is reduced. It actually remains an open question how and to what extent a chemical interaction between the filler and the matrix can affect κ_c. In the case of GO, where the graphene network is already damaged during oxidation, the covalent bonding of RGO to the matrix seems to be a logical strategy. However, reports on GO-based composites with the goal of improving κ are limited, and the results so far are not as promising as with GNPs [9, 215, 216]. There is an idea that hybrids of RGO–GNPs can be good candidates, as the chemical structure (and probably lattice parameters!) of GOs is somewhere between pure graphene and polar polymers [224]. Placing GO at the interface of graphene (or graphite) and polymers may reduce the interfacial resistance. However, this idea is yet to be examined.

10.2.4 Barrier Properties

Transport of liquid and gas molecules through a polymeric film is a vital phenomenon for many applications, such as membranes, protective coatings, chemically resistive polymers, etc. Engineering the transport of small molecules through a polymer is crucial for those applications. GO turns out to show fascinating transport properties, especially selective transport [225–229]. We will not discuss the details here, as they have been discussed in previous chapters. In the following, we consider GO as a disk-shaped filler in polymeric media, and its impact on barrier properties in general.

Imagine that a molecule tries to penetrate through a matrix, and the graphene sheets block its way [230, 231]. This makes a "tortuous path" for the diffusion of a penetrant [232]. This is the case when graphenic sheets have no affinity to the penetrant; otherwise even faster (even ultrafast!) diffusion of penetrant is possible [225]. In the case of a tortuous path, a simple analytical model suggests the highest impermeability by aligning the layered fillers perpendicular to the penetrant diffusion path (a "nematic-phase" morphology) [232] (Figure 10.14). In addition, the aspect ratio of the graphene sheets has a huge effect on the diffusion coefficient of composites, and an aspect ratio of 10^3 (ratio length to thickness) is shown to have high performance.

GO and its derivatives have a high chance to be used in composites for barrier applications as their orientation and size can be easily controlled. For instance, Kim *et al.* [230]

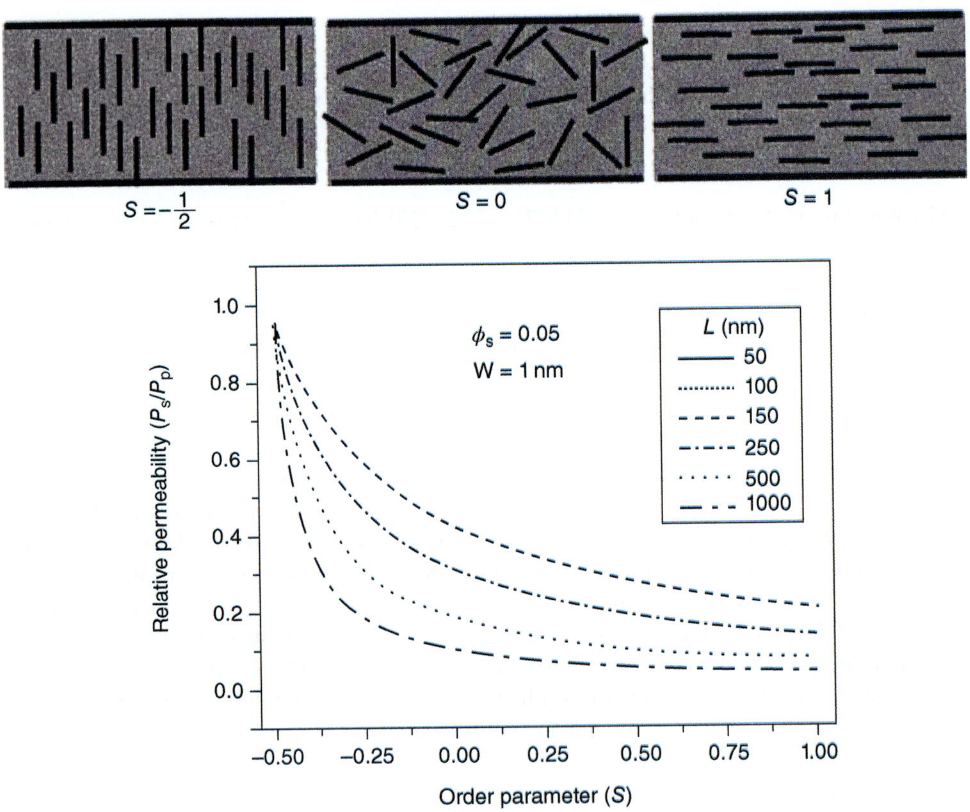

Figure 10.14 *Effect of sheet orientation and aspect ratio of nanoplatelets on the relative permeability in composites containing layered fillers. Illustrations (top) for three values of the order parameter, S = −½, 0 and 1. Relative permeability of composites at fixed content of filler increases with increasing filler orientation and aspect ratio (bottom). Reproduced from [232] with permission of ACS*

reported an 80-fold decrease in nitrogen permeability of a polyurethane matrix upon addition of 1.6 vol.% of functionalized GO. Recent reports also support the success of GO to enhance the barrier properties of polymer matrices [41, 103, 138, 230].

Apart from general-purpose applications as a gas barrier (such as protective coatings or packaging), there is a need for fabrication of transparent super gas barriers mainly for the organic light-emitting diode (OLED) industry and especially for flexible OLEDs [233, 234]. The current technology is based on ceramic-based fillers and these films are usually fragile and non-flexible. Graphene-based composites should meet the limit of about 10^{-5} cm^3 m^{-2} day^{-1} atm^{-1} for oxygen permeability to reach industrial acceptance [233, 234]. The design of such composites has challenged researchers for years, however; recent results have raised the hope that layered composites using 2D fillers including GO could serve as flexible super gas barriers [234–239] (Figure 10.15). Beside these breakthroughs,

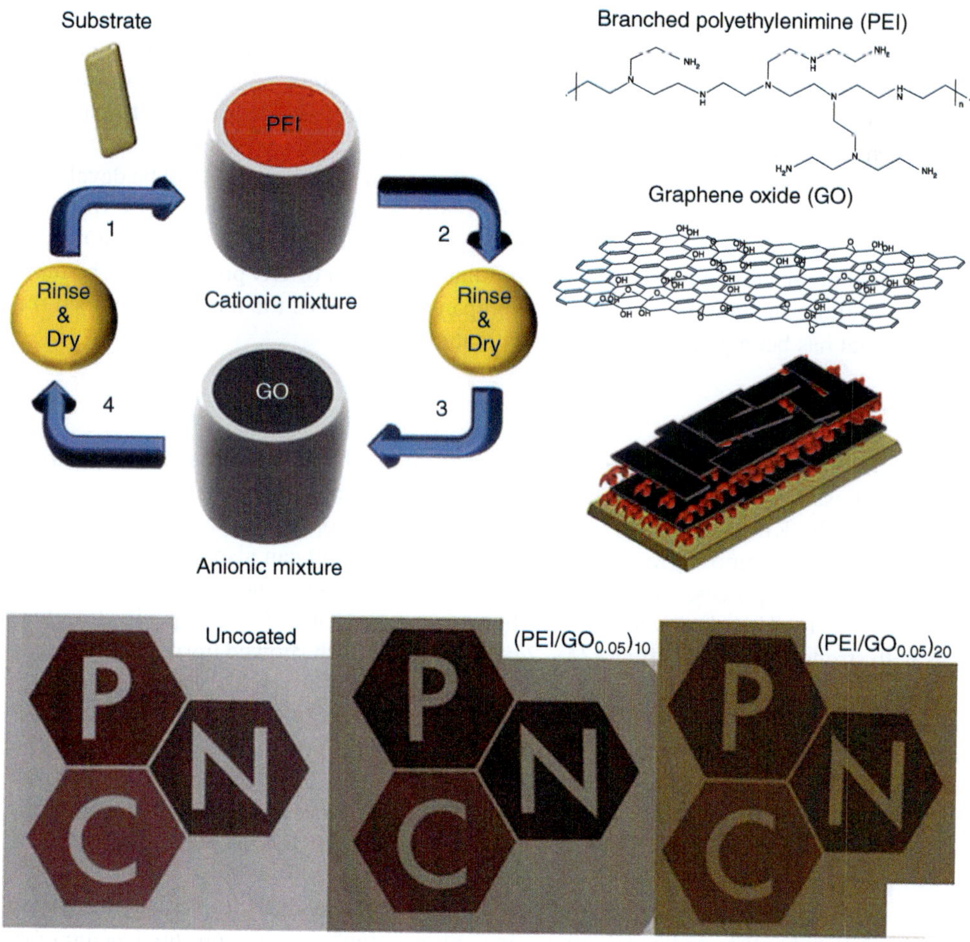

Figure 10.15 Developing a super gas barrier based on GO composites through layer-by-layer assembly on a plastic substrate resulting in a gas barrier with the performance of commercial ceramic gas barriers. Reproduced from [234] with permission of Wiley-VCH Verlag GmbH & Co

the fundamentals of molecular transport and even ions through atomically thin membranes challenge our current knowledge about the transport properties of materials [240, 241]. These lines of study need to be further pursued to complete our understanding of the graphene-based barrier mechanisms and to suggest appropriate guidelines.

10.3 Graphene Oxide or Graphene Oxides?

The most crucial factor to process any material for composite production, apart from its cost effectiveness, is the existence of a consistent, reliable and standardized method for the production. This is, in fact, the factor that severely hindered the growth of CNT application for use in different sectors for more than a decade. To this end, many research efforts have been devoted to understanding GO formation [242], and consequently its final chemistry and structure [60, 243]. So far, however, most studies in the field have been theoretical [244, 245], lacking the experimental verification and prone to simple models. As such, there is an immediate need to launch production guidelines for certain sets of material quality specifications. This baseline, if established, can, in principle, enable the optimization and real-time process control of GO production, leading to validated models for specific applications as required by different sectors.

To realize such a goal, a set of standard characterization procedures should be developed as an enabling platform to study the final properties of as-prepared GO structures. It should be noted that there are various factors governing GO production that determine GO structure, its final properties and consequently the standardization of GO production. The most common of those include the size of the as-prepared GO sheets, the medium in which the final material has been prepared and the purification process.

10.3.1 Size Effect

It has been shown that the size of GO has unprecedented effects on the final material structure properties, as it can be used to fine-tune its amphiphilicity (Figure 10.16). This effect, first observed by Kim *et al.* [246], opened up an avenue of opportunities for the formulation of GO-based composite materials, as the amphiphilic GO can also act as a molecular dispersing agent to process insoluble materials such as graphite and carbon nanotubes in water [247–251]. This discovery opened the way toward GO/CNT composite formulations with strong synergistic effect [248, 249, 251], and which showed unrivaled hydrogen storage capacity at room temperature [248]. Moreover, self-assembling of GO sheets at the liquid–liquid interface paved the way for environmentally benign processes for the production of high-quality polymer–graphene nanocomposites that are water-based (no organic solvent is employed) and soap-free (Figure 10.17) [252]. In these examples, the mixture of monomer, GO suspension and azobisisobutyronitrile (AIBN) can be emulsified using ultrasound to promote GO exfoliation and create emulsion droplets, which can be further polymerized [195, 253].

It should be noted that the dispersibility of GO in water and its hydrophilicity have been mainly attributed to the oxidized domains. In contrast, graphenic domains are hydrophobic [47, 246–248, 255]. With decreasing oxidation level of GO, the fraction of hydrophobic graphenic domains increases, rendering the structure less hydrophilic and leading to an

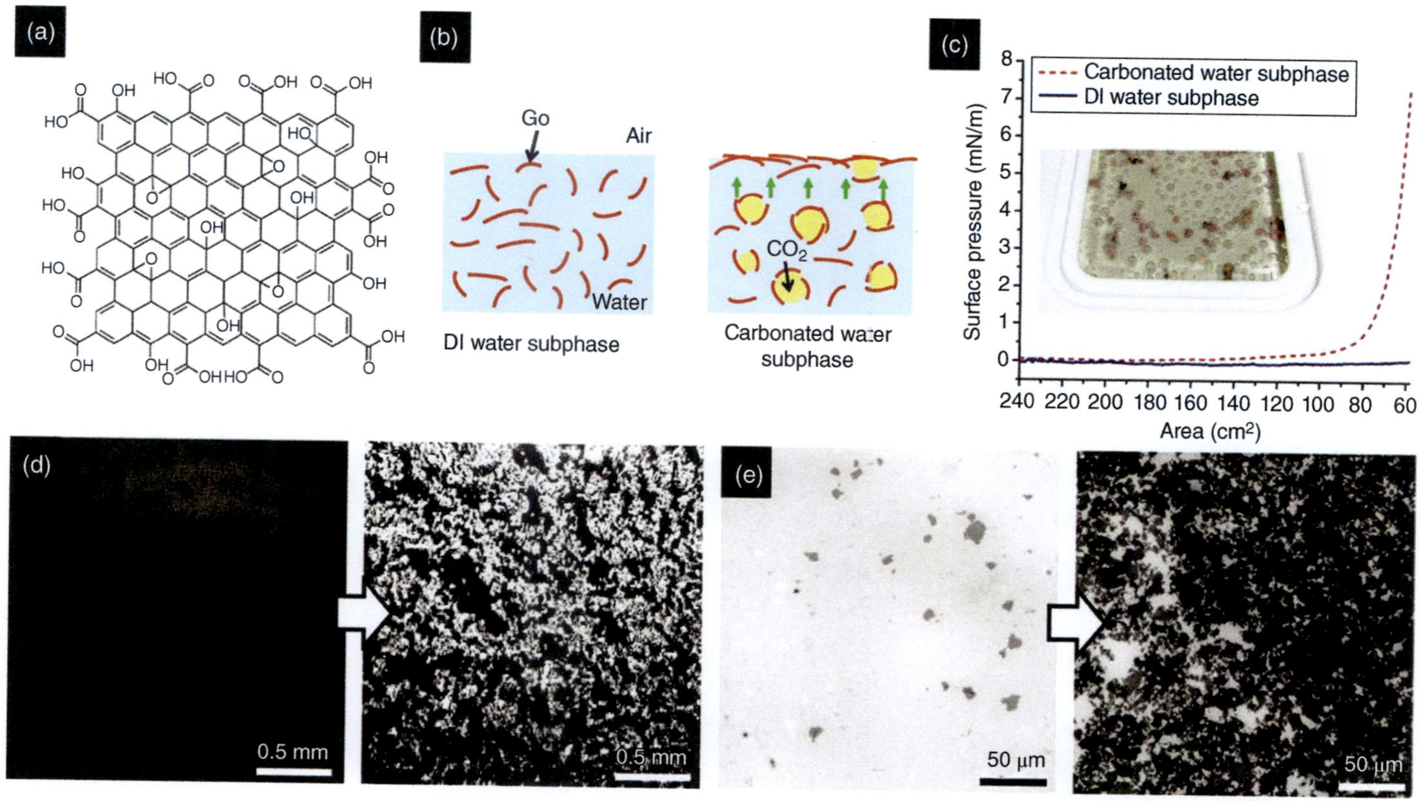

Figure 10.16 GO at the air – water interface. (a) GO can be viewed as a 2D molecular amphiphile, with hydrophobic π domains interspersed on its basal plane and hydrophilic – COOH groups on the edges, as shown in the structural model [47, 246]. (b) Schematic illustration showing the flotation of GO in carbonated water. GO is first captured by the rising CO_2 bubbles and then transported to the water surface. (c) Flotation experiment in a Langmuir–Blodgett (LB) trough, in which boiling stones were added to promote the evolution of bubbles, as shown in the inset. After flotation, the increased surface pressure (dashed red line) during isothermal compression indicates the presence of GO at the water surface. In contrast, the surface pressure of GO in deionized (DI) water remained nearly constant (solid blue line). Both (d) the in situ Brewster angle microscopy (BAM) images of the water surface and (e) the fluorescence quenching microscopy (FQM) images of GO sheets collected by dip coating show a massive increase of GO at the surface after flotation. Reproduced from [201] with permission of ACS

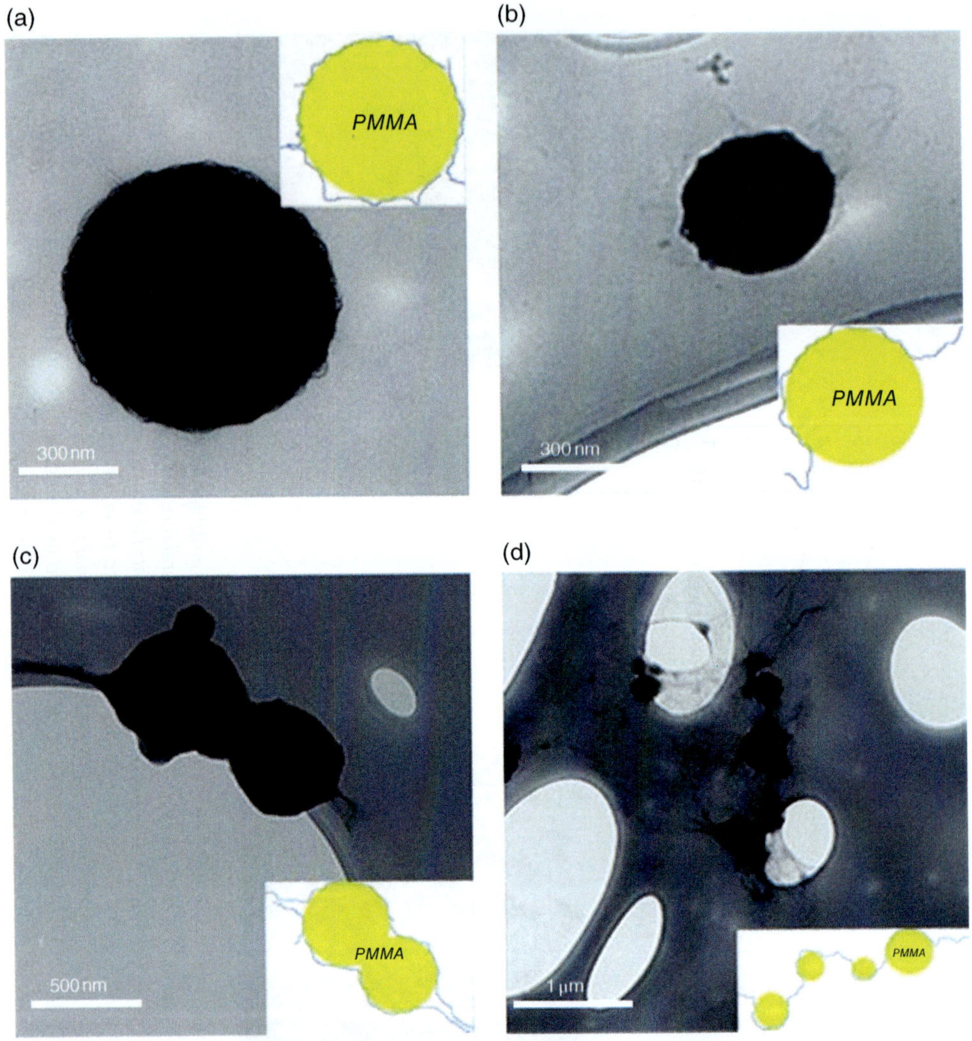

Figure 10.17 Transmission electron micrographs of a PMMA–GO colloidal hybrid showing various arrangements of polymer particle–GO nanolayers. Insets show speculative schemes of the colloidal hybrids. Reproduced from [254] with permission of Elsevier

amphiphilic structure, which can serve as both a molecular amphiphile and a colloidal surfactant, simultaneously [247, 248, 255, 256]. GO sheets with extended lateral size have also been shown to exhibit overall better mechanical properties [257], electrochemical capacitance [248], hydrogen storage properties [249] and electrical properties after reduction [257], arising from the lower degree of non-stoichiometric oxygen in their structure [249]. In contrast, the dominant hydrophilic nature of small GO sheets makes them an ideal material of choice for the preparation of GO–bioactive molecule composites for cellular delivery [258]. Nevertheless, this important finding, for the first time, demonstrated the possibility of the existence of different non-stoichiometric GO structures based on the size of the final GO sheets.

The size-dependent properties of GO sheets also played an important role in the discovery of GO liquid crystals [255, 259–261], and in subsequent processing of these materials in almost all industrially scalable fabrication methods [262], including but not limited to electrospraying, spray coating, a range of wet-spinning techniques [262–266], inkjet printing, 3D printing and dry spinning [262]. The discovery of GO liquid crystals then proved to be instrumental in guiding macroscopic, large-scale, material self-assembly at the atomic level through π–π stacking and hydrogen bonding interactions for composite formulations [254, 267–271].

10.3.2 Effect of Medium on GO Structure

The preparation of GO in organic solvents, first reported by Paredes *et al.* [272], facilitated the introduction of this material, beyond the original aqueous media solution-phase manipulation, to new composite formulations with water-sensitive metallic particles [273, 274], and water-insoluble polymers [250]. However, this brought up a new challenge regarding mainly the structure of the new material. It is pertinent to mention that, although all these materials are conveniently called GO, the characterization techniques reveal that their structures might be different. As an example, it was shown that the corresponding layer-by-layer distance (*d*-spacing) of each GO sample prepared in different organic solvents is varied consistent with the size of the organic solvent molecules (Figure 10.18).

XRD patterns of GO films prepared from dispersions in other organic solvents (*N*-methylpyrrolidone, NMP) also showed similar results [142], further suggesting that individual GO sheets in organic solvents are interlinked via a non-uniform network of hydrogen bonds mediated by oxygen functional groups and solvent molecules [250]. Thermogravimetric analysis (TGA) also showed that GO papers cast from organic solvents with high boiling points (154 °C for *N*-cyclohexyl-2-pyrrolidone (CHP) and 153 °C for dimethylformamide (DMF), respectively) exhibit higher decomposition temperatures compared to GO paper cast from water. In contrast, GO papers cast from GO dispersions in volatile solvents (acetone, tetrahydrofuran (THF) and ethanol) experienced weight loss at much lower temperatures. This behavior can pinpoint possible confinement of solvent molecules during film formation. This, fundamentally, proposes that the structure of the so-called GO in organic media is substantially different from that in water [142, 250]. Another interesting aspect, in this regard, is again the effect of the lateral size of the GO sheets. It has been shown that the extension of the size of the GO sheets to the range of a few to tens of micrometers can support the formation of fully dispersed, mostly monolayer GO sheets in polar aprotic solvents such as DMF with high stability [142, 250, 272–275], whereas small GO sheets cannot be exfoliated in such solvents [8, 276].

Nevertheless, the change in structure does not impose any significant problems regarding the performance of the final composite materials prepared from GO dispersions in organic solvents, at least from the mechanical properties point of view [250]. However, the application of GO dispersions in organic solvents should be investigated more fundamentally.

10.3.3 The Purification Process

The purification process is also another factor that is often neglected by groups working in this field. This process, although playing an important key role in the exfoliation of graphite oxide, has been mostly simplified in an effort to scale up production without

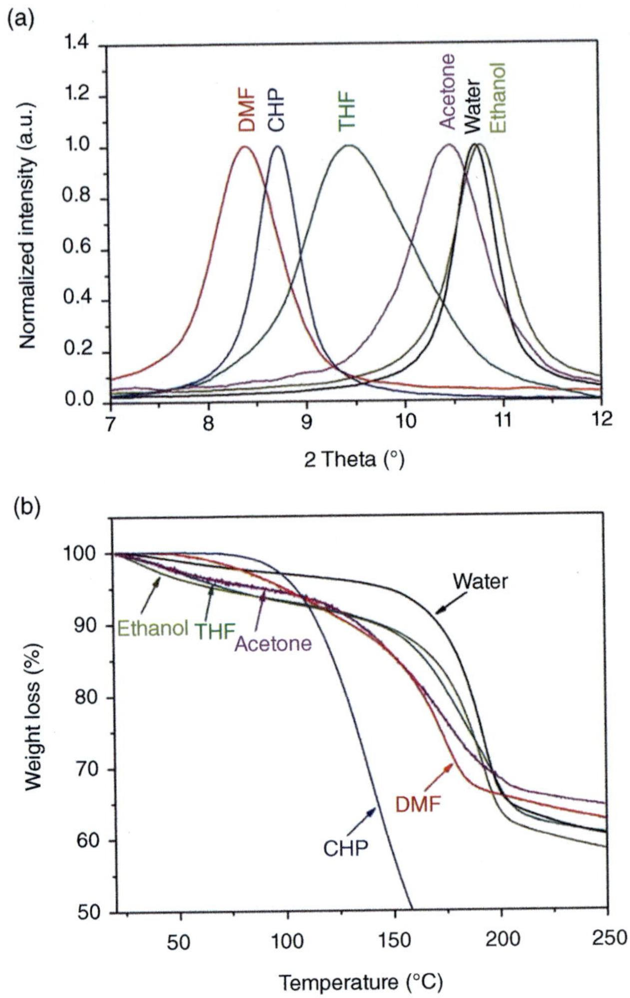

Figure 10.18 (a) XRD patterns and (b) TGA of GO films as a function of solvents. Reproduced from [250] with permission of American Chemical Society

undertaking the proper and adequate systematic investigation required to fully understand the fundamentals. Quenching and purification processes have been shown to significantly affect the structure and properties of GO [277]. Dimiev *et al.* [277] demonstrated that, in contrast to common belief, there is a threshold oxidation degree (TOD) and the TOD does not change significantly upon the addition of excess oxidizing agents (i.e. the variation of the oxidation level has some upper limit). Moreover, they showed that prolonged purification times are required if one wishes to remove the covalent sulfate moieties which can serve as a guideline for the scale-up of the GO production processes [277]. As such, the oxidation protocol should not contribute considerably to any substantial change in the final structure [277]. The type of the initial graphite, however, can affect the final structure, as different graphite types yield different GO flake lateral sizes [255, 257, 264].

10.3.4 Thermal Instability

GO is a highly energetic material with a high amount of stored chemical energy that can be readily released [278]. It has been demonstrated that, upon thermal reduction, GO is subjected to disproportionation reactions. During these reactions, some part of the carbon atoms, typically around 10% of C atoms, will be fully oxidized to CO_2, and the rest of the structure will be reduced to sp^2 graphenic products [278]. However, this introduces carbon vacancies in the basal planes, leading to a defective graphene structure [279] (Figure 10.19).

This, in essence, means that the "thermal reduction" of GO cannot lead to high-quality graphene, unless carbon atoms are added to the missing points through a carbonaceous atmosphere, or the arrangements of the carbon atoms are altered at elevated temperature (usually above 800 °C) [281, 282]. Another complication arising from the thermal instability of GO is the limited shelf-life of its solid form. Moreover, heating chunks of GO on a hotplate results in an explosive reaction as a direct consequence of the evolution of gaseous H_2O and CO_2 during the deoxygenation process (Figure 10.20).

These results suggest that the storage of large quantities of GO in its solid form can lead to fire hazards [278]. Additionally, it should be noted that, although GO is reported to have flame-retarding properties, the remains of any residue salts in the structure can make the material highly flammable, as shown by Krishnan *et al.* [278]. This imposes other problems, as it suggests the wet storage, transport, processing and distribution of GO sheets, to make them safe [283]. Nevertheless, this suggests some additional opportunities for composite making processes, as chemical reduction methods are not typically feasible for preparing RGO–polymer composites. Heating of the final composites is not a good solution, either, as typically most polymers are unstable at elevated temperatures. However, in these cases, flash photothermal irradiation can serve as the reduction method of choice for "reducing" the GO structure into RGO with minimal penalty effect on the surrounding polymer [278, 284, 285].

10.3.5 Health Issues

For the safe production, handling and application of any materials, it is essential to thoroughly assess the possibility of human exposure to that material through inhalation, digestion, skin contact, etc. However, as a direct result of the rapid development and growth of GO in different fields, the assessment of its potential health risks/hazards has not yet been methodically investigated. Nevertheless, with the introduction of more GO-based composites to the market, and the development of new research applications, many *in vitro* studies have been devoted to evaluate its potential biohazards [286, 287]. It has been reported that the toxicity of GO is concentration-dependent. It is suggested that the cytotoxicity stems from direct interactions between the cell membrane and GO nanosheets, leading to physical damage to the cell membrane [288]. However, it should be noted that the GO sheet sizes used here were all in the nanometer region. In contrast, in another study, it was shown that the size of the GO sheets plays an important role in its toxicity to A549 cells, a widely used model cell line for toxicity studies [289]. It was demonstrated that small GO sheets could cause the most serious oxidative stress to the cells, at a concentration as low as 10 $\mu g\ ml^{-1}$. In contrast, no meaningful difference between cells exposed to large GO and medium GO was found [289]. Interestingly, in another account, in which the low cytotoxicity of GO was reported, the size of the GO sheets was found to be in the micrometer region [290].

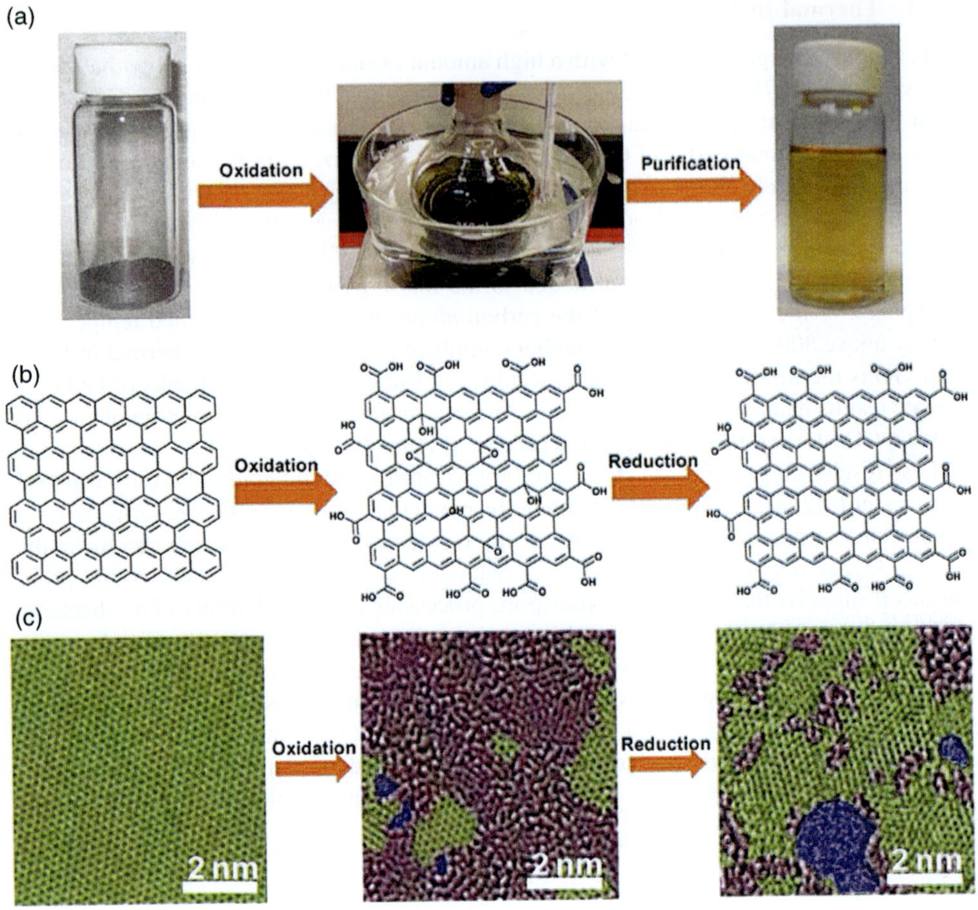

Figure 10.19 Synthesis, structural model and microstructures of GO sheets. (a) GO is typically synthesized by reacting graphite powder with strongly oxidizing agents such as $KMnO_4$ in concentrated H_2SO_4, followed by purification and exfoliation in water to yield a colloidal dispersion of single layers. (b) Structural model of graphene (left), GO (middle) and its reduction product RGO. GO is insulating due to broken conjugation in the basal plane. After reduction, RGO becomes conductive, but it is still a very defective structure compared to graphene. Reproduced with permission from Ref. [278]. Copyright 2012 Elsevier. (c) Color-coded high-resolution TEM images showing the atomic structures of graphene (left), GO (middle) and RGO (right). The green, purple and blue areas depict ordered graphitic sp^2 domains, disordered highly oxidized sp^3 domains, and holes on the sheet, respectively. Reproduced from [280] with permission of John Wiley & Sons

Unfortunately, apart from one study that has shown the cytotoxicity of both medium GO sheets and small GO sheets [291], for other studies regarding GO-induced cytotoxicity, genotoxicity and oxidative stress, the effect of size and consequently the structural difference of GO sheets with different sizes have not been investigated [226, 292–294]. However, the composites of GO with polyvinylpyrrolidone have been considered as an available

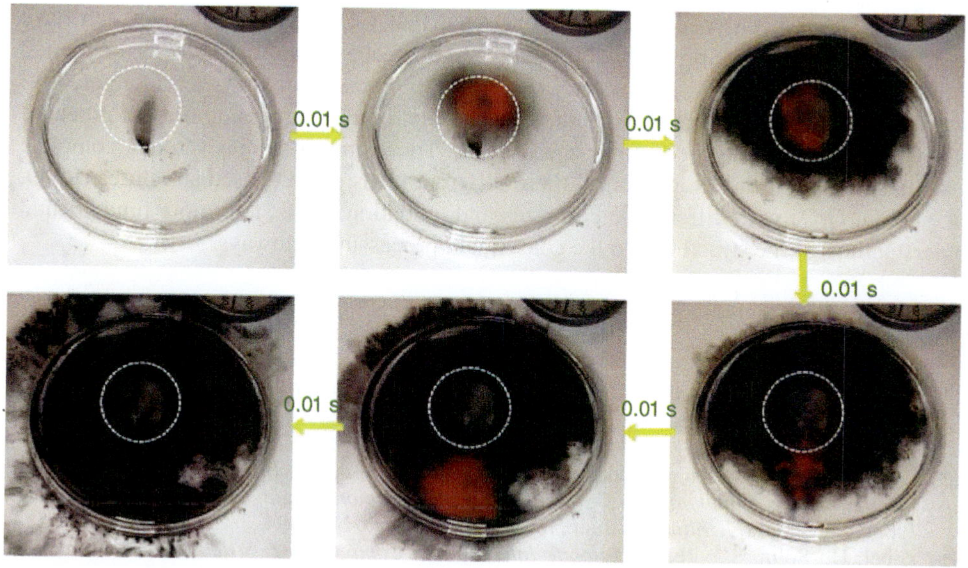

Figure 10.20 Snapshots showing the explosion of a piece of GO solid inside a Petri dish a few seconds after being placed on a hotplate preheated to 300 °C. Before explosion, water vapor was driven out of the GO paper and condensed on the top cover as highlighted by the white broken circle. Reproduced from [278] with permission of Elsevier

candidate as an immunoadjuvant to promote vaccine efficacy, just like the colloid aluminum hydroxide [295]. As such, the health effects of different structures of GO sheets is still unclear and should be more carefully investigated, which requires an interdisciplinary effort. Until such studies are performed, precaution in handling and application of GO in different sectors is suggested.

10.3.6 Environmental Impact

Recent research efforts have revealed the negative environmental impact of small and medium GO sheets on the wastewater microbial community, signifying the potential of GO to impede the crucial microbial functions needed in activated sludge processes, such as removal of organic matter and other nutrients from wastewater [296]. However, more studies in this regard should be conducted, as the environmental impact of GO is an important factor determining the application of this material in various fields.

10.4 Conclusion

Constructing multi-functional hybrid nanocomposites, which exhibit simultaneous improvement in mechanical, thermal, electrical, electrochemical and optical aspects with no or minimal environmental and health impact penalties, is an ambitious frontier research that is yet to be realized. In this regard, GO and its derivatives enjoy a unique tunable amphiphilic structure that gives them an edge in a wide range of composite making strategies. It has been successfully demonstrated that GO and its derivatives can have immense positive impact on the final properties of the matrix in almost all areas mentioned above, especially when the high lateral size of GO is preserved. However, there are still many fundamental questions that should be answered before harnessing the full potential of this exciting material for real-life applications. These questions range from a complete systematic study of the toxicity of all these different types of GO-based structures (RGO, tpGO, GO, GrO) with different lateral sizes and oxidation levels, to a more fundamental understanding of the structure and the consequent processing pathways.

References

[1] Novoselov, K.S.; Geim, A.K.; Morozov, S.; *et al.*, Electric field effect in atomically thin carbon films. *Science* **2004**, *306* (5696), 666–669.
[2] Geim, A.K.; Novoselov, K.S., The rise of graphene. *Nature Mater.* **2007**, *6* (3), 183–191.
[3] Geim, A.K., Nobel Lecture: Random walk to graphene. *Rev. Mod. Phys.* **2011**, *83* (3), 851.
[4] Pierson, H.O., *Handbook of Carbon, Graphite, Diamonds and Fullerenes: Processing, Properties and Applications*. William Andrew, Park Ridge, NJ, **1994**.
[5] Zabel, H.; Solin, S.A. (eds), *Graphite Intercalation Compounds I: Structure and Dynamics*. Springer Series in Materials Science, vol. *14*. Springer, Berlin, **1990**.
[6] Chung, D.D.L., Review: Graphite. *J. Mater. Sci.* **2002**, *37* (8), 1475–1489.
[7] Li, B.; Zhong, W.-H., Review on polymer/graphite nanoplatelet nanocomposites. *J. Mater. Sci.* **2011**, *46* (17), 5595–5614.
[8] Stankovich, S.; Dikin, D.A.; Piner, R.D.; *et al.*, Synthesis of graphene-based nanosheets via chemical reduction of exfoliated graphite oxide. *Carbon* **2007**, *45* (7), 1558–1565.
[9] Sadasivuni, K.K.; Ponnamma, D.; Thomas, S.; Grohens, Y., Evolution from graphite to graphene elastomer composites. *Prog. Polym. Sci.* **2014**, *39* (4), 749–780.
[10] Kroto, H.W.; Heath, J.R.; O'Brien, S.C.; *et al.*, C_{60}: buckminsterfullerene. *Nature* **1985**, *318* (6042), 162–163.
[11] Iijima, S., Helical microtubules of graphitic carbon. *Nature* **1991**, *354* (6348), 56–58.
[12] Thompson, B.C.; Fréchet, J.M.J., Polymer–fullerene composite solar cells. *Angew. Chem. Int. Ed.* **2008**, *47* (1), 58–77.
[13] Spitalsky, Z.; Tasis, D.; Papagelis, K.; Galiotis, C., Carbon nanotube–polymer composites: chemistry, processing, mechanical and electrical properties. *Prog. Polym. Sci.* **2010**, *35* (3), 357–401.
[14] Richard, C.; Balavoine, F.; Schultz, P.; *et al.*, Supramolecular self-assembly of lipid derivatives on carbon nanotubes. *Science* **2003**, *300* (5620), 775–778.
[15] Gogotsi, Y.; Presser, V., *Carbon Nanomaterials*, 2nd edn. CRC Press, Boca Raton, FL, **2013**.
[16] Chung, D., Exfoliation of graphite. *J. Mater. Sci.* **1987**, *22* (12), 4190–4198.
[17] Pan, Y.X.; Yu, Z.Z.; Ou, Y.C.; Hu, G.H., A new process of fabricating electrically conducting nylon 6/graphite nanocomposites via intercalation polymerization. *J. Polym. Sci. B: Polym. Phys.* **2000**, *38* (12), 1626–1633.
[18] Zurutuza, A.; Marinelli, C., Challenges and opportunities in graphene commercialization. *Nature Nanotechnol.* **2014**, *9* (10), 730–734.
[19] Feytis, A., The bright side of graphite. *Indust. Miner.* **2010, July** (*514*), 31–39.
[20] Aylsworth, J.W., Expanded graphite and composition thereof. US Patent 1137373, **1915**.
[21] Anderson, S.; Chung, D., Exfoliation of intercalated graphite. *Carbon* **1984**, *22* (3), 253–263.
[22] Sengupta, R.; Bhattacharya, M.; Bandyopadhyay, S.; Bhowmick, A.K., A review on the mechanical and electrical properties of graphite and modified graphite reinforced polymer composites. *Prog. Polym. Sci.* **2011**, *36* (5), 638–670.
[23] Li, J.; Kim, J.-K.; Sham, M.L., Conductive graphite nanoplatelet/epoxy nanocomposites: effects of exfoliation and UV/ozone treatment of graphite. *Scr. Mater.* **2005**, *53* (2), 235–240.
[24] Brodie, B.C., On the atomic weight of graphite. *Phil. Trans. R. Soc. Lond.* **1859**, *149*, 249–259.

[25] Olsen, L.C.; Seeman, S.E.; Scott, H.W., Expanded pyrolytic graphite: structural and transport properties. *Carbon* **1970**, *8* (1), 85–93.
[26] Dresselhaus, M.; Dresselhaus, G., Intercalation compounds of graphite. *Adv. Phys.* **1981**, *30* (2), 139–326.
[27] Weller, T.E.; Ellerby, M.; Saxena, S.S.; *et al.*, Superconductivity in the intercalated graphite compounds C_6Yb and C_6Ca. *Nature Phys.* **2005**, *1* (1), 39–41.
[28] Hannay, N.; Geballe, T.; Matthias, B.; *et al.*, Superconductivity in graphitic compounds. *Phys. Rev. Lett.* **1965**, *14* (7), 225.
[29] Schweizer, E., Analyse des Porphyrs von Kreuznach im Nahethale. *J. Prakt. Chem.* **1841**, *22* (1), 155–158.
[30] Jang, B.Z.; Zhamu, A., Processing of nanographene platelets (NGPs) and NGP nanocomposites: a review. *J. Mater. Sci.* **2008**, *43* (15), 5092–5101.
[31] McAllister, M.J.; Li, J.-L.; Adamson, D.H.; *et al.*, Single sheet functionalized graphene by oxidation and thermal expansion of graphite. *Chem. Mater.* **2007**, *19* (18), 4396–4404.
[32] Cai, M.; Thorpe, D.; Adamson, D.H.; Schniepp, H.C., Methods of graphite exfoliation. *J. Mater. Chem.* **2012**, *22* (48), 24992–25002.
[33] Viculis, L.M.; Mack, J.J.; Kaner, R.B., A chemical route to carbon nanoscrolls. *Science* **2003**, *299* (5611), 1361.
[34] Viculis, L.M.; Mack, J.J.; Mayer, O.M.; *et al.*, Intercalation and exfoliation routes to graphite nanoplatelets. *J. Mater. Chem.* **2005**, *15* (9), 974–978.
[35] Su, C.-Y.; Lu, A.-Y.; Xu, Y.; *et al.*, High-quality thin graphene films from fast electrochemical exfoliation. *ACS Nano* **2011**, *5* (3), 2332–2339.
[36] Usuki, A.; Kojima, Y.; Kawasumi, M.; *et al.*, Synthesis of nylon 6–clay hybrid. *J. Mater. Res.* **1993**, *8* (5), 1179–1184.
[37] Okada, A.; Usuki, A., Twenty years of polymer–clay nanocomposites. *Macromol. Mater. Eng.* **2006**, *291* (12), 1449–1476.
[38] Mack, J.J.; Viculis, L.M.; Ali, A.; *et al.*, Graphite nanoplatelet reinforcement of electrospun polyacrylonitrile nanofibers. *Adv. Mater.* **2005**, *17* (1), 77–80.
[39] Ajayan, P.M.; Schadler, L.S.; Braun, P.V., *Nanocomposite Science and Technology*. Wiley-VCH, Weinheim, **2003**.
[40] Krishnamoorti, R.; Vaia, R.A., *Polymer Nanocomposites: Synthesis, Characterization, and Modeling*. ACS Symposium Series 804. American Chemical Society, Washington, DC, **2002**.
[41] Mukhopadhyay, P.; Gupta, R.K., *Graphite, Graphene, and Their Polymer Nanocomposites*. CRC Press, Boca Raton, FL, **2013**.
[42] Choi, W.; Lee, J.-W., *Graphene: Synthesis and Applications*. CRC Press, Boca Raton, FL, **2012**.
[43] Boehm, H.P.; Clauss, A.; Fischer, G.; Hofmann, U., Surface properties of extremely thin graphite lamellae. In *Carbon: Proceedings of the Fifth Conference*, vol. *1*, pp. 73–80. Pergamon Press, Oxford, **1962**.
[44] Boehm, H.-P., Graphene – how a laboratory curiosity suddenly became extremely interesting. *Angew. Chem. Int. Ed.* **2010**, *49* (49), 9332–9335.
[45] Dreyer, D.R.; Ruoff, R.S.; Bielawski, C.W., From conception to realization: an historial account of graphene and some perspectives for its future. *Angew. Chem. Int. Ed.* **2010**, *49* (49), 9336–9344.
[46] Jeong, H.-K.; Lee, Y.P.; Lahaye, R.J.W.E.; *et al.*, Evidence of graphitic AB stacking order of graphite oxides. *J. Am. Chem. Soc.* **2008**, *130* (4), 1362–1366.
[47] Lerf, A.; He, H.; Forster, M.; Klinowski, J., Structure of graphite oxide revisited. *J. Phys. Chem. B* **1998**, *102* (23), 4477–4482.
[48] Cai, W.; Piner, R.D.; Stadermann, F.J.; *et al.*, Synthesis and solid-state NMR structural characterization of ^{13}C-labeled graphite oxide. *Science* **2008**, *321* (5897), 1815–1817.
[49] Gao, W.; Alemany, L.B.; Ci, L.; Ajayan, P.M., New insights into the structure and reduction of graphite oxide. *Nature Chem.* **2009**, *1* (5), 403–408.
[50] Matsuo, Y.; Miyabe, T.; Fukutsuka, T.; Sugie, Y., Preparation and characterization of alkylamine-intercalated graphite oxides. *Carbon* **2007**, *45* (5), 1005–1012.
[51] Bourlinos, A.B.; Gournis, D.; Petridis, D.; *et al.*, Graphite oxide: chemical reduction to graphite and surface modification with primary aliphatic amines and amino acids. *Langmuir* **2003**, *19* (15), 6050–6055.

[52] Stankovich, S.; Dikin, D.A.; Dommett, G.H.; *et al.*, Graphene-based composite materials. *Nature* **2006**, *442* (7100), 282–286.
[53] Kotov, N.A.; Dékány, I.; Fendler, J.H., Ultrathin graphite oxide–polyelectrolyte composites prepared by self-assembly: transition between conductive and non-conductive states. *Adv. Mater.* **1996**, *8* (8), 637–641.
[54] Kovtyukhova, N.I.; Ollivier, P.J.; Martin, B.R.; *et al.*, Layer-by-layer assembly of ultrathin composite films from micron-sized graphite oxide sheets and polycations. *Chem. Mater.* **1999**, *11* (3), 771–778.
[55] Novoselov, K.; Jiang, D.; Schedin, F.; *et al.*, Two-dimensional atomic crystals. *Proc. Natl. Acad. Sci. USA* **2005**, *102* (30), 10451–10453.
[56] Geim, A.; Grigorieva, I., Van der Waals heterostructures. *Nature* **2013**, *499* (7459), 419–425.
[57] Baughman, R.H.; Zakhidov, A.A.; de Heer, W.A., Carbon nanotubes – the route toward applications. *Science* **2002**, *297* (5582), 787–792.
[58] Ruoff, R., Graphene: calling all chemists. *Nature Nanotechnol.* **2008**, *3* (1), 10–11.
[59] Grossiord, N.; Loos, J.; Regev, O.; Koning, C.E., Toolbox for dispersing carbon nanotubes into polymers to get conductive nanocomposites. *Chem. Mater.* **2006**, *18* (5), 1089–1099.
[60] Dreyer, D.R.; Park, S.; Bielawski, C.W.; Ruoff, R.S., The chemistry of graphene oxide. *Chem. Soc. Rev.* **2010**, *39* (1), 228–240.
[61] Chen, D.; Feng, H.; Li, J., Graphene oxide: preparation, functionalization, and electrochemical applications. *Chem. Rev.* **2012**, *112* (11), 6027–6053.
[62] Lomeda, J.R.; Doyle, C.D.; Kosynkin, D.V.; *et al.*, Diazonium functionalization of surfactant-wrapped chemically converted graphene sheets. *J. Am. Chem. Soc.* **2008**, *130* (48), 16201–16206.
[63] Chua, C.K.; Pumera, M., Friedel–Crafts acylation on graphene. *Chemistry – Asian J.* **2012**, *7* (5), 1009–1012.
[64] Hamilton, C.E.; Lomeda, J.R.; Sun, Z.; *et al.*, Radical addition of perfluorinated alkyl iodides to multi-layered graphene and single-walled carbon nanotubes. *Nano Res.* **2010**, *3* (2), 138–145.
[65] Lee, S.H.; Dreyer, D.R.; An, J.; *et al.*, Polymer brushes via controlled, surface-initiated atom transfer radical polymerization (ATRP) from graphene oxide. *Macromol. Rapid Commun.* **2010**, *31* (3), 281–288.
[66] Fang, M.; Wang, K.; Lu, H.; *et al.*, Covalent polymer functionalization of graphene nanosheets and mechanical properties of composites. *J. Mater. Chem.* **2009**, *19* (38), 7098–7105.
[67] Collins, W.R.; Lewandowski, W.; Schmois, E.; *et al.*, Claisen rearrangement of graphite oxide: a route to covalently functionalized graphenes. *Angew. Chem.* **2011**, *123* (38), 9010–9014.
[68] Xu, Y.; Bai, H.; Lu, G.; *et al.*, Flexible graphene films via the filtration of water-soluble noncovalent functionalized graphene sheets. *J. Am. Chem. Soc.* **2008**, *130* (18), 5856–5857.
[69] Zhang, M.; Parajuli, R.R.; Mastrogiovanni, D.; *et al.*, Production of graphene sheets by direct dispersion with aromatic healing agents. *Small* **2010**, *6* (10), 1100–1107.
[70] Jankovsky, O.; Simek, P.; Klimova, K.; *et al.*, Towards graphene bromide: bromination of graphite oxide. *Nanoscale* **2014**, *6* (11), 6065–6074.
[71] Jeon, I.-Y.; Shin, Y.-R.; Sohn, G.-J.; *et al.*, Edge-carboxylated graphene nanosheets via ball milling. *Proc. Natl. Acad. Sci. USA* **2012**, *109* (15), 5588–5593.
[72] Dai, L., Functionalization of graphene for efficient energy conversion and storage. *Acc. Chem. Res.* **2012**, *46* (1), 31–42.
[73] Ajayan, P.M.; Tour, J.M., Materials science: nanotube composites. *Nature* **2007**, *447* (7148), 1066–1068.
[74] Yan, L.; Zheng, Y.B.; Zhao, F.; *et al.*, Chemistry and physics of a single atomic layer: strategies and challenges for functionalization of graphene and graphene-based materials. *Chem. Soc. Rev.* **2012**, *41* (1), 97–114.
[75] Hernandez, Y.; Nicolosi, V.; Lotya, M.; *et al.*, High-yield production of graphene by liquid-phase exfoliation of graphite. *Nature Nanotechnol.* **2008**, *3* (9), 563–568.
[76] Bergin, S.D.; Sun, Z.; Rickard, D.; *et al.*, Multicomponent solubility parameters for single-walled carbon nanotube–solvent mixtures. *ACS Nano* **2009**, *3* (8), 2340–2350.
[77] Lotya, M.; Hernandez, Y.; King, P.J.; *et al.*, Liquid phase production of graphene by exfoliation of graphite in surfactant/water solutions. *J. Am. Chem. Soc.* **2009**, *131* (10), 3611–3620.

[78] Islam, M.; Rojas, E.; Bergey, D.; *et al.*, High weight fraction surfactant solubilization of single-wall carbon nanotubes in water. *Nano Lett.* **2003**, *3* (2), 269–273.
[79] Coleman, J.N., Liquid-phase exfoliation of nanotubes and graphene. *Adv. Funct. Mater.* **2009**, *19* (23), 3680–3695.
[80] Zhou, K.G.; Mao, N.N.; Wang, H.X.; *et al.*, A mixed-solvent strategy for efficient exfoliation of inorganic graphene analogues. *Angew. Chem. Int. Ed.* **2011**, *50* (46), 10839–10842.
[81] Backes, C.; Hauke, F.; Hirsch, A., The potential of perylene bisimide derivatives for the solubilization of carbon nanotubes and graphene. *Adv. Mater.* **2011**, *23* (22–23), 2588–2601.
[82] Moore, V.C.; Strano, M.S.; Haroz, E.H.; *et al.*, Individually suspended single-walled carbon nanotubes in various surfactants. *Nano Lett.* **2003**, *3* (10), 1379–1382.
[83] Park, S.; An, J.; Potts, J.R.; *et al.*, Hydrazine-reduction of graphite- and graphene oxide. *Carbon* **2011**, *49* (9), 3019–3023.
[84] Kotov, N.A., Materials science: carbon sheet solutions. *Nature* **2006**, *442* (7100), 254–255.
[85] Editorial, Ten years in two dimensions. *Nature Nanotechnol.* **2014**, *9* (10), 725.
[86] Hodge, S.A.; Bayazit, M.K.; Coleman, K.S.; Shaffer, M.S.P., Unweaving the rainbow: a review of the relationship between single-walled carbon nanotube molecular structures and their chemical reactivity. *Chem. Soc. Rev.* **2012**, *41* (12), 4409–4429.
[87] Wang, X.; Zhi, L.; Müllen, K., Transparent, conductive graphene electrodes for dye-sensitized solar cells. *Nano Lett.* **2008**, *8* (1), 323–327.
[88] Stoller, M.D.; Park, S.; Zhu, Y.; *et al.*, Graphene-based ultracapacitors. *Nano Lett.* **2008**, *8* (10), 3498–3502.
[89] Liu, Z.; Robinson, J.T.; Sun, X.; Dai, H., PEGylated nanographene oxide for delivery of water-insoluble cancer drugs. *J. Am. Chem. Soc.* **2008**, *130* (33), 10876–10877.
[90] Eda, G.; Fanchini, G.; Chhowalla, M., Large-area ultrathin films of reduced graphene oxide as a transparent and flexible electronic material. *Nature Nanotechnol.* **2008**, *3* (5), 270–274.
[91] Wu, Z.; Chen, Z.; Du, X.; *et al.*, Transparent, conductive carbon nanotube films. *Science* **2004**, *305* (5688), 1273–1276.
[92] Dikin, D.A.; Stankovich, S.; Zimney, E.J.; *et al.*, Preparation and characterization of graphene oxide paper. *Nature* **2007**, *448* (7152), 457–460.
[93] Bae, S.; Kim, H.; Lee, Y.; *et al.*, Roll-to-roll production of 30-inch graphene films for transparent electrodes. *Nature Nanotechnol.* **2010**, *5* (8), 574–578.
[94] Zheng, Q.B.; Gudarzi, M.M.; Wang, S.J.; *et al.*, Improved electrical and optical characteristics of transparent graphene thin films produced by acid and doping treatments. *Carbon* **2011**, *49* (9), 2905–2916.
[95] Dettlaff-Weglikowska, U.; Skákalová, V.; Graupner, R.; *et al.*, Effect of $SOCl_2$ treatment on electrical and mechanical properties of single-wall carbon nanotube networks. *J. Am. Chem. Soc.* **2005**, *127* (14), 5125–5131.
[96] Arnold, M.S.; Green, A.A.; Hulvat, J.F.; *et al.*, Sorting carbon nanotubes by electronic structure using density differentiation. *Nature Nanotechnol.* **2006**, *1* (1), 60–65.
[97] Green, A.A.; Hersam, M.C., Solution phase production of graphene with controlled thickness via density differentiation. *Nano Lett.* **2009**, *9* (12), 4031–4036.
[98] Behabtu, N.; Lomeda, J.R.; Green, M.J.; *et al.*, Spontaneous high-concentration dispersions and liquid crystals of graphene. *Nature Nanotechnol.* **2010**, *5* (6), 406–411.
[99] Xu, Z.; Gao, C., Graphene chiral liquid crystals and macroscopic assembled fibres. *Nature Commun.* **2011**, *2*, 571.
[100] Jalili, R.; Aboutalebi, S.H.; Esrafilzadeh, D.; *et al.*, Scalable one-step wet-spinning of graphene fibers and yarns from liquid crystalline dispersions of graphene oxide: towards multifunctional textiles. *Adv. Funct. Mater.* **2013**, *23* (43), 5345–5354.
[101] Ericson, L.M.; Fan, H.; Peng, H.; *et al.*, Macroscopic, neat, single-walled carbon nanotube fibers. *Science* **2004**, *305* (5689), 1447–1450.
[102] Tasis, D.; Tagmatarchis, N.; Bianco, A.; Prato, M., Chemistry of carbon nanotubes. *Chem. Rev.* **2006**, *106* (3), 1105–1136.
[103] Kuilla, T.; Bhadra, S.; Yao, D.; *et al.*, Recent advances in graphene based polymer composites. *Prog. Polym. Sci.* **2010**, *35* (11), 1350–1375.

[104] Zhang, D.; Ryu, K.; Liu, X.; *et al.*, Transparent, conductive, and flexible carbon nanotube films and their application in organic light-emitting diodes. *Nano Lett.* **2006**, *6* (9), 1880–1886.
[105] Endo, M.; Muramatsu, H.; Hayashi, T.; *et al.*, Nanotechnology: "buckypaper" from coaxial nanotubes. *Nature* **2005**, *433* (7025), 476.
[106] Ramanathan, T.; Abdala, A.A.; Stankovich, S.; *et al.*, Functionalized graphene sheets for polymer nanocomposites. *Nature Nanotechnol.* **2008**, *3* (6), 327–331.
[107] Sanchez, V.C.; Jachak, A.; Hurt, R.H.; Kane, A.B., Biological interactions of graphene-family nanomaterials: an interdisciplinary review. *Chem. Res. Toxicol.* **2012**, *25* (1), 15–34.
[108] Akhavan, O.; Ghaderi, E., Toxicity of graphene and graphene oxide nanowalls against bacteria. *ACS Nano* **2010**, *4* (10), 5731–5736.
[109] Lacerda, L.; Bianco, A.; Prato, M.; Kostarelos, K., Carbon nanotubes as nanomedicines: from toxicology to pharmacology. *Adv. Drug Deliv. Rev.* **2006**, *58* (14), 1460–1470.
[110] Kostarelos, K.; Novoselov, K.S., Graphene devices for life. *Nature Nanotechnol.* **2014**, *9* (10), 744–745.
[111] Gedde, U.W., *Polymer Physics*. Springer, Dordrecht, **1995**.
[112] Cowie, J.M.G.; Arrighi, V., *Polymers: Chemistry and Physics of Modern Materials*, 3rd edn. CRC Press, Boca Raton, FL, **2008**.
[113] Chandrasekhar, P., *Conducting Polymers, Fundamentals and Applications: A Practical Approach*. Kluwer Academic, Dordrecht, **1999**.
[114] Tsai, S.W.; Hahn, H.T., *Introduction to Composite Materials*. Technomic, Lancaster, PA, **1980**.
[115] Yu, Z.; Li, L.; Zhang, Q.; *et al.*, Silver nanowire–polymer composite electrodes for efficient polymer solar cells. *Adv. Mater.* **2011**, *23* (38), 4453–4457.
[116] Yousefi, N.; Gudarzi, M.M.; Zheng, Q.; *et al.*, Self-alignment and high electrical conductivity of ultralarge graphene oxide–polyurethane nanocomposites. *J. Mater. Chem.* **2012**, *22* (25), 12709–12717.
[117] Yu, C.; Kim, Y.S.; Kim, D.; Grunlan, J.C., Thermoelectric behavior of segregated-network polymer nanocomposites. *Nano Lett.* **2008**, *8* (12), 4428–4432.
[118] Du, J.; Zhao, L.; Zeng, Y.; *et al.*, Comparison of electrical properties between multi-walled carbon nanotube and graphene nanosheet/high density polyethylene composites with a segregated network structure. *Carbon* **2011**, *49* (4), 1094–1100.
[119] Kim, H.; Abdala, A.A.; Macosko, C.W., Graphene/polymer nanocomposites. *Macromolecules* **2010**, *43* (16), 6515–6530.
[120] Park, Y.T.; Qian, Y.; Chan, C.; *et al.*, Epoxy toughening with low graphene loading. *Adv. Funct. Mater.* **2015**, *25* (4), 575–585.
[121] Rafiee, M.A.; Rafiee, J.; Wang, Z.; *et al.*, Enhanced mechanical properties of nanocomposites at low graphene content. *ACS Nano* **2009**, *3* (12), 3884–3890.
[122] Ning, H.; Li, J.; Hu, N.; *et al.*, Interlaminar mechanical properties of carbon fiber reinforced plastic laminates modified with graphene oxide interleaf. *Carbon* **2015**, *91*, 224–233.
[123] Walther, A.; Bjurhager, I.; Malho, J.-M.; *et al.*, Large-area, lightweight and thick biomimetic composites with superior material properties via fast, economic, and green pathways. *Nano Lett.* **2010**, *10* (8), 2742–2748.
[124] Pugno, N.M., Space elevator: out of order? *Nano Today* **2007**, *2* (6), 44–47.
[125] Chen, H.; Müller, M.B.; Gilmore, K.J.; *et al.*, Mechanically strong, electrically conductive, and biocompatible graphene paper. *Adv. Mater.* **2008**, *20* (18), 3557–3561.
[126] Lee, C.; Wei, X.; Kysar, J.W.; Hone, J., Measurement of the elastic properties and intrinsic strength of monolayer graphene. *Science* **2008**, *321* (5887), 385–388.
[127] Luong, N.D.; Hippi, U.; Korhonen, J.T.; *et al.*, Enhanced mechanical and electrical properties of polyimide film by graphene sheets via in situ polymerization. *Polymer* **2011**, *52* (23), 5237–5242.
[128] Gómez-Navarro, C.; Burghard, M.; Kern, K., Elastic properties of chemically derived single graphene sheets. *Nano Lett.* **2008**, *8* (7), 2045–2049.
[129] Suk, J.W.; Piner, R.D.; An, J.; Ruoff, R.S., Mechanical properties of monolayer graphene oxide. *ACS Nano* **2010**, *4* (11), 6557–6564.
[130] Gong, L.; Kinloch, I.A.; Young, R.J.; *et al.*, Interfacial stress transfer in a graphene monolayer nanocomposite. *Adv. Mater.* **2010**, *22* (24), 2694–2697.

[131] Cadek, M.; Coleman, J.N.; Ryan, K.P.; *et al.*, Reinforcement of polymers with carbon nanotubes: the role of nanotube surface area. *Nano Lett.* **2004**, *4* (2), 353–356.
[132] Coleman, J.N.; Khan, U.; Gun'ko, Y.K., Mechanical reinforcement of polymers using carbon nanotubes. *Adv. Mater.* **2006**, *18* (6), 689–706.
[133] Coleman, J.N.; Khan, U.; Blau, W.J.; Gun'ko, Y.K., Small but strong: a review of the mechanical properties of carbon nanotube–polymer composites. *Carbon* **2006**, *44* (9), 1624–1652.
[134] Young, R.J.; Kinloch, I.A.; Gong, L.; Novoselov, K.S., The mechanics of graphene nanocomposites: a review. *Compos. Sci. Technol.* **2012**, *72* (12), 1459–1476.
[135] Gong, L.; Young, R.J.; Kinloch, I.A.; *et al.*, Optimizing the reinforcement of polymer-based nanocomposites by graphene. *ACS Nano* **2012**, *6* (3), 2086–2095.
[136] Liang, J.; Huang, Y.; Zhang, L.; *et al.*, Molecular level dispersion of graphene into poly(vinyl alcohol) and effective reinforcement of their nanocomposites. *Adv. Funct. Mater.* **2009**, *19* (14), 2297–2302.
[137] Gudarzi, M.M.; Sharif, F., Enhancement of dispersion and bonding of graphene–polymer through wet transfer of functionalized graphene oxide. *Express Polym. Lett.* **2012**, *6* (12), 1017–1031.
[138] Yousefi, N.; Gudarzi, M.M.; Zheng, Q.; *et al.*, Highly aligned, ultralarge-size reduced graphene oxide/polyurethane nanocomposites: mechanical properties and moisture permeability. *Composites A* **2013**, *49*, 42–50.
[139] Rafiee, M.A.; Lu, W.; Thomas, A.V.; *et al.*, Graphene nanoribbon composites. *ACS Nano* **2010**, *4* (12), 7415–7420.
[140] Mallick, P.K., *Fiber-Reinforced Composites: Materials, Manufacturing, and Design*, 3rd edn. CRC Press, Boca Raton, FL, **2008**.
[141] Aboutalebi, S.H.; Gudarzi, M.M.; Zheng, Q.B.; Kim, J.-K., Spontaneous formation of liquid crystals in ultralarge graphene oxide dispersions. *Adv. Funct. Mater.* **2011**, *21* (15), 2978–2988.
[142] Gudarzi, M.M.; Moghadam, M.H.M.; Sharif, F., Spontaneous exfoliation of graphite oxide in polar aprotic solvents as the route to produce graphene oxide–organic solvents liquid crystals. *Carbon* **2013**, *64*, 403–415.
[143] Jalili, R.; Aboutalebi, S. H.; Esrafilzadeh, D.; *et al.*, Formation and processability of liquid crystalline dispersions of graphene oxide. *Mater. Horiz.* **2014**, *1* (1), 87–91.
[144] Crosby, A.J.; Lee, J.Y., Polymer nanocomposites: the "nano" effect on mechanical properties. *Polym. Rev.* **2007**, *47* (2), 217–229.
[145] Cote, L.J.; Kim, F.; Huang, J., Langmuir–Blodgett assembly of graphite oxide single layers. *J. Am. Chem. Soc.* **2009**, *131* (3), 1043–1049.
[146] Kulkarni, D.D.; Choi, I.; Singamaneni, S.S.; Tsukruk, V.V., Graphene oxide–polyelectrolyte nanomembranes. *ACS Nano* **2010**, *4* (8), 4667–4676.
[147] Ansari, S.; Kelarakis, A.; Estevez, L.; Giannelis, E.P., Oriented arrays of graphene in a polymer matrix by in situ reduction of graphite oxide nanosheets. *Small* **2010**, *6* (2), 205–209.
[148] Putz, K.W.; Compton, O.C.; Palmeri, M.J.; *et al.*, High-nanofiller-content graphene oxide–polymer nanocomposites via vacuum-assisted self-assembly. *Adv. Funct. Mater.* **2010**, *20* (19), 3322–3329.
[149] Xie, X.-L.; Mai, Y.-W.; Zhou, X.-P., Dispersion and alignment of carbon nanotubes in polymer matrix: a review. *Mater. Sci. Eng. R – Rep.* **2005**, *49* (4), 89–112.
[150] Kim, H.; Macosko, C.W., Processing–property relationships of polycarbonate/graphene composites. *Polymer* **2009**, *50* (15), 3797–3809.
[151] Aboutalebi, S.H.; Jalili, R.; Esrafilzadeh, D.; *et al.*, High-performance multifunctional graphene yarns: toward wearable all-carbon energy storage textiles. *ACS Nano* **2014**, *8* (3), 2456–2466.
[152] Xu, Y.; Hong, W.; Bai, H.; *et al.*, Strong and ductile poly(vinyl alcohol)/graphene oxide composite films with a layered structure. *Carbon* **2009**, *47* (15), 3538–3543.
[153] Xu, Z.; Gao, C., In situ polymerization approach to graphene-reinforced nylon-6 composites. *Macromolecules* **2010**, *43* (16), 6716–6723.
[154] Huang, X.; Qi, X.; Boey, F.; Zhang, H., Graphene-based composites. *Chem. Soc. Rev.* **2012**, *41* (2), 666–686.
[155] Suhr, J.; Victor, P.; Ci, L.; *et al.*, Fatigue resistance of aligned carbon nanotube arrays under cyclic compression. *Nature Nanotechnol.* **2007**, *2* (7), 417–421.

[156] Koratkar, N.A.; Suhr, J.; Joshi, A.; *et al.*, Characterizing energy dissipation in single-walled carbon nanotube polycarbonate composites. *Appl. Phys. Lett.* **2005**, *87* (6), 063102.
[157] Suhr, J.; Koratkar, N.; Keblinski, P.; Ajayan, P., Viscoelasticity in carbon nanotube composites. *Nature Mater.* **2005**, *4* (2), 134–137.
[158] Ash, B.J.; Siegel, R.W.; Schadler, L.S., Mechanical behavior of alumina/poly(methyl methacrylate) nanocomposites. *Macromolecules* **2004**, *37* (4), 1358–1369.
[159] Koratkar, N.; Wei, B.Q.; Ajayan, P.M., Carbon nanotube films for damping applications. *Adv. Mater.* **2002**, *14* (13–14), 997–1000.
[160] Espinosa, H.D.; Juster, A.L.; Latourte, F.J.; *et al.*, Tablet-level origin of toughening in abalone shells and translation to synthetic composite materials. *Nature Commun.* **2011**, *2*, 173.
[161] Jackson, A.; Vincent, J.; Turner, R., The mechanical design of nacre. *Proc. R. Soc. B: Biol. Sci.* **1988**, *234* (1277), 415–440.
[162] Bonderer, L.J.; Studart, A.R.; Gauckler, L.J., Bioinspired design and assembly of platelet reinforced polymer films. *Science* **2008**, *319* (5866), 1069–1073.
[163] Podsiadlo, P.; Kaushik, A.K.; Arruda, E.M.; *et al.*, Ultrastrong and stiff layered polymer nanocomposites. *Science* **2007**, *318* (5847), 80–83.
[164] Tang, Z.; Kotov, N.A.; Magonov, S.; Ozturk, B., Nanostructured artificial nacre. *Nature Mater.* **2003**, *2* (6), 413–418.
[165] Cui, W.; Li, M.; Liu, J.; *et al.*, Strong integrated strength and toughness artificial nacre based on dopamine cross-linked graphene oxide. *ACS Nano* **2014**, *8* (9), 9511–9517.
[166] Cheng, Q.; Wu, M.; Li, M.; *et al.*, Ultratough artificial nacre based on conjugated cross-linked graphene oxide. *Angew. Chem.* **2013**, *125* (13), 3838–3843.
[167] Zhang, M.; Huang, L.; Chen, J.; *et al.*, Ultratough, ultrastrong, and highly conductive graphene films with arbitrary sizes. *Adv. Mater.* **2014**, *26* (45), 7588–7592.
[168] Liu, Z.; Xu, Z.; Hu, X.; Gao, C., Lyotropic liquid crystal of polyacrylonitrile-grafted graphene oxide and its assembled continuous strong nacre-mimetic fibers. *Macromolecules* **2013**, *46* (17), 6931–6941.
[169] Wang, J.; Cheng, Q.; Tang, Z., Layered nanocomposites inspired by the structure and mechanical properties of nacre. *Chem. Soc. Rev.* **2012**, *41* (3), 1111–1129.
[170] Xu, Z.-H.; Li, X., Deformation strengthening of biopolymer in nacre. *Adv. Funct. Mater.* **2011**, *21* (20), 3883–3888.
[171] Wang, L.; Boyce, M.C., Bioinspired structural material exhibiting post-yield lateral expansion and volumetric energy dissipation during tension. *Adv. Funct. Mater.* **2010**, *20* (18), 3025–3030.
[172] Shyu, T.C.; Damasceno, P.F.; Dodd, P.M.; *et al.*, A kirigami approach to engineering elasticity in nanocomposites through patterned defects. *Nature Mater.* **2015**, *14* (8), 785–789.
[173] Gao, H.; Ji, B.; Jäger, I.L.; *et al.*, Materials become insensitive to flaws at nanoscale: lessons from nature. *Proc. Natl. Acad. Sci. USA* **2003**, *100* (10), 5597–5600.
[174] Pei, S.; Cheng, H.-M., The reduction of graphene oxide. *Carbon* **2012**, *50* (9), 3210–3228.
[175] Eda, G.; Chhowalla, M., Graphene-based composite thin films for electronics. *Nano Lett.* **2009**, *9* (2), 814–818.
[176] Kim, J.Y.; Lee, W.H.; Suk, J.W.; *et al.*, Chlorination of reduced graphene oxide enhances the dielectric constant of reduced graphene oxide/polymer composites. *Adv. Mater.* **2013**, *25* (16), 2308–2313.
[177] Raccichini, R.; Varzi, A.; Passerini, S.; Scrosati, B., The role of graphene for electrochemical energy storage. *Nature Mater.* **2015**, *14* (3), 271–279.
[178] Pumera, M.; Ambrosi, A.; Bonanni, A.; *et al.*, Graphene for electrochemical sensing and biosensing. *Trends Anal. Chem.* **2010**, *29* (9), 954–965.
[179] Yin, J.; Chang, R.; Shui, Y.; Zhao, X., Preparation and enhanced electro-responsive characteristic of reduced graphene oxide/polypyrrole composite sheet suspensions. *Soft Matter* **2013**, *9* (31), 7468–7478.
[180] Zhao, Y.; Song, L.; Zhang, Z.; Qu, L., Stimulus-responsive graphene systems towards actuator applications. *Energy Environ. Sci.* **2013**, *6* (12), 3520–3536.
[181] Zhang, J.; Song, L.; Zhang, Z.; *et al.*, Environmentally responsive graphene systems. *Small* **2014**, *10* (11), 2151–2164.

[182] Zakri, C.; Poulin, P., Phase behavior of nanotube suspensions: from attraction induced percolation to liquid crystalline phases. *J. Mater. Chem.* **2006**, *16* (42), 4095–4098.
[183] Bauhofer, W.; Kovacs, J.Z., A review and analysis of electrical percolation in carbon nanotube polymer composites. *Compos. Sci. Technol.* **2009**, *69* (10), 1486–1498.
[184] Li, J.; Kim, J.-K., Percolation threshold of conducting polymer composites containing 3D randomly distributed graphite nanoplatelets. *Compos. Sci. Technol.* **2007**, *67* (10), 2114–2120.
[185] Celzard, A.; McRae, E.; Deleuze, C.; *et al.*, Critical concentration in percolating systems containing a high-aspect-ratio filler. *Phys. Rev. B* **1996**, *53* (10), 6209–6214.
[186] Zhang, H.-B.; Zheng, W.-G.; Yan, Q.; *et al.*, The effect of surface chemistry of graphene on rheological and electrical properties of polymethylmethacrylate composites. *Carbon* **2012**, *50* (14), 5117–5125.
[187] Pham, V.H.; Dang, T.T.; Hur, S.H.; *et al.*, Highly conductive poly(methyl methacrylate) (PMMA)-reduced graphene oxide composite prepared by self-assembly of PMMA latex and graphene oxide through electrostatic interaction. *ACS Appl. Mater. Interfaces* **2012**, *4* (5), 2630–2636.
[188] Wu, C.; Huang, X.; Wang, G.; *et al.*, Highly conductive nanocomposites with three-dimensional, compactly interconnected graphene networks via a self-assembly process. *Adv. Funct. Mater.* **2013**, *23* (4), 506–513.
[189] Krishnan, K.; Ganguli, N., Large anisotropy of the electrical conductivity of graphite. *Nature* **1939**, *144* (3650), 667.
[190] Kimura, T.; Ago, H.; Tobita, M.; *et al.*, Polymer composites of carbon nanotubes aligned by a magnetic field. *Adv. Mater.* **2002**, *14* (19), 1380–1383.
[191] Li, J.; Wong, P.-S.; Kim, J.-K., Hybrid nanocomposites containing carbon nanotubes and graphite nanoplatelets. *Mater. Sci. Eng. A* **2008**, *483–484*, 660–663.
[192] Safdari, M.; Al-Haik, M., Electrical conductivity of synergistically hybridized nanocomposites based on graphite nanoplatelets and carbon nanotubes. *Nanotechnology* **2012**, *23* (40), 405202.
[193] Xu, Q.; Song, T.; Cui, W.; *et al.*, Solution-processed highly conductive PEDOT:PSS/AgNW/GO transparent film for efficient organic–Si hybrid solar cells. *ACS Appl. Mater. Interfaces* **2015**, *7* (5), 3272–3279.
[194] Liang, J.; Li, L.; Tong, K.; *et al.*, Silver nanowire percolation network soldered with graphene oxide at room temperature and its application for fully stretchable polymer light-emitting diodes. *ACS Nano* **2014**, *8* (2), 1590–1600.
[195] Bourgeat-Lami, E.; Faucheu, J.; Noel, A., Latex routes to graphene-based nanocomposites. *Polym. Chem.* **2015**, *6* (30), 5323–5357.
[196] Tkalya, E.; Ghislandi, M.; Alekseev, A.; *et al.*, Latex-based concept for the preparation of graphene-based polymer nanocomposites. *J. Mater. Chem.* **2010**, *20* (15), 3035–3039.
[197] Long, G.; Tang, C.; Wong, K.-W.; *et al.*, Resolving the dilemma of gaining conductivity but losing environmental friendliness in producing polystyrene/graphene composites via optimizing the matrix–filler structure. *Green Chem.* **2013**, *15* (3), 821–828.
[198] Pickering, S.U., CXCVI – Emulsions. *J. Chem. Soc., Trans.* **1907**, *91*, 2001–2021.
[199] Ramsden, W., Separation of solids in the surface-layers of solutions and "suspensions" (observations on surface-membranes, bubbles, emulsions, and mechanical coagulation). – Preliminary account. *Proc. R. Soc. Lond.* **1903**, *72* (477–486), 156–164.
[200] Cao, Y.; Zhang, J.; Feng, J.; Wu, P., Compatibilization of immiscible polymer blends using graphene oxide sheets. *ACS Nano* **2011**, *5* (7), 5920–5927.
[201] Kim, J.; Cote, L.J.; Kim, F.; *et al.*, Graphene oxide sheets at interfaces. *J. Am. Chem. Soc.* **2010**, *132* (23), 8180–8186.
[202] Kim, H.; Kobayashi, S.; AbdurRahim, M.A.; *et al.*, Graphene/polyethylene nanocomposites: effect of polyethylene functionalization and blending methods. *Polymer* **2011**, *52* (8), 1837–1846.
[203] Nair, S.T.; Vijayan, P.P.; Xavier, P.; *et al.*, Selective localisation of multi walled carbon nanotubes in polypropylene/natural rubber blends to reduce the percolation threshold. *Compos. Sci. Technol.* **2015**, *116*, 9–17.
[204] Paul, D.R.; Newman, S. (eds), *Polymer Blends*, vol. *1* Academic Press, New York, **1978**.

[205] Chen, Z.; Ren, W.; Gao, L.; *et al.*, Three-dimensional flexible and conductive interconnected graphene networks grown by chemical vapour deposition. *Nature Mater.* **2011**, *10* (6), 424–428.
[206] Jun, Y.-S.; Sy, S.; Ahn, W.; *et al.*, Highly conductive interconnected graphene foam based polymer composite. *Carbon* **2015**, *95*, 653–658.
[207] Zhao, J.; Zhang, G.-Y.; Shi, D.-X., Review of graphene-based strain sensors. *Chin. Phys. B* **2013**, *22* (5), 057701.
[208] Balandin, A.A.; Ghosh, S.; Bao, W.; *et al.*, Superior thermal conductivity of single-layer graphene. *Nano Lett.* **2008**, *8* (3), 902–907.
[209] Balandin, A.A., Thermal properties of graphene and nanostructured carbon materials. *Nature Mater.* **2011**, *10* (8), 569–581.
[210] Hermann, A.; Chaudhuri, T.; Spagnol, P., Bipolar plates for PEM fuel cells: a review. *Int. J. Hydrogen Energy* **2005**, *30* (12), 1297–1302.
[211] Marotta, E.E.; Mazzuca, S.J.; Norley, J., Thermal joint conductance for flexible graphite materials: analytical and experimental study. *IEEE Trans. Components Packaging Technol.* **2005**, *28* (1), 102–110.
[212] Han, Z.; Fina, A., Thermal conductivity of carbon nanotubes and their polymer nanocomposites: a review. *Prog. Polym. Sci.* **2011**, *36* (7), 914–944.
[213] Huxtable, S.T.; Cahill, D.G.; Shenogin, S.; *et al.*, Interfacial heat flow in carbon nanotube suspensions. *Nature Mater.* **2003**, *2* (11), 731–734.
[214] Alexeev, D.; Chen, J.; Walther, J.H.; *et al.*, Kapitza resistance between few-layer graphene and water: liquid layering effects. *Nano Lett.* **2015**, *15* (9), 5744–5749.
[215] Im, H.; Kim, J., Thermal conductivity of a graphene oxide–carbon nanotube hybrid/epoxy composite. *Carbon* **2012**, *50* (15), 5429–5440.
[216] Shahil, K.M.; Balandin, A.A., Graphene–multilayer graphene nanocomposites as highly efficient thermal interface materials. *Nano Lett.* **2012**, *12* (2), 861–867.
[217] Sarvar, F.; Whalley, D.C.; Conway, P.P., Thermal interface materials – a review of the state of the art. In *2006 1st Electronic System Integration Technology Conference*, Dresden, 5–7 September 2006, vol. 2, pp. 1292–1302. IEEE, Piscataway, NJ, 2006.
[218] Pollack, G.L., Kapitza resistance. *Rev. Mod. Phys.* **1969**, *41* (1), 48.
[219] Pop, E., Energy dissipation and transport in nanoscale devices. *Nano Res.* **2010**, *3* (3), 147–169.
[220] Renteria, J.D.; Ramirez, S.; Malekpour, H.; *et al.*, Strongly anisotropic thermal conductivity of free-standing reduced graphene oxide films annealed at high temperature. *Adv. Funct. Mater.* **2015**, *25* (29), 4664–4672.
[221] Xin, G.; Yao, T.; Sun, H.; *et al.*, Highly thermally conductive and mechanically strong graphene fibers. *Science* **2015**, *349* (6252), 1083–1087.
[222] Shtein, M.; Nadiv, R.; Buzaglo, M.; *et al.*, Thermally conductive graphene–polymer composites: size, percolation, and synergy effects. *Chem. Mater.* **2015**, *27* (6), 2100–2106.
[223] Du, F.-P.; Yang, W.; Zhang, F.; *et al.*, Enhancing the heat transfer efficiency in graphene–epoxy nanocomposites using a magnesium oxide–graphene hybrid structure. *ACS Appl. Mater. Interfaces* **2015**, *7* (26), 14397–14403.
[224] Tian, L.; Anilkumar, P.; Cao, L.; *et al.*, Graphene oxides dispersing and hosting graphene sheets for unique nanocomposite materials. *ACS Nano* **2011**, *5* (4), 3052–3058.
[225] Nair, R.R.; Wu, H.A.; Jayaram, P.N.; *et al.*, Unimpeded permeation of water through helium-leak-tight graphene-based membranes. *Science* **2012**, *335* (6067), 442–444.
[226] Qu, G.; Liu, S.; Zhang, S.; *et al.*, Graphene oxide induces toll-like receptor 4 (TLR4)-dependent necrosis in macrophages. *ACS Nano* **2013**, *7* (7), 5732–5745.
[227] Kim, H.W.; Yoon, H.W.; Yoon, S.-M.; *et al.*, Selective gas transport through few-layered graphene and graphene oxide membranes. *Science* **2013**, *342* (6154), 91–95.
[228] Mi, B., Graphene oxide membranes for ionic and molecular sieving. *Science* **2014**, *343* (6172), 740–742.
[229] Joshi, R.K.; Carbone, P.; Wang, F.C.; *et al.*, Precise and ultrafast molecular sieving through graphene oxide membranes. *Science* **2014**, *343* (6172), 752–754.
[230] Kim, H.; Miura, Y.; Macosko, C.W., Graphene/polyurethane nanocomposites for improved gas barrier and electrical conductivity. *Chem. Mater.* **2010**, *22* (11), 3441–3450.

[231] Choudalakis, G.; Gotsis, A., Permeability of polymer/clay nanocomposites: a review. *Eur. Polym. J.* **2009**, *45* (4), 967–984.
[232] Bharadwaj, R.K., Modeling the barrier properties of polymer–layered silicate nanocomposites. *Macromolecules* **2001**, *34* (26), 9189–9192.
[233] Priolo, M.A.; Gamboa, D.; Holder, K.M.; Grunlan, J.C., Super gas barrier of transparent polymer – clay multilayer ultrathin films. *Nano Lett.* **2010**, *10* (12), 4970–4974.
[234] Yang, Y.H.; Bolling, L.; Priolo, M.A.; Grunlan, J.C., Super gas barrier and selectivity of graphene oxide–polymer multilayer thin films. *Adv. Mater.* **2013**, *25* (4), 503–508.
[235] Su, Y.; Kravets, V.; Wong, S.; *et al.*, Impermeable barrier films and protective coatings based on reduced graphene oxide. *Nature Commun.* **2014**, *5*, 4843.
[236] Kim, J.; Song, S.H.; Im, H.G.; *et al.*, Moisture barrier composites made of non-oxidized graphene flakes. *Small* **2015**, *11* (26), 3124–3129.
[237] Wang, L.; Sasaki, T., Titanium oxide nanosheets: graphene analogues with versatile functionalities. *Chem. Rev.* **2014**, *114* (19), 9455–9486.
[238] Wong, M.; Ishige, R.; White, K.L.; *et al.*, Large-scale self-assembled zirconium phosphate smectic layers via a simple spray-coating process. *Nature Commun.* **2014**, *5*, 3589.
[239] Dou, Y.; Pan, T.; Xu, S.; *et al.*, Transparent, ultrahigh-gas-barrier films with a brick–mortar–sand structure. *Angew. Chem. Int. Ed.* **2015**, *54* (33), 9673–9678.
[240] Hu, S.; Lozada-Hidalgo, M.; Wang, F.C.; *et al.*, Proton transport through one-atom-thick crystals. *Nature* **2014**, *516* (7530), 227–230.
[241] Karnik, R.N., Materials science: breakthrough for protons. *Nature* **2014**, *516* (7530), 173–175.
[242] Dimiev, A.M.; Tour, J.M., Mechanism of graphene oxide formation. *ACS Nano* **2014**, *8* (3), 3060–3068.
[243] Eigler, S.; Hirsch, A., Chemistry with graphene and graphene oxide – challenges for synthetic chemists. *Angew. Chem. Int. Ed.* **2014**, *53* (30), 7720–7738.
[244] Sun, T.; Fabris, S., Mechanisms for oxidative unzipping and cutting of graphene. *Nano Lett.* **2012**, *12* (1), 17–21.
[245] Li, J.-L.; Kudin, K.N.; McAllister, M.J.; *et al.*, Oxygen-driven unzipping of graphitic materials. *Phys. Rev. Lett.* **2006**, *96* (17), 176101.
[246] Kim, J.; Cote, L.J.; Kim, F.; *et al.*, Graphene oxide sheets at interfaces. *J. Am. Chem. Soc.* **2010**, *132* (23), 8180–8186.
[247] Aboutalebi, S.H.; Chidembo, A.T.; Salari, M.; *et al.*, Comparison of GO, GO/MWCNTs composite and MWCNTs as potential electrode materials for supercapacitors. *Energy Environ. Sci.* **2011**, *4* (5), 1855–1865.
[248] Aboutalebi, S.H.; Aminorroaya-Yamini, S.; Nevirkovets, I.; *et al.*, Enhanced hydrogen storage in graphene oxide–MWCNTs composite at room temperature. *Adv. Energy Mater.* **2012**, *2* (12), 1439–1446.
[249] Jalili, R.; Aboutalebi, S.H.; Esrafilzadeh, D.; *et al.*, Organic solvent-based graphene oxide liquid crystals: a facile route toward the next generation of self-assembled layer-by-layer multifunctional 3D architectures. *ACS Nano* **2013**, *7* (5), 3981–3990.
[250] Qiu, L.; Yang, X.; Gou, X.; *et al.*, Dispersing carbon nanotubes with graphene oxide in water and synergistic effects between graphene derivatives. *Chem. Eur. J.* **2010**, *16* (35), 10653–10658.
[251] Gudarzi, M.M.; Sharif, F., Self assembly of graphene oxide at the liquid–liquid interface: a new route to the fabrication of graphene based composites. *Soft Matter* **2011**, *7* (7), 3432–3440.
[252] Kim, S.D.; Zhang, W.L.; Choi, H.J., Pickering emulsion-fabricated polystyrene–graphene oxide microspheres and their electrorheology. *J. Mater. Chem. C* **2014**, *2* (36), 7541–7546.
[253] Gudarzi, M.M.; Sharif, F., Molecular level dispersion of graphene in polymer matrices using colloidal polymer and graphene. *J. Colloid Interface Sci.* **2012**, *366* (1), 44–50.
[254] Aboutalebi, S.H.; Gudarzi, M.M.; Zheng, Q.B.; Kim, J.-K., Spontaneous formation of liquid crystals in ultralarge graphene oxide dispersions. *Adv. Funct. Mater.* **2011**, *21* (15), 2978–2988.
[255] Nakajima, T.; Matsuo, Y., Formation process and structure of graphite oxide. *Carbon* **1994**, *32* (3), 469–475.
[256] Kim, J.; Cote, L.J.; Huang, J., Two dimensional soft material: new faces of graphene oxide. *Acc. Chem. Res.* **2012**, *45* (8), 1356–1364.

[257] Lin, X.; Shen, X.; Zheng, Q.; *et al.*, Fabrication of highly-aligned, conductive, and strong graphene papers using ultralarge graphene oxide sheets. *ACS Nano* **2012**, *6* (12), 10708–10719.

[258] Hung, A.H.; Holbrook, R.J.; Rotz, M.W.; *et al.*, Graphene oxide enhances cellular delivery of hydrophilic small molecules by co-incubation. *ACS Nano* **2014**, *8* (10), 10168–10177.

[259] Kim, J.E.; Han, T.H.; Lee, S.H.; *et al.*, Graphene oxide liquid crystals. *Angew. Chem. Int. Ed.* **2011**, *50* (13), 3043–3047.

[260] Xu, Z.; Gao, C., Aqueous liquid crystals of graphene oxide. *ACS Nano* **2011**, *5* (4), 2908–2915.

[261] Dan, B.; Behabtu, N.; Martinez, A.; *et al.*, Liquid crystals of aqueous, giant graphene oxide flakes. *Soft Matter* **2011**, *7* (23), 11154–11159.

[262] Naficy, S.; Jalili, R.; Aboutalebi, S.H.; *et al.*, Graphene oxide dispersions: tuning rheology to enable fabrication. *Mater. Horiz.* **2014**, *1* (6), 326–331.

[263] Xu, Z.; Gao, C., Graphene chiral liquid crystals and macroscopic assembled fibres. *Nature Commun.* **2011**, *2*, 571.

[264] Aboutalebi, S.H.; Jalili, R.; Esrafilzadeh, D.; *et al.*, High-performance multifunctional graphene yarns: toward wearable all-carbon energy storage textiles. *ACS Nano* **2014**, *8* (3), 2456–2466.

[265] Jalili, R.; Aboutalebi, S.H.; Esrafilzadeh, D.; *et al.*, Scalable one-step wet-spinning of graphene fibers and yarns from liquid crystalline dispersions of graphene oxide: towards multifunctional textiles. *Adv. Funct. Mater.* **2013**, *10* (6), 5345–5354.

[266] Jalili, R.; Aboutalebi, S.H.; Esrafilzadeh, D.; *et al.*, Formation and processability of liquid crystalline dispersions of graphene oxide. *Mater. Horiz.* **2014**, *1* (1), 87–91.

[267] Islam, M.M.; Aboutalebi, S.H.; Cardillo, D.; *et al.*, Self-assembled multifunctional hybrids: toward developing high-performance graphene-based architectures for energy storage devices. *ACS Central Sci.* **2015**, *1* (4), 206–216.

[268] Chidembo, A.; Aboutalebi, S.H.; Konstantinov, K.; *et al.*, Globular reduced graphene oxide–metal oxide structures for energy storage applications. *Energy Environ. Sci.* **2012**, *5* (1), 5236–5240.

[269] Yousefi, N.; Gudarzi, M.M.; Zheng, Q.; *et al.*, Self-alignment and high electrical conductivity of ultralarge graphene oxide-polyurethane nanocomposites. *J. Mater. Chem.* **2012**, *22* (25), 12709–12717.

[270] Gudarzi, M.M.; Aboutalebi, S.H.; Yousefi, N.; *et al.*, Self-aligned graphene sheets–polyurethane nanocomposites. *MRS Proc.* **2011**, *1344* (1), mrss11-1344-y02-06.

[271] Chidembo, A.T.; Aboutalebi, S.H.; Konstantinov, K.; *et al.*, Liquid crystalline dispersions of graphene-oxide-based hybrids: a practical approach towards the next generation of 3D isotropic architectures for energy storage applications. *Part. Part. Syst. Charact.* **2014**, *31* (4), 463–473.

[272] Paredes, J.I.; Villar-Rodil, S.; Martínez-Alonso, A.; Tascón, J.M.D., Graphene oxide dispersions in organic solvents. *Langmuir* **2008**, *24* (19), 10560–10564.

[273] De Silva, K.S.B.; Aboutalebi, S.H.; Xu, X.; *et al.*, A significant improvement in both low- and high-field performance of MgB_2 superconductors through graphene oxide doping. *Scr. Mater.* **2013**, *69* (6), 437–440.

[274] Mustapić, M.; De Silva, K.S.B.; Aboutalebi, S.H.; *et al.*, Improvements in the dispersion of nanosilver in a MgB_2 matrix through a graphene oxide net. *J. Phys. Chem. C* **2015**, *119* (19), 10631–10640.

[275] Cai, D.; Song, M., Preparation of fully exfoliated graphite oxide nanoplatelets in organic solvents. *J. Mater. Chem.* **2007**, *17* (35), 3678–3680.

[276] Wang, Y.; Shi, Z.; Fang, J.; *et al.*, Graphene oxide/polybenzimidazole composites fabricated by a solvent-exchange method. *Carbon* **2011**, *49* (4), 1199–1207.

[277] Dimiev, A.; Kosynkin, D.V.; Alemany, L.B.; *et al.*, Pristine graphite oxide. *J. Am. Chem. Soc.* **2012**, *134* (5), 2815–2822.

[278] Krishnan, D.; Kim, F.; Luo, J.; *et al.*, Energetic graphene oxide: challenges and opportunities. *Nano Today* **2012**, *7* (2), 137–152.

[279] Erickson, K.; Erni, R.; Lee, Z.; *et al.*, Determination of the local chemical structure of graphene oxide and reduced graphene oxide. *Adv. Mater.* **2010**, *22* (40), 4467–4472.

[280] Izadi-Najafabadi, A.; Yasuda, S.; Kobashi, K.; *et al.*, Extracting the full potential of single-walled carbon nanotubes as durable supercapacitor electrodes operable at 4 V with high power and energy density. *Adv. Mater.* **2010**, *22* (35), E235–E241.

[281] Zheng, Q.; Ip, W.H.; Lin, X.; *et al.*, Transparent conductive films consisting of ultralarge graphene sheets produced by Langmuir–Blodgett assembly. *ACS Nano* **2011**, *5* (7), 6039–6051.

[282] Su, C.-Y.; Xu, Y.; Zhang, W.; *et al.*, Highly efficient restoration of graphitic structure in graphene oxide using alcohol vapors. *ACS Nano* **2010**, *4* (9), 5285–5292.

[283] Zurutuza, A.; Marinelli, C., Challenges and opportunities in graphene commercialization. *Nature Nanotechnol.* **2014**, *9* (10), 730–734.

[284] Cote, L.J.; Cruz-Silva, R.; Huang, J., Flash reduction and patterning of graphite oxide and its polymer composite. *J. Am. Chem. Soc.* **2009**, *131* (31), 11027–11032.

[285] Gilje, S.; Dubin, S.; Badakhshan, A.; *et al.*, Photothermal deoxygenation of graphene oxide for patterning and distributed ignition applications. *Adv. Mater.* **2010**, *22* (3), 419–423.

[286] Seabra, A.B.; Paula, A.J.; de Lima, R.; *et al.*, Nanotoxicity of graphene and graphene oxide. *Chem. Res. Toxicol.* **2014**, *27* (2), 159–168.

[287] Guo, X.; Mei, N., Assessment of the toxic potential of graphene family nanomaterials. *Food Drug Anal.* **2014**, *22* (1), 105–115.

[288] Hu, W.; Peng, C.; Lv, M.; *et al.*, Protein corona-mediated mitigation of cytotoxicity of graphene oxide. *ACS Nano* **2011**, *5* (5), 3693–3700.

[289] Chang, Y.; Yang, S.-T.; Liu, J.-H.; *et al.*, In vitro toxicity evaluation of graphene oxide on A549 cells. *Toxicol. Lett.* **2011**, *200* (3), 201–210.

[290] Yuan, J.; Gao, H.; Sui, J.; *et al.*, Cytotoxicity evaluation of oxidized single-walled carbon nanotubes and graphene oxide on human hepatoma HepG2 cells: an iTRAQ-coupled 2D LC-MS/MS proteome analysis. *Toxicol. Sci.* **2012**, *126* (1), 149–161.

[291] Chen, G.-Y.; Yang, H.-J.; Lu, C.-H.; *et al.*, Simultaneous induction of autophagy and toll-like receptor signaling pathways by graphene oxide. *Biomaterials* **2012**, *33* (27), 6559–6569.

[292] Lv, M.; Zhang, Y.; Liang, L.; *et al.*, Effect of graphene oxide on undifferentiated and retinoic acid-differentiated SH-SY5Y cells line. *Nanoscale* **2012**, *4* (13), 3861–3866.

[293] Wang, A.; Pu, K.; Dong, B.; *et al.*, Role of surface charge and oxidative stress in cytotoxicity and genotoxicity of graphene oxide towards human lung fibroblast cells. *J. Appl. Toxicol.* **2013**, *33* (10), 1156–1164.

[294] Kang, S.-M.; Kim, T.-H.; Choi, J.-W., Cell chip to detect effects of graphene oxide nanopellet on human neural stem cell. *J. Nanosci. Nanotechnol.* **2012**, *12* (7), 5185–5190.

[295] Zhi, X.; Fang, H.; Bao, C.; *et al.*, The immunotoxicity of graphene oxides and the effect of PVP-coating. *Biomaterials* **2013**, *34* (21), 5254–5261.

[296] Ahmed, F.; Rodrigues, D.F., Investigation of acute effects of graphene oxide on wastewater microbial community: a case study. *J. Haz. Mater.* **2013**, *256–257*, 33–39.

11

Toxicity Studies and Biomedical Applications of Graphene Oxide

Larisa Kovbasyuk and Andriy Mokhir

11.1 Introduction

Graphene oxide (GO) is a carbon-rich material that is derived from graphene. Similarly to the parent material, GO contains flat regions made of sp^2-hybridized carbon atoms. In contrast to graphene, it also contains non-flat regions and modified edges, which can be formally considered as products of the partial oxidation of the sp^2 system (Figure 11.1). Such non-flat GO sections carry a rich plethora of chemical fragments, including rather abundant epoxides, alcohols, carboxylic acids, carbonyl groups and sulfate esters, as well as a number of less abundant fragments and ions, whose role in GO properties relevant to its biological activity is often poorly understood. The presence of these groups explains the good solubility of GO in aqueous solutions at pH close to 7 and its substantially lower tendency to aggregation than that observed for graphene. Though GO has a lower area of flat, sp^2-hybridized sections, it seems to be sufficient to provide for the efficient interaction with biomolecules of different types, including small molecules and biopolymers such as nucleic acids, as well as with unnatural biologically active compounds, e.g. drugs and fluorescent dyes. Finally, GO exhibits substantial cell membrane permeability and relatively low toxicity both in cellular assays and *in vivo*. This combination of properties, which is rather unusual for carbon-rich materials, makes GO an interesting material for biomedical and medicinal applications.

Graphene Oxide: Fundamentals and Applications, First Edition. Edited by Ayrat M. Dimiev and Siegfried Eigler.

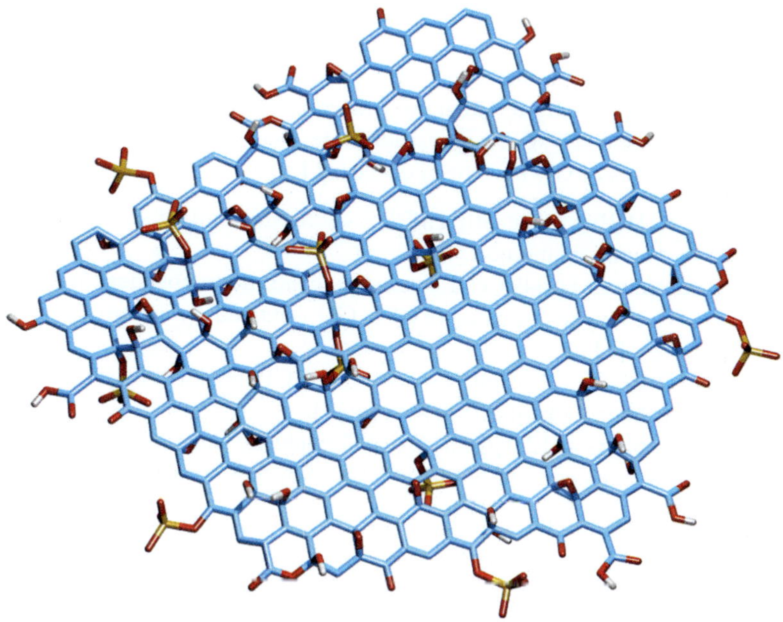

Figure 11.1 *A cartoon illustrating the presence of a variety of functional chemical groups, flat graphene-like regions and non-flat regions on the surface of graphene oxide (GO)*

11.2 Toxicity of Graphene Oxide

Graphene oxide is an amphiphilic material, which has an overall negative charge at physiological conditions. The charge can be reversed by covering GO with polycationic reagents, e.g. polymers or dendrimers. Correspondingly, in cells, GO can potentially interact with hydrophobic, positively charged and negatively charged surfaces, e.g. membranes, proteins and nucleic acids, thereby inducing toxicity. In this section, we will discuss known toxic effects of GO observed in cellular assays (*in vitro*) and *in vivo*, and, where possible, outline reasons for the toxicity. Biological effects of GO and analogous materials, including their cytotoxicity, have been previously reviewed [1–8].

Data on the toxicity of GOs in cellular assays found in the literature are often contradictory [1–9]. This is partially explained by the large number of parameters that have to be controlled to be able to compare the results obtained in different laboratories. In particular, the source of the starting materials as well as the method of synthesis and purification of GO affect the size, the number of sheets in the material, surface charge, oxidative state and the presence of low-molecular-weight impurities and different functional groups on the surface. Substantial efforts have to be invested to account for all of these parameters to obtain standardized GO materials. Unfortunately, this is not yet done routinely. Moreover, GO can interfere with cell viability assays, producing false positive results. For example, Macosko, Haynes and coworkers have observed that methylthiazolyldiphenyl-tetrazolium bromide (MTT), which is used as a reagent in the popular cell viability assay, is efficiently reduced in the presence of GO with the formation of a blue-colored product [9]. A product of the same color is produced when MTT is reduced in viable cells. Therefore, MTT-based assays will fail to indicate GO cytotoxicity. The same authors have found that another tetrazole-based reagent, the water-soluble tetrazolium salt WST-8, as well as trypan blue exclusion, allow for accurate estimation of the number of viable and dead cells [9].

GO toxicity in cells is usually moderate at low concentrations ($\leq 10\,\mu g\,ml^{-1}$). At higher doses, it is dependent upon GO size, aggregation state, oxygen content and surface charge. For example, toxic effects of GO have been observed for:

i. human fibroblast (HDF) cells ($>50\,\mu g\,ml^{-1}$) – decreasing cell adhesion, cell apoptosis; GO obtained by Hummers method [10];
ii. human lung carcinoma (A549) cell line – concentration-dependent increase of the amount of reactive oxygen species (ROS); GO obtained by Hummers method and fractionated by size [11];
iii. red blood cells (RBCs) ($>25\,\mu g\,ml^{-1}$) – hemolysis; GO obtained by Hummers method and sonicated to obtain GOs of different sizes [9];
iv. human skin fibroblasts ($\geq 12.5\,\mu g\,ml^{-1}$) – cell viability decreased; GO obtained by Hummers method and sonicated to obtain GOs of different sizes [9].

A number of other studies on the toxicity of GO, nano-GO (NGO) and related materials toward various cell lines have appeared recently [1–3, 12–19].

Interestingly, Fiorillo *et al.* have observed that GO inhibits the proliferative expansion of single cancer stem cells in the tumor-sphere assay [19]. The effect has been confirmed for six different cancer types, including breast, pancreatic, prostate, ovarian, lung cancer and glioblastoma. Surprisingly, GO has been found to be only weakly toxic to mature (non-stem) cancer cells. This is a significant result, since cancer stem cells are tumor-initiating cells, which are practically insensitive to conventional chemotherapy and radiation. The survival of a few cells of this type after treatment leads to tumor recurrence and distant metastasis.

Toxicity of GO *in vivo* depends on the experimental settings selected and the parameters investigated. For example, it has been found that NGO at a dose of $25\,mg\,kg^{-1}$ (injected *via* the tail vein) exhibits practically no toxicity for reproductive function of male mice [20], and GO-derived carrier of Stat3 siRNA is practically non-toxic in mice, as shown in studies with a mouse model of melanoma [21]. However, at $\sim 14\,mg\,kg^{-1}$, chronic toxicity of GO has been observed for Kunming mice [10], whereas oral exposure to a dose ~0.8 mg GO per day per mouse in the lactating period strongly delayed the development of offspring and caused many other negative effects in the development of mice [22]. Furthermore, a systematic study of Li *et al.* on the distribution and toxicity of NGO in C57BL/6 mice for three months after the exposure has revealed that NGO can be retained in the lungs, thereby resulting in acute lung injury and chronic pulmonary fibrosis [23].

11.3 On the Toxicity Mechanism

11.3.1 Membrane as a Target

Graphene is known to enter cells by the edge-first uptake mechanism, which can lead to membrane damage [24]. An analogous mechanism can be assumed for GO and other GO-derived materials, since they have graphene-like regions, including edges, whose extent depends on the C/O ratio and other factors. Other mechanisms of GO-induced membrane damage are possible [1–3]. The current literature on the subject indicates that the effect of GO on outer cellular membranes is strongly dependent upon the cell type. For example, Cao, Wang and coworkers have observed that the incubation of human alveolar adenocarcinoma A549 cells with GO at concentrations of up to 200 μg ml^{-1} does not significantly modulate the level of the extracellular lactate dehydrogenase (LDH) activity, which is a common marker for membrane damage [11]. Similar results have been obtained by Dai, Lu, Liu and coworkers, who studied the effect of GO on eyesight both *in vitro* and *in vivo* [17]. In particular, they have observed that the level of LDH did not exceed 8% in the *in vitro* assay with ARPE-19 cells (a cell line derived from human retinal pigment epithelium) incubated for a variable time (24–72 h) with variable GO concentrations (5–100 μg ml^{-1}). For comparison, ~2–3% LDH have been released from the untreated cells. Furthermore, Mullick Chowdhury *et al.* have studied the toxicity of oxidized graphene nanoribbons (O-GNR, width ~125–220 nm) stabilized with 1,2-distearoyl-*sn*-glycero-3-phosphoethanolamine-*N*-[amino(polyethylene glycol) (PEG-DSPE) in several selected cancer cell lines: cervical cancer cells HeLa and breast cancer cells SKBR3 and MCF-7 [25]. Upon the incubation of MCF-7 cells for 24 h with 0.4 mg ml^{-1} of O-GNR–PEG-DSPE (the highest concentration used), the cells release ~55% LDH compared to the LDH activity in the lysed cells. For SKBR3 cells, the effect was comparable. In a negative control experiment (cells not treated with anything), ~40% and ~55% LDH activity was observed for MCF-7 and SKBR3, respectively. These data indicate that the membrane is not significantly affected by the treatment of the breast cancer cells with O-GNR–PEG-DSPE. In contrast, the membrane of HeLa cells has been found to be substantially more sensitive: 95% LDH release in the presence of O-GNR–PEG-DSPE versus ~50% in its absence. Moreover, the membrane of RBCs has been found to be highly sensitive to GO. For example, Jiang and coworkers have investigated the toxic effects of GO and nitrogen-doped graphene quantum dots (N-GODs) on RBCs. By using infrared (IR) spectroscopy in combination with monitoring hemolysis, observing morphological changes and detecting the adenosine triphosphate (ATP) content of RBCs, they have confirmed that the GO materials were first adsorbed on the external part of the lipid bilayer of the RBC membrane, which led to its disintegration, hemolysis and aberrant forms [16]. Haynes and coworkers have found that hemolysis of RBCs was especially pronounced for GOs of small size [9]. In particular, pGO-30 with a hydrodynamic diameter $d = 324 \pm 17$ nm at 50 μg ml^{-1} induced hemolysis of >90% RBCs, whereas the usual GO obtained by the Hummers method ($d = 765 \pm 19$ nm) applied at the same concentration affected only ~25% RBCs. Finally, the membranes of a variety of bacterial cells have been found to be sensitive to graphene-based materials [26–28].

The interaction of GO with cellular membranes can be further modulated by proteins present in biological fluids, since some of them bind to the GO surface with high affinity. For example, serum albumins (SAs) are present in large quantities in blood and can potentially affect GO toxicity. One example of such an influence has been reported by Ge, Zhou and coworkers. By using electron microscopy, these authors have observed that bovine serum albumin (BSA) reduced the cell membrane permeation of GO, inhibited the cellular damage induced by GO and reduced its cytotoxicity [29]. Based on molecular dynamics studies, they have concluded that the protein–GO interaction weakens the GO–phospholipid interaction due to the reduction of the surface available for binding. In other work, the effect of GO on human serum albumin (HSA) properties has been reported by Ding *et al.* [30]. In particular, they have observed that GO inhibited the interaction of HSA with bilirubin. Thus, GO and serum albumins mutually affect the properties of each other.

11.3.2 Oxidative Stress

A number of reports confirm that GO treatment results in an increase in the amount of ROS in cells. The latter can be detected, for example, by using a variety of commercially available leuco-dyes, including dichlorodihydrofluorescein diacetate or dihydroethidium, in combination with flow cytometry or fluorescence microscopy. In particular, Chang *et al.* [11] have observed that incubation of A549 cells with GO induced a dose-dependent intracellular oxidative stress that leads to a slight loss of cell viability at high concentrations. Moreover, GO toxicity toward human multiple myeloma RPMI 8226 cells has been found to be closely associated with an elevated amount of ROS [14]. A similar effect has been observed by Lammel and Navas [31], who studied the influence of GO and carboxyl graphene (CXYG) on fish hepatoma cell line PLHC-1. For example, they found that graphene materials penetrated spontaneously through the cellular membrane and in the cytosol they interacted with mitochondrial and nuclear membranes. The treated PLHC-1 cells demonstrated significantly reduced mitochondrial membrane potential and increased ROS levels at $16\,\mu g\,ml^{-1}$ GO and CXYG (72h incubation). Other reports confirming the GO-induced oxidative stress in cellular assays have been reviewed elsewhere [1–3].

The data obtained in *in vitro* assays are supported by *in vivo* data. For example, the effects of prolonged exposure of the roundworm *Caenorhabditis elegans* to GO have been evaluated by Wu *et al.* [32]. *Caenorhabditis elegans* is especially well suited as a model organism for evaluation of the biological effects (including toxicity) of chemical compounds *in vivo*, since this organism is transparent and can be monitored/studied by using fluorescence imaging. Wu *et al.* [32] have found that prolonged exposure of this organism to $0.5–100\,mg\,l^{-1}$ of GO caused a negative effect on the functions of both primary (intestine) and secondary (neuron and reproductive organ) targeted organs. Interestingly, in the intestine, the production of ROS was detected, which correlated with the adverse effects observed. Furthermore, Li *et al.* [23] have proven that NGO-induced acute lung injury (ALI) and chronic pulmonary fibrosis were related to the oxidative stress and could be relieved with dexamethasone treatment, which is a steroid drug with anti-inflammatory properties. In another model organism, zebrafish, GO induced a significant hatching delay and cardiac edema during embryogenesis [33]. Moreover, its treatment led to the excessive production of ROS (e.g. hydroxyl radicals) and changes in the secondary structure of proteins.

The question of why GO induces oxidative stress in cells is currently being actively investigated. For example, Nie's group has reported that the ROS-generating ability of GOs in mouse embryo fibroblasts (MEFs) is dependent upon the oxidation degree of GOs [34]. In particular, the least oxidized GO exhibited the highest ROS-enhancing ability, which was explained by the conversion of less toxic H_2O_2 into highly toxic $HO^\bullet$ radicals in cells. The theoretical simulations by the same authors revealed the involvement of carboxyl groups and planar domains of GO in varying the energy barrier of the H_2O_2 reduction reaction. Furthermore, using a fluorogenic, DNA-based probe, Mokhir and colleagues have confirmed that GOs obtained either by the Hummers method or by the milder method first

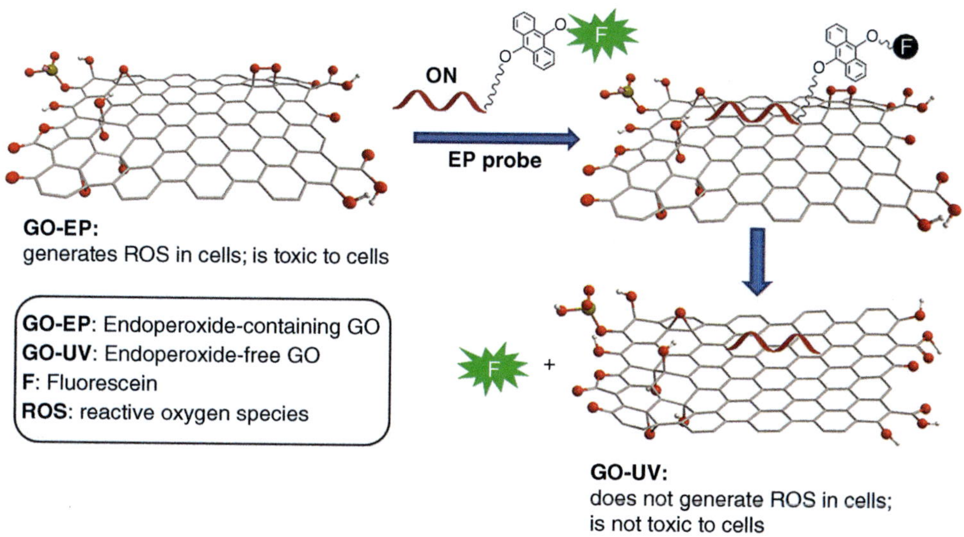

Figure 11.2 Detection of endoperoxides (EP) on the graphene oxide (GO–EP) surface using fluorogenic probes (EP probe) consisting of an oligonucleotide (ON), which binds strongly to the GO–EP, a reactive moiety (an anthracene derivative) and a fluorescent dye (fluorescein, F) [35]

reported by Eigler contained low amounts of surface-bound endoperoxides: one moiety per ~10^4 carbon atoms (Figure 11.2) [35].

These GOs were efficiently taken up by HeLa cells, which was accompanied by an increase in the intracellular ROS concentration and a decrease in cell viability. Interestingly, endoperoxide-free GOs, obtained by irradiation of the GOs with ultraviolet light of low power, were also taken up by the cells, but neither increased the intracellular ROS amount nor affected cell viability. These data allowed the authors to conclude that endoperoxides play an important role in the ROS-generating ability of GOs. Next, Chen and coworkers have investigated the effects of GO on T-lymphocytes and HSA [30]. In particular, they have observed that the treatment of T-lymphocytes with GO led to an increase in ROS generation, damage to DNA, cell apoptosis and limited suppression of the immune response of T-lymphocytes. Based on these data, they suggested that GO interacts directly with protein receptors, which inhibits their ligand binding ability, thereby leading to ROS-dependent passive apoptosis through the B-cell lymphoma-2 (Bcl-2) pathway.

11.3.3 Other Factors

New information about the toxicity of GO in relation to gene expression in cells has recently become available in the literature. These data may contribute to further understanding of the mechanism of GO toxicity *in vivo*. In particular, Wu *et al.* [36] have observed that mutations in several genes, including hsp-16.48, gas-1, sod-2, sod-3, aak-2 as well as isp-1 and clk-1, strongly affected translocation of GO into the body of *C. elegans*, its toxicity on both primary and secondary targeted organs compared with wild type, the intestinal permeability and the mean defecation cycle length. Furthermore, Wang and coworkers have investigated the role of micro-RNAs (miRNAs) in GO toxicity [37]. They have identified 23 up-regulated and eight down-regulated miRNAs in GO-treated *C. elegans*, and provided evidence to suggest that GO may reduce the lifespan of nematodes by affecting insulin/IGF (insulin-like growth factor) signaling, TOR (target of rapamycin) signaling as well as germline signaling pathways. Finally, the same authors have established the role of innate immunity in regulating chronic toxicity of GO in *C. elegans* [38].

11.4 Biomedical Applications of Graphene Oxide

11.4.1 Graphene Oxide in Treatment of Cancer and Bacterial Infections

In general, disease therapy relies on the selective action of a drug on disease-associated cells, biomolecules (e.g. enzymes, nucleic acids) or biochemical states (e.g. inflammation), which ideally occurs without affecting healthy organs and normal cells. The currently applied therapies for cancer treatment, including chemotherapy (using, for example, Pt(II)-based drugs, bleomycin and 5-fluorouracil) and radiotherapy, are not sufficiently cancer-cell-specific. Therefore, such treatments exhibit characteristic dose-limiting toxicities. Moreover, repeat treatments lead to the development of resistance. This partially explains why cancer is still one of the most common causes of death (together with cardiovascular disease) in developed countries. Therefore, the search for new approaches to cancer treatment is warranted. Targeted therapy is an advanced, recently introduced method, in which cancer-specific drugs (or prodrugs) are applied. GO is used in several approaches for cancer targeting, including photothermal and photodynamic therapy and as a nano-sized carrier to improve the cell membrane permeability of drugs and achieve their accumulation in tumors due to the enhanced permeability and retention (EPR) effect [6–8].

11.4.2 Photothermal Therapy

In photothermal therapy (PTT), disease-causing cells, including cancer cells in tumors and bacteria in wounds, are loaded with a reagent that absorbs near-infrared (NIR) light. Then subsequent exposure to NIR light heats up the system, inducing hyperthermia and thereby causing cell death. However, human tissues contain large amounts of hemoglobin and water, which strongly absorb visible and NIR light. To avoid unspecific heating of healthy tissues, for PTT, light is used that is practically not absorbed by the tissues: in the first biological window, 700–980 nm (BW1); and in the second biological window, 1000–1400 nm (BW2) [39]. Such light can penetrate through several centimeters of human tissue [40], whereas deeper located sites can be accessed by delivery of the light *via* optical fibers in combination with endoscopy [41]. Since the light beam can be focused on a specified area (e.g. tumor location) and its intensity (dose) can be easily controlled, PTT allows surgery-free tumor ablation practically without affecting healthy tissues.

Single-layered GO is suitable for PTT, since, in addition to its excellent water solubility, membrane permeability and stability, this material absorbs light in the NIR range [42–44]. It has been reported that the NIR absorptivity of GO can be improved by optimization of its size. For example, small GOs (less than 300 nm in size) absorb NIR light more efficiently than the conventional material. In particular, extinctions at 808 and 1200 nm have been found to be between about five- and eight-fold higher for small GOs [45]. The latter material is often called nano-GO (NGO) in the scientific literature.

GO of optimal size accumulates in tumors due to the EPR effect. Additionally, the accumulation can be achieved by decoration of GO with ligands that bind to cancer-specific receptors. These aspects will be discussed in the next subsection. A number of excellent reviews on PTT have been published, which cover the literature on the subject up to 2014 [1, 46, 47]. Here we will discuss only two selected reports, which demonstrate the applicability of GO-based materials as sensitizers for PTT of bacterial infections and cancer.

In 2013 bacterial infections affected over 48 million people in the USA and caused 80 deaths. They are especially dangerous for people with compromised immune systems and patients having extensive wounds following surgery [48]. Moreover, chronic infections are known to develop resistance against conventional organic drugs (antibiotics) and cause cancer: e.g. infection of *Helicobacter pylori* often leads to stomach cancer [49]. Therefore, novel antibacterial drugs are necessary. Wu and coworkers have explored the applicability of conventional GO obtained by the Hummers method in combination with PTT for the treatment of bacterial infection in wounds [50]. This is a rare example demonstrating the biological activity of conventional, chemically unmodified GO *in vivo*. The authors have conducted their investigation on healthy albino mice. Each mouse received three wounds, all of which were infected with *Staphylococcus aureus*: the first wound was left untreated; another one was exposed to the light from a Nd:YAG laser ($\lambda = 1064$ nm, 3 min irradiation, every day for 12 days); and the third wound was treated with GO and exposed to the same light under the same conditions. The healing of the wounds treated with both GO and the laser irradiation was accelerated in comparison to the control wounds. These data indicate that GO combined with PTT can potentially be used as an efficient and cheap alternative to antibiotics.

PTT in combination with GO or other NIR-absorbing nanomaterials has the potential to improve the side effects that occur during conventional chemotherapy. Moreover, when applied together, PTT and chemotherapy have been shown to exhibit synergistic effects [51]. For example, Guo and coworkers have prepared a hybrid material NGO–PEG–DOX containing NGO covalently modified with polyethylene glycol (PEG) residues and doxorubicin (DOX) [52]. The PEG fragment stabilizes NGO in the serum-containing media, enabling the application of this material *in vivo*. DOX is an anthracycline antitumor drug, which exhibits its activity due to binding of genomic DNA *via* intercalation. This fragment binds the NGO–PEG *via* non-covalent π–π interactions. The antitumor activity of NGO–PEG–DOX *in vivo* has been studied on a xenograft tumor mouse model, balb/c female mice, which carried tumors derived from murine mammary tumor cell line EMT6. In particular, the solution of the NGO construct was injected intravenously and the tumor was irradiated for 5 min (24 h post-injection) by a laser with $\lambda_{em} = 808$ nm (2 W cm^{-2}) focused on a 6 × 8 mm spot. The authors have reported a strong synergistic antitumor effect of DOX and PTT: the tumors were completely destroyed 30 days after the beginning of treatment. Interestingly, DOX alone exhibited substantially stronger side effects than the nano-sized construct NGO–PEG–DOX. These data demonstrate that PTT combined with chemotherapy can be a feasible approach for the improvement of current methods of cancer treatment.

11.4.3 Graphene Oxide as a Drug Carrier

In GO, every atom is exposed to the surface. Therefore, its surface area is very large (for graphene, 2600 m^2 g^{-1}). Correspondingly, one can densely load GO with cargo, e.g. drugs, cell-surface-directing fragments, nucleic acids and proteins. Though GO is soluble in water, it can aggregate in the presence of salts and components of serum. Therefore, this material is often chemically modified to improve its bioavailability, which includes either non-covalent (electrostatic or π–π interactions) or covalent modification [53]. For example, mixing GO with poly(ethyleneimine) (PEI) of different sizes (1.2–10 kDa) leads to facile formation of hybrid GO–PEI materials, which, in contrast to GO, are retained in the physiological solution and the serum-containing medium in the monomeric state. Moreover, such constructs have lower toxicity than free PEI [54]. In this case, the electrostatic interaction at neutral pH between positively charged PEI and negatively charged GO is the driving force for the GO–PEI formation. Furthermore, π–π interactions can be applied to modify GO and reduced GO (RGO). For example, planar aromatic molecules, such as porphyrins, pyrenes, perylenes and coronenes, have been used as anchors for attaching different functionalities to GO and RGO [55]. Since GO contains a variety of reactive functional groups, one often applies covalent chemistry to modify the GO surface. The most popular reactions include the formation of amide bonds, which typically occurs between a GO-bound –COOH group and a modifier-bound $-NH_2$ group [42, 53]. For example, the attachment of polymers like PEG, poly-L-lysine and polyacrylamide (PAA) to GO has been conducted by using such reactions.

11.4.3.1 Low-Molecular-Weight Drugs as a Cargo

Organic drugs containing extended π systems are common. For example, such compounds can act as intercalators of genomic DNA (e.g. anthracycline antibiotics), inhibit specific kinases (e.g. imatinib) and act as antimetabolites (e.g. methotrexate). Many such drugs are not well soluble in water. This problem can be solved by loading them onto well soluble GO-based materials. Moreover, GO–drug hybrids often enter the cells *via* pathways that are different from those of the free drugs. The same is the case for the removal of the drugs from the cell: hybrids may be retained in cells longer. Therefore, the activity of hybrids is often higher and they can be used to overcome the resistance of cells to particular drugs. Resistance is an important problem in the chemotherapy of cancer. It is developed as a result of repeated treatments with the same drug.

A number of reports were devoted to the transport of DOX and its analogs with the help of graphene-based carriers, as reviewed elsewhere [45–47, 53]. One such example has already been described in section 11.4.2 devoted to PTT [52]. Despite being a potent anticancer drug, DOX has an extended aromatic polycyclic structure and can, therefore, interact with sp^2-hybridized regions of GO due to π–π stacking interactions. This interaction is so strong that it is sufficient for the immobilization of DOX onto GO and no additional covalent attachment is usually required. Moreover, DOX is a fluorescent molecule that allows easy monitoring of its loading into cells by using either fluorescence microscopy or flow cytometry. In particular, Wang, Zhang and colleagues have used as a carrier NGO with sheet size below 100 nm and thickness between 0.8 and 1.5 nm, suggesting single- and double-layered structure [56]. By simple incubation of the components and removal of excess drug by centrifugation, high cargo loading could be achieved, 0.468 g DOX per 1 g NGO. Interestingly, the interaction strength between the NGO and DOX was strongly modulated by the pH of the solution. In particular, at neutral (pH 7.2) and basic (pH 9.0) conditions, less than 6.5% of the drug was released after standing for over 40 h in the correspondingly buffered phosphate saline (PBS) solutions. In contrast, at slightly acidic conditions (pH 5.0) ~15% of DOX was released within the same time. This effect has been attributed to the presence of hydrogen bonding between the functional groups on the NGO (–OH, –COOH) and those of DOX (–OH, $-NH_2$) in addition to the π–π stacking. The former bonds were expected to be destabilized under the acidic conditions. Since the microenvironment in tumors is often acidic, this property of NGO–DOX can be used for the selective release of DOX at the cancer-specific conditions, which should improve the therapeutic index of the nano-drug. Zhang and colleagues have explored whether NGO–DOX can reverse the DOX resistance of cells that overexpress multi-drug-resistant (MDR) gene [56]. In particular, they have tested the cytotoxicity of NGO–DOX toward DOX-sensitive MCF-7 and DOX-resistant MCF-7/ADR cell lines and compared the data obtained with the effects of free DOX. They have observed that NGO–DOX exhibited a toxicity toward MCF-7 cells (~1 $\mu g\,ml^{-1}$) that was comparable to DOX, but was substantially more toxic toward MCF-7/ADR than DOX: $IC_{50} \approx 1$ and 14 $\mu g\,ml^{-1}$ respectively. These data indicate that application of the NGO-based hybrid drug allows reversal of the resistance of MCF-7/ADR cells toward DOX. Other examples of graphene-based hybrids containing organic and metal-containing anticancer drugs, photosensitizers for photodynamic therapy (PDT) and receptor-targeting fragments include PEGylated reduced NGO carrying natural phenol resveratrol [57], unmodified NGO carrying bioactive flavonoid quercetin [58], GO-PEG loaded with chlorin e6 (Ce6) [59], RGD-motif-containing reduced NGO [60] and others [45–48, 53].

11.4.3.2 Oligonucleotide-Based Drugs as a Cargo

Nucleic acids play a central role in the storage of genetic information, protein synthesis and regulation. Due to the recent scientific and technological advancements in the field of genome wide sequencing, knowledge of the role of nucleic acids in cellular biology is rapidly expanding. For example, apart from messenger RNAs (mRNAs), ribosomal RNAs (rRNAs) and transfer RNAs (tRNAs), which have been known for a long time, many new RNA types have recently been discovered, and further discoveries of this type are certainly under way. They include micro-RNAs (miRNAs), pseudo-genes, circular RNAs, long non-coding RNAs and others. These biomolecules are termed non-coding RNAs (ncRNAs). Synthesis, processing, mode of action and targets of miRNAs are mostly well understood. They participate in the regulation of gene expression, regulation of protein synthesis and are often overexpressed or down-regulated in diseases (e.g. cancer) with respect to the normal state. However, functions of other ncRNAs are less well understood and are still actively being investigated.

Binders of intracellular RNAs can inhibit their biological activity (*via* antisense effect or RNA interference), which can help in the elucidation of the functions of newly discovered ncRNAs. Moreover, binders of the RNAs, which are fluorescently labelled with dyes and are responsive to the hybridization state (e.g. molecular beacons, MBs), can be used to monitor ncRNAs and mRNAs directly in cells. Oligonucleotides (ONs) are highly specific and strong binders of complementary RNA sequences. However, due to their polyanionic character, these reagents are not permeable through the cellular membrane. Moreover, they are not stable in cells due to their efficient cleavage by abundant intracellular endo- and exonucleases. To improve these properties, a number of chemically modified oligonucleotides have been prepared. They include phosphorothioate DNAs (PTOs), 2′-OMe RNAs, peptide nucleic acids (PNAs) and others. These compounds are more stable to nucleases than unmodified ONs. However, with few rare exceptions (e.g. PTOs), ON analogs are not cell-membrane-permeable.

Apart from inhibiting the RNA function, it can be required to upregulate (increase) the RNA amount in cells. This goal can be achieved by the introduction of circular, double-stranded nucleic acids (plasmids). The genes within the plasmids are expressed, producing the corresponding mRNAs and proteins, which can be used for determination of gene functions and design of cellular reporters, e.g. fluorescent proteins or luciferases. Moreover, required RNAs can also be introduced directly. However, in both cases the problem of cell membrane permeability exists. Typically, nucleic acids (plasmids, RNAs) and nucleic acid inhibitors (ONs, analogs of ONs, small interfering RNAs, siRNA) are brought into cells by reversible membrane permeabilization using streptolysin O (SLO), electroporation, transfection with positively charged oligomers or dendrimers and direct microinjection. These approaches are toxic or damage cells to some extent, are not applicable to all cell types and their possible applications *in vivo* are limited. Therefore, studies of new approaches for improving cell membrane permeation of nucleic acids are warranted.

GO-based materials can potentially evolve as true alternatives to the usually used transfection reagents. For example, it has already been demonstrated that they are applicable for transfection of plasmids and siRNAs into cells. In particular, Liu and coworkers have modified GO covalently with cationic PEI polymers of different sizes: 1.2 and 10 kDa [54]. They have found that both GO–PEI-1.2 kDa and GO–PEI-10 kDa induce the efficient transfection of a plasmid carrying enhanced green fluorescent protein (EGFP), which could be followed by monitoring the EGFP expression *via* fluorescence microscopy. In contrast, PEI-1.2 kDa itself is not functional. Though PEI-10 kDa is an efficient transfection agent, it has been found to be toxic. In contrast, the toxicity of the hybrid GO–PEI-10 kDa was reduced.

Zhang *et al.* [61] have designed a hybrid consisting of GO that was covalently modified with PEI-25 kDa and contained an siRNA targeting an mRNA of the Bcl-2 gene. The latter reagent was adsorbed onto the GO–PEI-25 kDa due to electrostatic interactions. The authors have confirmed that, in human cervical cancer cell line (HeLa cells) incubated with the GO–PEI-25 kDa–siRNA hybrid, which had the optimal ratio of nitrogen (proportional to the PEI) to phosphorus (proportional to the RNA) of 20, the expression of the Bcl-2 gene was suppressed down to ~30%. The inhibition observed using the PEI-25 kDa–siRNA associate under the same conditions was significantly weaker, ~60%. Additionally, the toxicity of the GO–PEI-25 kDa has been found to be negligible up to a concentration of 4 $ng\,l^{-1}$, whereas only ~50% of cells remained viable after their incubation with the same amount of PEI-25 kDa. Switching off the Bcl-2 gene was expected to overcome the MDR system of cancer cells and make them more sensitive to chemotherapeutic agents. To prove this hypothesis, the authors treated HeLa cells first with GO–PEI-25 kDa–siRNA, which was followed by DOX. In a control experiment, they used the hybrid containing a scrambled siRNA, which was not targeting any gene in the cells. The authors observed that the cells become more sensitive to DOX after the inhibition of the Bcl-2 gene by the siRNA-containing hybrid. These data confirm the applicability of GO as a carrier for the transport of siRNAs into cells. Moreover, Yin *et al.* [21] have demonstrated the delivery of a plasmid-based Stat3 siRNA in a mouse model of melanoma, which resulted in the significant inhibition of tumor growth without any toxicity.

Since GO has a large surface area, many different components can be introduced onto it simultaneously to obtain multi-functional drugs or prodrugs. A demonstration of this possibility has been described in the publication of Yang, Xiang, Chen and coworkers [62]. In particular, they have prepared PEGylated GO, which carried at the terminus of each PEG residue one folic acid (FA) fragment. FA was attached to direct (target) the hybrid to cancer cells overexpressing the FA receptor. Next, 1-pyrenemethylamine was adsorbed *via* strong non-covalent π–π interactions with flat regions of GO that provided an overall positive charge for the resulting construct. Finally, siRNA targeting a human telomerase reverse transcriptase (hTERT) gene was attached *via* electrostatic interactions. The authors have demonstrated that the obtained hybrid is not toxic to HeLa cells, but acts as a strong inhibitor of hTERT expression, as evidenced by monitoring the corresponding transcript and the protein.

Both plasmids and siRNAs act by catalytic mechanisms. Therefore, one molecule of the plasmid can generate many equivalents of mRNAs, whereas one molecule of siRNA can induce cleavage of many equivalents of mRNAs. Thus, delivery of even small amounts of these reagents in cells will cause dramatic changes of the concentration of the targeted nucleic acids. Correspondingly, experiments that rely on detection of gene expression from plasmids or target inhibition using siRNAs allow one to answer the question whether the carriers transport their cargo through the cell membrane or not. However, they do not provide an accurate estimate of how much cargo crosses the membrane and stays in the active form in the cell. It should be mentioned that the large proportion of ONs that cross the cellular membrane are trapped in intracellular compartments and remain inactive. A more accurate experiment for determination of the delivery efficiency would be the transport of a labelled ON, e.g. molecular beacons (MBs) or another hybridization-sensitive probe, which binds to its intracellular target and causes changes of the fluorescence of the probe. The latter reaction is stoichiometric and the fluorescence intensity is expected to correlate with the concentration of the target in the cell. Chen, Yang and coworkers have reported on the transport of a DNA-based MB containing Dabcyl as a quencher and Cy5 as a reporter fluorophore [63]. This probe was designed to bind a survivin mRNA. The authors have observed that the NGO protected the MB from nucleases and substantially reduced the background signal of the MB in the absence of its target. Moreover, the NGO brought this MB into cells where it bound the survivin mRNA, as indicated by the increase of the fluorescence intensity in cells. A construct containing NGO and control MB has been found to generate a 2.4-fold lower fluorescence signal in cells. These results confirm some selectivity of the survivin MB, which is, however, still not sufficient for practical applications. Therefore, further optimizations of the delivery of MBs or other hybridization-sensitive probes are warranted.

Aptamers are short ON sequences (RNA or DNA) that bind specifically small molecules. These reagents can be applied as drugs and for imaging of biomolecules both in cell-free conditions and directly in cells. For example, Li, Lin and coworkers have explored the possibility for the delivery of an ATP-aptamer (ON) to cells using GO [64]. To be able to monitor the reagent delivery as well as its binding to intracellular ATP, the authors have labelled the aptamer with a fluorescent dye (Fl) to obtain ON-Fl. The GO-bound ON-Fl remained weakly fluorescent due to strong quenching of the Fl with the GO. However, upon binding of ATP, the aptamer was folded with the formation of the structure, which does not have the high affinity to GO. Therefore, the aptamer was released from the GO in the presence of ATP, which was reflected in the dequenching of the fluorescence of the

ON-Fl. Thus, the fluorescence of ON-Fl/GO correlates with the concentration of ATP in solution. Li, Lin and coworkers have observed this behavior both in cell-free assays and in cells, thereby confirming that GO is a suitable carrier of aptamers.

11.5 Bioanalytical Applications

Graphene oxide has been used in a variety of electrochemical and optical assays for the detection of biomolecules and xenobiotics, as reviewed elsewhere [65]. In this section we will concentrate on the detection of nucleic acids by using fluorescence-based approaches, which is mainly due to the current research interests of the authors of this chapter.

Oligonucleotides (ONs) interact strongly with GO. In particular, Maheshwari, Liu and coworkers have studied in detail the binding of fluorescein-labeled oligonucleotides (Fl-ONs) of different length with GO [66]. The GO used in that work was obtained by a modified Hummers method, including oxidation with potassium persulfate and phosphorus pentoxide at 90 °C. The binding was studied by fluorescence spectroscopy monitoring the fluorescence quenching upon the addition of the GO to the Fl-ONs. The authors observed that the binding efficiency decreased with increasing ON length (12- to 32-mers have been investigated). Moreover, the kinetics of this interaction has been found to be substantially faster for shorter strands (12- to 24-mers) than for the longer one (36-mer). Since both the nucleobases of ONs and the flat regions of the GO are planar aromatics, they are able to get engaged in π–π interactions with each other analogously to those found between single-walled nanotubes (SWNTs) and nucleic acids [67]. However, this does not seem to be the only factor defining the affinity of the GO to ONs. For example, it has been observed that the binding of Fl-ONs to GO was strongly salt-dependent, which allowed the formation of salt bridges GO^{n-}–Na^{+}–ON^{m-} between the reacting partners to be suggested. Furthermore, the GO–ON interaction is highly sensitive to the hybridization state of the nucleic acids. For example, single-stranded nucleic acids bind GO efficiently, whereas folded ones, including double-stranded nucleic acids, quadruplexes [68] or aptamers [64] bound to their target molecules, do not bind GO. This property of GO has been used in a number of bioanalytical applications for detection of nucleic acids and, in combination with aptamers, for the detection of small molecules.

In early assays, a single-stranded ON containing a fluorophore was loaded onto GO or NGO, leading to fluorophore quenching (Figure 11.3). Adding a complementary nucleic acid to the resulting mixture caused the formation of dsDNA, its release from the GO surface and, consequently, fluorophore dequenching (pathway A, Figure 11.3).

For example, in this way Yang and coworkers have detected a DNA with an HIV1 sequence [69]. Ai and coworkers have recently optimized this assay for detection of single mismatches in nucleic acids [70]. However, since GO binding to single-stranded nucleic acids is in general strong and weakly sequence-specific, an alternative activation mechanism can occur. In particular, a mismatched nucleic acid can replace the probe from the GO surface by an unspecific interaction with the GO that leads to a strong background signal (false positive, pathway B in Figure 11.3). For example, Yang and coworkers have observed [69] that an MHIV1 target, containing a single mismatch, also enhances the fluorescence of the solution containing the probe–GO construct. In the latter case, the signal was only about half that observed in the presence of the fully matching target. Recently, Liu and coworkers have provided experimental evidence that the unspecific probe displacement

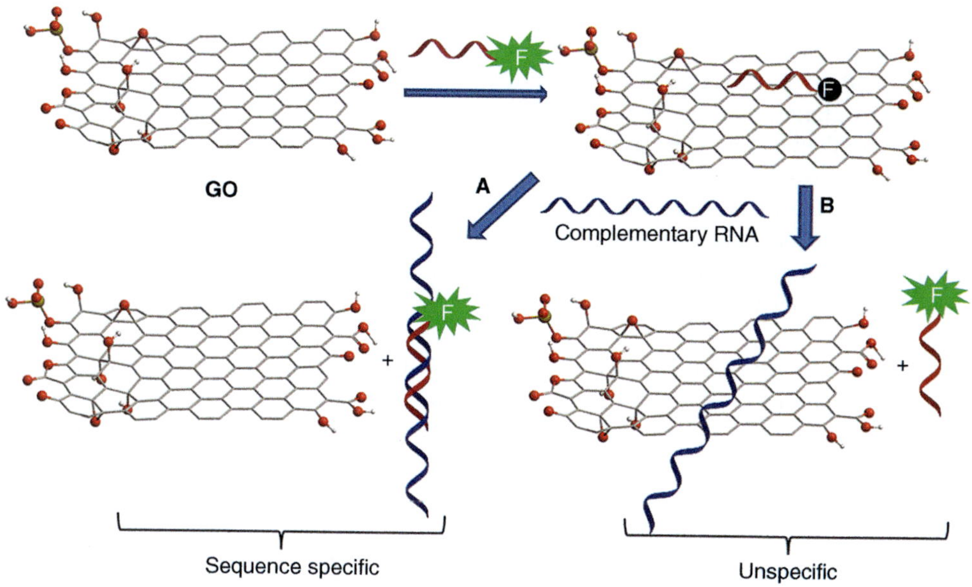

Figure 11.3 An approach for detection of nucleic acids by making use of quenching and ON-binding properties of GO. Binding of the probe via pathway A leads to the sequence-specific fluorescence increase, whereas when pathway B is realized the fluorescence increase is not specific [66, 69, 70]

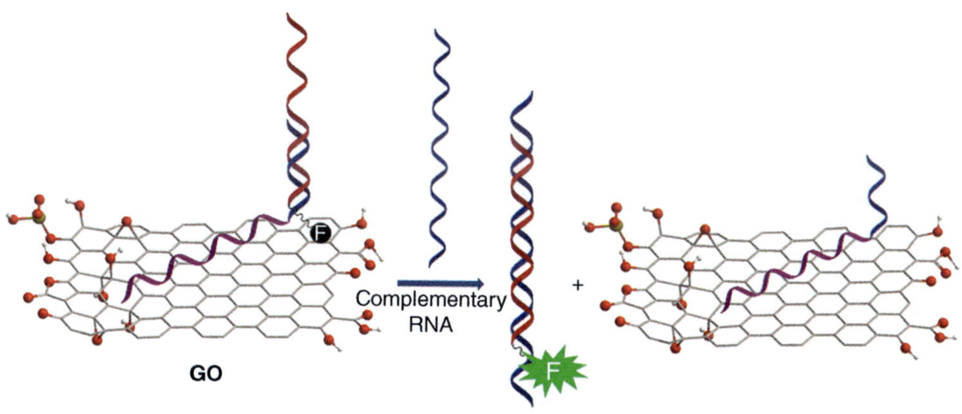

Figure 11.4 An improved method of nucleic acid analysis using GO as a quencher and probe binder. A toehold sequence of the probe was selected [72]

from the GO followed by the hybridization of the probe with the target in solution is a major pathway of activation of the probe–GO sensors. They have also confirmed that only a small portion of the target is hybridized to the probe, whereas the majority of it remains bound to the GO [71]. The solution to this problem has been suggested by the research groups of Kitamura and Ihara, who separated a target-binding DNA sequence from the GO-binding element (Figure 11.4) [72]. The latter can in principle be any GO-binding chemical moiety that has a substantially stronger affinity toward GO than any analyte present in solution. In the original work, DNA dA_{20} (magenta colored in Figure 11.4) has been selected as a GO-binding sequence and a toehold construct, containing a double-stranded part covalently connected to the dA_{20} and a single-stranded DNA overhang for anchoring the target. The reporter fluorophore in the resulting sensor has been placed in the proximity to the GO and was, therefore, strongly quenched. In this case the target cannot displace the sensor (dA_{20}–toehold sequence) from the GO. Rather, it hybridizes with one of the strands of the toehold sequence, thereby replacing another strand and leading to the fluorescence dequenching. In this system, the target-induced fluorescence enhancement with respect to that obtained in the presence of a mismatched target has been found to be about 7-fold higher than the same parameter for the parent system with a simple, single-stranded probe.

Another important factor to account for in GO-based DNA detection assays is the level of oxidation of GO (C/O ratio). This factor has not been considered in earlier reports. For example, Nguyen and coworkers have reported that the C/O ratio strongly affects the fluorescence quenching ability of GO as well as its affinity toward single-stranded ONs [73].

Acknowledgments

We thank Dr. Siegfried Eigler for drawing the GO structure shown in Figure 11.1 and Serghei Chercheja for drawing the GO structures shown in Figures 11.2, 11.3 and 11.4.

References

[1] Bianco, A., Graphene: safe or toxic? Two faces of the medal. *Angew. Chem. Int. Ed.* **2013**, *52*, 4986–4997.
[2] Nezakati, T.; Cousins, B.G.; Seifalian, A.M., Toxicology of chemically modified graphene-based materials for medical application. *Arch. Toxicol.* **2014**, *88*, 1987–2012.
[3] Jachak, A.C.; Creighton, M.; Giu, Y.; *et al.*, Biological interactions and safety of graphene materials. *MRS Bull.* **2012**, *37* (12), 1307–1313.
[4] Guo, X.; Mei, N., Assessment of the toxic potential of graphene family nanomaterials. *J. Food. Drug. Anal.* **2014**, *22* (1), 105–115.
[5] Seabra, A.B.; Paula, A.J.; de Lima, R.; *et al.*, Nanotoxicity of graphene and graphene oxide. *Chem. Res. Toxicol.* **2014**, *27* (2), 159–168.
[6] Goenka, S.; Sant, V.; Sant, S., Graphene-based nanomaterials for drug delivery and tissue engineering. *J. Control. Release* **2014**, *173*, 75–88.
[7] Krishna, K.V.; Ménard-Moyon, C.; Verma, S.; Bianco, A., Graphene-based nanomaterials for nanobiotechnology and biomedical applications. *Nanomedicine* **2013**, *8* (10), 1669–1688.
[8] Skoda, M.; Dudek, I.; Jarosz, A.; Szukiewicz, D., Graphene: one material, many possibilities – application difficulties in biological systems. *J. Nanomater.* **2014**, *2014*, 890246.
[9] Liao, K.-H.; Lin, Y.-S.; Macosko, C.W.; Haynes, C.L., Cytotoxicity of graphene oxide and graphene in human erythrocytes and skin fibroblasts. *ACS Appl. Mater. Interfaces* **2011**, *3*, 2607–2615.
[10] Wang, K.; Ruan, J.; Song, H.; *et al.*, Biocompatibility of graphene oxide. *Nanoscale Res. Lett.* **2011**, *6*, 8.
[11] Chang, Y.; Yang, S.-T.; Liu, J.-H.; *et al.*, In vitro toxicity evaluation of graphene oxide on A549 cells. *Toxicol. Lett.* **2011**, *200*, 201–210.
[12] Nguyen, T.H.D.; Lin, M.; Mustapha, A., Toxicity of graphene oxide on intestinal bacteria and caco-2 cells. *J. Food Protect.* **2015**, *78* (5), 996–1002.

[13] Jaworski, S.; Sawosz, E.; Kutwin, M.; *et al.*, In vitro and in vivo effects of graphene oxide and reduced graphene oxide on glioblastoma. *Int. J. Nanomed.* **2015**, *10*, 1585–1596.
[14] Wang, Y.; Wu, S.; Zhao, X.; *et al.*, In vitro toxicity evaluation of graphene oxide on human RPMI 8226 cells. *Biomed. Mater. Eng.* **2014**, *24* (6), 2007–2013.
[15] Chng, E.L.K.; Chua, C.K.; Pumera, M., Graphene oxide nanoribbons exhibit significantly greater toxicity than graphene oxide nanoplatelets. *Nanoscale* **2014**, *6* (18), 10792–10797.
[16] Wang, T.; Zhu, S.; Jiang, X., Toxicity mechanism of graphene oxide and nitrogen-doped graphene quantum dots in RBC revealed by surface-enhanced infrared absorption spectroscopy. *Toxicol. Res.* **2015**, *4* (4), 885–894.
[17] Yan, L.; Wang, Y.; Xu, X.; *et al.*, Can graphene oxide cause damage to eyesight? *Chem. Res. Toxicol.* **2012**, *25* (6), 1265–1270.
[18] Wojtoniszak, M.; Chen, X.; Kalenczuk, R.J.; *et al.*, Synthesis, dispersion, and cytocompatibility of graphene oxide and reduced graphene oxide. *Colloids Surfaces B: Biointerfaces* **2012**, *89*, 79–85.
[19] Fiorillo, M.; Verre, A.F.; Iliut, M.; *et al.*, Graphene oxide selectively targets cancer stem cells, across multiple tumor types: implications for non-toxic cancer treatment, *via* "differentiation-based nano-therapy". *Oncotarget* **2015**, *6* (6), 3553–3562.
[20] Liang, S.; Xu, S.; Zhang, D.; *et al.*, Reproductive toxicity of nanoscale graphene oxide in male mice. *Nanotoxicity* **2015**, *9* (1), 92–105.
[21] Yin, D.; Li, Y.; Lin, H.; *et al.*, Functional graphene oxide as a plasmid-based Stat3 siRNA carrier inhibits mouse malignant melanoma growth in vivo. *Nanotechnology* **2013**, *24*, 105012.
[22] Fu, C.; Liu, T.; Li, L.; *et al.*, Effects of graphene oxide on the development of offspring mice in lactation period. *Biomaterials* **2015**, *40*, 23–31.
[23] Li, B.; Yang, J.; Huang, Q.; *et al.*, Biodistribution and pulmonary toxicity of intratracheally instilled graphene oxide in mice. *NPG Asia Materials* **2013**, *5*, e44.
[24] Li, Y.; Yuan, H.; von dem Bussche A.; *et al.*, Graphene microsheets enter cells through spontaneous membrane penetration at edge asperities and corner sites. *Proc. Natl. Acad. Sci. USA* **2013**, *110* (30), 12295–12300.
[25] Mullick Chowdhury, S.; Lalwani, G.; Zhang, K.; *et al.*, Cell specific cytotoxicity and uptake of graphene nanoribbons. *Biomaterials* **2013**, *34* (1), 283–293.
[26] Liu, S.; Zeng, T.H.; Hofmann, M.; *et al.*, Antibacterial activity of graphite, graphite oxide, graphene oxide, and reduced graphene oxide: membrane and oxidative stress. *ACS Nano* **2011**, *5*, 6971–6980.
[27] Akhavan, O.; Ghaderi, E., Toxicity of graphene and graphene oxide nanowalls against bacteria. *ACS Nano* **2010**, *4*, 5731–5736.
[28] Chen, J.; Wang, X.; Han, H., A new function of graphene oxide emerges: inactivating phytopathogenic bacterium *Xanthomonas oryzae pv. oryzae*. *J. Nanopart. Res.* **2013**, *15*, 1658–1671.
[29] Duan, G.; Kang, S.; Tian, X.; *et al.*, Protein corona mitigates the cytotoxicity of graphene oxide by reducing its physical interaction with cell membrane. *Nanoscale* **2015**, *7* (37), 15214–15224.
[30] Ding, Z.; Zhang, Z.; Ma, H.; Chen, Y., In vitro hemocompatibility and toxic mechanism of graphene oxide on human peripheral blood T lymphocytes and serum albumin. *ACS Appl. Mater. Interfaces* **2014**, *6* (22), 19797–19807.
[31] Lammel, T.; Navas, J.M., Graphene nanoplatelets spontaneously translocate into the cytosol and physically interact with cellular organelles in the fish cell line PLHC-1. *Aquatic Toxicol.* **2014**, *150*, 55–65.
[32] Wu, Q.; Yin, L.; Li, X.; *et al.*, Contributions of altered permeability of intestinal barrier and defecation behavior to toxicity formation from graphene oxide in nematode *Caenorhabditis elegans*. *Nanoscale* **2013**, *5* (20), 9934–9943.
[33] Chen, Y.; Ren, C.; Ouyang, S.; *et al.*, Mitigation in multiple effects of graphene oxide toxicity in zebrafish embryogenesis driven by humic acid. *Environ. Sci. Technol.* **2015**, *49* (16), 10147–10154.
[34] Zhang, W.; Yan, L.; Li, M.; *et al.*, Deciphering the underlying mechanisms of oxidation-state dependent cytotoxicity of graphene oxide on mammalian cells. *Toxicol. Lett.* **2015**, *237*, 61–71.
[35] Pieper, H.; Chercheja, S.; Eigler, S.; *et al.*, Endoperoxides revealed as origin of toxicity of graphene oxide. *Angew. Chem. Int. Ed.* **2016**, *55* (1), 405–407.

[36] Wu, Q.; Zhao, Y.; Li, Y.; Wang, D., Molecular signals regulating translocation and toxicity of graphene oxide in the nematode *Caenorhabditis elegans*. *Nanoscale* **2014**, *6* (19), 11204–11212.
[37] Wu, Q.; Zhao, Y.; Zhao, G.; Wang, D., microRNAs control of in vivo toxicity from graphene oxide in *Caenorhabditis elegans*. *Nanomedicine* **2014**, *10* (7), 1401–1410.
[38] Wu, Q.; Zhao, Y.; Fang, J.; Wang, D., Immune response is required for the control of in vivo translocation and chronic toxicity of graphene oxide. *Nanoscale* **2014**, *6* (11), 5894–5906.
[39] Jaque, D.; Martínes Maestro, L.; del Rosal, B.; *et al.*, Nanoparticles for photothermal therapies. *Nanoscale* **2014**, *6*, 9494–9530.
[40] Helmchen, F.; Denk, W., Deep tissue two-photon microscopy. *Nature Meth.* **2005**, *2*, 932–940.
[41] Olivo, M.; Fu, C.Y.; Raghavan, V.; Lau, W.K., New frontier in hypericin-mediated diagnosis of cancer with current optical technologies. *Ann. Biomed. Eng.* **2012**, *49* (2), 460–473.
[42] Lai, Q.; Zhu, S.; Luo, X.; *et al.*, Ultraviolet–visible spectroscopy of graphene oxides. *AIP Advances* **2012**, *2*, 032146.
[43] Saxena, S.; Tyson, T.A.; Shukla, S.; *et al.*, Investigation of structural and electronic properties of graphene oxide. *Appl. Phys. Lett.* **2011**, *99*, 013104.
[44] Li, M.; Yang, X.; Ren, J.; *et al.*, Using graphene oxide high near-infrared absorbance for photothermal treatment of Alzheimer's disease. *Adv. Mater.* **2012**, *24*, 1722–1728.
[45] Sun, X.; Liu, Z.; Welsher, K.; *et al.*, Nano-graphene oxide for cellular imaging and drug delivery. *Nano Res.* **2008**, *1* (3), 203–212.
[46] Liu, J.; Cui, L.; Losic, D., Graphene and graphene oxide as new nanocarriers for drug delivery applications. *Acta Biomater.* **2013**, *9*, 9243–9257.
[47] Orecchiono, M.; Cabizza, R.; Bianco, A.; Delogu, L.G., Graphene as cancer theranostic tool: progress and future challenges. *Theranostics* **2015**, *5* (7), 710–723.
[48] Centers for Disease Control and Prevention, *Estimates of foodborne illness in the United States.* See http://www.cdc.gov/foodborneburden/trends-in-foodborne-illness.html (accessed 11 May 2016).
[49] Kim, W.; Moss, S.F., The role of *Heliobacter pylori* in the pathogenesis of gastric malignancies. *Oncol. Rev.* **2008**, *2* (3), 131–140.
[50] Khan, M.S.; Abdelhamid, H.N.; Wu, H.-F., Near infrared (NIR) laser mediated surface activation of graphene oxide nanoflakes for efficient antibacterial, antifungal and wound healing treatment. *Colloids Surfaces B: Biointerfaces* **2015**, *127*, 281–291.
[51] Hauck, T.S.; Jennings, T.L.; Yatsenko, T.; *et al.*, Enhancing the toxicity of cancer chemotherapeutics with gold nanorod hyperthermia. *Adv. Mater.* **2008**, *20*, 3832–3838.
[52] Zhang, W.; Guo, Z.; Huang, D.; *et al.*, Synergistic effect of chemo-photothermal therapy using PEGylated graphene oxide. *Biomaterials* **2011**, *32*, 8555–8561.
[53] Shi, S.; Chen, F.; Ehlerding, E.B.; Cai, W., Surface engineering of graphene-based nanomaterials for biomedicinal applications. *Bioconjug. Chem.* **2014**, *25*, 1609–1619.
[54] Feng, L.; Zhang, S.; Liu, Z., Graphene based gene transfection. *Nanoscale* **2011**, *3*, 1252–1257.
[55] Eigler, S.; Hirsch, A., Chemistry with graphene and graphene oxide – challenges for synthetic chemists. *Angew. Chem. Int. Ed.* **2014**, *53*, 7720–7738.
[56] Wu, J.; Wang, Y.; Yang, X.; *et al.*, Graphene oxide used as a carrier for adriamycin can reverse drug resistance in breast cancer cells. *Nanotechnology* **2012**, *23*, 355101.
[57] Chen, J.; Liu, H.; Zhao, C.; *et al.*, One-step reduction and PEGylation of graphene oxide for photothermally controlled drug delivery. *Biomaterials* **2014**, *35*, 4986–4995.
[58] Rahmanian, N.; Hamishehkar, H.; Dolatabadi, J.E.N., Nano graphene oxide: a novel carrier for oral delivery of flavonoids. *Colloids Surfaces B: Biointerfaces* **2014**, *123*, 331–338.
[59] Tian, B.; Wang, C.; Zhang, S.; *et al.*, Photothermally enhanced photodynamic therapy delivered by nano-graphene oxide. *ACS Nano* **2011**, *5* (9), 7000–7009.
[60] Robinson, J.T.; Tabakman, S.M.; Liang, Y.Y.; *et al.*, Ultrasmall reduced graphene oxide with high near-infrared absorbance for photothermal therapy. *J. Am. Chem. Soc.* **2011**, *133*, 6825–6831.
[61] Zhang, L.; Lu, Z.; Zhao, Q.; *et al.*, Enhanced chemotherapy efficacy by sequential delivery of siRNA and anticancer drugs using PEI-grafted graphene oxide. *Small* **2011**, *7* (4), 460–464.
[62] Yang, X.; Niu, G.; Cao, X.; *et al.*, The preparation of functionalized graphene oxide for targeted intracellular delivery of siRNA. *J. Mater. Chem.* **2012**, *22*, 6649–6654.

[63] Lu, C.-H.; Zhu, C.-L.; Li, J.; *et al.*, Using graphene to protect DNA from cleavage during cellular delivery. *Chem. Commun.* **2010**, *46*, 3116–3118.
[64] Wang, Y.; Li, Z.; Hu, D.; *et al.*, Aptamer/graphene oxide nanocomplex for in situ molecular probing in living cells. *J. Am. Chem. Soc.* **2010**, *132*, 9274–9276.
[65] Tang, L:, Wang, Y.; Li, J., The graphene/nucleic acid nanobiointerface. *Chem. Soc. Rev.* **2015**, *44*, 6954–6980.
[66] Wu, M.; Kempaiah, R.; Huang, P.-J.J.; *et al.*, Adsorption and desorption of DNA on graphene oxide studied by fluorescently labeled oligonucleotides. *Langmuir* **2011**, *27*, 2731–2738.
[67] Zheng, M.; Jagota, A.; Semke, E.D.; *et al.*, DNA-assisted dispersion and separation of carbon nanotubes. *Nature Mater.* **2003**, *2*, 338–342.
[68] Wang, H.; Chen, T.; Wu, S.; *et al.*, A novel biosensing strategy for screening G-quadruplex ligands based on graphene oxide sheets. *Biosens. Bioelectron.* **2012**, *34*, 88–93.
[69] Lu, C.-H.; Yang, H.-H.; Zhu, C.-L.; *et al.*, A graphene platform for sensing biomolecules. *Angew. Chem. Int. Ed.* **2009**, *48*, 4785–4787.
[70] Huang, Y.; Yang, H.Y.; Ai, Y., DNA single-base mismatch study using graphene oxide nanosheets-based fluorometric biosensors. *Anal. Chem.* **2015**, *87*, 9132–9136.
[71] Liu, B.; Sun, Z.; Zhang, X.; Liu, J., Mechanisms of DNA sensing on graphene oxide. *Anal. Chem.* **2013**, *85* (16), 7987–7993.
[72] Miyahata, T.; Kitamura, Y.; Futamura, A.; *et al.*, DNA analysis based on toehold-mediated strand displacement on graphene oxide. *Chem. Commun.* **2013**, *49*, 10139–10141.
[73] Hong, B.J.; An, Z.; Compton, O.C.; Nguyen, S.B., Tunable biomolecular interaction and fluorescence quenching ability of graphene oxide: application to "turn-on" DNA sensing in biological media. *Small* **2012**, *8* (16), 2469–2476.

12
Catalysis

Ioannis V. Pavlidis

12.1 Introduction

The catalytic potential of graphene oxide (GO) and graphite oxide is nothing new. The reactivity of graphite oxide as a solid-phase heterogeneous catalyst was cited in 1962 for the first time [1], catalyzed on the production of HBr, just four years after Hummers' publication on the method of production of GO [2]. It is astonishing that the authors were able to estimate that 100 carbon atoms in the layer plane catalyze the formation of 10 HBr molecules per hour and that the carbon atoms at the periphery functionalized as oxides are the active sites. Moreover, they were able to estimate that the GO consists of two or three layers, with about 80% carbon and 20% oxygen atoms, and most of the oxygen was identified as bound hydroxyl and ether groups. However, it took more than 40 years for carbocatalysis to draw the attention that it deserves. In the meantime, two Nobel Prizes were awarded for research in the field (the 1996 Nobel Prize in Chemistry for the discovery of fullerenes, and the 2010 Nobel Prize in Physics for the groundbreaking experiments regarding the two-dimensional material graphene). Just recently the catalytic properties of GO have started to be harnessed for chemical reactions, and the potential of GO and its derivatives in synthetic chemistry still remains unexplored to a great extent.

As its name indicates, GO is related to graphene, in that they are both two-dimensional (2D) carbon nanomaterials. However, despite this basic similarity, these materials differ significantly in their intrinsic properties. Graphene is composed of sp^2-hybridized carbon atoms, while GO comprises a mixture of sp^2- and sp^3-hybridized carbon atoms, with the latter being functionalized with oxygen. This particular difference has a profound impact on the properties of the final material [3], as will be discussed later, while it provides a lot of groups

Graphene Oxide: Fundamentals and Applications, First Edition. Edited by Ayrat M. Dimiev and Siegfried Eigler.

acting as active sites. It is known that the performance of a catalyst is influenced by the nature of the material, the concentration and the accessibility of the active sites [4]. Taking this into account, the catalytic potential of GO is far greater than that of pristine graphene. GO is used already in several oxidative reactions, but also in polymerization reactions and Friedel–Crafts or Michael additions, just to mention some of the reactions that GO can catalyze. However, there are still some obstacles to surpass due to the difficulty of the not fully understood structure, and thus the elucidation of the catalytic mechanism is difficult in most cases. These topics are going to be discussed in detail in the following sections.

12.2 Graphene Oxide Properties

In order to elaborate the catalytic potential of GO, one needs to understand the unique properties that GO and similar or derived materials possess. It is a common trend that researchers in the field mention vaguely its unique properties in the introductions of their publications; however, in most cases, they never define the distinctive property of the material that makes it proper for the specific reaction or use. In this section, the most interesting properties of GO – concerning its catalytic behavior, which is the focus of the present chapter – are going to be discussed.

In order to be fair about the criticism raised in the introduction concerning the length of time needed to rediscover GO as a catalyst, there is a huge barrier that hindered its exploitation: the structure of GO is still under debate. Several structural models have been proposed [5–8], with the one proposed by Lerf *et al.* [9] being the most widely accepted. This model proposes that there are two kinds of regions: an aromatic region with non-oxidized benzene rings; and regions with oxidized aliphatic six-membered rings that have epoxy and hydroxyl functionalities on both sides of the monolayer. Of importance is the slightly distorted configuration of the hydroxylated carbon centers, which results in wrinkling of the layers, and thus increasing the distance between the lattice planes. The grid terminates with carboxyl or hydroxyl groups. However, even if this model was universally accepted (which it is to a great extent), it does underline an uncertainty about the level of oxidation, something that relies heavily on the preparation method.

Apart from the structure of GO, there are significant properties that need to be taken into account when GO is used as a catalyst. The most important properties for the catalytic activity are the following.

- *Carbon-to-oxygen (C/O) atomic ratio.* The abundance of oxygen atoms in the material reflects the number of functional groups that were implanted on the carbon grid upon oxidation of the parent material. As mentioned in the description of the structure, there is a significant variation, in terms of functional groups, as well as their localization. The C/O ratio can vary between 2 : 1 and 3 : 1 depending on the manufacturing method [2, 10], while further thermal and/or chemical treatment can increase the ratio to 14 : 1, by decreasing the oxygen concentration due to reduction [11, 12]. The strict reproducibility of the C/O ratio, as well as the quality of the functional groups incorporated, is of utmost importance, as in most cases these functional groups are the ones responsible for the catalytic activity of GO, as discussed later. For this reason, synthesis techniques like the ones proposed by Compton *et al.* [13], which allows the fine tuning of the C/O ratio, and Marcano *et al.* [14], which allows high oxidation rate (up to 70%), are of extreme importance.

- *High surface area.* One of the major benefits of nanomaterials, which makes them so appealing not only for catalysis but also for other applications, is their high specific surface area, as usually determined by the Brunauer–Emmett–Teller (BET) method. This provides a high interaction area, compared to bulk materials, and – in combination with the abundance of functional groups on the surface of GO – a high proportion of catalytically active moieties per unit weight of material. The theoretical surface area of monolayer graphene is $2630\,m^2\,g^{-1}$ [15]. At the same time, the one calculated for GO is $890\,m^2\,g^{-1}$, while the experimentally determined value in aqueous solution is $736.6\,m^2\,g^{-1}$ [16]. The apparent value is in line with the theoretical one, once the agglomeration level is taken into account. The surface area can be increased again by reduction of the material, at the expense of the oxygen functionalities. For instance, activated reduced GO (RGO) films have a much higher specific area, of about $2400\,m^2\,g^{-1}$ [12], quite close to the theoretical value for monolayer graphene. Extremely high surface areas of up to $3100\,m^2\,g^{-1}$ were reported for microwave-exfoliated graphite oxide, activated after treatment with KOH [15]. The authors suggest that the fine-tuning of the surface area relies on the ratio of KOH to graphite oxide. Although it seems a paradox, higher surface areas compared to graphene are possible, by introducing heptagons and octagons into the carbon lattice, thus disrupting the typical six-membered rings – the sheet acquires a negative curvature and adopts a saddle shape [17].
- *Conductivity/capacitance.* The conductivity of GO and its derivatives is one of the properties that are easily tunable [18]. The fine-tuning of these properties is highly desirable for applications such as the design of supercapacitors and fuel cells. The high electrical conductivity and electron mobility of graphene results from a very small effective mass. The electrons behave as massless particles in the graphene lattice, traveling with a speed of about $10^6\,m\,s^{-1}$, granting an exceptional in-plane conductivity of about $20000\,S\,cm^{-1}$ [19] and an intrinsic mobility limit of $2\times10^5\,cm^2V^{-1}\,s^{-1}$, significantly higher compared to any silicon conductor [20]. Due to the disruption of the sp^2 bonding network, the conductivity of GO (of C/O ratio 2 : 1) is quite low, at the level of some $\mu S\,cm^{-1}$ [4, 21]; thus GO is often described as an insulator. The removal of the oxygen atoms via reduction can regain a significant portion of the conductivity, up to $5880\,S\,m^{-1}$ [12]. The removal of the oxygen functionalities of GO can significantly increase its capacitance, to levels comparable to those for pristine graphite or graphene (about $100–300\,F\,g^{-1}$) [22]. As a result of the fine-tuning of the C/O ratio, as well as the functionalization, GO and its derivatives are appealing materials for energy storage or supercapacitor fabrication.

12.3 Oxidative Activity

12.3.1 Oxidation Reactions of GO

Although the fact that pristine graphene can be used as a catalyst, using the delocalized π electron system to drive complexation reactions [23], the oxygen functionalities on the surface of GO make it much more appealing as a catalyst. These functional groups facilitate the catalysis of oxidation reactions, either by using other oxidizing agents, or by using their own oxygen atoms, as will be discussed later in this chapter. The reactivity is attributed to the high acidity of GO (pH 4.5 in aqueous media at 0.1 mg ml^{-1}) [11], and thus GO is often described as acting as a Lewis acid. Due to the nature of the functional groups of GO, most of the first studies concerning carbocatalysis with graphitic compounds focused on redox processes [24]. The reactions are performed at mild conditions, highlighting the potential of GO and its derivatives as catalysts. However, there are also some drawbacks in the use of GO as a catalyst. For instance, the catalytic efficiency in several cases is not high, requiring high loading (up to 400%) in typical oxidative reactions, while the non-stoichiometric and inhomogeneous nature of GO hinder the understanding of the detailed mechanism, thus making it more difficult to design novel reactions [25]. A significant part of the studies on the catalytic reactivity of GO has been performed by Dreyer, Bielawski and their coworkers, who evaluated the substrate spectrum of GO, as well as optimizing the processes for some of these reactions [26–31].

One of the typical examples of the oxidative reactivity of GO is the oxidation of benzyl alcohol to benzaldehyde (Table 12.1, entry 1), attaining conversions over 90% at 100 °C, and at the same time avoiding over-oxidation to benzoic acid [27]. Conversions to benzoic acid were observed mostly at higher temperatures (>100 °C). The non-preferred by-product could be avoided by performing the reaction at lower temperatures and using higher catalyst loadings to compensate for the lower productivity. Apart from benzyl alcohol, GO was found to be able to oxidize a broad spectrum of alcohols to the corresponding aldehydes or ketones [27]. Some of the compounds synthesized are presented in Table 12.1 (entries 1–6). Aromatic groups on the substrate are not a necessity for the catalytic activity of GO, as it also exhibits high catalytic activity toward cyclohexanol, a cyclic but not aromatic substrate. More interestingly, GO seems to be a selective catalyst, as it oxidized 2-thiophenemethanol into the respective aldehyde, but without any detectable sulfur oxidation (despite the fact that GO can oxidize sulfur compounds, as discussed later).

However, the oxidative potential of GO is not restricted only to alcohols. In the same work [27], the authors highlighted that GO can oxidize unsaturated hydrocarbons, as exemplified by the production of benzil from *cis*-stilbene, via a Wacker–Tsuji type of oxidation (Table 12.1, entry 7). The optimization showed that, for increased product yield, high GO loading up to 400% is needed and moderately high temperatures of about 100–120 °C, as minimal conversion was observed at lower temperatures (<60 °C) [30]. At temperatures over 120 °C, the conversion also decreases, something that the authors attribute to the thermal degradation/reduction of GO [26]. Interestingly, GO was inert toward *trans*-stilbenes, showcasing its selectivity as a catalyst. GO could also catalyze the oxidation of substituted *cis*-stilbenes with generally high yields (between 38% and 68%), once the substrates had electron-donating or moderately electron-withdrawing groups (Table 12.1, entries 8–11). Modest yield (25%) was observed for a substrate substituted with a strong electron-withdrawing nitro group, such as 4,4′-dinitro-*cis*-stilbene (Table 12.1, entry 12) [30]. Despite the fact that the yields attained are far from perfect, the catalytic potential of GO was exemplified in a broad set of substrates, and it seems to have potential as a substitute for metal catalysts, like selenium(IV) oxide, or chromate-containing catalysts, for the oxidation of stilbenes and 1,2-disubstituted olefins [30]. The low cost of GO and its availability are two of the advantages that could favor its use compared to metal catalysts.

Table 12.1 *Oxidative reactions catalyzed by GO. The efficiency is presented either in conversion [c] or yield [y], as described in the literature*

Entry	Substrate	Product	Efficiency	Ref.	Entry	Substrate	Product	Efficiency	Ref.
1	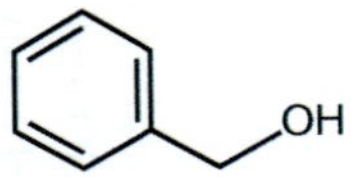	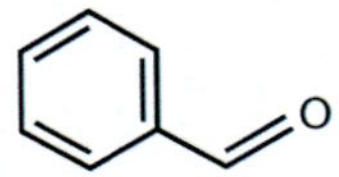	>90% [c]	[27]	14	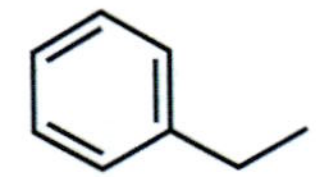	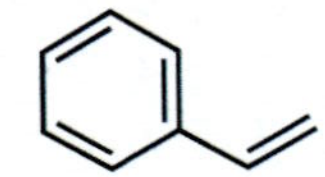	~40% [y]	[34]
2	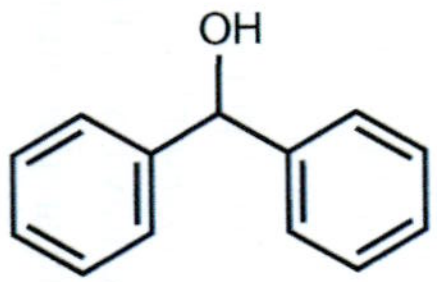	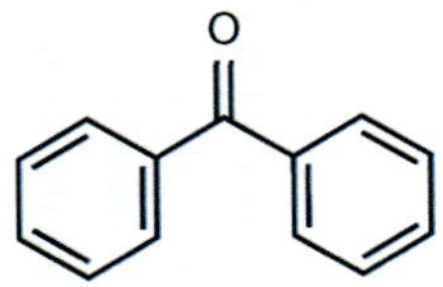	>98% [c]	[27]	15	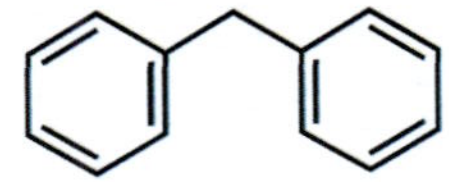	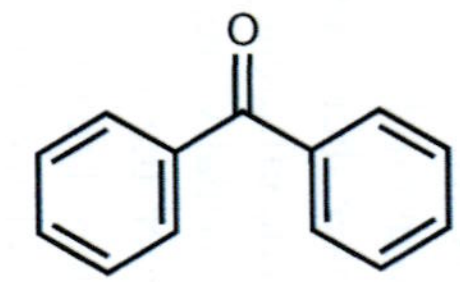	72% [y]	[30]
3	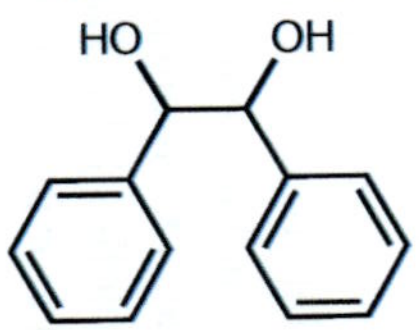	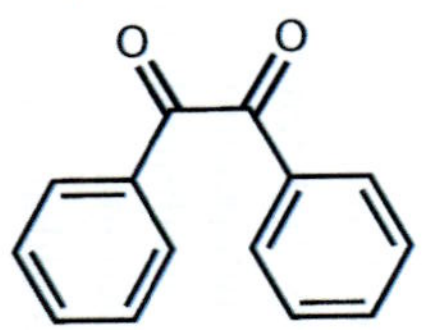	96% [c]	[27]	16	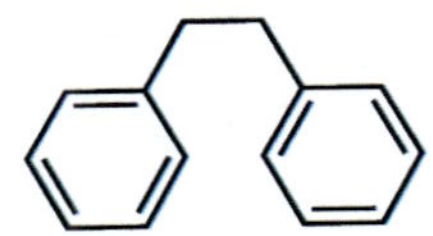	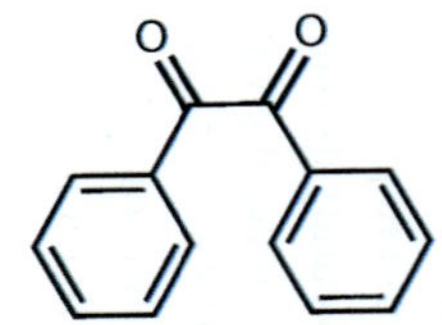	17% [y]	[30]
4	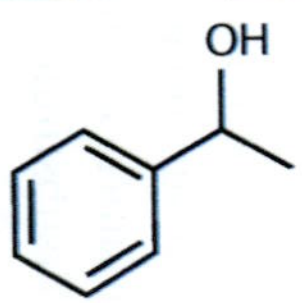	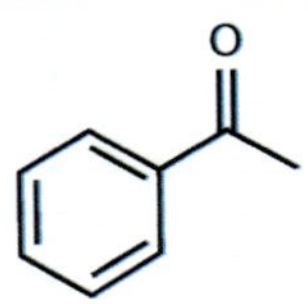	26% [c]	[27]	17	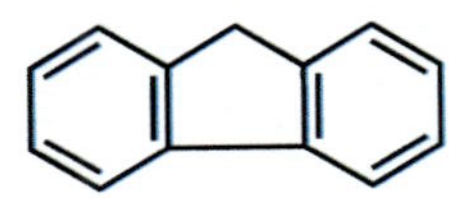	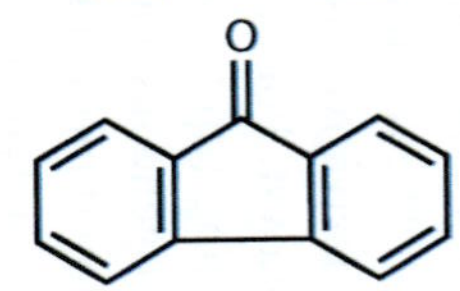	59% [y]	[30]
5	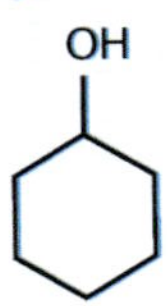	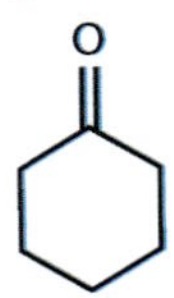	>98% [c]	[27]	18	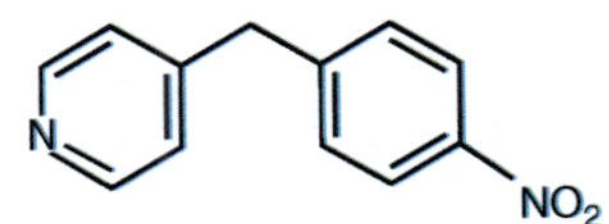	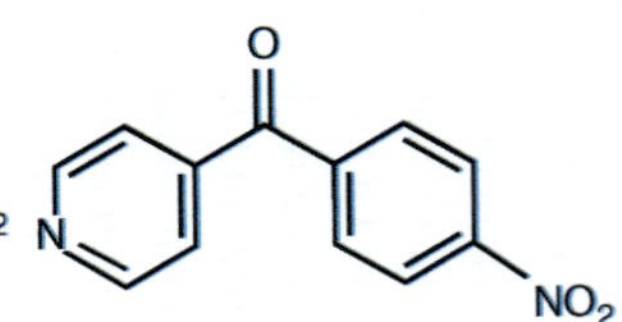	80% [y]	[30]

6	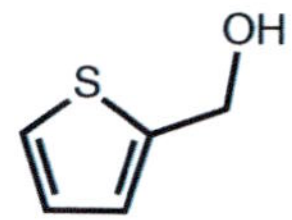	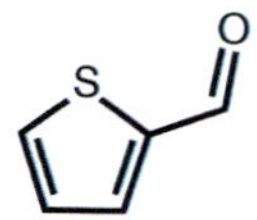	18% [c]	[27]	19	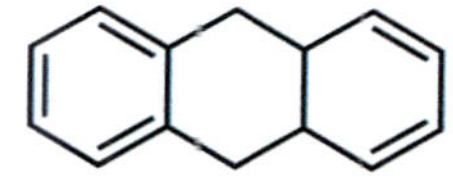	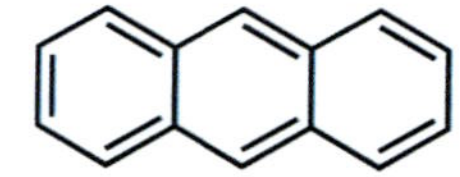	49% [y]	[30]
							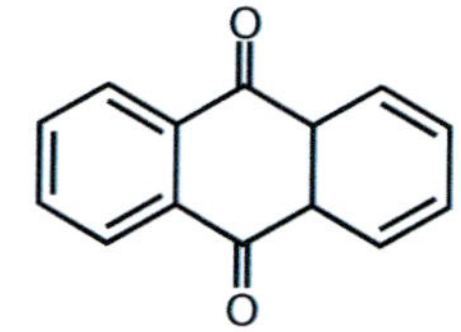	31% [y]	
7	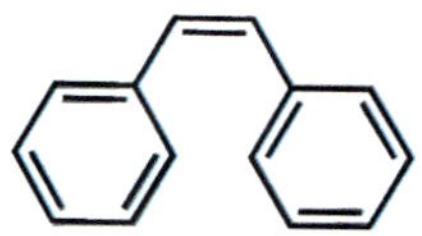	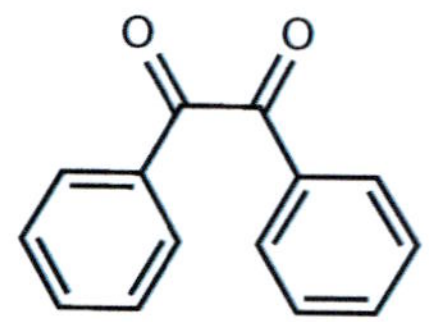	68% [y]	[30]	20	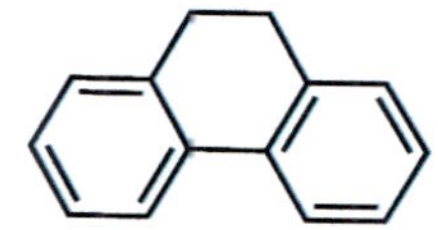	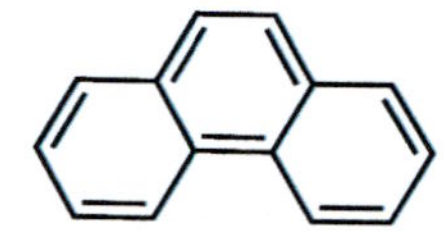	85% [y]	[30]
8	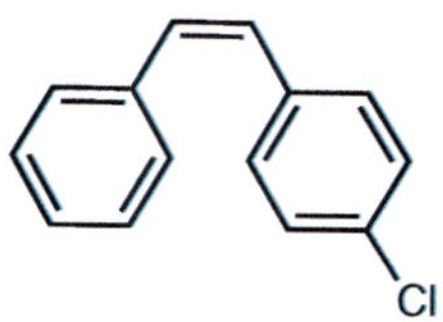	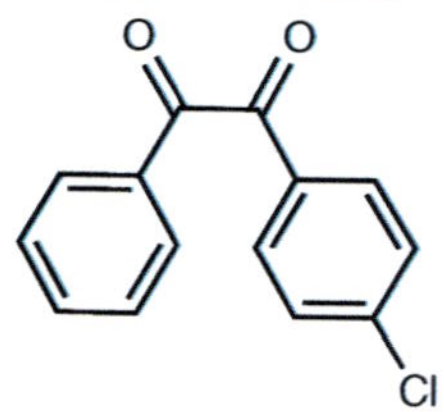	50% [y]	[30]	21	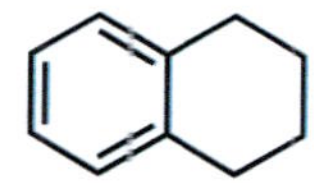	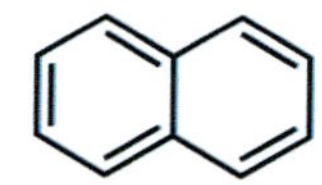	24% [y]	[30]
9	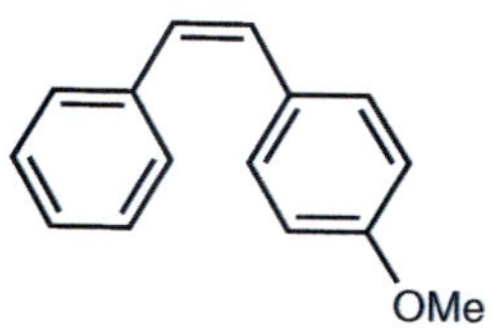	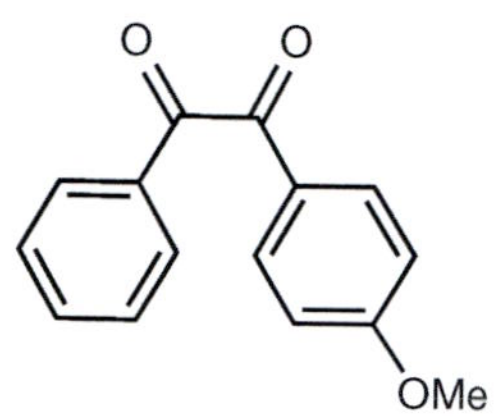	38% [y]	[30]	22	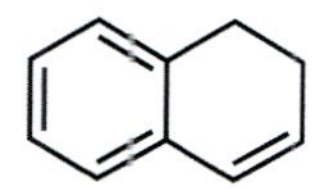	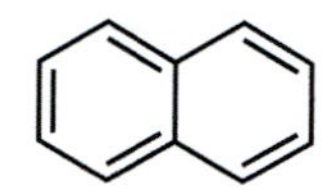	26% [y]	[30]

(*Continued*)

Table 12.1 *(Continued)*

Entry	Substrate	Product	Efficiency	Ref.	Entry	Substrate	Product	Efficiency	Ref.
10	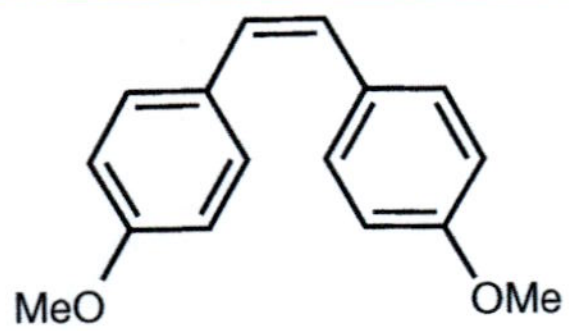	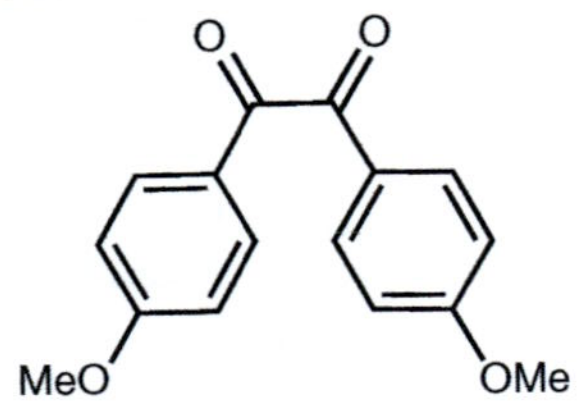	48% [y]	[30]	23	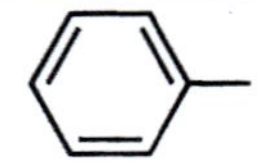	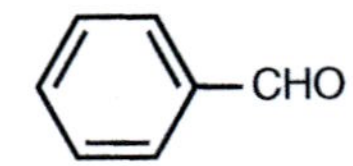	6% [y]	[30]
11	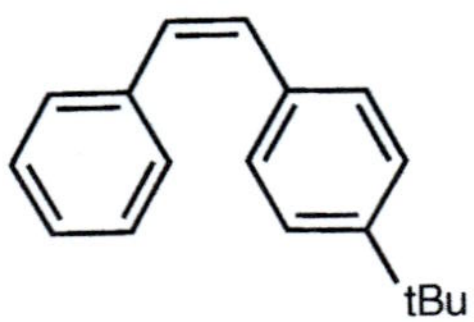	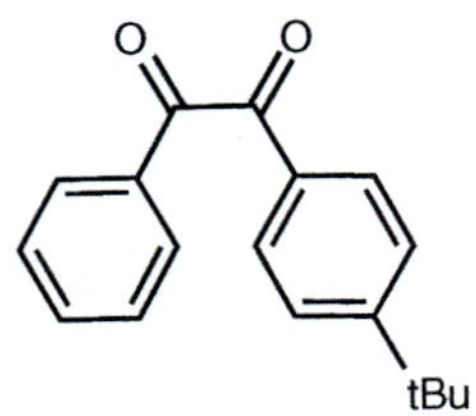	68% [y]	[30]	24	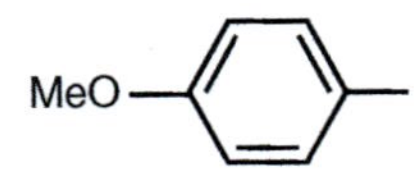	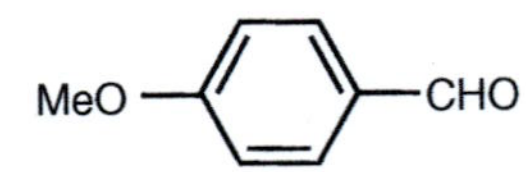	16% [y]	[30]
12	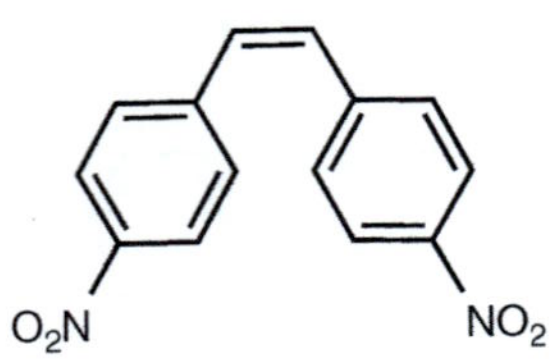	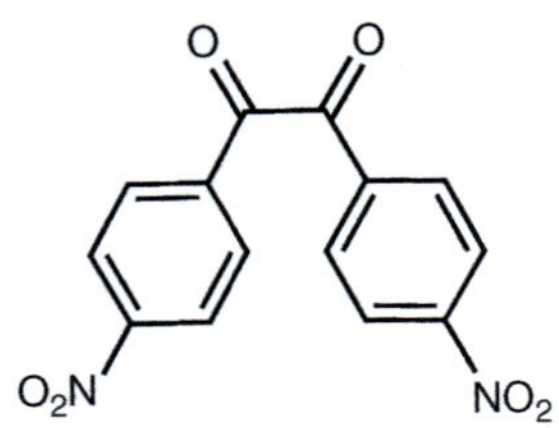	25% [y]	[30]	25	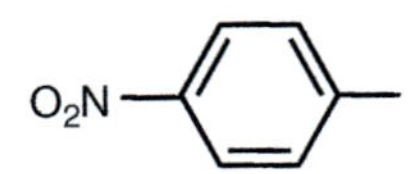	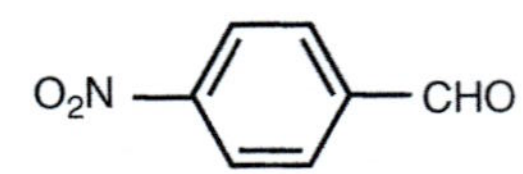	4% [y]	[30]
13	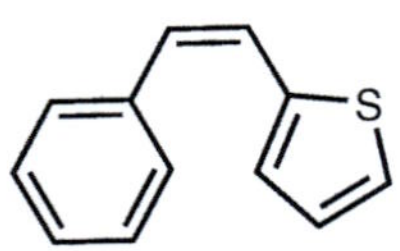	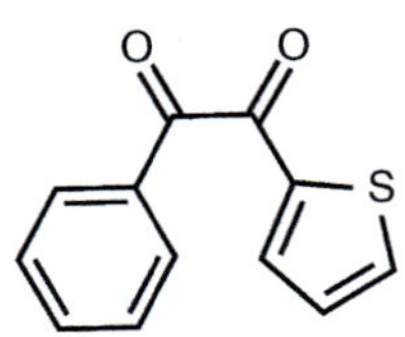	12% [y]	[30]					

In the aforementioned works, it was suggested that GO does not act as the oxidant, but it activates O_2 molecules, which in turn act as the final oxidant in these reactions. This hypothesis was based on the observation that the reaction does not proceed in a nitrogen atmosphere, and the oxygen content of the material remains intact after it catalyzed the reaction. Interestingly, the reaction was also not catalyzed by graphite or RGO, even in the presence of oxygen, highlighting the role of the functional groups on the surface of GO. It is suggested that the oxygen-rich groups that are abundant on the surface of GO promote the adsorption of O_2 [32]. However, there are also cases where GO acts not only as a catalyst but also as a co-substrate, implying a different catalytic mechanism. GO catalyzes the hydration of alkynes and aryl alkynes to the respective methyl ketones in solvent-free systems at ambient temperatures – a very interesting reaction and an alternative to the established oxymercuration reaction. However, in this case, a significant decrease of the oxygen content is observed, providing hints on the consumption of surface-bound oxygen during the reaction [27]. Although a competing thermal degradation process can occur at high loadings of GO [33], the diminishing of oxygen functionalities in catalytic reactions at ambient temperatures suggests that probably GO is also used as a co-substrate or oxidant.

In order to expand the study of the substrate scope, the catalytic efficiency of GO was tested toward more challenging substrates such as hydrocarbons that possess sp^3-hybridized C–H bonds. The oxidative dehydrogenation of hydrocarbons (ODH) is catalyzed by nucleophilic oxygen atoms on the carbon surface, located on prismatic edges of stacked graphene sheets or at defect surfaces, such as in the case of GO [35]. Most works in the field focus on the production of styrene via oxidative dehydrogenation of ethylbenzene, which is an industrially interesting reaction [35]. After further investigation of the specific reaction, it was found that the quinone groups of the GO are the ones responsible for the catalytic activity [34]. It needs to be mentioned, however, that this specific reaction is not challenging, as the stable conjugated system of styrene drives the reaction [4]. On the contrary, ODH of light alkanes has low selectivity and it is more challenging, as the C–H bond produced is weaker than the one present in the substrate [36]. GO is an efficient catalyst also for these substrates. For some substrates, the result of the oxidation was the corresponding ketone (Table 12.1, entries 15–18), while in other cases dehydrogenative aromatization was observed (Table 12.1, entries 20–22) [30]. Several substrates with benzylic methylene groups were successfully converted (Table 12.1, entries 23–25). When compounds with double activated benzylic positions were used as substrates, the corresponding ketone was formed as a product in good yields. The efficiency of GO as a catalyst for ODH reactions and its potential as a catalyst in industrial processes is highlighted by a very recently published patent [31]. The inventors suggest a process for the production of propene from propane, at 1 atm pressure, but high temperatures (400 °C). Using 250 mg of catalyst, only 15% conversion was observed after 48 h, but there were hints that equilibrium was not reached and that higher yields could be attained, to grant the system more efficiency. It is expected that the process can be more efficient than that described so far.

The elucidation of the catalytic mechanism from Tang and Cao [37] could help to achieve higher yields and push the barriers of industrial application. First-principles calculations were used to explore the plausible mechanisms for the oxidative dehydrogenation of propane to propene. As presented in Figure 12.1, two epoxide groups at the right distance on GO are needed, in order to activate the C–H bond of the substrate. In this proposed mechanism, the proton from one carbon of the substrate is initially abstracted by the first epoxide group,

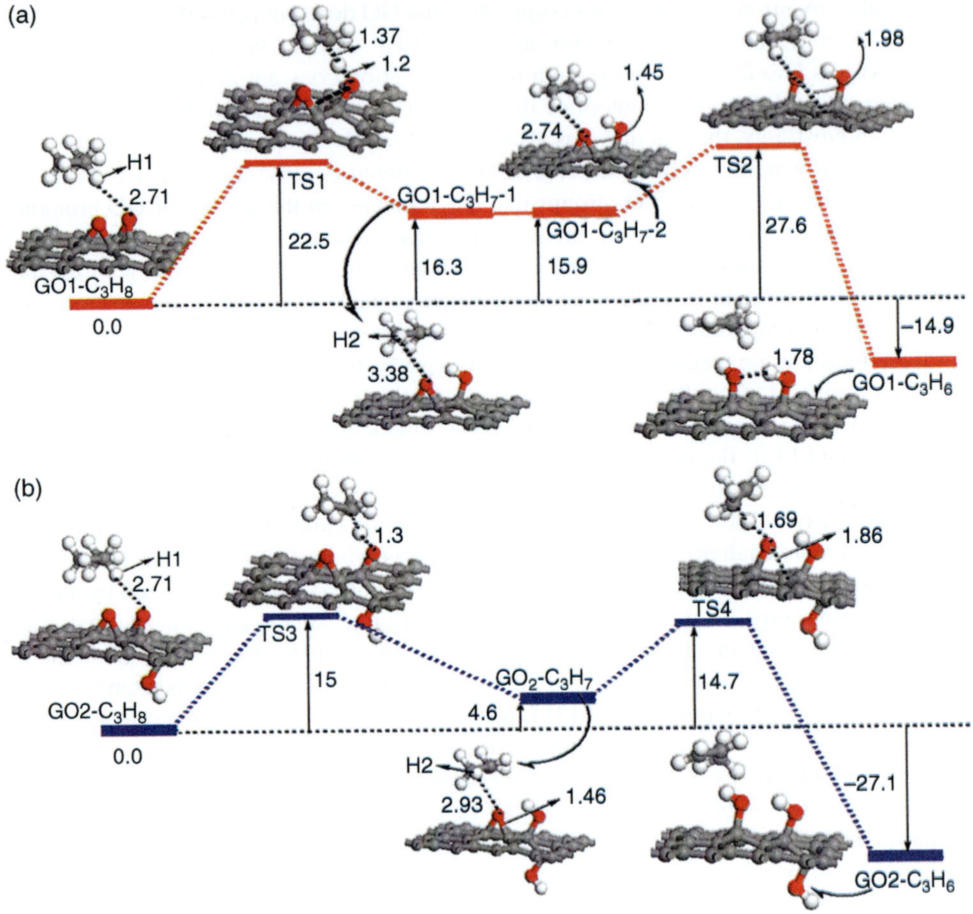

Figure 12.1 Relative energy profiles for ODH of propane to propene on (a) GO1 and (b) GO2 (GO1 with one added hydroxyl group at the opposite side). All energies (kcal mol^{-1}) in (a) and (b) are relative to the reactants GO1–C_3H_8 and GO2–C_3H_8, respectively, and the optimized configurations (distances, Å) of species involved in ODH are shown. Reproduced from [37] with permission of RSC

yielding the first transition state (TS1) with the intermediate radical C_3H_7. Then, the second epoxide abstracts the second proton from the radical, in order to release the final product. Propene is physically adsorbed on the GO surface after its release. In order to regenerate the GO catalyst, the hydroxyl groups formed on the surface of the GO can form H_2O and epoxy or a carbonyl group due to their physical vicinity [38].

Supporting the proposed mechanism, the calculations led to the conclusion that GO with modified oxygen-containing groups can afford high catalytic activity [37]. More specifically, it was calculated that the presence of hydroxyl groups around the epoxy groups, which act as the catalytic centers of the material, can enhance the C–H bond activation of the propane by lowering the energy barrier (Figure 12.1b). Thus, it was suggested that an external electric field can be applied to facilitate the diffusion of the oxygen functional groups on the surface, and subsequently orient them better for catalysis. This should lead to increased reactivity of GO toward oxidative dehydrogenation of propane.

The various oxidative reactions that are catalyzed by GO can be combined in tandem one-pot reactions to provide more complex compounds. In such an example, alcohol oxidation and alkyl hydration were combined to produce chalcones. Electron-rich and electron-poor aryl alkenes or methyl ketones were coupled with aryl alcohols or aldehydes to form the desired products, in acceptable isolated yields (>60%) [28]. For instance, the simultaneous oxidation of 4-phenylbenzyl alcohol and 4-ethynyl-1,1′-biphenyl leads to the spontaneous condensation of the products. It has been suggested that GO acts in the aforementioned reaction as an acid catalyst, in a Claisen–Schmidt condensation [39]. Another condensation reaction that is catalyzed by GO is the condensation of pyrroles with dialkyl ketones [40]. Typically, this reaction is catalyzed by Lewis acids [41]. However, GO was proven to be an efficient catalyst, changing also the product specificity depending on the organic solvent in which the reaction is performed. Another advantage of GO catalysis is the fact that GO is a heterogeneous catalyst, which facilitates the isolation of the products. For instance, in the previous example, the synthesized chalcones are soluble in CH_2Cl_2, and thus they can be easily separated from the hydrophilic GO, which would remain in the aqueous phase [28].

It needs to be highlighted that in all these reactions high loadings of the GO were used (in most cases a GO loading of 200%); however, the authors suggest that the low cost of GO as a catalyst – compared to classic transition metal catalysts – compensates for the higher loading. It remains to be proven if the downstream processing of such amounts would be economically affordable and that GO will find its way to industrial applications. So far there are some examples that show the potential of the application of GO as a catalyst. For instance, the selective oxygenation of acrolein to acrylic acid is described with no oxidation of the double bond. This reaction was performed quite efficiently using surface oxygen-functionalized graphitic carbon. The production rate exceeded 26 mmol g^{-1} h^{-1} [35].

Graphene oxide can also catalyze the oxidative coupling of amines to imines, with excellent yields (up to 98%) and, particularly in this case, low catalyst loading (5 wt.%) [25]. The high activity in this case seems to arise from the synergistic effect of the carboxylic acid groups and unpaired electrons that are located at the edge defects, facilitated by the sequential base–acid treatment of GO. The base reduction of GO generates nanovoids that are surrounded by conjugated domains. Thus, even more edge defects are generated in the center of the sheet and thus more carboxylic functionalities can form. Also, more unpaired electrons can be found [25]. The carbocatalyst is not reduced upon the reaction and can be used for several cycles. Using trapping molecules, the authors were able to prove the formation of superoxide radicals, leading to the conclusion that hydrogen peroxide is formed as an intermediate, as usually occurs in aerobic oxidation reactions. The suggested mechanism is presented at Figure 12.2.

12.3.2 Oxidation of Sulfur Compounds

GO and its derivatives have been successfully used for the oxidation of sulfur compounds. GO foam and dispersions of GO in aqueous medium were used to oxidize SO_2 to SO_3 in an environmentally friendly way, avoiding noble metal catalysts [42]. The oxidation of the

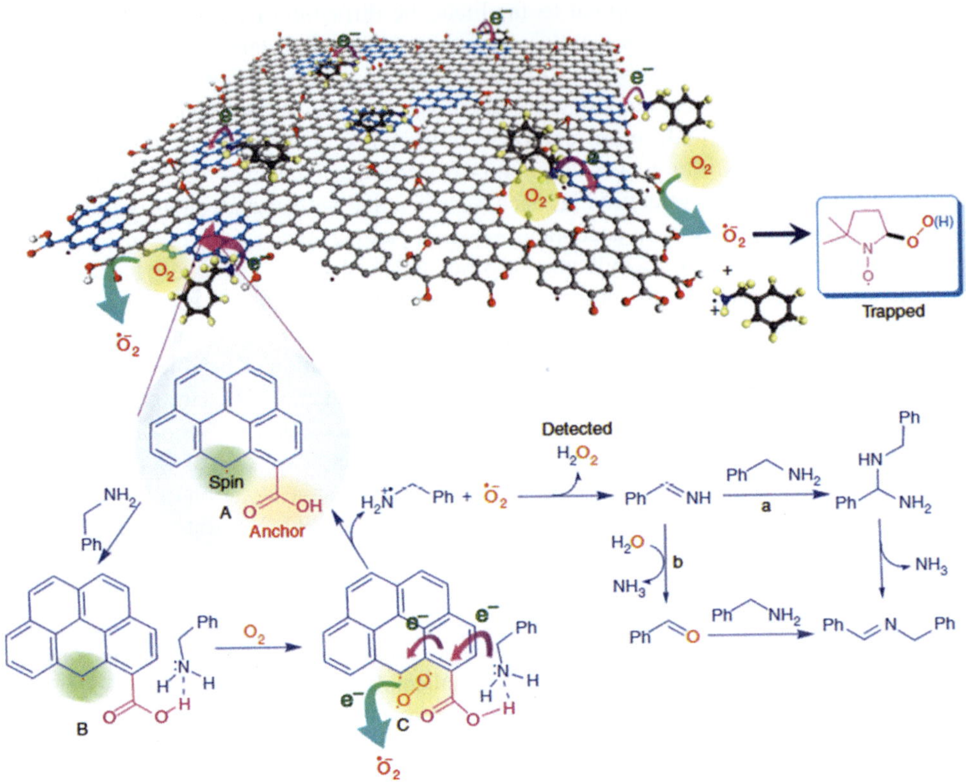

Figure 12.2 Schematic representation of the oxidation mechanism of basified GO-catalyzed oxidative coupling of primary amines. Reproduced from [25] with permission of Nature Publishing Group

gaseous SO_2 proceeds using both molecular oxygen and the oxygen functionalities of the GO as oxidant, resulting in SO_3, which in water transforms spontaneously to sulfuric acid. The authors base their suggestion that GO acts as an oxidant on two macroscopic observations: (i) a gradual change of color, from brown to black; and (ii) precipitation, which was assigned to increase of hydrophobicity due to the reduction. Both indications lead to the conclusion that the functional groups are consumed and the GO is gradually reduced.

The application of GO for the oxidation of sulfur compounds was also exemplified in the oxidation of thioanisole, an industrially interesting reaction used for the desulfurization of liquid fuels. GO is a perfect catalyst for this reaction, as high yields are attained at room temperature, with a catalyst-to-substrate weight ratio of about 0.11 [43]. Interestingly, the final oxidant in this case was not the GO itself, but the H_2O_2 supplied. It was proven that the reaction proceeds via $^{\bullet}OH$ radicals, formed from the homolytic GO-mediated cleavage of H_2O_2. The free radicals produced are then able to oxidize the organosulfur compounds [43]. Interestingly, the three-dimensional (3D) GO foam is a more efficient catalyst compared to the 2D GO, reaching conversions of about 85%, while the 2D material had only 65% conversion under the same conditions. In addition to that, the selectivity of the 3D GO was better compared to the 2D counterpart, as it produced the sulfoxide in high excess, compared to the over-oxidized sulfone (about 90% of the total products, in most cases). Although the mechanism of the selectivity is unclear, a hypothesis was proposed that the 3D GO is a milder catalyst than the 2D GO, as it undergoes reduction to some extent during the formation of the 3D foam. On the contrary, 2D GO has more oxygen functional groups on its surface, thus being more reactive and leading to over-oxidation of the substrate. The 3D GO foam was proved to be an efficient catalyst also for the oxidation of diphenyl sulfide, although yielding lower conversions (about 45% in the same time interval and under the same conditions), probably due to steric hindrance. It is still unclear if steric hindrance or partial reduction is responsible for the lower activity. However, Dreyer and coworkers could reach an isolated yield of 86% with 2D GO [29]. In fact, in this specific work, the potential of GO as oxidant was exemplified with a set of several sulfides, always leading to the corresponding sulfoxide, with high yields (>51%).

More than that, GO has a good potential for use as an oxidant in the oxidation of thiols [29]. First, GO was used for the oxidation of thiophenol, in order to produce diphenyl disulfide quantitatively in only 10 min, which is a significantly lower reaction time compared to the reaction with metal catalysts, and with a moderate catalyst loading of 60%. After proving the benefit of the use of GO as a catalyst in this reaction, the oxidation of several aliphatic and aromatic thiols was showcased [29]. In all cases, a fast reaction was observed with high conversion rates (>75% in less that 30 min). The deoxygenation of GO was confirmed, so the catalyst was used also as an oxidant, hindering its reuse.

12.3.3 Functionalized Materials

Apart from unmodified GO, its derivatives can also exhibit interesting oxidizing activities. GO has a variety of oxygen-rich functional groups, which can be used for further functionalization in order to incorporate interesting groups for catalysis. Herein, some examples are presented to show the potential of functionalized GO.

Carboxyl-modified GO (GO-COOH) has intrinsic peroxidase-like activity, catalyzing the oxidation of 3,3,5,5-tetramethylbenzidine (TMB) in the presence of hydrogen peroxide [44]. The catalytic activity of this material is dependent on pH and temperature, while the hydrogen peroxide concentration also affects the activity. Interestingly, the catalytic efficiency of GO-COOH was higher than that observed for the same reaction using horseradish peroxidase, despite the fact that they share the same "ping-pong" catalytic mechanism. As long as the catalytic activity was affected linearly by the H_2O_2 concentration, the material was used for the detection of peroxide released from enzymes. For instance, a glucose oxidase released hydrogen peroxide from the oxidation of glucose, which was subsequently detected by GO-COOH. A glucose detector was fabricated in this way, with a detection limit of 1 μM and a linear range of 1–20 μM [44]. Taking advantage of the same principle, other detectors could be fabricated (e.g. for the detection of uric acid or xanthine), once the proper oxidase is used. Compared to the enzymatic electrodes that use horseradish peroxidase for the detection of hydrogen peroxide, GO-COOH is more stable and much cheaper to produce. Concerning the catalytic mechanism, it was suggested that TMB is adsorbed on the surface of the GO-COOH and donates the lone-pair electrons of the amino groups to the GO-COOH, which increases the electron density and mobility [45]. This increases the Fermi level and therefore the electrochemical potential from the lowest unoccupied molecular orbital of H_2O_2; this leads to the acceleration of the electron transfer from GO-COOH to H_2O_2.

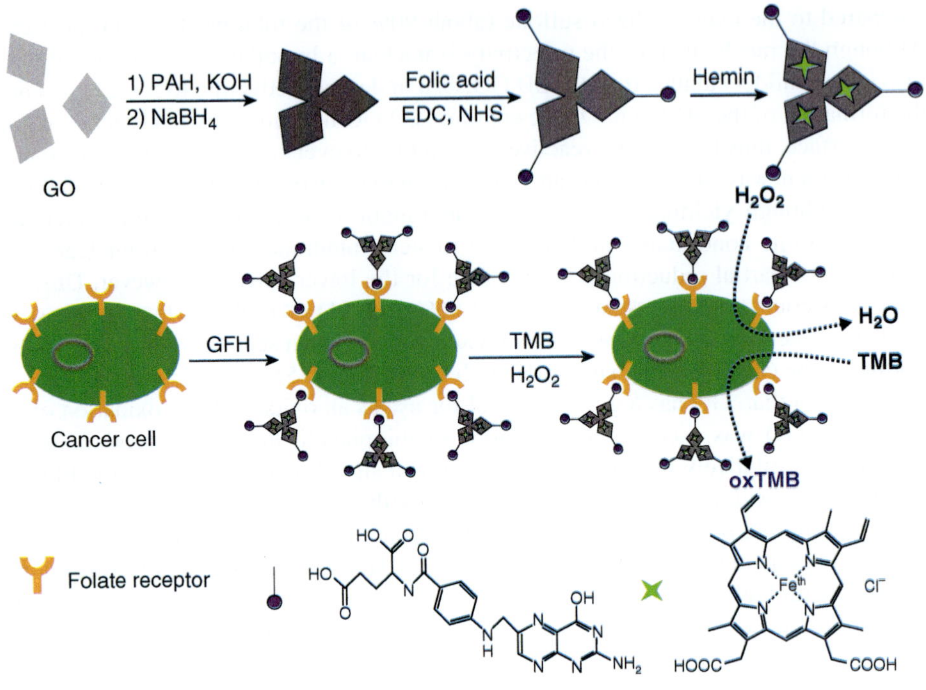

Figure 12.3 *Illustration of graphene–hemin composite (GFH) preparation and cancer cell detection by using target-directed GFH. Reproduced from [49] with permission of RSC*

While carboxyl functional groups are also present in pristine GO, other functional groups can be incorporated. For instance, RGO was sulfonated using 4-benzenediazonium sulfonate, in order to perform hydrolytic reactions [46]. The functionalized GO was tested for the hydrolysis of ethyl acetate. Sulfuric acid was a better catalyst, but the reuse of sulfonated GO and the ease of separation makes it an interesting alternative for hydrolysis or esterification reactions.

In a more sophisticated approach, the intrinsic peroxidase activity of GO was utilized to prepare a detector of cancer cells [47]. Taking into consideration the fact that folate receptors are overexpressed in several types of cancer cells, GO functionalized with folic acid was prepared, in order to be recognized by the cells. At the same time, the nanomaterial was also functionalized with hemin to increase the catalytic activity, especially at physiological values of pH, where the peroxidase activity of GO-COOH is negligible [44]. Hemin functionalization is facilitated by effective binding via π–π interactions [48], while a synergistic effect is observed for the activity of the hybrid system (Figure 12.3) [49]. This detector was able to detect as few as 1000 cells.

12.4 Polymerization

The acidity of GO is exploited by using this nanomaterial as a Lewis or Brønsted acid catalyst for the formation of polymers. The mechanism of polymerization may vary and, to date, three different ways of preparing polymers have been reported. As the GO is active toward olefins, an acid-initiated polymerization process of n-butyl vinyl ether in a solvent-free system was established [50], which is typical for the first described mechanism. The polymerization is a rapid and exothermic reaction, which without optimization was completed after 14 h with 100% conversion of the monomer, yielding a polymer with an average molecular weight of 5.2 kDa (about 50 monomers combined) and a polydispersity index (PDI) of 9.42. After optimization, the reaction time could be significantly decreased to 4 h, while polymers with average molecular weight up to 8.1 kDa were obtained. Interestingly, and in contrast to the oxidative reactivity toward olefins, the loading of the heterogeneous catalyst was quite low (0.1–5.0 wt.%). The GO does not seem to act as a co-substrate, as the catalyst could be used consecutively five times while retaining about 90% of its initial conversion rate. As in the case of the oxidative reactivity toward olefins, neither pristine graphite nor RGO were able to catalyze the polymerization reaction. The potential of GO as a heterogeneous acid catalyst for the polymerization of olefins via cationic pathways was exemplified also with other substrates. For example, *N*-vinylcarbazole was polymerized as efficiently as n-butyl vinyl ether, with completion of the reaction after 4 h and an average molecular weight of 1.9 kDa. Styrene was also polymerized using GO as a catalyst; however, despite the high conversion of the monomer (>90%), only low-molecular-weight oligomers were formed (about four monomers per molecule).

GO was also used for the polymerization of benzyl alcohol, in order to produce poly(phenylene methylene) (PPM). PPM is usually produced using strong acids [51], yielding highly branched products due to the presence of multiple reactive sites on the aromatic ring [52]. The polymerization takes place through a step-growth dehydrative polymerization reaction, the second polymerization mechanism catalyzed by GO, which is presented in Figure 12.4. In the end, a highly branched product is yielded with an average molecular weight of 2.3 kDa and a PDI of 1.26 [53]. The process is quite straightforward and requires only 10 wt.% of GO; the water produced creates a second phase and can be easily removed, while the GO can be separated from the polymer by extraction in organic solvents. In this case of polymerization reactions, the reduction of GO is observed, probably due to its use as oxidant, as in the case of the oxidation of benzyl alcohol.

The potency of GO to catalyze polymerization reactions is not restricted to substrates that were transformed also in oxidative reactions. GO can catalyze the ring-opening polymerization of several lactones and lactams, yielding the respective polyesters and polyamides, using the third mechanism of polymerization described. Using only 2.5 wt.% GO, poly(ε-caprolactone) was produced with an average molecular weight of 5.1 kDa and a PDI of 2.1, with yields up to 91% [54]. Although polymers up to 50 kDa were produced using metal catalysts, their fine-tuning is challenging, and in most cases a comparable product is

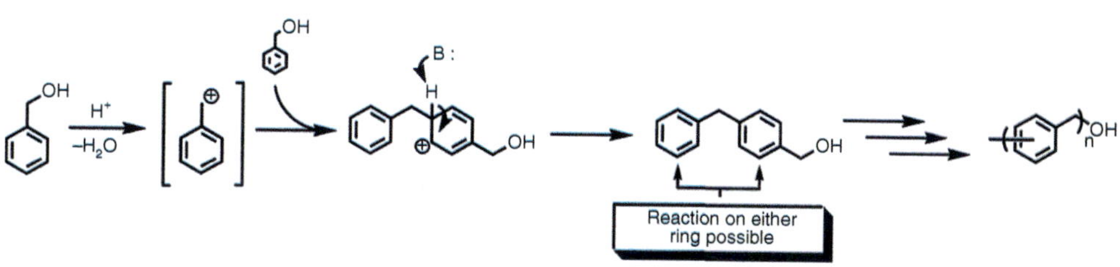

Figure 12.4 Proposed structure and formation of poly(phenylene methylene) synthesized via an acid-catalyzed dehydration process [51]. Reproduced from [53] with permission of ACS

generated (typically less than 5 kDa) [55]. Optimization of the process did not lead to the increase of the polymer size and thus the authors concluded that the polymer reaches an equilibrium chain length under the given conditions [54]. GO is an efficient catalyst for the polymerization of other lactones also. Poly(δ-valerolactone) with an average molecular weight of 10.2 kDa and a PDI of 1.6 was prepared with a yield of 86.2%, using only 2.5 wt.% of GO. The ring-opening polymerization reactivity of GO can also be expanded to lactams, for the preparation of aliphatic polyamides. However, the acidity and the hygroscopicity of GO is a hindering factor, as, for instance the water can lead to premature termination. To surpass this problem, GO basified with triethylamine (TEA) was prepared for such reactions. This functionalization retained the oxygen functional groups on the surface of the GO, granting the material neutrality while retaining the oxygen functional groups. Nylon-6 of about 15 kDa could be produced from the polymerization of ε-caprolactam, at a yield of 70%, using this TEA-treated GO [54]. A really interesting aspect of this work is that, during the catalysis, GO adopted a different conformation and transformed to multi-walled fullerenes during the catalysis, a fact that probably also affected the efficiency of these reactions.

12.5 Oxygen Reduction Reaction

Fuel cells are electrochemical devices that convert hydrogen and oxygen to water and heat in order to produce energy. In a simple form, a fuel cell has an anode and a cathode, with electrolyte in between. At the anode, supplied hydrogen reacts with the catalyst, producing a proton and an electron. The proton passes to the electrolyte, while the electron travels through a circuit to produce current. In the cathode, oxygen reacts with the hydrogen cation and the electron, to form water and heat. The oxygen reduction reactions (ORRs) can follow two distinctive pathways: either the four-electron transfer pathway, or the peroxidase pathway, as described in Table 12.2. The strong bond between the atoms of molecular oxygen makes the ORRs kinetically very sluggish, and therefore efficient catalysts for the reduction of oxygen are needed for the fabrication of the cathode [56]. Usually, metal catalysts are used (especially Pt); however, the diminishing supply of such metals used in industrial processes creates a need for novel materials to substitute them. More than the high price and the diminishing availability, there are also other serious disadvantages that come with the use of metal catalysts, such as their CO poisoning and sintering [4].

Originally, GO derivatives were used as support materials for the fabrication of electrodes. A fuel cell was prepared with a hybrid catalyst composed of Pt/RGO as the cathode material, and Pt dispersed on carbon black as the anode. The partially discharged Pt/RGO-based fuel cell delivered up to 161 mW cm^{-2}, significantly more than the analog without RGO as support material, highlighting the benefit of RGO usage [57].

Table 12.2 Reaction pathways of ORR in aqueous electrolytes

	Alkaline medium	Acidic medium
Four-electron transfer pathway	$O_2 + 2H_2O + 4e^- \rightarrow 4OH^-$	$O_2 + 4H^+ + 4e^- \rightarrow 2H_2O$
Peroxidase pathway	$O_2 + H_2O + 2e^- \rightarrow HO_2^- + OH^-$	$O_2 + 2H^+ + 2e^- \rightarrow H_2O_2$
	$HO_2^- + H_2O + 2e^- \rightarrow 3OH^-$	$H_2O_2 + 2H^+ + 2e^- \rightarrow 2H_2O$

To use 2D carbon materials for ORR applications, the electroneutrality of the sp^2-hybridized carbons of graphene or pristine graphite needs to be broken, but at the same time the π electrons of the lattice need to be used efficiently [58]. The abundant functional groups of GO make this carbon nanomaterial really appealing as a catalyst for ORRs. GO is favorable for the design of cathodes, as the abundant oxygen-containing groups facilitate the adsorption and subsequent reduction of O_2 [32]. The electrocatalytic activity was also decreased significantly when the oxygen functionalities were removed by thermal treatment, suggesting that these are responsible for facilitating the electrocatalysis of O_2.

The atoms that disrupt electroneutrality do not necessarily need to be oxygen atoms. At this point, a clear distinction between doping and functionalization is needed, as in the literature there seems to be some confusion between the two terms. Doping implies the *substitution* of sp^2-hybridized carbon atoms for a heteroatom. However, not all heteroatoms can be conjugated. For instance, phosphorus atoms are not coplanar, so they cannot be conjugated on the lattice, while oxygen atoms have π electrons. However, none of them are "graphitic" or "quaternary" in GO and thus we are speaking about functionalization [59, 60]. Taking this into consideration, the materials will be referred to as in the original publications, subject to reservations, due to the insufficient characterization in some cases, which does not allow us to distinguish between functionalization and doping.

S-doped graphenes, with a sulfur content of 1.3–1.5%, were produced by annealing GO with benzyl disulfide [61]. In another work, poly(diallyldimethylammonium chloride) (PDDA) was used to functionalize RGO [62]. These materials exhibited a better catalytic activity (9.3 mA cm^{-2} at −0.8 V for the S-doped graphene; 8 mA cm^{-2} at −1.0 V for the PDDA/RGO) than the commercial Pt/C catalyst (4.5 mA cm^{-2} at −0.8 V). The onset potentials for both carbon-based nanomaterials were similar and the number of electrons transferred was about 3.8, underlying the superior catalytic efficiency of the S-doped graphene compared to the widely used metal catalysts. However, the advantages do not stop only at catalytic efficiency. Both modified GO materials were stable in 3 M methanol or 0.1 M KOH, exhibiting no crossover effects, while PDDA/RGO is also tolerant to CO poisoning [62]. In an effort to explain the beneficial effect of PDDA/RGO, Wang and coworkers suggested that the PDDA charges the surface positively, thus increasing the diffusion of oxygen to the surface of the cathode [62]. The functionalized RGO was superior to graphene, as graphene had an electron transfer number of only 1.5, while at the same voltage PDDA/RGO could transfer 3.5 electrons.

In another work, Lee and coworkers suggested that ORR activity is increased after reduction of GO with hydrazine [63], albeit this seems in contrast to what was suggested before about the disruption of electroneutrality. It was shown that the reduction of GO with hydrazine catalyzes also the incorporation of nitrogen atoms on the graphitic layer, with a mechanism that until now is not fully understood [64]. The nitrogen content is much lower than the oxygen content of the pristine material (GO); however, the nitrogen atoms facilitate the oxygen reduction when inserted into carbon-based nanomaterials. The elucidation of the mechanism (Figure 12.5) [65] revealed that the edge structures lower the oxygen adsorption barrier and the first electron transfer barrier significantly, which are rate-limiting factors for ORRs. However, this effect is also observed for GO. The nitrogen doping improves the ORR performance by further enhancing the first electron transfer rate, as well as the preference for the four-electron reduction pathway. Thus nitrogen inserted next to the edge carbon (N0) was proposed to be the main active site in N-doped RGO, compared to nitrogens that are further from the edge.

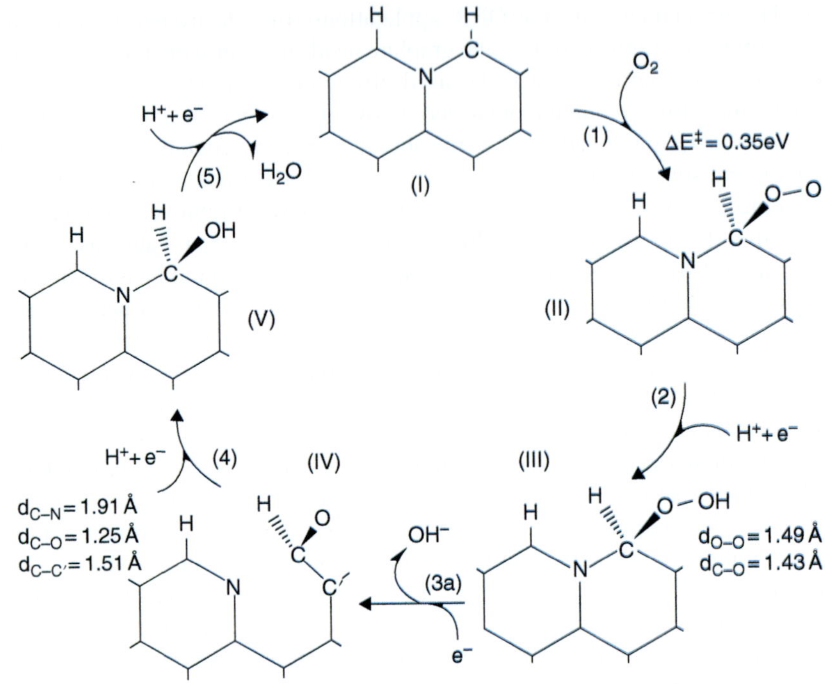

Figure 12.5 The proposed ORR catalytic cycle for the N0 structure, showing only the catalytically active part of the graphene nanoribbon edge. Reproduced from [65] with permission of RSC

The advantage of GO and other graphene derivatives as cathode catalysts was exemplified in fuel cells, but the application of these materials is not restricted to those. GO in a mixture with carbon nanotubes was tested as the cathode in aprotic aqueous hybrid electrolyte Li–air batteries, with the mixture of GO and carbon nanotubes in the ratio 1 : 4 yielding the best results [32]. To underline the significance of the oxygen-bearing functional groups, thermal treatment was applied and the C/O ratio increased, diminishing at the same time the efficiency of the cathode. This result highlights the fact that the presence of oxygen atoms on the carbon lattice seems to be beneficial, and supports the theory of better adsorption of O_2 on such nanomaterials.

A correlated application in the energy field is the use of 2D carbon-based nanomaterials as supercapacitors. In this case, the material needs to have the ability to store electrons, which can be released upon discharge. Supercapacitors are advantageous compared to the conventional lithium-ion batteries, as they have large power delivery capability, fast charge and discharge rates and long life cycles [66]. Their major problem is low energy density; however, novel nanomaterials can overcome this by their high specific surface area. For instance, in the case of single-layer graphene, the maximum electric capacitance was estimated at ~550 $F\,g^{-1}$ [67]. For this application, RGO is a better material compared to as-prepared GO, due to its surface area and the capacity to store electrons and discharge them in a non-Faradaical way. For this reason, several protocols for the reduction of GO were investigated in order to find the optimum for capacitor fabrication [22]. The production method of the RGO significantly alters the performance of the material. As just as one example of this differentiation, two works will be mentioned. On the one hand, thermally processed GO (tpGO) obtained by annealing at 200 °C exhibited a maximum specific capacity of 315 F g^{-1}, one of the highest values reported so far [68]. On the other hand, RGO prepared with hydrazine had a maximum specific capacity of only 205 F g^{-1}, but was stable for over 1200 charge–discharge cycles, maintaining 90% of its initial capacitance [69]. The use of imidazole-based ionic liquids facilitated the reuse of the electrodes, and electrodes in butyl methyl imidazolium with hexafluorophosphate or tetrafluoroborate retained their initial specific capacity even after 3000 cycles [70].

12.6 Friedel–Crafts and Michael Additions

The Friedel–Crafts reaction is an important reaction for the formation of new C–C bonds on aromatic systems, which is usually catalyzed by Lewis acids. GO was successfully used to catalyze the Friedel–Crafts addition of indoles to α,β-unsaturated ketones [71]. The existing methods require metal catalysts that yield low turnovers and need stoichiometric amounts of the reagents, while the downstream processing is cumbersome [72, 73]. GO was used as catalyst to perform the indole addition to methyl vinyl ketone. High isolated yields of over 90% were attained within 3 h with only 20 wt.% of catalyst in an H_2O/THF (tetrahydrofuran) system, while the increase of the catalyst loading to 50% significantly reduced the reaction time needed [71]. Interestingly, the addition took place only on the β-position, while the NH moiety was not protected. The superior catalytic activity of GO was exemplified by comparing its activity with the performance of several other catalytic systems. Several substituted indoles were tested and the complete conversion was monitored. At latest after 8 h the reaction was complete, without the formation of polymers of dimerization products. The only exception was the case of α-substituted indoles, where no conversion was detected. The reaction could also be performed with the much less reactive electron-deficient substituted nitrostyrenes. However, to get moderate yields, a rather long reaction time was required. In all these reactions, GO does not seem to be reduced, and it could be reused up to five times without losing any of its activity, reflecting high recovery rates.

The catalytic efficiency of GO as a phase transition catalyst in a Michael addition reaction was tested for the nucleophilic addition of 2,4-pentanedione to *trans*-β-nitrostyrene, and was compared to an established catalyst for this reaction, namely 18-crown-6 ether [74]. In general, GO was a better catalyst than the crown ether, which was attributed to the affinity to the cation, making the hydroxide a stronger base in an organic phase. The method was further explored with other substrates, such as several substituted *trans*-β-olefins and 1,3-dicarbonyl compounds, yielding fairly good conversions but only a mediocre enantiomeric excess of ~7% [74].

GO can also be used to form new carbon–heteroatom bonds via Michael addition. The catalytic activity was highlighted for the addition of amines to activated alkenes via aza-Michael addition [75]. It needs to be stated that the model reactions that were selected for the study needed no activation and also proceed under the suggested experimental conditions within 35 min, without the addition of GO. However, GO could significantly speed up the addition of diethylamine to acrylonitrile, while RGO was almost unreactive. The

GO facilitated the reaction without getting reduced and retaining its catalytic efficiency even after nine runs. The GO can be used also for the addition of other primary and secondary amines to several α,β-unsaturated compounds, such as acrylamide, methyl acrylate, acrylonitrile, etc., to yield the corresponding β-amino compound, without any indication of by-products.

12.7 Photocatalysis

Graphene oxide with an appropriate oxidation level can be used to catalyze water splitting under ultraviolet (UV) or visible irradiation, for stable H_2 generation [76]. The hydrophilic character of GO facilitates good dispersion in aqueous media. Despite the fact that GO is reduced through the process, the energy level of the conduction band edge is high enough to supply overpotential for H_2 generation and thus the productivity of hydrogen is not diminished [76]. The reactivity of GO with sunlight is so high that GO can undergo spontaneous photolysis, yielding CO_2 and RGO [77]. It is of utmost importance to tame this reactivity for the benefit of some interesting reactions. For instance, GO was used with Rose Bengal to facilitate the photo-oxidation of tertiary amines [78]. Although the GO by itself is a very poor catalyst (yielding only 10%), it has synergistic effects with Rose Bengal, an organic dye that has a strong absorption at about 550 nm. With this hybrid the photo-oxidation of tertiary amines is driven to completion. The photo-reactivity of GO with Rose Bengal was proven in a diverse set of cyclic tertiary amines and trialkyl amines, yielding the respective α-cyanide. In general, the potential of graphite and graphene materials for photocatalysis is recognized and possible applications (for instance, in solar cells) are expected to emerge in the years to come.

12.8 Catalytic Activity of Other Layered Carbon-Based Materials and Hybrid Materials of GO

12.8.1 Non-Functionalized Carbon-Based Nanomaterials

Despite the fact that graphite and graphene consist only of sp^2-hybridized carbon, they can also exhibit some catalytic activity. A very interesting review about the biological and environmental applications of graphene and some of its derivatives was recently published [79]. Here, only some examples of catalytically active graphite will be mentioned to highlight the potential of this layered carbon-based material as a catalyst. Graphite is known to catalyze the cleavage of alkyl or aromatic ethers using acyl halides to yield the corresponding ester [80]. It was suggested that the cationic intermediates are stabilized by π–π interactions and the reaction is catalyzed through a Lewis acid type of mechanism [80]. This suggestion is also supported by the fact that graphite can efficiently catalyze (yield up to 99%) an effect similar to the one that graphite has in alkylation of aromatic compounds and primary alcohols, which furnished different diphenyl methanes [81]. However, Schaetz *et al.* [24] suggest that the reactions are S_N1-type reactions, as graphite was not able to cleave the primary and secondary alkyl ethers. Moreover, graphite is able to catalyze the [4+2]-cycloaddition reaction of anthracene with various electron-deficient dienophiles, facilitated by microwave irradiation [82]. In this case the advantage of graphite is its high thermal conductivity of 19 W cm^{-1} K^{-1} at 300 K [83]. Finally, graphite can catalyze the reduction of nitroarenes, using hydrazine hydrate as the terminal reductant [84].

12.8.2 Hybrid Catalysts and Alternative Applications

Although the main focus of this book is GO, the catalytic applications of hybrid materials made with GO or its derivatives will also be discussed. Some hybrid materials were already described in the previous subsection. Interesting properties of this nanomaterial can be combined with other catalysts, leading to new hybrid catalysts. In this subsection, some indicative works in important fields of research will be discussed to highlight the potential of such systems. However, there are far more research reports in each field available that describe the hybrids in detail.

12.8.2.1 Metal Clusters

GO and other layered carbon-based nanomaterials are perfect supports for the deposition of metallic catalysts to prepare heterogeneous hybrid catalysts. The production of such hybrids, as well as their applications, have been well reviewed [56, 85–87]. In general, GO and other functionalized carbon-based materials are preferred for such applications, as they are better for the attachment of metal or metal oxide particles, compared to the chemically inert graphene. The formation of hybrids is supported by the defect sites of GO [56]. For instance, palladium nanoparticles dispersed on GO were able to catalyze a Suzuki–Miyaura cross-coupling reaction, with a turnover frequency (TOF) of more than 39 000 per hour [88]. This excellent TOF value is a result of the high surface area of GO, something that would not be possible with other carbon nanomaterials, even with graphene, due to the high agglomeration of the individual sheets. However, one needs to be clear that the active catalyst is the transition metal, while GO is used as a support only.

In a similar work studying the same reaction, microwaves were used for the preparation of the hybrid catalyst, in order to disperse a Pd salt on the GO surface and at the same time to reduce the GO, subsequently forming a Pd/RGO catalyst [89]. In this case, the TOF was up to 108 000 h^{-1}, assisted by microwave irradiation during the reaction. The higher TOF value was attributed to the high concentration of well-dispersed Pd on RGO. As dispersion seems to be the limiting factor, an alternative approach was applied for the preparation of the hybrid catalyst, by using a pulsed laser of 532 nm for irradiation of the GO dispersion in the presence of a palladium source [89]. The hybrid catalyst had the best TOF value reported so far, with 230 000 h^{-1}. However, it needs to be stated that again microwave irradiation was used during the reaction to facilitate the dispersion.

Recently, a composite material was prepared with CdSe quantum dots of several sizes, varying from 2 to 4 nm diameter, and GO, in order to be used on conducting glass electrodes. The difference in the size of the quantum dots allows the modulation of the photoresponse [90]. However, questions about the applicability of RGO for photocatalytic reactions arose, as the RGO is vulnerable to OH$^{\bullet}$ attack [91]. Hence, it should be stated that these radicals are the primary oxidant generated when TiO_2 (which is used for these solar cells) are excited with UV in aqueous media.

Carbon nanomaterial could also be used as an electron donor. Nitrogen-doped graphene was used as an electron donor for the CO hydrogenation of light olefins with an iron catalyst [92]. The nitrogen content could be fine-tuned, up to 16 wt.%, something that leads to higher conversions, as the doping with nitrogen atoms produces more defect sites on the sheet and allows a better distribution of the iron catalyst. The reaction could not take place without the use of the metal catalyst, proving that the graphene derivative did not possess any catalytic activity. However, iron does not possess selectivity for light olefins; the attribute probably derives from the N atoms of the graphene.

Another field for GO–metal catalyst hybrids is the fabrication of glucose biosensors. Previously, electrochemical glucose sensors have been developed that incorporate enzymes, such as glucose oxidase, for the detection of glucose. However, the use of enzymatic detectors still has several drawbacks. The main problems are related to the stability of the enzymes, the low reproducibility and the influence of oxygen limitation in the vicinity of the enzyme and its active site. More sophisticated systems have been developed in recent years, but this significantly raises the cost. Thus, in the last few years research has focused toward the production of non-enzymatic electrochemical sensors [93]. In one of these works, RGO nanosheets were used as a matrix for the deposition of $Ni(OH)_2$ nanoplates [94]. In alkaline media, the following reaction is observed:

$$\mathrm{Ni(OH)_2 + OH^- \rightarrow NiO(OH) + H_2O + e^-}$$

$$\mathrm{NiO(OH) + glucose \rightarrow Ni(OH)_2 + glucolactone}$$

Although RGO does not directly contribute to the catalytic mechanism, the authors claim that the catalytic current is much higher when RGO is present. Synergistic effects of the two materials are proposed, in addition to the increased conductivity and surface area of the composite film on the electrode [95]. This hybrid catalyst possesses a low detection limit of 0.6 mM with a linear range of 2.0 to 3.1 mM in alkaline media [94]. In another biosensor application, nanocomposites of RGO and PtNi alloy nanoparticles (NPs) have been reported [96]. Under physiological conditions, the response current of the sensor is linear to glucose concentrations up to 35 mM, with a sensitivity of 20.42 $\mu A\,cm^{-2}\,mM^{-1}$ at a substantially negative potential (i.e. 0.35 V). Compared to PtNi alloy NPs, PtNi–chemically RGO and PtNi–SWNT (single-walled nanotube) nanocomposites, the PtNi–electrochemically RGO modified electrodes exhibit smaller electron transfer resistances with a larger electrochemically active surface area, which makes it an ideal electrode material for electrocatalytic application.

12.8.2.2 Hybridization with Nucleic Acids

Graphene materials and especially GO have a high affinity toward single-stranded nucleotides. Although this application is not catalytic, the affinity is used to manufacture detectors. For instance, hemin–graphene hybrid materials were prepared via π–π stacking to detect single nucleotide polymorphisms [47]. The principle of the detector is as follows. Graphene has significantly higher affinity to single-stranded DNA (ssDNA) compared to double-stranded DNA (dsDNA). When graphene interacts with ssDNA, no precipitation is observed, while with the dsDNA significant precipitation is observed. The single nucleotide polymorphism gives only a small precipitation and can be differentiated from the two other states. As the hemin–graphene hybrid possesses peroxidase activity, the activity on the supernatant – after the interaction and centrifugation – is directly related to the amount of the hybrid catalyst that remained soluble. A simple colorimetric assay can be used to evaluate the activity and thus detect the polymorphism.

The same principle was used in later work to detect pathogenic DNA, using GO in this case [97, 98]. The GO can be deposited on a glassy carbon electrode and the interaction with DNA molecules changes the behavior in experiments of electrochemical impedance spectroscopy and differential pulse voltammetry. With the proper design of oligonucleotides, it is possible to detect the hepatitis B virus genes, without the need for any fluorescent molecule, or any labeling [97]. This is due to the fact that silver nanoclusters, which are functionalized with ssDNA of several pathogenic viruses (like the hepatitis B virus gene, the immunodeficiency virus gene and the syphilis gene), are intrinsically fluorescent [98]. However, when the ssDNA/Ag hybridizes with GO, then the fluorescence is quenched. In the presence of the virus in the sample, the ssDNA/Ag could form a dsDNA, and thus lose the affinity to GO; this results in an increase of the fluorescence.

A more specific colorimetric immunoassay was developed for the detection of prostate cancer through the identification of prostate specific antigen (PSA) [99]. GO functionalized with a secondary antibody was used in this assay to label magnetic beads (MBs) functionalized with the primary anti-PSA antibody. Once the antibodies interact with PSA, the complexes can be isolated from the solution using a magnetic field. Then GO can be tested for peroxidase activity by the oxidation of hydroquinone with the use of H_2O_2 (Figure 12.6). As in all previous assays described, the authors underline the advantages of GO due to the higher stability and lower cost compared to enzymatic detectors based on horseradish peroxidase.

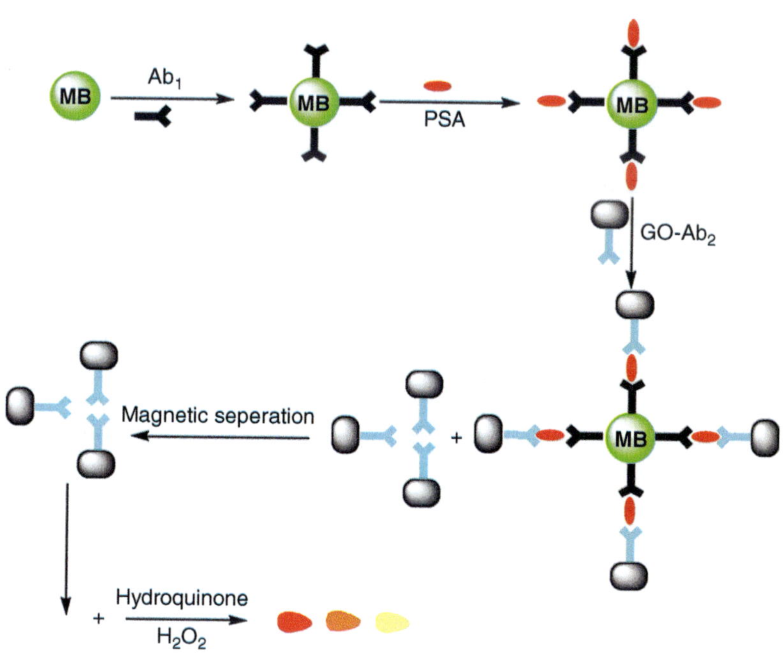

Figure 12.6 Schematic representation of the immunoassay procedure for the detection of prostate cancer. Reproduced from [99] with permission of Elsevier

The approach is applicable also for RNA molecules. GO can distinguish between single- and double-stranded RNA, an attribute that was used to develop a fluorometric method for the detection of single nucleotide changes, with the use of RNA cleaving DNAzyme [100]. The method requires three steps: (i) the specific DNAzyme cleaves the target mutant RNA, and RNA fragments are produced; (ii) these fragments are hybridized with a fluorescent-probe-labeled DNA, to prepare an RNA–DNA duplex with no single-strand areas; and (c) GO is used to quench the fluorescence of the single-stranded long wild-type RNA. The mutant fragment in this case is totally hybridized and thus does not interact with GO. Although the high affinity leads to quantitative results, the method is highly specific, requiring the development of a specific DNAzyme, as well as the production of fluorescent-probe-labeled DNA, something that definitely cannot be used in a high-throughput manner. However, this method underlines the potential of such approaches for fast and accurate tests for the detection of single nucleotide polymorphisms.

12.8.2.3 Enzyme–GO Hybrids

Graphene oxide and its derivatives can be used efficiently as immobilization matrices for the anchoring of biocatalysts, like enzymes. However, not only do they provide the support and facilitate the reuse of the biocatalyst, but also its intrinsic properties are important for the process to enhance the catalytic behavior in several cases. Although the field of enzyme immobilization on GO for biocatalytic processes is still in progress, recent developments have been summarized in a recently published review [101]. The applications of these hybrid bioconjugates are mainly focused on the development of biosensors, on biofuel cells, the degradation of pollutants and some applications in proteomic analysis. However, careful evaluation of the enzyme–nanomaterial interactions is expected to give a new perspective on the field in the next few years. For instance, it is known that GO can facilitate the wiring of the redox centers of proteins to electrodes for the preparation of biosensors, bioelectronics or biofuel cells [102]. The electron transfer was facilitated for several enzymes, such as cytochrome c [103], horseradish peroxidase and myoglobin [102]. This enhancement does not arise from structural changes, and the further functionalization of the GO did not lead to further improvement, underlining the significance of the oxygen-rich surface for the electron transfer. The peroxidase activity of GO was also validated in the oxidation of pyrogallol, assisted by hemoglobin. Hydrogels of GO and hemoglobin were formulated as hybrid catalysts and they were stable in several organic solvents. These hydrogels possessed higher catalytic activity than each component separately (GO or hemoglobin), underlining the beneficial effect that the immobilization of enzymes have on GO [104]. This finding further underlines the potential of GO as immobilization support for redox enzymes, and thus it is expected that more research on redox enzyme immobilization for biosensing applications will emerge in the next few years.

12.9 Outlook

Carbocatalysis has attracted attention over the last few years. GO and its derivatives are versatile catalysts that can be used in several processes, mostly related to redox reactions. In particular, GO is quite reactive, due to the abundant functional groups on its surface, serving as active sites, combined with the high surface area and the hydrophilicity of the material. These properties allow its dispersion in aqueous media. From all the aforementioned applications that were described in this chapter, it seems that GO and all 2D carbon-based nanomaterials have the potential to play a major role in carbocatalysis in the years to come.

However, there are some pitfalls in the field that need to be addressed in future studies, in order to facilitate the exponential growth of the field. One of the major points is that the preparation of GO is prone to metal contamination. In several cases, manganese and other metals were found at low, but still detectable, levels (ppb level) [27, 30]. Despite their low concentration, it cannot be excluded that these metal traces contribute to the catalytic reactivity of the final material [105]. Thus, the first point to be taken into consideration is to identify the metal contaminants, in order to know the exact composition of the final material. The second barrier has to do with the unelucidated structure of GO [106]. Without a universal model and the defined structure of the catalyst, the field lacks reproducibility, mechanism understanding, as well as proper comparison between different studies. Techniques for the production of GO that heavily disrupt the carbon lattice should be avoided, and more mild and reproducible techniques should be used. Recently, a new synthesis procedure for GO that preserved the honeycomb lattice and only caused a minor number of σ hole defects (as low as 0.01%) was established [107]. Full advantage of such techniques should be taken, in order to be able to exploit the GO materials and their properties to a greater extent. Moreover, the suggestion to use an electric field for beneficial orientation of the functional groups on the surface of the 2D material may lead to significant improvement of the catalytic efficiency of these materials [38, 108], something that has not been well studied until now.

From all the aforementioned, it is highlighted that the key for the application of GO in synthetic chemistry and other catalytic applications is the careful characterization of the material. This step is crucial for the reproducibility of the experiments, and it is expected to lead faster to new and more efficient processes that will reach industrial application soon.

References

[1] Boehm, H.P.; Clauss, A.; Fischer, G.; Hofmann, T., Surface properties of extremely thin graphite lamellae. In *Carbon: Proceedings of the Fifth Conference*, vol. *1*, pp. 73–80. Pergamon Press, Oxford, **1962**.

[2] Hummers, W.S.; Offeman, R.E., Preparation of graphitic oxide. *J. Am. Chem. Soc.* **1958**, *80* (6), 1339.

[3] Dreyer, D.R.; Todd, A.D.; Bielawski, C.W., Harnessing the chemistry of graphene oxide. *Chem. Soc. Rev.* **2014**, *43* (15), 5288–5301.

[4] Machado, B.F.; Serp, P., Graphene-based materials for catalysis. *Catal. Sci. Technol.* **2012**, *2* (1), 54–75.

[5] Hofmann, U.; Holst, R., Über die Säurenatur und die Methylierung von Graphitoxyd. *Ber. Dtsch. Chem. Ges.* **1939**, *72*, 754–771.

[6] Ruess, G., Über das Graphitoxyhydroxyd (Graphitoxyd). *Monatsh. Chem.* **1947**, *76* (3–5), 381–417.

[7] Scholz, W.; Boehm, H.P., Betrachtungen zur Struktur des Graphitoxids. *Z. Anorg. Allg. Chem.* **1969**, *369*, 327–339.

[8] Nakajima, T.; Yoshiaki, M., Formation process and structure of graphite oxide. *Carbon* **1994**, *32* (3), 469–475.

[9] Lerf, A.; He, H.; Forster, M.; Klinowski, J. Structure of graphite oxide revisited. *J. Phys. Chem. B* **1998**, *102* (23), 4477–4482.
[10] Szabó, T.; Berkesi, O.; Forgó, P.; *et al.*, Evolution of surface functional groups in a series of progressively oxidized graphite oxides. *Chem. Mater.* **2006**, *18* (11), 2740–2749.
[11] Yang, D.; Velamakanni, A.; Bozoklu, G.; *et al.*, Chemical analysis of graphene oxide films after heat and chemical treatments by X-ray photoelectron and micro-Raman spectroscopy. *Carbon* **2009**, *47* (1), 145–152.
[12] Zhang, L.L.; Zhao, X.; Stoller, M.D.; *et al.*, Highly conductive and porous activated reduced graphene oxide films for high-power supercapacitors. *Nano Lett.* **2012**, *12* (4), 1806–1812.
[13] Compton, O.C.; Jain, B.; Dikin, D.A.; *et al.*, Chemically active reduced graphene oxide with tunable C/O ratios. *ACS Nano* **2011**, *5* (6), 4380–4391.
[14] Marcano, D.C.; Kosynkin, D.V.; Berlin, J.M.; *et al.*, Improved synthesis of graphene oxide. *ACS Nano* **2010**, *4* (8), 4806–4814.
[15] Zhu, Y.; Murali, S.; Stoller, M.D.; *et al.*, Carbon-based supercapacitors produced by activation of graphene. *Science* **2011**, *332* (6037), 1537–1541.
[16] Montes-Navajas, P.; Asenjo, N.G.; Santamaría, R.; *et al.*, Surface area measurement of graphene oxide in aqueous solutions. *Langmuir* **2013**, *29* (44), 13443–13448.
[17] Ruoff, R., Perspective: A means to an end. *Nature* **2012**, *483* (7389), S42–S42.
[18] Jung, I.; Dikin, D.A.; Piner, R.D.; Ruoff, R.S., Tunable electrical conductivity of individual graphene oxide sheets reduced at "low" temperatures. *Nano Lett.* **2008**, *8* (12), 4283–4287.
[19] Wei, D.; Kivioja, J., Graphene for energy solutions and its industrialization. *Nanoscale* **2013**, *5* (21), 10108–10126.
[20] Chen, J.-H.; Jang, C.; Xiao, S.; *et al.*, Intrinsic and extrinsic performance limits of graphene devices on SiO_2. *Nature Nanotechnol.* **2008**, *3* (4), 206–209.
[21] Punckt, C.; Muckel, F.; Wolff, S.; *et al.*, The effect of degree of reduction on the electrical properties of functionalized graphene sheets. *Appl. Phys. Lett.* **2013**, *102*, 023114.
[22] Kuila, T.; Mishra, A.K.; Khanra, P.; *et al.*, Recent advances in the efficient reduction of graphene oxide and its application as energy storage electrode materials. *Nanoscale* **2013**, *5* (1), 52–71.
[23] Sarkar, S.; Niyogi, S.; Bekyarova, E.; Haddon, R.C., Organometallic chemistry of extended periodic π-electron systems: hexahapto-chromium complexes of graphene and single-walled carbon nanotubes. *Chem. Sci.* **2011**, *2* (7), 1326–1333.
[24] Schaetz, A.; Zeltner, M.; Stark, W.J., Carbon modifications and surfaces for catalytic organic transformations. *ACS Catal.* **2012**, *2* (6), 1267–1284.
[25] Su, C.; Acik, M.; Takai, K.; *et al.*, Probing the catalytic activity of porous graphene oxide and the origin of this behaviour. *Nature Commun.* **2012**, *3*, 1298.
[26] Dreyer, D.R.; Park, S.; Bielawski, C.W.; Ruoff, R.S., The chemistry of graphene oxide. *Chem. Soc. Rev.* **2010**, *39* (1), 228–240.
[27] Dreyer, D.R.; Jia, H.-P.; Bielawski, C.W., Graphene oxide: a convenient carbocatalyst for facilitating oxidation and hydration reactions. *Angew. Chem. Int. Ed.* **2010**, *49* (38), 6813–6816.
[28] Dreyer, D.R.; Bielawski, C.W., Carbocatalysis: heterogeneous carbons finding utility in synthetic chemistry. *Chem. Sci.* **2011**, *2* (7), 1233–1240.
[29] Dreyer, D.R.; Jia, H.-P.; Todd, A.D.; *et al.*, Graphite oxide: a selective and highly efficient oxidant of thiols and sulfides. *Org. Biomol. Chem.* **2011**, *9* (21), 7292–7295.
[30] Jia, H.-P.; Dreyer, D.R.; Bielawski, C.W., C–H oxidation using graphite oxide. *Tetrahedron* **2011**, *67* (24), 4431–4434.
[31] Bielawski, C.W.; Dreyer, D.R.; Miller, R., Production of propene. US Patent US2015/025289 A1, 2015.
[32] Wang, S.; Dong, S.; Wang, J.; *et al.*, Oxygen-enriched carbon material for catalyzing oxygen reduction towards hybrid electrolyte Li–air battery. *J. Mater. Chem.* **2012**, *22* (39), 21051–21056.
[33] Zhu, Y.; Stoller, M.D.; Cai, W.; *et al.*, Exfoliation of graphite oxide in propylene carbonate and thermal reduction of the resulting graphene oxide platelets. *ACS Nano* **2010**, *4* (2), 1227–1233.
[34] Pereira, M.F.R.; Orfao, J.J.M.; Figueiredo, J.L., Oxidative dehydrogenation of ethylbenzene on activated carbon catalysts: 2. Kinetic modelling. *Appl. Catal. Gen.* **2000**, *196* (1), 43–54.

[35] Frank, B.; Blume, R.; Rinaldi, A.; *et al.*, Oxygen insertion catalysis by sp^2 carbon. *Angew. Chem. Int. Ed.* **2011**, *50* (43), 10226–10230.
[36] Frank, B.; Zhang, J.; Blume, R.; *et al.*, Heteroatoms increase the selectivity in oxidative dehydrogenation reactions on nanocarbons. *Angew. Chem. Int. Ed.* **2009**, *48* (37), 6913–6917.
[37] Tang, S.; Cao, Z., Site-dependent catalytic activity of graphene oxides towards oxidative dehydrogenation of propane. *Phys. Chem. Chem. Phys.* **2012**, *14*, 16558–16565.
[38] Tang, S.; Zhang, S., Adsorption of epoxy and hydroxyl groups on zigzag graphene nanoribbons: insights from density functional calculations. *Chem. Phys.* **2012**, *392* (1), 33–45.
[39] Hamwi, A.; Marchand, V. Some chemical and electrochemical properties of graphite oxide. *J. Phys. Chem. Solids* **1996**, *57* (6–8), 867–872.
[40] Singh Chauhan, S.M.; Mishra, S., Use of graphite oxide and graphene oxide as catalysts in the synthesis of dipyrromethane and calix[4]pyrrole. *Molecules* **2011**, *16* (12), 7256–7266.
[41] Shao, S.; Wang, A.; Yang, M.; *et al.*, Synthesis of *meso*-aryl substituted calix[4]pyrroles. *Synth. Commun.* **2001**, *31* (9), 1421–1426.
[42] Long, Y.; Zhang, C.; Wang, X.; *et al.*, Oxidation of SO_2 to SO_3 catalyzed by graphene oxide foams. *J. Mater. Chem.* **2011**, *21* (36), 13934–13941.
[43] Gonçalves, G.A.B.; Pires, S.M.G.; Simões, M.M.Q.; *et al.*, Three-dimensional graphene oxide: a promising green and sustainable catalyst for oxidation reactions at room temperature. *Chem. Commun.* **2014**, *50* (57), 7673–7676.
[44] Song, Y.; Qu, K.; Zhao, C.; *et al.*, Graphene oxide: intrinsic peroxidase catalytic activity and its application to glucose detection. *Adv. Mater.* **2010**, *22* (19), 2206–2210.
[45] Song, Y.; Wei, W.; Qu, X., Colorimetric biosensing using smart materials. *Adv. Mater.* **2011**, *23* (37), 4215–4236.
[46] Ji, J.; Zhang, G.; Chen, H.; *et al.*, Sulfonated graphene as water-tolerant solid acid catalyst. *Chem. Sci.* **2011**, *2* (3), 484–487.
[47] Guo, Y.; Deng, L.; Li, J.; *et al.*, Hemin – graphene hybrid nanosheets with intrinsic peroxidase-like activity for label-free colorimetric detection of single-nucleotide polymorphism. *ACS Nano* **2011**, *5* (2), 1282–1290.
[48] Xu, Y.; Zhao, L.; Bai, H.; *et al.*, Chemically converted graphene induced molecular flattening of 5,10,15,20-tetrakis(1-methyl-4-pyridinio)porphyrin and its application for optical detection of cadmium(II) ions. *J. Am. Chem. Soc.* **2009**, *131*, 13490–13497.
[49] Song, Y.; Chen, Y.; Feng, L.; *et al.*, Selective and quantitative cancer cell detection using target-directed functionalized graphene and its synergetic peroxidase-like activity. *Chem. Commun.* **2011**, *47* (15), 4436–4438.
[50] Dreyer, D.R.; Bielawski, C.W., Graphite oxide as an olefin polymerization carbocatalyst: applications in electrochemical double layer capacitors. *Adv. Funct. Mater.* **2012**, *22* (15), 3247–3253.
[51] Montaudo, G.; Finocchiaro, P.; Caccamese, S.; Bottino, F., Polycondensation of benzyl chloride and its derivatives: a study of the reaction at different temperatures. *J. Polym. Sci.* **1970**, *8* (9), 2475–2490.
[52] Lenz, R.W., *Organic Chemistry of Synthetic High Polymers*. Interscience, New York, **1967**.
[53] Dreyer, D.R.; Jarvis, K.A.; Ferreira, P.J.; Bielawski, C.W., Graphite oxide as a dehydrative polymerization catalyst: a one-step synthesis of carbon-reinforced poly(phenylene methylene) composites. *Macromolecules* **2011**, *44* (19), 7659–7667.
[54] Dreyer, D.R.; Jarvis, K.A.; Ferreira, P.J.; Bielawski, C.W., Graphite oxide as a carbocatalyst for the preparation of fullerene-reinforced polyester and polyamide nanocomposites. *Polym. Chem.* **2012**, *3* (3), 757–766.
[55] Arbaoui, A.; Redshaw, C., Metal catalysts for ε-caprolactone polymerisation. *Polym. Chem.* **2010**, *1* (6), 801–826.
[56] Liang, Y.; Li, Y.; Wang, H.; Dai, H., Strongly coupled inorganic/nanocarbon hybrid materials for advanced electrocatalysis. *J. Am. Chem. Soc.* **2013**, *135* (6), 2013–2036.
[57] Seger, B.; Kamat, P.V. Electrocatalytically active graphene–platinum nanocomposites. Role of 2-D carbon support in PEM fuel cells. *J. Phys. Chem. C* **2009**, *113* (19), 7990–7995.
[58] Su, D.S.; Perathoner, S.; Centi, G. Nanocarbons for the development of advanced catalysts. *Chem. Rev.* **2013**, *113* (8), 5782–5816.

[59] Strelko, V.V., Role of carbon matrix heteroatoms at synthesis of carbons for catalysis and energy applications. *J. Energy Chem.* **2013**, *22* (2), 174–182.
[60] Wang, D.-W.; Su, D., Heterogeneous nanocarbon materials for oxygen reduction reaction. *Energy Environ. Sci.* **2014**, *7* (2), 576.
[61] Yang, Z.; Yao, Z.; Li, G.; *et al.*, Sulfur-doped graphene as an efficient metal-free cathode catalyst for oxygen reduction. *ACS Nano* **2012**, *6* (1), 205–211.
[62] Wang, S.; Yu, D.; Dai, L.; *et al.*, Polyelectrolyte-functionalized graphene as metal-free electrocatalysts for oxygen reduction. *ACS Nano* **2011**, *5* (8), 6202–6209.
[63] Lee, K.R.; Lee, K.U.; Lee, J.W., *et al.*, Electrochemical oxygen reduction on nitrogen doped graphene sheets in acid media. *Electrochem. Commun.* **2010**, *12* (8), 1052–1055.
[64] Park, S.; An, J.; Potts, J.R.; *et al.*, Hydrazine-reduction of graphite- and graphene oxide. *Carbon* **2011**, *49* (9), 3019–3023.
[65] Kim, H.; Lee, K.; Woo, S.I.; Jung, Y., On the mechanism of enhanced oxygen reduction reaction in nitrogen-doped graphene nanoribbons. *Phys. Chem. Chem. Phys.* **2011**, *13* (39), 17505–17510.
[66] Dong, L.; Chen, Z.; Yang, D.; Lu, H., Hierarchically structured graphene-based supercapacitor electrodes. *RSC Advances* **2013**, *3* (44), 21183–21191.
[67] Zhai, Y.; Dou, Y.; Zhao, D.; *et al.*, Carbon materials for chemical capacitive energy storage. *Adv. Mater.* **2011**, *23* (42), 4828–4850.
[68] Ye, J.; Zhang, H.; Chen, Y.; *et al.*, Supercapacitors based on low-temperature partially exfoliated and reduced graphite oxide. *J. Power Sources* **2012**, *212*, 105–110.
[69] Wang, Y.; Shi, Z.; Huang, Y.; Ma, Y.; Chen, M.; Chen, Y., Supercapacitor devices based on graphene materials. *J. Phys. Chem. C* **2009**, *113* (30), 13103–13107.
[70] Chen, Y.; Zhang, X.; Zhang, D.; Ma, Y., High power density of graphene-based supercapacitors in ionic liquid electrolytes. *Mater. Lett.* **2012**, *68*, 475–477.
[71] Vijay Kumar, A.; Rama Rao, K., Recyclable graphite oxide catalyzed Friedel–Crafts addition of indoles to α,β-unsaturated ketones. *Tetrahedron Lett.* **2011**, *52* (40), 5188–5191.
[72] Gu, Y.; Ogawa, C.; Kobayashi, J.; *et al.*, A heterogeneous silica-supported scandium/ionic liquid catalyst system for organic reactions in water. *Angew. Chem. Int. Ed.* **2006**, *45* (43), 7217–7220.
[73] Hagiwara, H.; Sekifuji, M.; Hoshi, T.; *et al.*, Sustainable conjugate addition of indoles catalyzed by acidic ionic liquid immobilized on silica. *Synlett* **2008**, *2008* (4), 608–610.
[74] Kim, Y.; Some, S.; Lee, H., Graphene oxide as a recyclable phase transfer catalyst. *Chem. Commun.* **2013**, *49* (50), 5702–5704.
[75] Verma, S.; Mungse, H.P.; Kumar, N.; *et al.*, Graphene oxide: an efficient and reusable carbocatalyst for aza-Michael addition of amines to activated alkenes. *Chem. Commun.* **2011**, *47* (47), 12673–12675.
[76] Yeh, T.-F.; Syu, J.-M.; Cheng, C.; *et al.*, Graphite oxide as a photocatalyst for hydrogen production from water. *Adv. Funct. Mater.* **2010**, *20* (14), 2255–2262.
[77] Hou, W.-C.; Chowdhury, I.; Goodwin, D.G.; *et al.*, Photochemical transformation of graphene oxide in sunlight. *Environ. Sci. Technol.* **2015**, *49* (6), 3435–3443.
[78] Pan, Y.; Wang, S.; Kee, C.W.; *et al.*, Graphene oxide and Rose Bengal: oxidative C–H functionalisation of tertiary amines using visible light. *Green Chem.* **2011**, *13* (12), 3341–3344.
[79] Sreeprasad, T.S.; Pradeep, T., Graphene for environmental and biological applications. *Int. J. Mod. Phys. B* **2012**, *26* (21), 1242001.
[80] Suzuki, Y.; Matsushima, M.; Kodomari, M., Graphite-catalyzed acylative cleavage of ethers with acyl halides. *Chem. Lett.* **1998**, *27* (4), 319–320.
[81] Sereda, G.A.; Rajpara, V.B.; Slaba, R.L. The synthetic potential of graphite-catalyzed alkylation. *Tetrahedron* **2007**, *63* (34), 8351–8357.
[82] Garrigues, B.; Laporte, C.; Laurent, R.; *et al.*, Microwave-assisted Diels–Alder reaction supported on graphite. *Liebigs Ann.* **1996**, *1996* (5), 739–741.
[83] Klemens, P.G.; Pedraza, D.F., Thermal conductivity of graphite in the basal plane. *Carbon* **1994**, *32* (4), 735–741.
[84] Han, B.H.; Shin, D.H.; Cho, S.Y., Graphite catalyzed reduction of aromatic and aliphatic nitro compounds with hydrazine hydrate. *Tetrahedron Lett.* **1985**, *26* (50), 6233–6234.
[85] Krishnan, D.; Kim, F.; Luo, J.; *et al.*, Energetic graphene oxide: challenges and opportunities. *Nano Today* **2012**, *7* (2), 137–152.

[86] Samy El-Shall, M., Heterogeneous catalysis by metal nanoparticles supported on graphene. In *Graphene: Synthesis, Properties and Phenomena*, eds C.N.R. Rao and A.K.Sood, pp. 303–338. Wiley-VCH, Weinheim, **2013**.

[87] Julkapli, N.M.; Bagheri, S., Graphene supported heterogeneous catalysts: an overview. *Int. J. Hydrogen Energy* **2015**, *40* (2), 948–979.

[88] Scheuermann, G.M.; Rumi, L.; Steurer, P.; *et al.*, Palladium nanoparticles on graphite oxide and its functionalized graphene derivatives as highly active catalysts for the Suzuki–Miyaura coupling reaction. *J. Am. Chem. Soc.* **2009**, *131* (23), 8262–8270.

[89] Siamaki, A.R.; Khder, A.E.R.S.; Abdelsayed, V.; *et al.*, Microwave-assisted synthesis of palladium nanoparticles supported on graphene: a highly active and recyclable catalyst for carbon–carbon cross-coupling reactions. *J. Catal.* **2011**, *279* (1), 1–11.

[90] Krishnamurthy, S.; Kamat, P.V., CdSe–graphene oxide light-harvesting assembly: size-dependent electron transfer and light energy conversion aspects. *ChemPhysChem* **2014**, *15* (10), 2129–2135.

[91] Radich, J.G.; Krenselewski, A.L.; Zhu, J.; Kamat, P.V., Is graphene a stable platform for photocatalysis? Mineralization of reduced graphene oxide with UV-irradiated TiO_2 nanoparticles. *Chem. Mater.* **2014**, *26* (15), 4662–4668.

[92] Chen, X.; Deng, D.; Pan, X.; *et al.*, N-doped graphene as an electron donor of iron catalyst for CO hydrogenation to light olefins. *Chem. Commun.* **2015**, *51*, 217–220.

[93] Wang, G.; He, X.; Wang, L.; *et al.*, Non-enzymatic electrochemical sensing of glucose. *Microchim. Acta* **2013**, *180* (3–4), 161–186.

[94] Zhang, Y.; Xu, F.; Sun, Y.; *et al.*, Assembly of $Ni(OH)_2$ nanoplates on reduced graphene oxide: a two dimensional nanocomposite for enzyme-free glucose sensing. *J. Mater. Chem.* **2011**, *21* (42), 16949–16954.

[95] Liu, C.; Wang, K.; Luo, S.; *et al.*, Direct electrodeposition of graphene enabling the one-step synthesis of graphene–metal nanocomposite films. *Small* **2011**, *7* (9), 1203–1206.

[96] Gao, H.; Xiao, F.; Ching, C.B.; Duan, H., One-step electrochemical synthesis of PtNi nanoparticle–graphene nanocomposites for non-enzymatic amperometric glucose detection. *ACS Appl. Mater. Interfaces* **2011**, *3* (8), 3049–3057.

[97] Muti, M.; Sharma, S.; Erdem, A.; Papakonstantinou, P., Electrochemical monitoring of nucleic acid hybridization by single-use graphene oxide-based sensor. *Electroanalysis* **2011**, *23* (1), 272–279.

[98] Liu, X.; Wang, F.; Aizen, R.; *et al.*, Graphene oxide/nucleic acid-stabilized silver nanoclusters: functional hybrid materials for optical aptamer sensing and multiplexed analysis of pathogenic DNAs. *J. Am. Chem. Soc.* **2013**, *135* (32), 11832–11839.

[99] Qu, F.; Li, T.; Yang, M., Colorimetric platform for visual detection of cancer biomarker based on intrinsic peroxidase activity of graphene oxide. *Biosens. Bioelectron.* **2011**, *26* (9), 3927–3931.

[100] Hong, C.; Kim, D.-M.; Baek, A.; *et al.*, Fluorescence-based detection of single-nucleotide changes in RNA using graphene oxide and DNAzyme. *Chem. Commun.* **2015**, *51* (26), 5641–5644.

[101] Pavlidis, I.V.; Patila, M.; Bornscheuer, U.T.; *et al.*, Graphene-based nanobiocatalytic systems: recent advances and future prospects. *Trends Biotechnol.* **2014**, *32* (6), 312–320.

[102] Zuo, X.; He, S.; Li, D.; Peng, C.; Huang, Q.; Song, S.; Fan, C., Graphene oxide-facilitated electron transfer of metalloproteins at electrode surfaces. *Langmuir* **2010**, *26* (3), 1936–1939.

[103] Patila, M.; Pavlidis, I.V.; Diamanti, E.K.; *et al.*, Enhancement of cytochrome c catalytic behaviour by affecting the heme environment using functionalized carbon-based nanomaterials. *Process Biochem.* **2013**, *48* (7), 1010–1017.

[104] Huang, C.; Bai, H.; Li, C.; Shi, G., A graphene oxide/hemoglobin composite hydrogel for enzymatic catalysis in organic solvents. *Chem. Commun.* **2011**, *47* (17), 4962–4964.

[105] Leadbeater, N.E., Cross coupling: when is free really free? *Nature Chem.* **2010**, *2*, 1007–1009.

[106] Eigler, S.; Hirsch, A., Chemistry with graphene and graphene oxide – challenges for synthetic chemists. *Angew. Chem. Int. Ed.* **2014**, *53* (30), 7720–7738.

[107] Eigler, S.; Enzelberger-Heim, M.; Grimm, S.; *et al.*, Wet chemical synthesis of graphene. *Adv. Mater.* **2013**, *25* (26), 3583–3587.

[108] Sun, P.; Zheng, F.; Wang, K.; *et al.*, Electro- and magneto-modulated ion transport through graphene oxide membranes. *Sci. Rep.* **2014**, *4*, 6798.

13

Challenges of Industrial-Scale Graphene Oxide Production

Sean E. Lowe and Yu Lin Zhong

13.1 Introduction

It is no longer a question of *whether* graphene can transform industries, but a question of *when*. The answer comes down to when graphene can be produced on an industrial scale, at low cost and with controllable quality. There are multiple routes toward producing graphene industrially, each route producing a somewhat different graphene product [1–4]. For high-end electronics industries, the chemical vapor deposition (CVD) method is the most suited, as it allows precise control of graphene defects and layers but, more importantly, the process can be easily integrated into the lucrative semiconductor industries. For industries that require massive amounts of graphene, such as energy storage, energy conversion and coatings, the scalable top-down approaches employing wet-chemical exfoliation of graphite seem to be most promising [2]. Wet-chemical exfoliation makes use of the abundant raw material graphite and has potentially low production costs. These are pivotal considerations for replacing currently employed materials, particularly those that are already inexpensive. Out of the wet-chemical methods, the graphite oxide route is by far the most popular. This is due to its ability to produce single-layer graphene oxide (GO) in high yield and with excellent dispersibility in water, which facilitates graphite oxide processing and use in downstream applications.

In this chapter, we discuss the technical and economic aspects of large-scale GO production. While there has been robust discussion of the scientific aspects of GO in this volume and by others [5–13], there have been few in-depth reviews about large-scale production

Graphene Oxide: Fundamentals and Applications, First Edition. Edited by Ayrat M. Dimiev and Siegfried Eigler.

from an industrial perspective. The current chapter aims in part to fill this gap, highlighting some of the key scientific, engineering and economic challenges that will be faced by large-scale GO producers. Firstly, to provide an economic backdrop, an introduction to the GO market is provided in section 13.2. Secondly, the major methods of GO synthesis are outlined in section 13.3. The key considerations and challenges associated with large-scale GO production are then delineated in section 13.4. These challenges range from sourcing appropriate graphite starting material to cost-effectively producing GO and then storing and handling the product. Finally, some future directions for industrial-scale GO production are considered in section 13.5.

13.2 Scope and Scale of the Graphene Market

Only a decade on from Novoselov and Geim's breakthrough isolation and characterization of graphene [14–16], the graphene industry is still nascent, with different types of graphene seeing different rates of commercialization and market uptake. The scale of graphene production is small relative to other industrial materials, which may be manufactured in tens of thousands of tonnes per year or more. For example, graphene production is orders of magnitude lower than its carbon nanomaterial counterparts, carbon fibers and carbon nanotubes (CNTs), as shown in Figure 13.1.

In 2014, Ren and Cheng [4] reported on the global growth of large-scale graphene production with targeted industrial applications (Table 13.1). Examining Table 13.1, it is not surprising that the large-scale production of CVD graphene film (up to 200 000 m^2 per year) is progressing rapidly, with well-defined quality benchmarks (e.g. conductivity and transparency), given its application in the fast-paced and lucrative electronics market. This growth is aided by the fact that the most common transparent conductive film currently

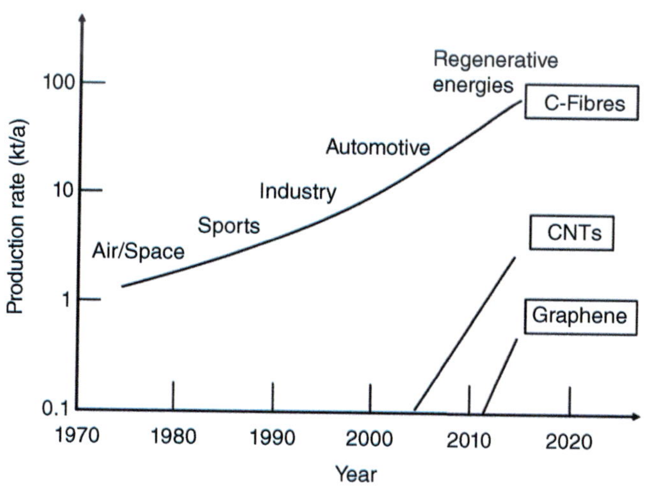

Figure 13.1 Annual production rates of carbon fibers, carbon nanotubes and graphene, 1970–2014. The industries in which carbon fibers found a market and approximate uptake times are also plotted. Adapted from [17] with permission from John Wiley & Sons

Table 13.1 *Synthesis method, product, production capacity and main application products of several large graphene manufacturers. Reproduced from [4] with permission from Nature Publishing Group, Copyright 2014*

Graphene manufacturer	Basic synthesis method	Basic product	Production capacity	Main application products
Angstron Materials (USA)	Liquid exfoliation	Pristine nano-graphene platelets Thickness: <100 nm Carbon content: ≥95% (for product under three layers in average thickness)	300 tonnes yr^{-1}	Graphene/silicon anode materials
Thomas Swan (UK)	Liquid exfoliation	Graphene nanoplatelets Thickness: 5–7 layers (average) Size: 0.5–2 μm Sheet resistance: 10 ± 5 Ω/ϒ (25 μm film, equal to 27–80 S cm^{-1}) Impurity: ~8 wt.%	kg day^{-1} (September 2014)	—
Vorbeck Materials (USA)	Oxidation–thermal exfoliation	Functionalized graphene Thickness: mostly 1–3 layers Morphology: wrinkled	40 tonnes yr^{-1} (October 2012)	Conductive inks and coatings, graphene rubbers, flexible battery straps
Sixth Element Materials (China)	Oxidation–exfoliation–reduction	Graphene oxide, reduced GO Thickness: ≤10 nm Carbon content: ≥95% Electrical conductivity: ≥2 S cm^{-1} (for electrically conductive type product)	100 tonnes yr^{-1} (November 2013)	Mechanical/thermally enhanced composites, anticorrosion coatings
XG Sciences (USA)	Intercalation–exfoliation	Graphene nanoplatelets Thickness: 2–15 nm (average) Electrical and thermal conductivity: 3800 S cm^{-1} and 500 W mK^{-1}, respectively (for 30 μm thick graphene paper product)	80 tonnes yr^{-1} (August 2012)	Graphene/silicon anode composite, supercapacitor electrode materials, conductive inks and coatings, graphene paper (for thermal spreading, and electrically conductive applications, and so on)

Ningbo Morsh (China)	Intercalation–expansion–exfoliation	Graphene nanoplatelets Average thickness: 3 nm	300 tonnes yr^{-1} (December 2013)	Graphene conductive additives and graphene-coated current collector for lithium-ion batteries, conductive inks, heat-radiating coatings, anticorrosion coatings, thermally/electrically conductive master batch
Deyang Carbonene Tech (China)	Intercalation–expansion–exfoliation	Graphene sheets (few-layer) Thickness: ≤10 layers Electrical conductivity: ~1000 S cm^{-1} (for ~15 μm thick membrane)	1.5 tonnes yr^{-1} (October 2012) 300 tonnes yr^{-1} (2017–2019)	Battery materials, thermal management materials, conductive inks, conductive anticorrosion coatings
Bluestone Global Tech (USA)	CVD	Graphene films on Cu, SiO_2/Si, polyethylene terephthalate (PET) Film on PET: available maximum size 8 × 10 $inch^2$, <30 **Ω/Υ** at >85% transmittance (excluding substrate), <800 **Ω/Υ** at 95% transmittance (excluding substrate) Film on SiO_2/Si: available maximum size 4 inch wafer, monolayer >95%, average Hall mobility: 2000–4000 $cm^2\ V^{-1}\ s^{-1}$	—	Field-effect transistors, touch panels
2D Carbon Tech (China)	CVD	Graphene films on Cu, SiO_2/Si, glass, PET Film on PET: available maximum size 450 × 550 mm^2, monolayer >90%, 200–400 **Ω/Υ** at >85% transmittance (including substrate)	30 000 m^2 (May 2013) 200 000 m^2 (December 2014)	Touch panels
Wuxi Graphene Film (China)	CVD	Graphene films on Cu, PET Film on PET: ~600 **Ω/Υ** at >97% transmittance (excluding substrate)	80 000 m^2 (December 2013)	Touch panels, touch sensors (5 million pieces, December 2013)
Power Booster (China)	CVD	Graphene films on Cu, PET Film on Cu: available maximum size 7.5 m^2 (2013) Film on PET: 50–140 **Ω/Υ** at 95.5% transmittance	—	Touch panels

used, indium tin oxide (containing 74% indium), is costly and has limited supply. Table 13.1 also highlights another class of graphene, graphene nanoplatelets, which are capable of being produced at large scale (up to 300 tonnes per year), typically via an intercalation–expansion–exfoliation process. Such a process is akin to the age-old technology of expandable graphite combined with a breaking-down process to yield graphene nanoplatelets. One concern is whether these graphene nanoplatelets exhibit properties closer to graphene or graphite. In contrast, there are fewer large companies producing GO, despite the fact that GO is the most processable and versatile graphene precursor. Commercialization of GO is not as straightforward as CVD graphene, as GO is usually further processed or used in conjunction with other materials, rather than as an end-product for a market with specific demand.

Graphene oxide is likely to find applications in composite materials and polymer blends [9, 18], conductive inks [19], energy storage [20] and other applications where defect-free graphene is not required [3]. In many cases, GO will need to undergo further transformation depending on the application, such as reduction for conductivity, film forming for desalination membranes, bioconjugation for bio-related applications, etc. Hence, GO is expected to be produced in a way that caters to the demands of a variety of industrial sectors. As with other industrial chemicals that are produced at different grades, purities and mixtures, there is a remarkable range of GO qualities that can be produced, including variations in flake sizes, oxidation degree, porosity and purity. It will be a significant advantage for GO if all these qualities can be controllably produced – but, if not, it will be a significant setback.

13.3 Overview of Graphene Oxide Synthesis

Graphene oxide is a single layer of oxidized graphene, typically obtained from the dispersion of bulk graphite oxide in neutral or alkaline water, assisted by stirring or, more often, sonication. It is noteworthy that even the simple process of dispersing bulk graphite oxide in water may lead to GO with varying flake sizes. Although there are applications involving the direct use of bulk graphite oxide, such as the formation of thermally processed GO (tpGO) [21], most applications seek to capitalize on the large surface area available with single-layer GO. Over time, synthetic methods for GO synthesis have become safer, more environmentally friendly and more scalable.

The first example of graphite oxide synthesis was demonstrated by Brodie in 1859 [22] when bulk graphite was oxidized via potassium perchlorate in fuming nitric acid. However, this process is very hazardous due to the formation of highly explosive chlorine dioxide gas when potassium perchlorate is added to concentrated acid. In 1898, Staudenmaier [23] made improvements to the Brodie method by replacing about two-thirds of the fuming nitric acid with concentrated sulfuric acid and adding potassium perchlorate in aliquots over four days. This modification reduces the explosion hazard, increases the overall oxidation degree of the graphite oxide and allows the production to be carried out in a single reaction vessel as compared to Brodie's approach. A revolutionary method for graphite oxide production was developed by Charpy [24] and later by Hummers and Offeman in 1958 [25], where one weight equivalent of graphite is oxidized by three weight equivalents of potassium permanganate and a half weight equivalent of sodium nitrate in concentrated sulfuric acid. The Hummers method is currently the most widely adopted one for the production of graphite oxide, as it eliminates the evolution of explosive gas and, more importantly, reduces the total reaction time to just a few hours or less.

A typical process based on the Hummers or comparable method is shown in Figure 13.2. In this process, molecules of sulfuric acid (H_2SO_4) are thought to first intercalate between

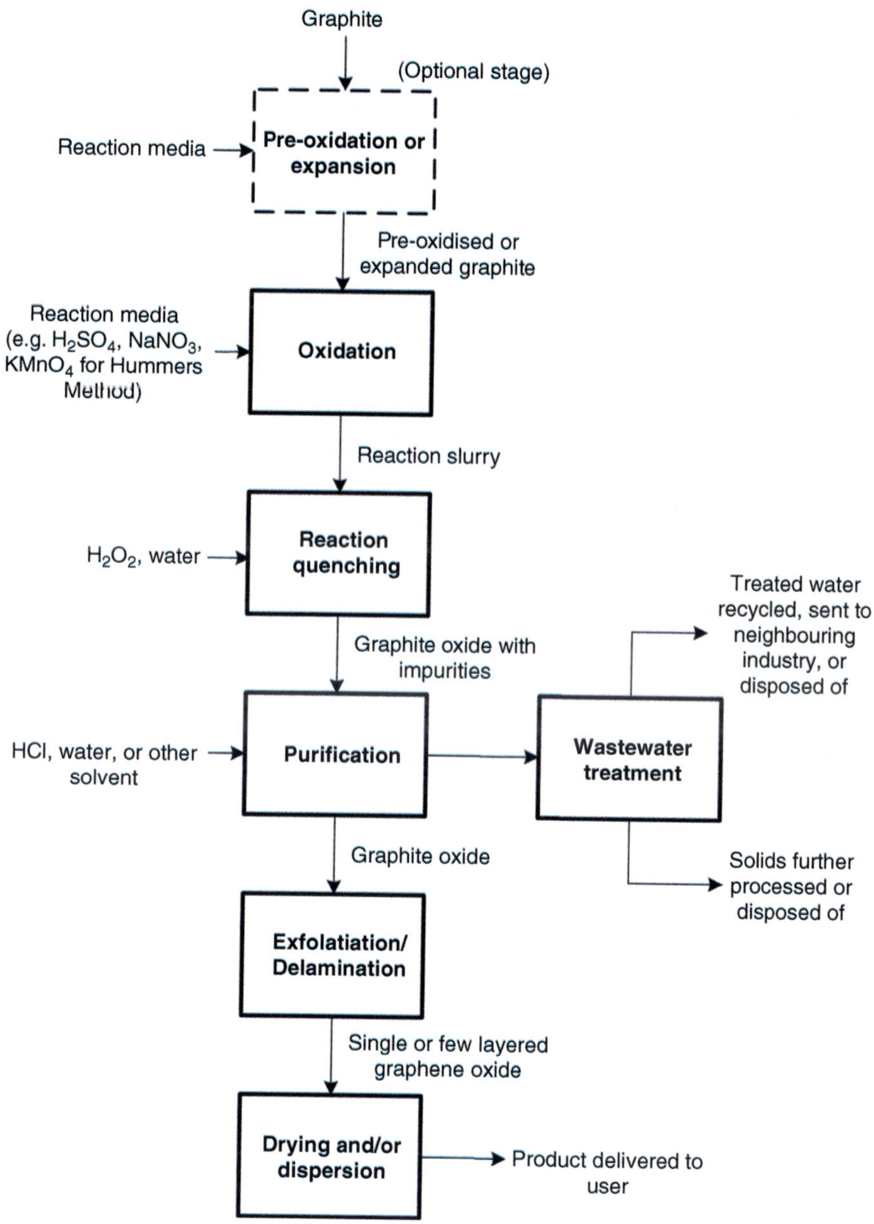

Figure 13.2 *Block flow diagram of a possible process for the production of GO*

the stacked layers of graphene making up graphite [7, 26]. An oxidant, potassium permanganate ($KMnO_4$), and sodium nitrate ($NaNO_3$) are also added to the reaction mixture. In concentrated sulfuric acid, potassium permanganate reacts to form manganese heptoxide (Mn_2O_7), a strong oxidant thought to be the active species in graphene oxidation, although other species such as the permanganate ion may also play a role [26]. Because of the intercalation of the graphite with sulfuric acid, the oxidant mixture has access to the space between the graphene layers. It can then react with the graphene, adding functional groups such as hydroxyl (–OH), epoxide (–O–) and carboxylic acids (–COOH) to form GO. Because in some cases the oxidation procedure does not lead to complete conversion of the graphite to graphite oxide, a pre-oxidation step may be added before the main oxidation (e.g. using P_2O_5 and $K_2S_2O_8$ in H_2SO_4) [27].

In the work-up procedure, the reaction is quenched with water and hydrogen peroxide (H_2O_2). The water reacts with the potassium heptoxide to form $HMnO_4$, effectively stopping the oxidation reaction. The hydrogen peroxide further reacts with manganese oxides in the mixture, reducing them to manganese ions. In a subsequent purification process, manganese and other ions in the solution are removed, along with any unoxidized or underoxidized graphite. The graphite oxide also needs to be exfoliated to single-layered or few-layered GO. This is generally accomplished by mechanical means (e.g. stirring or sonication), although thermal expansion is also possible [28]. At the conclusion of the process, the product is prepared for further use by drying or dispersion in water or other media.

13.4 Challenges of Graphene Oxide Production

There are several challenges that a potential large-scale producer of GO may face. Although the Hummers method should be scalable for industrial production, there are nonetheless some scientific questions that should be addressed before a large-scale operation is put into place (e.g. the optimal parameters for graphite exfoliation and GO storage). In addition to these technical issues, large-scale production brings with it a host of economic considerations, such as the need to acquire large quantities of starting materials at low cost. The various technical and economic challenges associated with each stage of the process, shown in Figure 13.2, are discussed below.

13.4.1 Graphite Sources

The graphite source is a particularly important consideration in the process, as it will affect not only the cost of the operation, but also the subsequent processing steps required and the final product quality.

Graphene oxide can be synthesized from naturally occurring graphite, as well as from synthetic graphite. Three forms of natural graphite are mined: vein graphite, flake graphite and amorphous/microcrystalline graphite. Vein graphite is the purest ore, with the highest carbon content before treatment (>90%). However, vein graphite is currently only produced by two mines in Sri Lanka with limited capacity [29], making up only 1% of the world's total natural graphite production [30].

Flake graphite has been the main source of natural graphite used in GO research and development. The world's largest producer of flake graphite is China, although increasing demand from the electronics industry has encouraged exploration in Canada, Brazil, Australia and elsewhere [31]. After mineral extraction, flake graphite is typically purified by mechanical separation and flotation to remove hetroatom contamination [29]. It is then supplied in dry form in various mesh sizes and purity.

World production capacity of flake graphite is increasing, following a 2011 exploration boom, and there are not expected to be supply shortages in the near future [30]. After a peak in graphite prices in 2011–2012, prices for flake graphite have been trending downward [32]. In 2014, prices ranged from \$750 to \$1350 per tonne, based on the carbon percentage, flake size, delivery location and quantity purchased [33]. There is speculation that graphite prices could increase with contracting supply from China [34], and rising prices may pose a challenge to GO commercialization.

Graphene oxide can also be synthesized from amorphous graphite (also known as microcrystalline graphite) [35]. Amorphous graphite is the cheapest and most abundant form of the mineral. As its name implies, amorphous graphite lacks the large, stacked lamellar graphene sheets characteristic of flake graphite [29, 36]. There is some evidence that the disordered stacking in amorphous graphite may impede intercalation by the commonly used intercalating agent, sulfuric acid (see Figure 13.3) [35].

Expanded graphite (EG) is also a suitable precursor for GO synthesis [35, 37]. Expanded graphite is natural graphite that has been pretreated with intercalating agents and heat, such that its specific volume is greatly increased. Using expanded graphite can assist in the oxidation process [35], but has the disadvantage of requiring an additional step and cost for graphite expansion.

In addition to natural graphite, synthetic graphite may also be used to produce GO [28, 38, 39]. Synthetic graphite is formed by heating unstructured carbon to very high temperatures

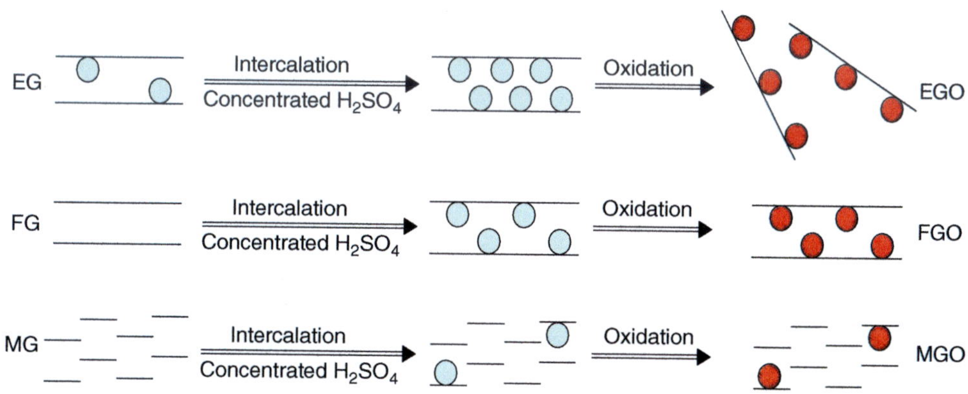

Figure 13.3 *Intercalation and oxidation of three forms of graphite: expanded graphite (EG), flake graphite (FG) and microcrystalline graphite (MG). EG has been pretreated with sulfuric acid, and is therefore most readily intercalated and oxidized in the Hummers process. In contrast, the microcrystalline structure of MG appears to impede intercalation/oxidation, leading to a greater amount of under-oxidized graphite in a given time period. Reproduced from [35]: "Effect of graphite precursor on oxidation degree, hydrophilicity and microstructure of graphene oxide", Hu, X.; Yu, Y.; Zhou, J.; Song, L.,* Nano **2014**, *9 (3) 1450037. Copyright 2014, World Scientific*

(greater than 2500 °C). Synthetic graphite has a somewhat less crystalline structure than naturally occurring graphite [29], which will affect the kinetics of the oxidation reaction. While the batch-to-batch reliability of synthetic graphite can be expected to be higher than that of mined graphite, synthetic graphite is significantly more expensive. Synthetic graphite prices are generally at least 50% higher than comparable grades of flake graphite [33], and can be higher for more processed forms.

Thus, the type and size of the graphite used in the process affects the course of the oxidation and exfoliation reactions. All else being equal, the graphite feed will affect the extent of the reaction (see section 13.4.2.3), the number of layers in the product and the degree of oxidation [28, 35]. In addition, defects occurring in mined graphite can influence the quality and properties of the final material. Therefore, the selection of graphite needs to take into account not only the cost, but also the starting materials' effect on the entire process and final product.

13.4.2 Reaction Conditions

There are several challenges associated with the graphite oxidation process: economically sourcing reaction media, health and safety issues, challenges in increasing yield/conversion, and challenges in achieving reproducible high product quality. However, for the most part, these challenges are not intractable, and recent work has moved to address them.

13.4.2.1 Selecting Cost-Effective Reaction Media

To produce GO from graphite, chemicals need to be sourced to assist with intercalation and oxidation. With some exceptions, most of the chemicals required for the process are widely and cheaply available.

In the original Hummers method, sulfuric acid is used as an intercalating agent, with potassium permanganate as the oxidant and sodium nitrate facilitating the oxidation process. Most modern variations of this method continue to rely primarily on the use of sulfuric acid for its intercalating effect, although phosphoric acid (H_3PO_4) may optionally be added [40]. Fortunately, these intercalating chemicals (sulfuric acid and phosphoric acid) are readily available on an industrial scale. Indeed, the primary component of the reaction mixture, sulfuric acid, is one of the most heavily traded chemicals in the world [41]. At $75–80/tonne in April 2015 [42], it will be a relatively inexpensive material input.

From an industrial perspective, the use of potassium permanganate poses its own challenges. Firstly, the chemical can be hazardous, as discussed in section 13.4.2.2. Secondly, the use of potassium permanganate will likely be costly. In 2008–2009, potassium permanganate sold at an average price of $2.16/lb ($4768/tonne) from US producers [43]. Potassium permanganate can also be sourced at significantly lower cost from China, but imports to the USA have been limited as part of anti-dumping measures [44, 45]. Given that the typical modified Hummers process uses a potassium permanganate to graphite mass ratio of 3 : 1 [10], it can be expected that the oxidant will be a significant component of operating costs. Sourcing affordable oxidant will be important to reducing the overall cost of GO, making it cost competitive with alternative industrial materials.

Alternatives to potassium permanganate have also recently been examined [44, 46]. Notably, Gao and colleagues [44] have shown that potassium ferrate (K_2FeO_4) in sulfuric acid has promise. Highly oxidized, single-layer GO could be produced with shorter reaction times, fewer dangerous waste by-products (e.g. no heavy metal manganese produced), and without the need to cool the reactor. They further showed that the reaction media could be recycled at least 10 times. However, despite this plethora of benefits, potassium ferrate is not currently produced on an industrial scale. Prices are on the order of magnitude of $2/gram [47]. Moreover, potassium ferrate decomposes quickly in water, which could lead to its decomposition in the process, particularly if the reaction medium is recycled. These challenges will need to be overcome in order to use this attractive alternative oxidant for large-scale GO production.

13.4.2.2 Managing Hazards

There are at least two sources of hazards characteristic of the Hummers method: the toxic and explosive gas evolving from sodium nitrate; and the explosion risk arising from the use of potassium permanganate in sulfuric acid.

- *Evolution of dangerous nitrogenous gas.* The addition of sodium nitrate, $NaNO_3$, can be a safety issue. Adding it to the reaction can lead to the evolution of toxic and hazardous gases, NO_2 and N_2O_4. This gas can be managed, although a more recent strategy is to prevent it entirely. It has been shown that $NaNO_3$ can be omitted from the reaction mixture because of the strong intercalating nature of sulfuric acid and the oxidizing nature of potassium permanganate. The strategy was first demonstrated by Tour and colleagues [40] and has since been employed by others [37, 48, 49]. Removing sodium nitrate has the added advantage of eliminating the need to purify the wastewater of sodium and nitrate ions.
- *Explosion risk from manganese heptoxide.* When potassium permanganate is added to the solution of sulfuric acid, a dangerous heptoxide is formed [50] according to the following reaction:

$$2MnO_4^- + 2H^+ \rightarrow Mn_2O_7 + H_2O$$

Manganese heptoxide is an extremely oxidizing compound, which can readily react and combust with most organic compounds at temperatures above 55 °C [51]. Further, potassium heptoxide itself can explode at temperatures greater than 95 °C. At the same time, the oxidation reaction is highly exothermic, and reaction mixtures can increase to 60 °C without cooling [52].

Therefore, appropriate temperature control systems, alarms and other safety mechanisms are essential to reduce the risk of explosion. For example, reactors can be jacketed with cooling water and the flow rate of cooling water controlled appropriately. Risk may be further mitigated by staggering the addition of reactants across multiple reactors, reducing the exothermic load at each reactor (see Figure 13.4).

13.4.2.3 Increasing Yield/Conversion

In the conventional Hummers process, the oxidation reaction does not go to completion, and there is typically unoxidized or under-oxidized graphite remaining in the reactor. It is desirable to increase the conversion of graphite to GO. Firstly, this maximizes use of input materials and reduces waste. Secondly, achieving 100% conversion would eliminate the need to separate out unoxidized graphite in the subsequent purification.

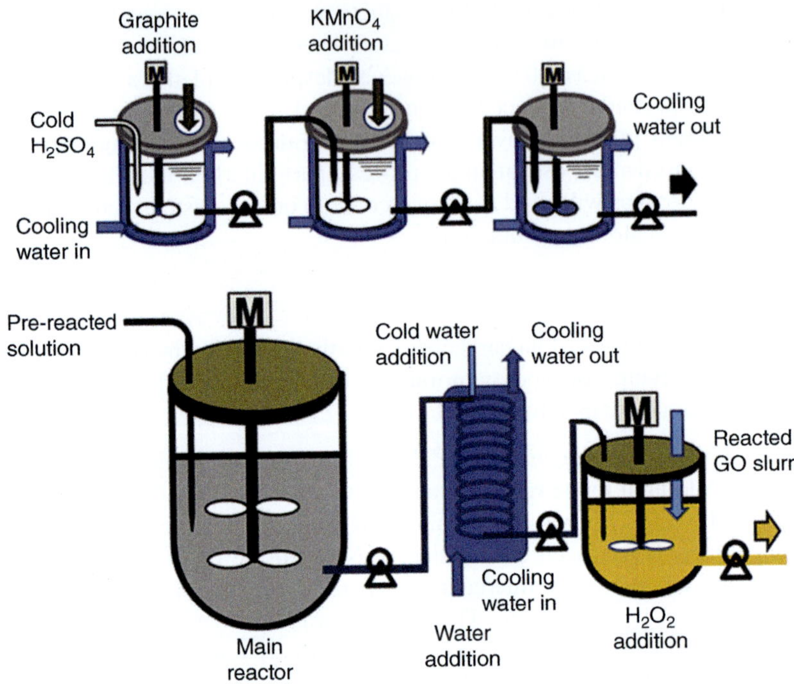

Figure 13.4 Schematic diagram of a continuous process for the oxidation of graphite. In this process, graphite and sulfuric acid are first premixed in a continuous stirred-tank reactor (CSTR). Potassium permanganate is added in a second CSTR, followed by further cooling in a third CSTR. This mixture is pumped to the main oxidation reactor. After cooling in a heat exchanger, the reaction is quenched with hydrogen peroxide in a final CSTR. Reproduced from [53] with permission from Elsevier, Copyright 2013

It is difficult to achieve complete conversion, although it can be achieved by adding a pre-oxidation step [27]. Conversion may also be increased by increasing the reaction time to several hours or by sequentially adding a large excess of $KMnO_4$. None of these procedures, however, is particularly desirable from an industrial standpoint, as they would require increased reagents, significantly larger equipment in a continuous process (to achieve larger residence times), or additional unit operations altogether.

One strategy that may be effective, however, is changing the size of the input graphite particles. Chen *et al.* [49] showed that using flake graphite with sizes less than 20 μm in a modified Hummers process could lead to complete oxidation of the graphite within 30 min. They showed that the reaction is mass-transfer-limited, and, with larger flakes (e.g. 100 μm), 30 min was not enough time for the oxidant to completely permeate the graphite.

In an alternative approach, mass transfer can be improved by changing the mixing regime within the oxidation reactor. Park *et al.* [54] designed a reactor with a rotating inner cylinder which created a Couette–Taylor flow pattern (see Figure 13.5). Using 50 μm flake graphite, they found the increased turbulent mixing allowed for complete conversion and the production of highly oxidized GO within 30 min.

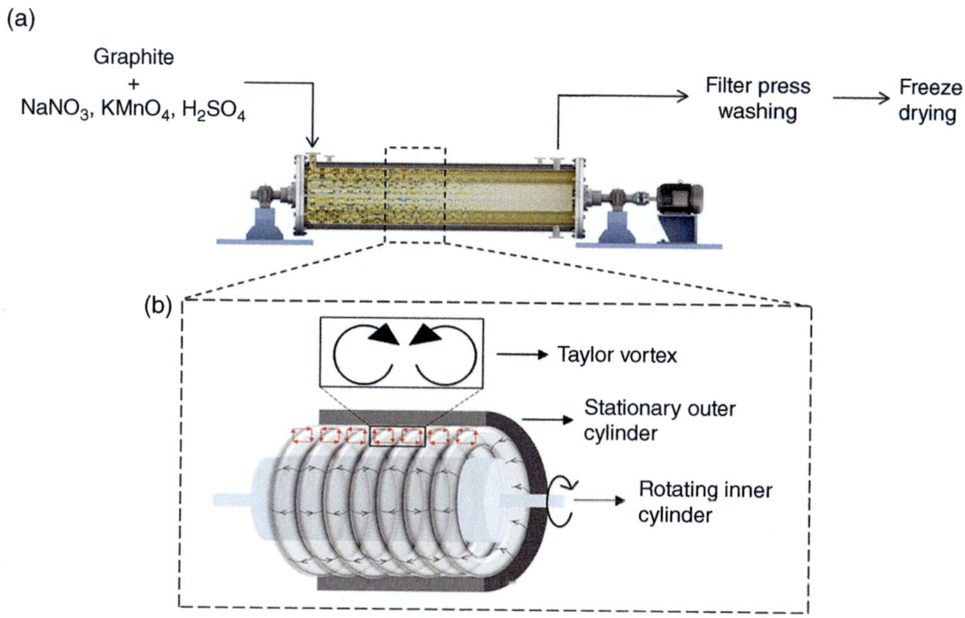

Figure 13.5 *Illustration of the oxidation reaction of graphite flakes in a Couette–Taylor flow reactor. (a) Schematic diagram of the Couette–Taylor flow reactor system. (b) Conceptual diagram of the vortex structure in the Couette–Taylor reactor. Reproduced from [54] with permission from Elsevier, Copyright 2015*

13.4.2.4 Tuning Product Quality

The reaction conditions will have an effect on the quality of the GO produced [1]. For example, the amount of oxidant used and the reaction time will affect the degree of oxidation and the number of layers that can be exfoliated [1, 55]. In many cases, the reaction conditions of a large-scale Hummers process may be modified and optimized to achieve the desired product.

However, there are some forms of GO made using synthetic procedures that present more challenges to scale up. For example Eigler *et al.* [56] describe GO with a more pristine, undamaged structure. They found that the reaction conditions of the traditional Hummers method can introduce hole defects in the GO lattice. These defects arise because some of the graphene's carbon in fact is removed in the form of carbon dioxide. This can change the carbon arrangement in the lattice from hexagonal to pentagonal or other forms, rendering the holes irreparable. Eigler *et al.* showed that this CO_2 formation could be minimized by using longer oxidation times (16 h as opposed to 30 min), and maintaining temperatures below 10 °C. However, there would be added challenges to economically scale up this procedure, given the increased reaction time, increased oxidant costs and the need for process cooling. Similar difficulties would be faced by other improved forms of GO [40] that rely upon longer reaction times under milder conditions. Even though there may be some trade-offs between GO quality and production costs, the target applications for such low-defect, but processable, GO are typically high-end (e.g. transparent conductive electrodes). Hence, the increased production cost is still acceptable (compared to the cost of CVD production) if its performance can truly fulfil industry requirements.

13.4.3 Work-up and Purification

At the end of the oxidation process, a number of impurities will be present in the reaction vessel, including an array of ions (i.e. K^+, H^+, Mn^{2+} and SO_4^{2-}) and any unoxidized graphite. Purification is important to ensure stability and safety of the final product. In fact, many downstream applications will not tolerate even small levels of impurities. For instance, while GO is normally chemically inert, the presence of potassium ions can increase the flammability of the product. During a reduction reaction, for example, GO can violently combust in the presence of air and potassium [57]. In some cases, removing sulfur contamination will be crucial. For example, for some catalytic applications, sulfur can contaminate catalytic sites [52].

Conventionally, the oxidation reaction has been quenched and the product washed with water, and water is the main work-up medium discussed here. However, it should be noted that Dimiev *et al.* [58] demonstrated that using other solvents for purification can have a significant effect on the structure and chemical make-up of GO. Using an organic medium such as methanol or trifluoroacetic acid led to a significantly different GO product (termed "pristine graphite oxide") than that produced using water purification. Pristine graphite oxide had a higher sulfate content, more vacancy defects and a more abundant presence of epoxide groups. The authors propose that water may mediate the rearrangement of functional groups on the graphene surface. Given that the work-up solvent will affect the nature of the product itself, research on the chemistry and engineering of GO produced using organic solvents is warranted.

Because of the very large volumes of solution to process after the oxidation reaction, the requirements for high purity and the unique chemical properties of GO, purification is often seen as the major challenge for industrial production. There are several parts of the work-up/purification process that could pose challenges for the large-scale production of GO: delamination, ion removal, the removal of unoxidized graphite and drying.

13.4.3.1 Exfoliation and Delamination

A potential challenge to large-scale production is determining the optimal conditions for the exfoliation of graphite oxide to GO. After the oxidation reaction, the raw product will typically be graphite oxide (i.e. multiple, stacked layers of GO). To delaminate the graphite oxide to single- or few-layered GO, laboratory procedures typically involve stirring [59], centrifugation [40] or sonication steps [28, 60]. These steps generally are carried out for very long times, often overnight. It does not appear, however, that harsh mechanical treatment is necessary to delaminate graphite oxide. Zhou and Liu [59] demonstrated that, with gentle shaking overnight, very large GO sheets (up to 200 μm) could be exfoliated after a modified Hummers reaction. They suggest that harsher ultrasonication treatment may in fact break up GO sheets.

Thus, before embarking on large-scale production of GO, it will be necessary to fully understand the effect of the delamination method on product quality. Very long exfoliation treatments could be costly and rate-limiting for the process, and so the exfoliation step should be optimized.

13.4.3.2 Ion Removal

In order to remove the large amounts of ions present in the reaction solution, GO is typically either washed using filtration or centrifugation or is subjected to dialysis. Dialysis, can take several weeks, requires frequent exchange of the dialysis medium and will likely represent a bottleneck in an industrial-scale GO process [52].

Filtration, however, can be accomplished in less time. There are multiple protocols for GO filtration and washing. However, typically, the metal ions are first washed out using HCl, followed by thorough rinsing with water. Both dead-end filtration and crossflow filtration may be used in GO purification to this end. In dead-end filtration, the fluid flows directly against a filter; whereas in crossflow filtration, the solution flows in a direction parallel to a filter. In both filtration arrangements, higher pressure on the filter cake side forces solution/filtrate through the filter.

The filter press is a common piece of equipment used to accomplish dead-end filtration and which has been used for GO purification [53, 54]. In this set-up, the GO slurry is first pumped into chambers within the filter press. HCl or water is then pumped into the chamber, and this liquid is extruded through cloth/membrane filter walls as the walls of the chamber are pressed together.

A challenge for GO filtration is that GO tends to gelate in aqueous media. This can greatly complicate certain types of filtration (e.g. gravity and vacuum-assisted filtration), as the filter cake can block the flow of solvent. One method of overcoming this is to use acetone instead of water after washing with HCl, as graphite oxide does not gelate in acetone [12, 57].

A second solution proposed by Tölle *et al.* [52] involves the use of crossflow filtration. In this arrangement, a graphite oxide slurry is first washed with HCl using dead-end filtration. In subsequent water washes, crossflow filtration is used. Graphite oxide is pumped continuously through a cylindrical membrane unit (see Figure 13.6). Wastewater is pushed outwards from the center of the filtration unit by a pressure gradient. The flow of the GO stream through the filtration unit serves to constantly move graphite oxide away from filter pores, reducing pore blockage and minimizing fouling in general. Fresh wash water is added to the system to make up for the lost wastewater. The solution is pumped through the crossflow filter until impurities in the wastewater reach an acceptably low level.

13.4.3.3 Removing Unoxidized Graphite

Depending on the graphite inputs and reaction conditions, some of the starting graphite may remain unoxidized after the reaction. This presents a further separation challenge. This problem can be precluded by choosing the process parameters such that all the starting materials are oxidized [49, 54]. If this is not possible, a variety of separation techniques are available to remove the unoxidized graphite. On a laboratory scale, centrifugation is often used to separate GO from unoxidized graphite based on density. While a vast array of different industrial batch and continuous centrifuge designs are available [61], high-speed centrifuges tend to have limited capacities and can be costly.

An alternative strategy may involve a settling tank where the denser unoxidized graphite flakes will eventually settle down to the bottom of the tank. The rate of settling will largely depend on the viscosity of the GO slurry/solution and the size of the unoxidized graphite

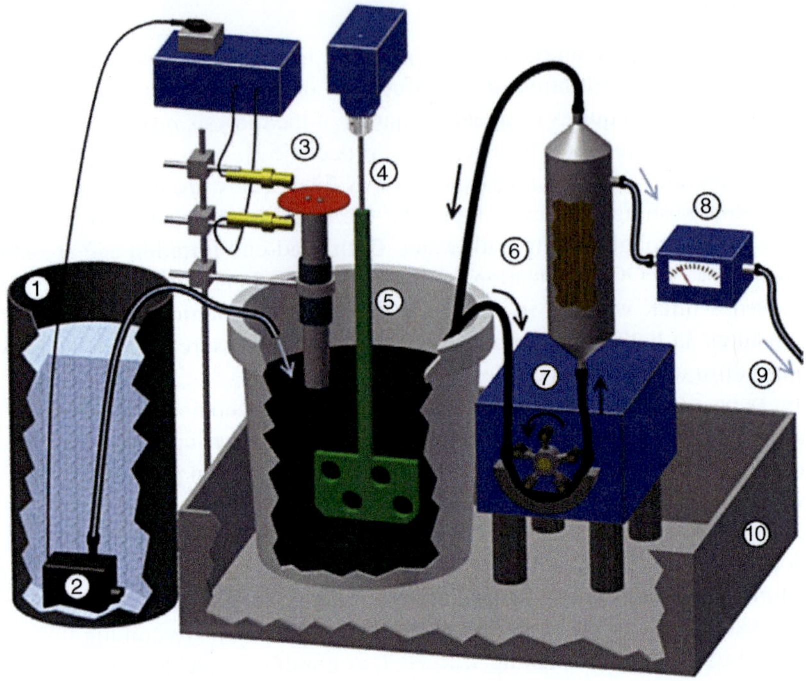

Figure 13.6 A GO purification pilot plant, enabling GO purification by automated crossflow filtration: (1) wash water reservoir, (2) pump for wash water feed, (3) control of filling level and controller for water feed pump, (4) blade stirrer with mixing holes, (5) vessel containing GO dispersion, (6) hollow-fiber crossflow filtration membrane cartridge, (7) peristaltic pump for GO circulation, (8) on-line ion conductivity measurement for monitoring the ion content of the wastewater, (9) wastewater stream, and (10) containment. Reproduced from [52] with permission from Elsevier, Copyright 2015

flakes. It is noteworthy that the product, GO, will precipitate out and settle if the pH of the solution is too low or the salt concentration is too high. Hence, the ideal stage for removing unoxidized graphite might be after the ion/salt purification stage.

13.4.3.4 Drying

With the bulk of the liquid removed from the GO, the product needs to be dried if it is to be stored in solid form. Typically, GO is freeze-dried in order to create a dry powder. Vacuum freeze-drying, however, is expensive on an industrial scale [61]. Moreover, Peng *et al.* [44] report that commercially available freeze-dried GO is very difficult to disperse, which limits its downstream applications. The freeze-drying process may lead to partial restacking of the GO sheets, creating aggregates that are difficult to disrupt during re-dispersion. Spray drying appears to avoid this problem [44]. Industrial-scale spray dryers are available and can be cost-effective. For this application, however, special pumping equipment will be required to feed the spray dryer's atomizer with the GO filter cake [61].

13.4.3.5 Wastewater Treatment

At the conclusion of the purification process, the wastewater will contain K^+, H^+, Mn^{2+} and SO_4^{2-} ions, and, if $NaNO_3$ was used during the oxidation, also Na^+ and NO_3^- ions. High salt concentrations and acidity mean that the wastewater cannot be released to the environment or recycled without treatment.

Typically, ionic wastewater streams are treated using chemical precipitation [62]. In this strategy, counter-ions that form insoluble compounds with existing ions in the solution are added. The insoluble precipitate is then filtered or otherwise separated off. For example, to remove sulfate ions (SO_4^{2-}), slaked lime, $Ca(OH)_2$, can be added to the solution. Calcium combines with the sulfate ions, leading to the formation of insoluble calcium sulfate, $CaSO_4$. Similarly, under neutral pH conditions, sodium carbonate can be added to precipitate manganese to manganese carbonate, $MnCO_3$. In this case, a precipitation strategy would need to be carefully designed in order to minimize unit operations and added chemicals.

Care should also be taken in increasing the pH excessively. Mn^{2+} can be precipitated by adding excess base (e.g. OH^- ions from KOH) [48] at a pH of 10, resulting in the formation of insoluble manganese tetroxide, Mn_3O_4. However, manganese tetroxide dust is highly neurotoxic and carcinogenic. It is generally recommended that processes are designed to avoid the compound unless a suitable downstream use can be found [63, 64].

Process designers should explore different opportunities for recycling and reusing spent materials, consistent with general waste management principles [65]. For example, reclaiming and reselling precipitated compounds could reduce landfill costs and provide a small profit.

Given the challenges of separating a multi-component mixture of salts, an alternative strategy is neutralization of the solution followed by evaporation. Evaporation of the wastewater will lead to the crystallization of $MnSO_4$, KOH and other salts, which can then be recycled or disposed of. Given the fact that a very large amount of water is required for purification of GO and fresh water is limited in many countries, having the right wastewater recycling system installed is decidedly one of the major considerations for the construction of a GO production plant.

13.4.4 Storage, Handling and Quality Control

There may be distinct challenges associated with the storage, handling and quality control of GO. Graphene oxide may be hazardous in some cases, and the product can be unstable at ambient conditions, increasing the need for good quality control.

13.4.4.1 Storage and Handling Challenges

While purified GO is not flammable, it becomes highly combustible in the presence of salt contaminants (notably potassium) [12]. Even without contamination, GO can be thermally reduced, which under certain conditions can cause explosion. When slowly heated under inert gas to approximately 140 °C, GO will explode [66]. In air, GO will explode when heated to around 300 °C [12]. Given this explosion risk, GO needs to be treated as a potentially hazardous material.

There are also challenges related to GO's chemical stability. While it was initially thought that GO was stable after the reaction and work-up, it has recently been found

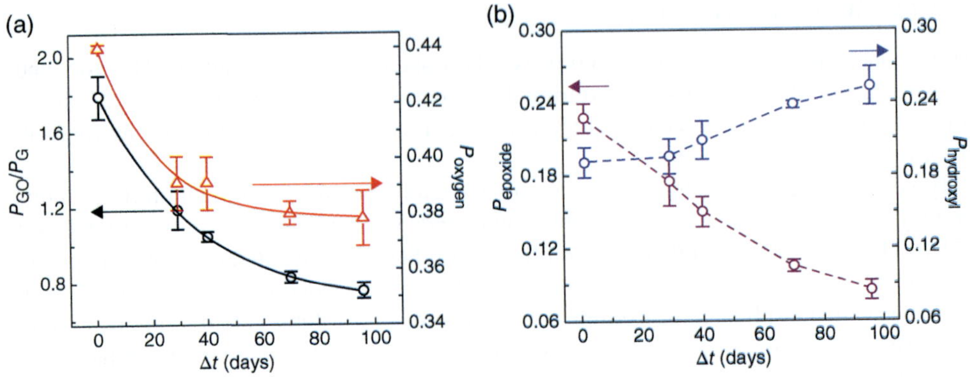

Figure 13.7 Chemical change of GO films over 100 days deduced from X-ray photoelectron spectroscopy (XPS) data. Symbols show experimental values and statistical error bars. (a) P_{GO} represents the hydroxyl/epoxide bonded carbon as a proportion of all carbon in the sample, while P_G represents the proportion of C–C bonded carbon. P_{GO}/P_G (black circles) is therefore a measure of the oxygen-bound carbon in the sample. P_{oxygen} (red triangles) is the oxygen-to-carbon ratio, estimated from the O 1s/C 1s intensity ratio, corrected by a Scofield relative sensitivity factor of 2.93. (b) Fraction of epoxide ($P_{epoxide}$; purple circles) and hydroxyl ($P_{hydroxyl}$; blue circles) species relative to the amount of C in GO. Reproduced from [67] with permission from Nature Publishing Group, Copyright 2012

that dried GO is only metastable at ambient conditions and will in fact chemically change during storage. Kim *et al.* [67] showed that multi-layer GO spontaneously reduces at room temperature, reaching a quasi-equilibrium state after approximately 35 days. Over time, the number of epoxide groups decreases, the number of hydroxyl groups increases, and the overall oxygen-to-carbon ratio (O/C) falls (see Figure 13.7). Further, Chua and Pumera [68] found that multi-layered and single-layered GO degradation could be observed over a period of 30 weeks. The degradation of single-layered GO was a function of exposure to air, while that of multi-layered GO was a function of light exposure. For example, GO exposed to air saw the O/C ratio fall from 0.46 to 0.40 at 30 weeks, again with a concomitant reduction in the number of epoxide groups. This degradation was not observed when the single-layer GO samples were kept under nitrogen.

In solution, GO appears to be stable over short time frames even when heated to 40 °C (but not in the presence of NaOH) [69]. However, no long-term assessments of the stability of GO in solution have been reported. Graphene oxide's post-oxidation chemical rearrangement is thought to be water-mediated [7, 58]. It therefore might be expected that storage in water would have an effect on the product's stability.

Lack of stability has clear implications for industrial production. Changes in the chemical composition of GO will affect its properties in downstream applications. Therefore, future work is required to fully understand the mechanism and time course of degradation. This will allow GO producers to develop strategies to ensure product consistency. Suitable quality control measures, such as routine product testing, will need to form part of this strategy.

13.4.4.2 Quality Control Measures

As with any chemical process, it will be important to ensure that the process is operating smoothly and that the finished product is of an acceptable quality. Ideally, on-line monitoring would be available to allow plant operators to quickly identify and troubleshoot problems as they evolve, thereby minimizing product loss [61, 70]. Simple on-line analysis might include measurements of the pH in the various reactors or measurement of the wastewater stream conductivity (abnormally low conductivity may indicate low ion content in the wastewater stream, which would suggest that there may be problems with the purification). Further, portholes can be installed in reactors to allow for visual inspection of the process stream. When the oxidation reaction is quenched with hydrogen peroxide, the reaction mixture becomes yellow, indicating that the material has been oxidized; operators may periodically check for this color change.

The final product will also need to be checked at on-site or independent analytical laboratories before distribution. There are a variety of analytical techniques for characterizing GO, which could be used to assess product quality [7, 71] (see Chapter 3 for a detailed discussion of these methods). Fourier-transform infrared spectroscopy (FTIR) is a rapid measure, which can be used to detect the functional groups in the GO molecules, thus confirming the quality of the sample. X-ray photoelectron spectroscopy (XPS) can indicate the degree of oxygenation, as can ultraviolet–visible spectroscopy. Raman spectroscopy can further confirm the identity and indicate the degree of defects. Tapping mode atomic force microscopy (AFM) can be used to determine the thickness of the sample, and therefore the number of stacked layers. AFM, along with other forms of microscopy, such as tunneling electron microscopy (TEM) and scanning electron microscopy (SEM), can be used to assess the overall morphology and lateral dimensions of the GO sheets. Some combination of these analytical techniques may be chosen based on quality requirements and cost/economic factors.

13.5 Concluding Remarks and Future Directions

In the field of polymer science, a single type of polymer (e.g. polyethylene) can be produced at various molecular weights and regiospecificity with different side/end groups to suit a variety of applications. This is aided by the extensive polymer chemistry available to precisely polymerize and tailor speciality monomers into the desired polymeric material in a bottom-up fashion. Unfortunately, a polymerizable "graphene monomer" via wet chemistry is difficult to achieve, and the size of such synthetic graphene is expected to be limited, as seen in the case of solvothermal growth [72]. The beauty of GO is that it is essentially a functional two-dimensional macromolecule with good dispersibility in water and many other solvents, thereby a highly processable intermediate to reduced GO, graphene nanohybrids and graphene nanocomposites. The utility of the oxidation route is that it can produce an array of single-layer GO types, from sheets with a nanoscale lateral size and porous sheets to the more elusive large diameter sheets with low irreparable hole defects and controlled oxygen functional groups.

Given the promising outlook for graphite supply, the top-down approach of graphite oxidation and exfoliation is undeniably one of the most promising methods for producing graphene-based products. It is the only scalable wet-chemical method capable of producing single-layer GO, and reduced GO, in high yield. This is in contrast to other methods, which typically produce a mixture of few-layered graphene or graphene nanoplatelets. The challenges of industrial-scale GO production are traditionally associated with the hazards involving strong corrosive acids and explosive oxidants, but these hazards can be overcome with modern industrial chemical processes. The control of oxidation degree and other critical properties of GO are currently being addressed, with promising results. The less recognized issues associated with purification and waste management are not obvious when operating at laboratory scale, but escalate to become the major challenges at the industrial scale. Moreover, seemingly minor details such as the shelf-life, packaging and delivery method have to be optimized to suit the needs of targeted industries.

Indeed, given that the market for GO is nascent, producers will need to play an active role in defining the format in which GO is produced and supplied. Among the many potential product types, GO may be supplied in dry form as powders or as films, in aqueous or organic solution, or as intermediary products such as graphene/polymer dispersions. As various market niches for GO are found, the establishment of industry standards for GO products will further encourage uptake of this new material [3]. However, some graphene producers are nonetheless operating in great secrecy, without providing details about their products, let alone their production processes. This is unsustainable, as it will be difficult for end-user industries to integrate these products reliably into existing applications or to make improvements to future products. Hence, there is a pressing need for full transparency on the materials characterization and quality control used for GO products.

There are multiple challenges for the industrial-scale production of GO, which can also be viewed as avenues for innovation and growth. By waiting for the "killer application" to drive the production of GO, there is the risk that production will be limited to highly specialized markets, with producers relying on one particular grade of graphite to make a small range of products. On the other hand, new GO producers can seek to be versatile and define the market for their products. There is the possibility of treating graphite as both a basic raw material and a precursor to a large range of new materials – analogous to the wide range of products obtained from crude oil. In this case, producers are responsible for sorting out the variety of low-cost graphite feedstocks and adding the capability of producing a greater diversity of products catering to a wider breadth of industries. Similarly, the apparent challenges of wastewater management and polluting by-products likely present opportunities for recycling and reuse. Minimizing cost and environmental impacts from waste products will be an important area of innovation for GO producers. Quite fittingly, it will probably not be long before the advancement in graphene technology (e.g. filtration, desalination, etc.) can be used to solve some of its own production challenges.

References

[1] Wan, X.; Huang, Y.; Chen, Y., Focusing on energy and optoelectronic applications: a journey for graphene and graphene oxide at large scale. *Acc. Chem. Res.* **2012**, *45* (4), 598–607.

[2] Zhong, Y.L.; Tian, Z.; Simon, G.P.; Li, D., Scalable production of graphene via wet chemistry: progress and challenges. *Mater. Today* **2015**, *18* (2), 73–78.

[3] Zurutuza, A.; Marinelli, C., Challenges and opportunities in graphene commercialization. *Nature Nanotechnol.* **2014**, *9* (10), 730–734.

[4] Ren, W.; Cheng, H.-M., The global growth of graphene. *Nature Nanotechnol.* **2014**, *9* (10), 726–730.
[5] Ferrari, A.C.; Bonaccorso, F.; Fal'ko, V.; *et al.*, Science and technology roadmap for graphene, related two-dimensional crystals, and hybrid systems. *Nanoscale* **2015**, *7* (11), 4598–4810.
[6] Kim, J.; Cote, L.J.; Huang, J., Two dimensional soft material: new faces of graphene oxide. *Acc. Chem. Res.* **2012**, *45* (8), 1356–1364.
[7] Dreyer, D.R.; Todd, A.D.; Bielawski, C.W., Harnessing the chemistry of graphene oxide. *Chem. Soc. Rev.* **2014**, *43* (15), 5288–5301.
[8] Eigler, S.; Enzelberger-Heim, M.; Grimm, S.; *et al.*, Wet chemical synthesis of graphene. *Adv. Mater.* **2013**, *25* (26), 3583–3587.
[9] Gambhir, S.; Jalili, R.; Officer, D.L.; Wallace, G.G., Chemically converted graphene: scalable chemistries to enable processing and fabrication. *NPG Asia Mater.* **2015**, *7*, e186.
[10] Zhu, Y.; James, D.K.; Tour, J.M., New routes to graphene, graphene oxide and their related applications. *Adv. Mater.* **2012**, *24* (36), 4924–4955.
[11] Eigler, S.; Hirsch, A., Chemistry with graphene and graphene oxide – challenges for synthetic chemists. *Angew. Chem. Int. Ed.* **2014**, *53* (30), 7720–7738.
[12] Krishnan, D.; Kim, F.; Luo, J.; *et al.*, Energetic graphene oxide: challenges and opportunities. *Nano Today* **2012**, *7* (2), 137–152.
[13] Galande, C.; Gao, W.; Mathkar, A.; *et al.*, Science and engineering of graphene oxide. *Part. Part. Syst. Charact.* **2014**, *31* (6), 619–638.
[14] Novoselov, K.S.; Geim, A.K.; Morozov, S.; *et al.*, Electric field effect in atomically thin carbon films. *Science* **2004**, *306* (5696), 666–669.
[15] Novoselov, K.; Jiang, D.; Schedin, F.; *et al.*, Two-dimensional atomic crystals. *Proc. Natl. Acad. Sci. USA* **2005**, *102* (30), 10451–10453.
[16] Novoselov, K.; Geim, A.K.; Morozov, S.; *et al.*, Two-dimensional gas of massless Dirac fermions in graphene. *Nature* **2005**, *438* (7065), 197–200.
[17] Greil, P., Perspectives of nano-carbon based engineering materials. *Adv. Eng. Mater.* **2015**, *17* (2), 124–137.
[18] Stankovich, S.; Dikin, D.A.; Dommett, G.H.B.; *et al.*, Graphene-based composite materials. *Nature* **2006**, *442* (7100), 282–286.
[19] Huang, L.; Huang, Y.; Liang, J.; *et al.*, Graphene-based conducting inks for direct inkjet printing of flexible conductive patterns and their applications in electric circuits and chemical sensors. *Nano Res.* **2011**, *4* (7), 675–684.
[20] Pumera, M., Graphene-based nanomaterials for energy storage. *Energy Environ. Sci.* **2011**, *4* (3), 668–674.
[21] Acik, M.; Lee, G.; Mattevi, C.; *et al.*, Unusual infrared-absorption mechanism in thermally reduced graphene oxide. *Nature Mater.* **2010**, *9* (10), 840–845.
[22] Brodie, B.C., On the atomic weight of graphite. *Phil. Trans. R. Soc. Lond.* **1859**, *149*, 249–259.
[23] Staudenmaier, L., Verfahren zur Darstellung der Graphitsäure. *Ber. Dtsch. Chem. Ges.* **1898**, *31* (2), 1481–1487.
[24] Charpy, G., Sur la formation de l'oxyde graphitique et la définition du graphite. *C. R. Hebd. Séances Acad. Sci.* **1909**, *148*, 920–923.
[25] Hummers, W.S.; Offeman, R.E., Preparation of graphitic oxide. *J. Am. Chem. Soc.* **1958**, *80* (6), 1339.
[26] Dreyer, D.R.; Park, S.; Bielawski, C.W.; Ruoff, R.S., The chemistry of graphene oxide. *Chem. Soc. Rev.* **2010**, *39* (1), 228–240.
[27] Kovtyukhova, N.I.; Ollivier, P.J.; Martin, B.R.; *et al.*, Layer-by-layer assembly of ultrathin composite films from micron-sized graphite oxide sheets and polycations. *Chem. Mater.* **1999**, *11* (3), 771–778.
[28] Wu, Z.-S.; Ren, W.; Gao, L.; *et al.*, Synthesis of high-quality graphene with a pre-determined number of layers. *Carbon* **2009**, *47* (2), 493–499.
[29] Wissler, M., Graphite and carbon powders for electrochemical applications. *J. Power Sources* **2006**, *156* (2), 142–150.
[30] Industrial Minerals, *Graphite: Market brief.* See http://www.indmin.com/Graphite.html#/MarketBrief (accessed 13 May 2016).

[31] Technology Metals Research, *TMR Advanced Graphite Projects index*. See http://www.techmetals research.com/metrics-indices/tmr-advanced-graphite-projects-index/ (accessed 13 May 2016).

[32] Moores, S., *Flake graphite: Supply, demand, prices – GMP securities*. GMP Securities Workshop, London, 10 June 2014. See http://www.slideshare.net/sdmoores/flake-graphite-supply-demand-prices-gmp-securities (accessed 13 May 2016).

[33] Industrial Minerals, *Pricing database, graphite*. See http://www.indmin.com/PricingDatabase.html (accessed 24 June 2015).

[34] Moores, S., *Has China reached peak graphite?* Mining.com. See http://www.mining.com/web/has-china-reached-peak-graphite/ (accessed 13 May 2016).

[35] Hu, X.; Yu, Y.; Zhou, J.; Song, L., Effect of graphite precursor on oxidation degree, hydrophilicity and microstructure of graphene oxide. *Nano* **2014**, *9* (3), 1450037.

[36] Asbury Carbons, *Amorphous graphite*. See http://asbury.com/technical-presentations-papers/materials-in-depth/amorphous-graphite/ (accessed 13 May 2016).

[37] Sun, L.; Fugetsu, B., Mass production of graphene oxide from expanded graphite. *Mater. Lett.* **2013**, *109*, 207–210.

[38] Hantel, M.M.; Kaspar, T.; Nesper, R.; *et al.*, Partially reduced graphite oxide for supercapacitor electrodes: effect of graphene layer spacing and huge specific capacitance. *Electrochem. Commun.* **2011**, *13* (1), 90–92.

[39] Burress, J.W.; Gadipelli, S.; Ford, J.; *et al.*, Graphene oxide framework materials: theoretical predictions and experimental results. *Angew. Chem. Int. Ed.* **2010**, *49* (47), 8902–8904.

[40] Marcano, D.C.; Kosynkin, D.V.; Berlin, J.M.; *et al.*, Improved synthesis of graphene oxide. *ACS Nano* **2010**, *4* (8), 4806–4814.

[41] Muller, T.L., Sulfuric acid and sulfur trioxide. In *Kirk–Othmer Encyclopedia of Chemical Technology*. John Wiley & Sons, Inc., New York, **2000**.

[42] Tonyan, D., US sulphuric acid import prices up. *ICIS Chemical Business* **2015**, 13 April (3534), 32.

[43] U.S. International Trade Commission, *Potassium permanganate from China*. Investigation no. 731-TA-125 (Third Review), Publ. 4183. U.S. International Trade Commission, Washington, DC, 2010.

[44] Peng, L.; Xu, Z.; Liu, Z.; *et al.*, An iron-based green approach to 1-h production of single-layer graphene oxide. *Nature Commun.* **2015**, *6*, 5716.

[45] International Trade Administration, *Potassium permanganate from the People's Republic of China: Preliminary results of antidumping duty administrative review; 2013*. Department of Commerce, US Government, Washington, DC, **2015**.

[46] Wojtoniszak, M.; Mijowska, E., Controlled oxidation of graphite to graphene oxide with novel oxidants in a bulk scale. *J. Nanopart. Res.* **2012**, *14* (11), 1248.

[47] Coons, R., New production process slashes potassium ferrate price. *Chem. Week* **2010**, *172* (8), 33.

[48] Chen, J.; Yao, B.; Li, C.; Shi, G., An improved Hummers method for eco-friendly synthesis of graphene oxide. *Carbon* **2013**, *64*, 225–229.

[49] Chen, J.; Li, Y.; Huang, L.; *et al.*, High-yield preparation of graphene oxide from small graphite flakes via an improved Hummers method with a simple purification process. *Carbon* **2015**, *81*, 826–834.

[50] Pisarczyk, K., Manganese compounds. In *Kirk–Othmer Encyclopedia of Chemical Technology*. John Wiley & Sons, Inc., New York, **2000**.

[51] Simon, A.; Dronskowski, R.; Krebs, B.; Hettich, B., The crystal structure of Mn_2O_7. *Angew. Chem. Int. Ed.* **1987**, *26* (2), 139–140.

[52] Tölle, F.J.; Gamp, K.; Mülhaupt, R., Scale-up and purification of graphite oxide as intermediate for functionalized graphene. *Carbon* **2014**, *75*, 432–442.

[53] Lee, S.; Eom, S.H.; Chung, J.S.; Hur, S.H., Large-scale production of high-quality reduced graphene oxide. *Chem. Eng. J.* **2013**, *233*, 297–304.

[54] Park, W.K.; Kim, H.; Kim, T.; *et al.*, Facile synthesis of graphene oxide in a Couette–Taylor flow reactor. *Carbon* **2015**, *83*, 217–223.

[55] Zhang, L.; Li, X.; Huang, Y.; *et al.*, Controlled synthesis of few-layered graphene sheets on a large scale using chemical exfoliation. *Carbon* **2010**, *48* (8), 2367–2371.

[56] Eigler, S.; Enzelberger-Heim, M.; Grimm, S.; *et al.*, Wet chemical synthesis of graphene. *Adv. Mater.* **2013**, *25* (26), 3583–3587.
[57] Kim, F.; Luo, J.; Cruz-Silva, R.; *et al.*, Self-propagating domino-like reactions in oxidized graphite. *Adv. Funct. Mater.* **2010**, *20* (17), 2867–2873.
[58] Dimiev, A.; Kosynkin, D.V.; Alemany, L.B.; *et al.*, Pristine graphite oxide. *J. Am. Chem. Soc.* **2012**, *134* (5), 2815–2822.
[59] Zhou, X.; Liu, Z., A scalable, solution-phase processing route to graphene oxide and graphene ultralarge sheets. *Chem. Commun.* **2010**, *46* (15), 2611–2613.
[60] Stankovich, S.; Dikin, D.A.; Piner, R.D.; *et al.*, Synthesis of graphene-based nanosheets via chemical reduction of exfoliated graphite oxide. *Carbon* **2007**, *45* (7), 1558–1565.
[61] Green, D.W.; Perry, R.H., *Perry's Chemical Engineers' Handbook*, 8th edn. McGraw-Hill, New York, **2007**.
[62] Guyer, H.H., *Industrial Processes and Waste Stream Management*. John Wiley & Sons, Inc., New York, **1998**.
[63] Hathaway, G.J., Proctor, N.H., *Proctor and Hughes' Chemical Hazards of the Workplace*, 5th edn. Wiley-Interscience, Hoboken, NJ, **2004**.
[64] Vincoli, J.W., *Risk Management for Hazardous Chemicals*. CRC Press, Boca Raton, FL, **1996**.
[65] European Commission, *Directive 2008/98/EC on waste (Waste Framework Directive)*. See http://ec.europa.eu/environment/waste/framework/ (accessed 13 May 2016).
[66] Qiu, Y.; Guo, F.; Hurt, R.; Külaots, I., Explosive thermal reduction of graphene oxide-based materials: mechanism and safety implications. *Carbon* **2014**, *72*, 215–223.
[67] Kim, S.; Zhou, S.; Hu, Y.; *et al.*, Room-temperature metastability of multilayer graphene oxide films. *Nature Mater.* **2012**, *11* (6), 544–549.
[68] Chua, C.K.; Pumera, M., Light and atmosphere affect the quasi-equilibrium states of graphite oxide and graphene oxide powders. *Small* **2015**, *11* (11), 1266–1272.
[69] Eigler, S.; Grimm, S.; Hof, F.; Hirsch, A., Graphene oxide: a stable carbon framework for functionalization. *J. Mater. Chem. A* **2013**, *1* (38), 11559–11562.
[70] Sinnott, R.K., *Coulson & Richardson's Chemical Engineering*, vol. *6*, *Chemical Engineering Design*, 2nd edn. Pergamon Press, Oxford, **1993**.
[71] Soldano, C.; Mahmood, A.; Dujardin, E., Production, properties and potential of graphene. *Carbon* **2010**, *48* (8), 2127–2150.
[72] Qu, D.; Zheng, M.; Zhang, L. G.; *et al.*, Formation mechanism and optimization of highly luminescent n-doped graphene quantum dots. *Sci. Rep.* **2014**, *4*, 5294.

Vocabulary

graphene oxide (GO)
epoxide
tertiary alcohol
ketone
carboxylic group
lactol
carbonyl
phenol
quinone
gem-diol
geminal diol
vic-diol
vicinal diol
formation mechanism
reaction mechanism
production
Brodie method
Staudenmaier method
Hummers method
Charpy–Hummers approach
Kovtyukhova modification
Marcano modification
chemical properties
nucleophilic attack
intercalation compounds of graphite
stage-1 GIC
chemical structure
Thiele model
Hofmann model
Nakajima model
Ruess model
Clauss model
Scholz–Boehm model
Lerf–Klinowski model
Szabó–Dékány model
Dimiev–Tour model
structural model
nuclear magnetic resonance (NMR)
infrared spectroscopy (IR)
Fourier-transform infrared spectroscopy (FTIR)
X-ray photoelectron spectroscopy (XPS)
Raman spectroscopy
microscopy
scanning electron microscopy (SEM)
atomic force microscopy (AFM)
transmission electron microscopy (TEM)
high-resolution transmission electron microscopy (HRTEM)
inflection point
acidity
cation exchange capacity
titration
reverse titration
intercalation compound of graphite oxide
deflagration
colloid
diazomethane
functionalization
electrophilic reactions
nucleophilic addition reactions
graphene
X-ray diffraction (XRD)
transition form
domain
Onsager theory
anisotropic
isotropic
lyotropic
nematic phase

Graphene Oxide: Fundamentals and Applications, First Edition. Edited by Ayrat M. Dimiev and Siegfried Eigler.

birefringence
polarized light
dynamic shear
steady shear
storage moduli
loss moduli
Boltzmann's constant
Brownian diffusion coefficient
Bingham model
Herschel–Bulkley model
Ostwald–de Waele model
linear viscoelastic region
dynamic frequency sweep
viscosity
Kerr coefficient
Kerr effect
shear thinning
dynamic structural model
thermal decomposition
thermogravimetric analysis
decarboxylation
oxo-functionalized graphene
density of defects
poly(diallyldimethylammonium chloride)
ethyl disulfide
N-isopropylacrylamide
sigmatrope rearrangement
"click" reaction
esterification
nascent hydrogen
edge functional groups
covalent functionalization
amide
deconvolution
physisorption
pyrazole
reduction
point defects
reducing agent
electrochemical reduction
hydrazine
hydrazone
sodium borohydride
hydrogen iodide
electrical conductivity
elemental analysis
contact resistance
lithography
sheet resistance
graphene oxide nanoribbons
deoxygenation
thermal reduction
decomposition
redox reaction
degradation
disproportionation
carbon dioxide
carbon monoxide
oxidation state
annealing
molecular dynamics (MD) simulations
thermally processed graphene oxide (tpGO)
Langmuir–Blodgett technique
diazonium chemistry
diazonium salts
aryl cation
azido-hydroxyl graphene
density functional theory (DFT)
semiconductor
bandgap
transistor device
Schottky contact model
cathode
anode
electrolyte
separator
solid–electrolyte interface (SEI)
irreversible capacity loss (ICL)
gravimetric capacity
volumetric capacity
specific capacity
surface area
cycling rate
energy density
power density
electrochemical impedance spectroscopy (EIS)
polymer–graphite composites
interfacial interaction
exfoliation
carbon nanotubes
N-methylpyrrolidone

dimethyl sulfoxide
GO–polymer composite
mechanical properties
mixture law
efficiency coefficient
volume fraction
filler
matrix
continuum model
Young's modulus
modulus
reinforcement
confinement
stress transfer
misalignment
agglomeration
interface
thermoplastic polymers
thermoset polymers
nacre
percolation
scaling law
percolation threshold
aspect ratio
anisotropy
segregated network
immiscible polymers
polymer blend
thermal conductivity
Kapitza resistance (R_K)
interfacial resistance
graphite nanoplatelets (GNPs)
barrier properties
diffusion coefficient
self-assembling
catalysis
benzyl alcohol
benzaldehyde
benzoic acid
unsaturated hydrocarbons
Wacker–Tsuji oxidation
oxidative dehydrogenation
Claisen–Schmidt condensation
condensation reaction
oxidative coupling
superoxide radicals
fuel cell
oxygen reduction reaction
Friedel–Crafts reaction
photolysis
photo-oxidation
GO hybrid materials
palladium nanoparticles
nucleotides
detector
antibody
biocatalyst
commercialization

Index

Graphene Oxide: Fundamentals and Applications, First Edition. Edited by Ayrat M. Dimiev and Siegfried Eigler.

Printed and bound by CPI Group (UK) Ltd, Croydon, CR0 4YY

17/09/2026

14964303-0005